Pathogenicity and Drug Resistance of Human Pathogens

Saif Hameed • Zeeshan Fatima
Editors

Pathogenicity and Drug Resistance of Human Pathogens

Mechanisms and Novel Approaches

Springer

Editors
Saif Hameed
Amity Institute of Biotechnology
Amity University Haryana
Manesar, Gurugram, India

Zeeshan Fatima
Amity Institute of Biotechnology
Amity University Haryana
Manesar, Gurugram, India

ISBN 978-981-32-9448-6 ISBN 978-981-32-9449-3 (eBook)
https://doi.org/10.1007/978-981-32-9449-3

This Springer imprint is published by the registered company Springer Nature Singapore Pte Ltd.
The registered company address is: 152 Beach Road, #21-01/04 Gateway East, Singapore 189721, Singapore

Dedicated to our "Mentor"

Prof. Rajendra Prasad "RP"

प्रो० राकेश भटनागर
कुलपति
Prof. Rakesh Bhatnagar Ph.D.
FNA, FASc, FNASc
Vice-Chancellor

काशी हिन्दू विश्वविद्यालय
Banaras Hindu University
(Established by Parliament by Notification No. 225 of 1916)
Varanasi – 221005 (INDIA)
Phone : 91-542-2368938, 2368339
Fax : 91-542-2369100, 2369951
E-mail : vc@bhu.ac.in
Website : www.bhu.ac.in

Foreword

I am delighted to endorse the book entitled *Pathogenicity and Drug Resistance of Human Pathogens: Mechanisms and Novel Approaches* being published by Springer Nature. The book carries the latest research and developments with focus on the recent advancements and modern trends in the field of human infectious diseases. The book is an integrated approach to widely disseminate the current research topics and frontier areas of human infections when the problem of antimicrobial resistance has surfaced alarmingly around the globe.

I am delighted to see that this book covers four human pathogens sectioned as Bacterial, Fungal, Protozoal, and Viral infectious diseases which are the major cause of human infections. I appreciate the efforts taken up by Drs. Saif Hameed and Zeeshan Fatima, young faculty members of Amity University Haryana working in the area of infectious diseases since a decade. Encouragingly, editors have successfully managed to assemble chapters from eminent veterans working in diverse areas of infectious microbes and touching upon most of the perceivable title points that provide completeness of the offering they made.

The book will also provide an opportunity to the students, faculty, research scholars, and policy makers to gain the knowledge and plan their prospective research. This piece of work is timely and an important addition in the human resource that will address the issue of antimicrobial resistance for the betterment of human health nationally and internationally.

"A step toward Swasth Bharat"

I wish the book a great success in the field.

(Rakesh Bhatnagar)

Prof. (Dr.) Seyed E. Hasnain
Professor, JH Institute of Molecular Medicine, and
Vice Chancellor, Jamia Hamdard,
New Delhi, India

Foreword

Hosts and pathogens have coevolved over millions of years, and with the passage of time, bacteria have modified their machinery to further improve the virulence mechanisms to target host defense systems. Multidisciplinary approaches have produced a plethora of new data for antibacterial therapy. This book represents a collection of chapters authored by domain experts who have summarized the most important mechanisms and novel approaches involved in the pathogenicity and drug resistance of human pathogens.

The chapters are categorized into five sections: Pathogenicity and Drug Resistance in *Mycobacterium tuberculosis*, *Candida* Infections and Therapeutic Strategies, Malaria Parasite Biology, Emerging Viral Diseases, and Translational Research Involving Human Microbes. The topics and their sequence of appearance present a logical framework for the readers to appreciate the strategies bacteria employ at the cellular level. Defensive virulence mechanisms that pathogens have adopted to survive within a multicellular host and therapies that have been used to target bacteria and their virulence mechanisms have been elaborated. Furthermore, the authors have comprehensively described the most relevant and updated information on molecular features of pathogenesis, the understanding of host-microbe interactions and microbial communities and how they can be exploited to develop new therapeutic strategies, the current and future perspective of nano-biotechnology to combat bacterial pathogenesis, and the novel drug targets.

This book also includes the effect of antimicrobial resistance on bacterial virulence and host fitness and covers the relationship between virulence and resistance, the different genetic mechanisms and how immune responses are modulated by host factors, and the role of ecological niche.

This book not only presents a compilation of articles on resistance and virulence but also potential therapeutic consequences, including new antimicrobials and natural compounds. Moreover, each chapter provides an informative review and offers comments on modification of current therapies to further stop the infection process.

The authors deserve to be complimented for bringing out this volume which will be useful for students, clinical scientists, and particularly those looking for new ideas to counter emerging problems of drug resistance.

Seyed E. Hasnain

Preface

"Drug resistant infection threatens end of the road for antibiotics with the prospect of moving towards post-antibiotic era."

Alexander Fleming, the scientist who discovered penicillin, himself shared his vision as a cautionary tale when collecting the Nobel Prize for his seminal work in 1945, predicting a time when antibiotics could become less useful due to frequent or improper use. Mankind has fought a long way in pursuit of diagnostics and therapeutics of diseases since antiquity. Several breakthrough attempts in researches have been made in the last couple of centuries. Since then, even the most dreadful diseases are not without some hope of cure today due to development of modern researches that allows efficient therapeutics. However, according to the WHO fact sheet, "Many of the medical breakthroughs of the last century could be lost through the spread of antimicrobial resistance." The book *Pathogenicity and Drug Resistance of Human Pathogens: Mechanisms and Novel Approaches* comprises topics pertaining to the advancements in the field of infectious disease research. Infectious diseases with emergence of drug resistance are thorn in the flesh before the scientific community since ages and have claimed enormous number of lives around the globe falling short of therapeutics. This book is therefore committed to give an overview of state of the art as well as upcoming trends to promote discussion about scientific, technological, or educational advances related to infectious diseases against pathogenic microbes. The book is comprehensive in scope and concise in approach to ensure vital information pertaining to the advancements and current researches related to human pathogens. Collectively, 20 chapters in the book cover 4 pathogens comprising bacterial, fungal, protozoal, and viral infectious diseases which are the major cause of human pathophysiology. It is a systematic compilation of four classes of microbes at a single platform that summarizes microbial pathogenesis, drug resistance, diagnostics, vaccines, and novel therapeutic aspects of various microbial life-threatening diseases. The contributors are stalwarts and eminent veterans working in diverse areas of infectious microbes from various research institutes and universities.

The first section covers the dreadful disease, tuberculosis, which is among the top ten causes of death worldwide. Multidrug-resistant TB (MDR-TB) remains a public health crisis and a health security threat. Ending the TB epidemic by 2030 is among the health targets of sustainable development goals of the WHO. Invasive fungal infections (IFI) particularly caused by *Candida* spp. are another major cause of mortality (fourth most common) in hospital-acquired infections. Thus, the second section deals with diagnosis and treatment strategies of *Candida*-mediated IFI. Similarly, malaria is another life-threatening disease caused by protozoan parasites transmitted through bites of female anopheles mosquitoes. The disease although is curable and preventable still attracts about 28% of the research funding according to the WHO and thus forms the basis of the third section of this book. The following section deals with some critical viral diseases such as hepatitis and Ebola. Hepatitis C in particular is a liver disease which can cause both acute and chronic diseases even leading to liver cirrhosis or liver cancer. About 75% of the viral carriers are found in Asia Pacific region, thus representing significant occupational hazard for health workers. Likewise, Ebola viral disease formerly known as Ebola hemorrhagic fever is a rare but severe and often fatal illness in humans. Additionally, in this section, we have included a special chapter on "prions" from an eminent contributor that focuses on various aspects of human prion diseases including epidemiology, clinical features, pathogenesis, treatment, and prevention. The last section in particular deals with thematic issues integrating vast areas of biotechnology. For instance, chapters pertaining to antimicrobials derived from microbes itself. We have chapter highlighting the significance of human microbiome and its homeostasis and how nanobiotechnology could be beneficial in combating microbial pathogenesis.

The book intends to bridge between advances across broad range of cutting-edge research and application-based domain. It will provide a forum to foster academic exchange among researches across different fields of microbial research. The chapters are presented in an impeccable manner to update and provide comprehensive knowledge. They are gated not only to the researchers but to help the physician, medical student, and nurse practitioner to understand, diagnose, and treat common infectious diseases and also shall serve interest for the public health. The recent explosion in infectious diseases is accelerating the pace of research and development in all areas of medical sciences. Thus, the aim and scope of this book is to clearly illustrate ideas on diverse ongoing researches in the field of infectious diseases and current scenario across a wide subject spectrum.

We are grateful for the blessings and constant motivation from Dr. Ashok K. Chauhan, President RBEF, and Dr. Aseem K. Chauhan, Additional President RBEF and Chancellor, Amity University Haryana (AUH) and Amity University Rajasthan (AUR). Sincere thanks to Dr. P.B. Sharma, VC, AUH; Maj. Gen. B.S. Suhag, DVC, AUH; and Dr. Padmakali Banerjee, PVC, AUH for their overall support to enhance our academic rigor. We are grateful to our esteemed contributors for their worthy and timely contributions without which this compilation would not become a ready reference for the researchers in this field. Patience and support during the manuscript preparation from Dr. Bhavik Sawhney, Springer Nature, is deeply acknowledged. Last but not the least, we feel privileged and proud to dedicate this piece of work to our mentor Prof. Rajendra Prasad, a scholar of scientific aptitudes, administrative qualities, and amiable disposition. We are sincerely grateful to him for his guidance both professionally and even personally.

Saif Hameed
Zeeshan Fatima

Contents

Editors and Contributors

About the Editors

Saif Hameed is currently Associate Professor and Research Coordinator at Amity Institute of Biotechnology, Amity University, Haryana. He did his Bachelors from University of Delhi and Masters from Jamia Hamdard with distinction in 2003 and 2005, respectively. He completed his doctoral studies in Life Sciences from Jawaharlal Nehru University in 2010, where he also received CSIR Research Associateship for the one-year postdoctoral work. He also worked as Visiting Scholar in Institut für Mikrobiologie, Heinrich-Heine-Universität, Düsseldorf, Germany, in 2008. He has received Young Scientist award from Science and Engineering Research Board, Department of Science and Technology, New Delhi, in 2012. He is actively engaged in research in the field of infectious diseases related to multidrug resistance (MDR) in pathogenic fungi. He has around 40 peer-reviewed papers to his credit in both international and national journals of repute, 2 books, 14 book chapters, and 21 popular articles with 2 of them as "cover story." He has successfully convened four national conferences as Organizing Secretary. He has guided 4 Ph.D. students for doctoral degree and supervised 25 UG- and PG-level students for their research projects.

Zeeshan Fatima is currently working as Associate Professor at Amity Institute of Biotechnology, Amity University, Haryana. She did her Bachelors and Masters from Banaras Hindu University in 2000 and 2002, respectively. She earned her doctoral degree in Biochemistry from Aligarh Muslim University in 2008. She has held research positions under nationally and internationally funded research projects which also include her Research Associateship at BHU and JNU and postdoctoral training from University of Cincinnati, Ohio, USA, in 2010. She has received two Young Scientist awards under Fast Track and Women Scientist Schemes from the Science and Engineering Research Board, Department of Science and Technology, New Delhi, in 2012. She has also bagged regular projects for funding from Board of Research in Nuclear Sciences (BRNS), BARC Mumbai and Ministry of AYUSH, New Delhi. She is actively engaged in research in the field of infectious diseases and particularly on the aspect of multidrug resistance in human pathogen *Mycobacterium tuberculosis* and *Candida albicans*. She has 2 books and around 35 peer-reviewed papers to her credit in both international and national journals of repute. She participated in several international and national conferences and received various accolade in the form of best paper awards. She has successfully convened three national conferences as Organizing Secretary and organized many guest lectures. She has supervised 4 Ph.D. students and guided more than 20 UG- and PG-level students for their research projects.

Contributors

Mohit Agarwal Mahatma Gandhi Medical College and Hospital, Jaipur, India

Malik M. Ahmad Department of Agriculture, Integral Institute of Agricultural Science and Technology (IIAST), Integral University, Lucknow, India

Salman Ahmad Department of Agriculture, Integral Institute of Agricultural Science and Technology (IIAST), Integral University, Lucknow, India

Reem Al Jindan Department of Microbiology, College of Medicine, Imam Abdulrahman Bin Faisal University, Dammam, Saudi Arabia

Zainab Al Salem Department of Epidemic Disease Research, Institute of Research and Medical Consultations (IRMC), Imam Abdulrahman Bin Faisal University, Dammam, Saudi Arabia

Mohd Asad International Centre for Genetic Engineering and Biotechnology, New Delhi, India

Dibyendu Banerjee Molecular Structural Biology Division, CSIR-Central Drug Research Institute, Lucknow, India

Rabiya Basri Department of Plant Protection, Aligarh Muslim University, Aligarh, India
Department of Bioengineering, Faculty of Engineering, Integral University, Lucknow, India

Ashima Bhaskar Signal transduction laboratory, National Institute of Immunology, New Delhi, India

Danishuddin School of Computation and Integrated Sciences, Jawaharlal Nehru University, New Delhi, India

Gagan Dhawan Department of Biomedical Science, Acharya Narendra Dev College, University of Delhi, New Delhi, India

Ved Prakash Dwivedi International Centre for Genetic Engineering and Biotechnology, New Delhi, India

Zeeshan Fatima Amity Institute of Biotechnology, Amity University Haryana, Manesar, Gurugram, India

Hemant K. Gautam Microbial Biotechnology Laboratory, CSIR-Institute of Genomics and Integrative Biology, New Delhi, India

Ayush Gupta Department of Microbiology, All India Institute of Medical Sciences, Bhopal, India

Jeena Gupta Department of Medical Laboratory Sciences, Division of Research and Development, Lovely Professional University, Phagwara, India

Shraddha Gupta Department of Microbiology, Vallabhbhai Patel Chest Institute, University of Delhi, New Delhi, India

Saif Hameed Amity Institute of Biotechnology, Amity University Haryana, Manesar, Gurugram, India

Md. Imtaiyaz Hassan Centre for Interdisciplinary Research in Basic Sciences, Jamia Millia Islamia, New Delhi, India

Jawed Iqbal Multidisciplinary Centre for Advanced Research and Studies, Jamia Millia Islamia, New Delhi, India

Khursheed ul Islam Multidisciplinary Centre for Advanced Research and Studies, Jamia Millia Islamia, New Delhi, India

Tushar Jain Molecular Structural Biology Division, CSIR-Central Drug Research Institute, Lucknow, India

Vishwa Mohan Katoch NASI-ICMR Chair on Public Health Research, Rajasthan University of Health Sciences (RUHS), Jaipur, Rajasthan, India

S. M. Paul Khurana Amity Institute of Biotechnology, Amity University Haryana, Manesar, Gurugram, India

Chanchal Kumar Department of Microbiology, Vallabhbhai Patel Chest Institute, University of Delhi, New Delhi, India

Narender Kumar Amity Institute of Biotechnology, Amity University Haryana, Manesar, Gurugram, India

Pradeep Kumar International Centre for Genetic Engineering and Biotechnology, New Delhi, India

Sushil Kumar Molecular Structural Biology Division, CSIR-Central Drug Research Institute, Lucknow, India

Deepika Kundu Department of Biochemistry, Maharshi Dayanand University, Rohtak, Haryana, India

Sabiha Majid Department of Biochemistry, Government Medical College, Srinagar, J&K, India

Pawan Malhotra International Centre for Genetic Engineering and Biotechnology, New Delhi, India

Nikhat Manzoor Department of Biosciences, Jamia Millia Islamia, New Delhi, India

Saika Manzoor Department of Clinical Biochemistry/Biochemistry, University of Kashmir, Srinagar, J&K, India

Taj Mohammad Centre for Interdisciplinary Research in Basic Sciences, Jamia Millia Islamia, New Delhi, India

Asif Mohmmed International Centre for Genetic Engineering and Biotechnology, New Delhi, India

Azhar Muneer International Centre for Genetic Engineering and Biotechnology, New Delhi, India

Vinay Kumar Nandicoori Signal Transduction Laboratory, National Institute of Immunology, New Delhi, India

Ahmad Abu Turab Naqvi Centre for Interdisciplinary Research in Basic Sciences, Jamia Millia Islamia, New Delhi, India

Monika Narwal International Centre for Genetic Engineering and Biotechnology, New Delhi, India

Ritu Pasrija Department of Biochemistry, Maharshi Dayanand University, Rohtak, Haryana, India

Fouzia Rashid Department of Clinical Biochemistry/Biochemistry, University of Kashmir, Srinagar, J&K, India

Sumit Rathore Department of Biotechnology, All India Institute of Medical Sciences, New Delhi, India

Ashok Rattan Pathkind Labs, Gurugram, India

Suriya Rehman Department of Epidemic Disease Research, Institute of Research and Medical Consultations (IRMC), Imam Abdulrahman Bin Faisal University, Dammam, Saudi Arabia

Syed Beenish Rufai Department of Microbiology, All India Institute of Medical Sciences, Bhopal, India

Munindra Ruwali Amity Institute of Biotechnology, Amity University Haryana, Manesar, Gurugram, India

Sandeep Sharma Department of Medical Laboratory Sciences, Division of Research and Development, Lovely Professional University, Phagwara, India

Kamal Shrivastava Department of Microbiology, Vallabhbhai Patel Chest Institute, University of Delhi, New Delhi, India

Rahul Shukla Department of Pharmaceutics, National Institute of Pharmaceutical Education and Research (NIPER-Raebareli), Lucknow, Uttar Pradesh, India

Mohammad Haris Siddiqui Department of Agriculture, Integral Institute of Agricultural Science and Technology (IIAST), Integral University, Lucknow, India
Department of Bioengineering, Faculty of Engineering, Integral University, Lucknow, India

Indu Singh Department of Biomedical Science, Acharya Narendra Dev College, University of Delhi, New Delhi, India
Microbial Biotechnology Laboratory, CSIR-Institute of Genomics and Integrative Biology, New Delhi, India

Sarman Singh Department of Microbiology, All India Institute of Medical Sciences, Bhopal, India

Shweta Singh International Centre for Genetic Engineering and Biotechnology, New Delhi, India

Vandana Thakur International Centre for Genetic Engineering and Biotechnology, New Delhi, India

Mandira Varma-Basil Department of Microbiology, Vallabhbhai Patel Chest Institute, University of Delhi, New Delhi, India

Part I

Pathogenicity and Drug Resistance in *Mycobacterium tuberculosis*

Molecular Basis of Drug Resistance in Mycobacteria

Vishwa Mohan Katoch

Abstract

Currently, about two dozen drugs belonging to 15 classes are being used to treat tuberculosis, leprosy, and various nontuberculous mycobacterial diseases. Over the years, wealth of information has accumulated on molecular targets and their alterations related to the mechanisms of susceptibility/resistance to these antimycobacterial drugs. Intrinsic mechanism(s) involving permeability, expression of various enzymes, and efflux pumps appear relevant for isoniazid (INH), ethambutol (EMB), fluroquinolones (FQs), macrolides, and tetracyclines (TET). Important gene targets/loci identified for susceptibility/resistance to various antimycobacterial drugs include, Rifampicin resistance-determining (RRD) region of *rpoB* for Rifampicin; katG, inhA, kasA, ndh, oxyR-ahpC, *furA-katG*, *dfrA*, mabAfor INH; inhA promoter, katG, ethA, ethR, mshA, ndh, inhA for Ethionamide/Prothionamide; pncA gene or upstream, RpsA and pyrazinoic acid (POA) efflux for Pyrazinamide; embB, embC for Ethambutol; ribosomal protein S12/16S rRNA. *gid*B locus, 5′ untranslated region of the transcriptional activator whiB7, eis (Rv2416c) and tap (Rv1258c)) for various aminoglycosides; interface involving 23S rRNA helix 69 and 16S rRNA helix, tlyA, rrs for cyclic peptide antibiotics; rv0678, MmpL5, MAB 2299c, MAB 1483, and MAB 0540 for clofazimine; *atpE*, mmpR for bedaquiline; upregulation of Mmpl5 for cross resistance between clofazimine and bedaqquiline; Gly81Ser and Gly81Asp for delaminid; byrrl, *rplC* and *rrl* genes, 23S rRNA, rplD, efflux pumps *lmrS* and *MmpL9*for linezolid*;* 16 different gene targets for cycloserine; gyrA, gyrBfor FQs; 23S rRNA gene, erm 39, hemerythin-like protein for macrolides; FolP1 for sulphones; *thy*A gene, *folC* for p-aminosalicylic acid PAS*;* L,D-transpeptidases for beta lactam antibioticsand for Tetracyclines -Tet M determinant which codes

V. M. Katoch (✉)
NASI-ICMR Chair on Public Health Research, Rajasthan University of Health Sciences (RUHS), Jaipur, Rajasthan, India
e-mail: vishwamohankatoch18@gmail.com

S. Hameed, Z. Fatima (eds.), *Pathogenicity and Drug Resistance of Human Pathogens*,
https://doi.org/10.1007/978-981-32-9449-3_1

for a ribosomal protection protein and Tet B efflux determinant, WhiB7-independent tetracycline-inactivating monooxygenase, MabTetX (MAB1496c) have been shown to be important. Despite this vast information, understanding appears to be incomplete for most of these agents.

Keywords
Molecular mechanisms · Antimycobacterial drugs · Resistance · Gene targets

Mycobacteria are a large family of members having pathogenicity characteristics raging from obligatory pathogens such as *M.tuberculosis (Mtb)*, *M.leprae* to several opportunistic pathogens and mostly those species/subspecies still recognized as saprophytes present in the soil, water and environment. Of the more than 212 species/subspecies of mycobacteria reported so far [1], hardly one-fourth have been known to cause disease in humans and other animals [2]; others are mainly saprophytes. Mycobacteria other than *M. tuberculosis* and *M. leprae* are popularly known as nontuberculous mycobacteria (NTMs). These NTMs are broadly classified into slow and rapid growing mycobacteria which is a therapeutically meaningful grouping. While rapidly growing mycobacteria (RGM) can be treated with antibiotics commonly used to treat other pyogenic infections, the treatment for slow-growing mycobacteria is often long and faces challenges such as persistence and drug resistance. Drugs described to be effective against *M. leprae*/*M. lepromatosis* are rifampicin (RIF), dapsone (DDS), clofazimine (CLF), clarithromycin (CLA), and fluoroquinolones such as ofloxacin and moxifloxacin [3, 4]. For drug-sensitive tuberculosis, also RIF remains a core drug, to be combined with isoniazid (INH), ethambutol (EMB), and pyrazinamide (PZA) [5]. There are several others used as second-line drugs which include aminoglycosides (streptomycin, kanamycin, amikacin, etc.), quinolones (levofloxacin and moxifloxacin), linezolid, PAS, CLF, ethionamide (ETH)/prothionamide (PTH), etc. New drugs such as bedaquiline and delaminid have recently appeared on the horizon to treat drug-resistant tuberculosis [5, 6]. As per the latest WHO guidelines [6], the drugs recommended for treatment of drug-resistant TB are levofloxacin (Lfx)/moxifloxacin (Mfx), bedaquiline (Bdq), and linezolid (Lzd) (group A); clofazimine (CLF/Cfz) and cycloserine (Cs)/terizidone (Trd) (group B); and ethambutol (EMB/E), delaminid (Dlm), pyrazinamide (PZA/Z), imipenem-cilastatin (Ipm-Cln)/meropenem (Mpm), amikacin(Am)/streptomycin (S), ethionamide (Eto)/prothionamide (Pto/PTH), and p-aminosalicylic acid (PAS) (group C); all group A drugs plus one of group B are recommended to be included in MDR-TB regimens. Group C drugs are to be included if drugs from group A and B cannot be used for some reason (WHO 2019). For NTM, RIF, rifabutin, CLF, amikacin, linezolid, new-generation macrolides (azithromycin/clarithromycin), and quinolones are recommended to treat infections due to slow-growing mycobacteria, whereas tetracyclines (doxycycline and minocycline), sulphonamides, cephalosporins, and macrolides like azithromycin/clarithromycin are commonly used to treat infections due to rapidly growing mycobacteria [2, 7].

Historically, chemotherapeutic agents as well as antibiotics active against tuberculosis started becoming available in the 1940s. Soon after, drug resistance was also reported and has progressively become a barrier to treat all cases effectively. Because of this challenge, the efforts to understand the mechanism(s) of susceptibility/resistance in mycobacteria have gained priority. Studies on mechanisms have addressed the issues such as penetration of that particular drug to its site of action, conversion of drug to its active form, interference of drug/or its active ingredient of sensitive enzyme/protein, secondary effects which result from the effect of drug on its primary sites, bacteriostatic/bactericidal action, tertiary effects, inactivation of drug by bacteria, and irrelevant effects of drug on the organism [8]. Drug resistance in mycobacteria may be to one agent (mono-resistance), to two or more (polyresistance), to combinations of Rifampicin and Isoniazid (referred to as multidrug resistance, MDR), Extensively drug-resistant (XDR) TB is resistant to the most effective second-line therapeutic drugs used commonly to treat MDR TB: fluoroquinolones and at least one of three injectable second-line drugs used to treat TB (amikacin, kanamycin, or capreomycin) and total resistance. Definitions are likely to be modified as drugs like kanamycin/capreomycin are no longer listed in the new list for DR TB [6].

In the earlier years, mechanistic studies for drug action were mainly focused on levels and activities of enzymes, proteins, and lipids [8]. That knowledge is important and will ever remain relevant. During the last 25 years, rapid advances have been made in understanding the molecular basis of action of drug resistance in mycobacteria. The knowledge generated on molecular mechanisms of resistance in *Mycobacterium tuberculosis* has been reviewed by [9–14]. Flandrois et al. [15] have developed a structured data base (MUBII-TB-DB) of the mutations occurring at seven loci of major therapeutic relevance in tuberculosis management. This system also provides interpretation of the mutations in biological and therapeutic terms and has scope to evolve further by the addition of newly described mutations. In another significant development [16], the Relational Sequencing TB Data Platform (ReSeqTB) consortium has critically reviewed and published a comprehensive list of globally prevalent mutations based on the strength of their association with drug resistance. Such studies in the case of *Mycobacterium leprae* are limited and quite fragmented in case of other NTMs.

There have been quite a few interesting publications on molecular evolution of drug resistance in mycobacteria including *M. tuberculosis*. Most of the studies have observed that drug resistance has evolved locally over a long period of history. The first whole genome-based analysis of the emergence of drug resistance among clinical isolates of *M. tuberculosis* [17] showed that the ancestral precursor of the LAM4 XDR outbreak strain in Tugela Ferry gained mutations to first-line drugs at the beginning of the antibiotic era. The restricted international transmission of drug-resistant TB suggests that containment efforts at the level of individual countries/regions are likely successful and should receive priority [18]. Studies on the coevolving groups published by Vats and Shanker [19] have provided an important approach toward finding solutions of multidrug resistance problem through coevolution analysis of proteins which in turn may help to develop new drug regimen

(s) against pathogens including *M. tuberculosis.* Mioto et al. [20] have carried out an exhaustive genotype-phenotype correlation analysis as a part of a comprehensive systematic review to develop a standardized analytical approach for interpreting resistance- associated mutations for rifampicin, isoniazid, ofloxacin/levofloxacin, moxifloxacin, amikacin, kanamycin, capreomycin, streptomycin, ethionamide/prothionamide, and pyrazinamide. Mutation frequencies in resistant and susceptible isolates in this analysis were calculated using novel statistical measures to classify mutations as high, moderate, minimal, or indeterminate confidence for predicting resistance. It will be logical to conclude that messages from this data cannot be generalized without carrying out actual in-depth studies on isolates from different parts of the world.

Resistance Due to Common Mechanisms like Permeability and Efflux Pumps

Mycobacteria like other unicellular organisms possess a variety of efflux pumps with important physiological roles to pump out toxic metabolites/any other substances that will be inimical to their survival. Kashyap et al. [21] have focused on intrinsic mechanism(s) of bacterial drug resistance via expression of various enzymes and efflux pumps that are responsible for the loss of activity of the therapeutic agents. Nasiri et al. [22] have also published a comprehensive review of intrinsic mechanisms associated with or responsible for drug resistance in mycobacteria.

Several investigators have investigated the role of efflux pumps in drug resistance in mycobacteria. Gupta et al. [23] observed increased activities of other eight efflux pump genes (Rv1819, Rv2209, Rv2459, Rv2477c, Rv2688, Rv2846, Rv2994, and Rv3728) in multidrug-resistant isolates. After confirmation of differential expression of these genes by real-time reverse transcription polymerase chain reaction, it was observed that a simultaneous overexpression of efflux pump genes Rv2459, Rv3728, and Rv3065 was associated with resistance to the combination of isoniazid and ethambutol [24]. Overexpression of two of these genes (Rv2477 and Rv2209) has also been observed with ofloxacin stress in *M. tuberculosis*. Singh et al. [25] reported the involvement of active efflux pumps of both major facilitator super family (MFS) family (inhibited by CCCP and DNP) and ATP-binding cassette (ABC) transporters (inhibited by verapamil) in the development of OFL resistance in *M. tuberculosis* isolates. Coelho et al. [26] using the approach of detection of active efflux by real-time fluorometry showed that all strains presented intrinsic efflux activity that contributes to the overall resistance which can be inhibited in the presence of the EIs. Sandhu and Akhter [27] proposed a ligand-binding mechanism responsible for peristaltic movements in the channels leading to the drug efflux.

Molecular Mechanism(s) Associated with/Contributing to Resistance Against Different Groups of Antimycobacterial Drugs

The molecular mechanism(s) identified to be responsible for/contributing to resistance in *M. tuberculosis*, *M. leprae*, and several clinically important nontuberculous mycobacteria (NTM) against different important groups of antimycobacterial drugs are summarized below.

Rifamycins

Rifampicin (RIF), a semisynthetic rifamycin derivative of naturally occurring antibiotic rifamycin B, was introduced in 1972 as an antituberculosis agent [14]. While mycobacteria in general are susceptible to RIF, susceptibility varies among different species as well as within the same species and has been ascribed to penetration of drug into bacteria rather than to variation in sensitivity to RNA polymerase [8]. The mode of action of RIF in *M. tuberculosis* is by binding to the β-subunit of the RNA polymerase, inhibiting the elongation of messenger RNA [28]. It has been observed that in more than 95% of *M. tuberculosis* isolates resistant to RIF, there are mutations in the so-called hotspot region of 81-bp spanning codons 507–533 of the *rpoB* gene. This region is also known as the rifampicin resistance-determining (RRD) region [10]. Siddiqi et al. [29] have also reported several novel mutations in the rpo beta locus in Mtb strains from different parts of India. In majority of the studies, mutations in codons 516, 526, and 531 have been observed to be commonly associated with RIF resistance [11].

Srivastava et al. [30] have observed that isolates of *M. tuberculosis* resistant at 10 and 40 μg/ml had either D516V, H526Y mutations or unknown mutations. Most (85.71%) isolates resistant at clinically relevant levels (64,128 μg/ml) exhibited double, triple, or more "R"-type mutations (R(2(D516V)), R(4a(H526Y)), R(4b (H526D)), R(5(S531L))) as well as unknown mutations. In some other studies also mutations in some codons (e.g., 518 or 529) have been associated with low-level resistance to rifampicin, but those strains are still susceptible to other rifamycins, such as rifabutin or rifalazil [31, 32]. Mutations in codon 531 have been observed to be important in recent Indian studies as well [33]. It has been observed that upon RIF binding, the S531L mutant exhibits a disordering of the RIF-binding interface, which effectively reduces the RIF affinity. In contrast, the H526Y mutation reshapes the RIF-binding pocket, generating significant steric conflicts that essentially prevent any RIF binding [34]. Another study has emphasized the role of acquisition of compensatory mutations in *rpoA* and *rpoC*, encoding the α and β′ subunits of RNA polymerase, in rifampicin-resistant strains with mutations in *rpoB* [35].

Leprosy

Rifampicin is a very important drug for treatment of leprosy as well, and development of resistance to RIF in leprosy is a matter of grave concern. Honore and Cole [36] concluded that the RIF resistance in lepromatous leprosy cases who had received RIF monotherapy was due to mutations in the rpoB gene which encodes for beta subunit of RNA polymerase of the *Mycobacterium leprae*. Williams et al. [37] also studied the molecular mechanisms of rifampin resistance in *M. leprae*, *Mycobacterium avium*, and *Mycobacterium africanum* and observed that mutations in the rpoB gene similar to that documented for *M. tuberculosis* were associated with rifampicin resistance in these species as well. Cambau et al. [38] reported mutations in rpoB locus of *M.leprae* from biopsies from relapse cases. Zhang et al. [39] reported that a missense mutation at codon 516 in the rpoB gene of *Mycobacterium leprae* conferred rifampin resistance. Vedithi et al. [40] have again concluded that the presence of point mutations within the rifampin resistance-determining (RRD) regions of rpoB gene plays a vital role in conferring resistance to RIF in leprosy. Hasanoor Reja et al. [41] have observed that missense mutation in CAC codon brings about a glutamic acid to histidine change in the amino acid sequence of RNA polymerase beta subunit at the position 442 (Glu442His), a region specific for rifampicin interaction, and this might be responsible for unresponsiveness to RIF. Mutations have also been reported outside the RRD region of *M. leprae* [42]. This study also reported that *M. leprae* strain having the new mutation at codon 442 Gln-His was found to be sensitive to all the three drugs; however, strains having additional mutations at 424 Val-Gly and 438 Gln-Val were resistant to RIF by mouse foot pad (MFP) assay. Recent data from WHO Antimicrobial Resistance Surveillance network has provided a global perspective of mutations present across the world [43].

NTM

Rifampicin is also an effective antimycobacterial drug against several nontuberculous mycobacteria including *M. kansasii* [2, 7]. As discussed in the preceding paragraph, molecular mechanism(s) of resistance in NTM also appear to be similar to *M. tuberculosis* and *M. leprae* [37]. In another study, all mutations detected (in codons 513, 526, and 531) of *M. kansasii* were the same as those previously described in rifampin-resistant *M. tuberculosis* isolates [44].

Rifamycin Congeners

As resistance mechanisms found in clinical settings may also occur in natural environments, Peek et al. [45] have postulated that bacteria could have evolved to produce rifamycin congeners active against clinically relevant resistance phenotypes. These authors surveyed soil metagenomes and identified a tailoring enzyme-rich family of gene clusters encoding biosynthesis of rifamycin congeners (kanglemycins, Kangs) with potent in vivo and in vitro activity against the most common clinically relevant Rif^R mutations. Unlike Rifs, Kangs function through a mechanism that includes interfering with 5′-initiating substrate binding [45]. These observations seem to have great significance for identifying alternate sites/ compounds to combat resistance against rifamycin compounds.

Isoniazid (INH)

Isoniazid (INH) is a central component of drug regimens used worldwide to treat tuberculosis. The antimycobacterial activity of INH was discovered in 1952, and almost as soon as its activity was published, the INH-resistant *Mycobacterium tuberculosis* strains were also identified. INH and its structural analog and second-line anti-TB drug ethionamide (ETH) are prodrugs. INH is activated by the catalase-peroxidase KatG, while ETH is activated by the monooxygenase EthA.

Various studies from 1954 onward [8, 46, 47] show that catalase-peroxidase(s) possessed by mycobacteria have crucial role in susceptibility/resistance to INH. Loss of this catalase activity due to a single mutation step has been observed to be linked to emergence of resistance to INH. It has been interpreted that this plays a role in its conversion to active drug and effect of mycolic acid biosynthesis [8].

It is widely recognized that the major mechanism of INH resistance is mutation in katG, encoding the activator of INH. Various studies have also identified resistance-associated mutations in katG, inhA, kasA, ndh, and the oxyR-ahpC intergenic region. Single nucleotide polymorphisms in other genes in isoniazid-resistant clinical isolates of *M. tuberculosis*, including *kasA* and the *oxyR-ahpC* and *furA-katG* intergenic regions, have also been reported [48]. In the review by [48], more than 75% of all INH-resistant isolates were observed to have a mutation in the katG gene. One specific KatG variant, S315T, has been reported to be present in most of INH-resistant clinical isolates of *M. tuberculosis*. The identification of a mutation in the katG gene of some of the clinical isolates showed four mutations, i.e., C1061T, G1261 A, G1388T, and G2161A, which correspond to the amino acid substitutions T354I, G421S, R463L, and V721M, respectively [49]. Ramasubban et al. [50] have reported that His276Met, Gln295His, and Ser315Thr mutations in the katG gene result in decreased stability and flexibility of the protein at INH-binding residues leading to impaired enzyme function.

The second important mechanism of INH resistance has been ascribed to mutation in the promoter region of inhA (c-15t), which results in inhA overexpression. Mutations in the inhA open reading frame and promoter region have been observed to be important mechanism of resistance to ETH, found more often in ETH-resistant clinical isolates than mutations in the activator of ETH. Besides this other mechanisms of resistance to INH and ETH include expression changes of the drugs' activators, redox alteration, drug inactivation, and efflux pump activation [51].

While mutations of the katG gene in *M. tuberculosis* have been identified as a major INH resistance mechanism, several other responsible genes/loci incriminated, yet the molecular basis for the resistance is unclear in 10–20% INH-resistant clinical isolates [49, 52]. Efflux pump-like mechanisms may be contributing partially as observed by [24], who reported that increased transcription of jefA also leads to increased resistance to ethambutol and isoniazid in *M. tuberculosis* via efflux pump-like mechanism.

Mutations in *dfrA* gene have also been linked with INH resistance; however, results are variable. It has also been observed that isoniazid-NADP adduct causes inhibition of the dihydrofolate reductase (DfrA) in *M. tuberculosis*, which led to thinking that mutations in *dfrA* may be possibly playing a role in resistance to INH [53]. However, others have failed to identify any mutation in *dfrA* associated with resistance to isoniazid [54].

Mutations in mabA [55], G279D mutant types [56], and several other novel loci [52] and novel mutations in KatG [50] have also been linked to INH resistance. As multiple mechanisms have been reported to be contributing to development of INH resistance in *M. tuberculosis*, their overall clinical significance needs to be determined by multicentric studies.

Ethionamide/Prothionamide

Ethionamide (ETH) and prothionamide (PTH) are thioamide drugs having structure resembling INH. Both are useful second-line antituberculosis drugs. They are bactericidal and act by inhibition of mycolic acid through specific action on inhA product enoyl-acyl carrier protein reductase. A systematic review published by [57] concludes that efficacy of both is similar with some studies showing PTH to be slightly better. Initially it was reported that the action of ETH is similar to INH [58]. However, notable distinction was observed by lack of cross resistance, and strains resistant to ETH were observed to be sensitive to INH and vice versa [59]. In most of the isolates, resistance to INH and ETH has been explained by mutations in the inhA promoter and in other genes including katG, ethA, ethR, mshA, ndh, and inhA [60]. A significant proportion of ETH-susceptible MDR-TB isolates also harbor mutations in katG; some show mutation in ethA, mshA, and inhA [60]. These studies as well as others discussed in the section on INH show that common and distinct molecular resistance mechanisms exist for INH and ETH/PTH.

Pyrazinamide (PZA)

Pyrazinamide (PZA) is a valuable anti-TB drug which also acts against persister/dormant bacilli. PZA has been thought to act after its conversion to pyrazinoic acid by bacilli possessing particular amidase [61]. Despite initial success, its exact mechanism remained unclear even after 25 years after that [8]. With the application of sequencing techniques, wealth of data on mutations associated with resistance to

PZA has been generated. Scorpio and Zhang [62] reported that transformation of PZA-resistant strains with a functional pncA gene restored PZase activity and PZA susceptibility. In other studies also mutations in the pyrazinamidase (PZAse) encoding gene, pncA, have been observed to be the main cause of pyrazinamide (PZA) resistance in *Mycobacterium tuberculosis* [63, 64]. In silico models [65] provided interesting explanation for the binding interaction of PZA with PZase. Singh et al. [63] reviewed the mechanisms of PZA resistance and concluded that majority (72–97%) of PZA-resistant isolates of *M. tuberculosis* exhibit mutations in their pncA gene or upstream area leading to loss of PZase activity. A wide diversity of pncA mutations scattered along the entire length of pncA gene have been linked to PZA resistance by Huy et al. [66]. However, PZA-resistant isolates with normal PZase activity and wild-type pncA sequences have also been reported in several studies which indicate that alternate mechanisms of PZA resistance exist. Yang et al. [67] have observed that PZA resistance is associated with RpsA mutations.

Another important observation is that pyrazinoic acid (POA) efflux rate is the basis of the PZA susceptibility, and its quantitative measurement by Wayne's test has been found to be a highly sensitive and specific predictor of PZA resistance. Based on biological considerations, the POA efflux rate is directly determined by the PZAse activity, the level of pncA expression, and the efficiency of the POA efflux pump system. These results suggest that tests which rely on pncA mutations or PZAse activity are likely to be less predictive of real PZA resistance than tests which measure the rate of POA efflux [68]. More recently, [69] have identified several efflux proteins involved in PZA resistance. It is thus clear that multiple molecular mechanisms contribute to susceptibility/resistance to PZA in *Mycobacterium tuberculosis*.

Ethambutol (EMB)

Ethambutol (EMB) was first introduced in the treatment of TB in 1966 and continues to be part of the current first-line regimen to treat the disease. Ethambutol is bacteriostatic against multiplying bacilli and is also useful to treat NTM disease due to MAC, *M. kansasii*, *M. marinum*, and others. Initial studies in *M. smegmatis* model showed that EMB causes declumping [70]. It was also observed to have effect on lipid and cell wall of mycobacteria [71, 72], but exact role remained unclear for a long period after that [8]. Takayama and Kilburn [73] showed that EMB interferes with the biosynthesis of arabinogalactan in the cell wall. After publication of important reports of EMB activity on cell wall lipids and arabinogalactan biosynthesis, interesting data on genes associated with these pathways has also been reported [74]. Ramaswamy et al. [75] observed that most *M. tuberculosis* isolates (68%) with resistance-associated mutations in a single gene had nucleotide changes in embB, a gene encoding an arabinosyltransferase involved in cell wall biosynthesis. The majority of these mutations resulted in amino acid replacements at position 306 or 406 of embB. Several other studies also show that phenotypic resistance to ethambutol is only partly explained by mutations in the *embB* 306 codon [14, 76]. Overall about 30% ethambutol-resistant strains do not have any mutation in embB indicating the need to identify other possible mechanisms of resistance to this drug [14]. Jadaun et al. [77] have reported that while embB306 mutations were commonly associated with EMB resistance, mutation at codon 270 of the embC gene was also observed to contribute to high-level EMB resistance in some Mtb isolates. Additional resistance-conferring mutations have been reported in the *embB* gene at codons 354, 406, and 497 [78]. There is clearly a lot of scope for searching for other molecular mechanisms for EMB resistance.

Aminoglycosides

Aminoglycosides have played a pivotal role in the treatment of tuberculosis and other mycobacterial infections. These act on mycobacteria through common mechanisms which are linked to penetration, anaerobiasis, effects linked to pH, effects on protein synthesis by misreading of mRNA, antagonize the associationof mRNA with ribosomal RNA subunits, inhibition of polypeptide synthesis and slow down of chain elongation. These mechanisms have been described in different reviews beginning from [79] to [8]. In the case of *Mycobacterium tuberculosis*, a high degree of streptomycin resistance has been demonstrated to be ribosomal, whereas low resistance appeared to be due to reduced penetrance [80, 81].

High level resistance to Streptomycin was long back reported to be due to mutations in gene strA (rpsL), encoding a component of 30S ribosomal subunit [82], mutation in this gene does not cause resistance to neomycin, kanamycin, viomycin or capreomycin. While kanamycin and neomycin also bind to 30S subunit of ribosomes, the sites are different. Kanamycin has little effect on streptomycin binding to ribosomes [83]. High-level resistance to kanamycin and neomycin can occur by single mutation step in nek gene, which does not result in resistance to streptomycin but resistance to viomycin and capreomycin [82].

Finken et al. [84] have observed that mutations responsible for aminoglycoside resistance either lead to amino acid changes in ribosomal protein S12 or alter the primary structure of the 16S rRNA. The 16S rRNA region mutated perturbs, a pseudoknot structure in a region which has been linked to ribosomal S12 protein. Overall, mutations in rpsL and rrs have been observed to be responsible for streptomycin resistance in 60–70% of Mtb isolates [14, 85], suggesting the need to search for additional mechanisms of streptomycin resistance.

Over the years ***gid*B** locus has also emerged as another important gene associated with streptomycin resistance. Okamoto et al. [86] were first to identify gidB as a new streptomycin resistant locus and uncovered a resistance mechanism that is mediated by loss of a conserved m(7)G modification in 16S rRNA. Subsequently, [87] reported microbiological evidence for the contribution of gidB in streptomycin resistance. Afterward, [88] identified eight novel sense mutations and four novel missense mutations in gidB.

Rominski et al. [89] reported that *M. abscessus* expresses two distinct AG resistance determinants, AAC (2′) and Eis2, which confer clinically relevant drug resistance. Viljoen et al. [90] observed that MmpLs contribute to streptomycin resistance mechanisms in mycobacteria. There is need to investigate these mechanisms in different clinically relevant mycobacteria before making any projections on usefulness of these interesting observations.

Sharma et al. [91] using a proteomic analysis approach identified several proteins differentially regulated in streptomycin (SM)-resistant isolates of *M. tuberculosis*. These are encoded by several genes, namely, Rv0350, Rv0440, Rv1240, Rv3075c, Rv2971, Rv3028c, Rv2145c, Rv2031c, and Rv0569. In silico docking analysis showed significant interactions of SM with essential (Rv0350, Rv0440, and Rv2971) and nonessential (Rv1240, Rv3075c, and Rv2031c) genes. The proteomic analysis of *M. tuberculosis* isolates resistant to second-line drugs (Kanamycin and Amikacin) showed that the genes/proteins involved in iron metabolism and the two hypothetical proteins (Rv3867 and Rv3224) may be playing some role in contributing resistance to second-line drugs [92].

Ahn and Kim [93] suggest that the Rv3168 mediates kanamycin resistance in *M. tuberculosis*, likely through phosphotransferase targeting of kanamycin. Reeves et al. [94] have also identified eight independent mutations in the 5′ untranslated region of the transcriptional activator whiB7 that confer low-level resistance to both aminoglycosides. The mutations lead to 23- to 145-fold increases in whiB7 transcripts and subsequent increased expression of both eis (Rv2416c) and tap (Rv1258c). Increased expression of eis confers kanamycin resistance in these mutants, while increased expression of tap, which encodes an efflux pump, has been associated with low-level streptomycin resistance.

Cyclic Peptide Antibiotics

Capreomycin and the structurally similar compound viomycin are cyclic peptide antibiotics which are known to be active against *Mycobacterium tuberculosis*, including multidrug-resistant strains. Maus et al. [95] have reported that mutation of tlyA, encoding a putative rRNA methyltransferase, confers capreomycin resistance. It has also been observed that mutation of the tlyA gene confers capreomycin and viomycin resistance in *Mycobacterium tuberculosis*. Maus et al. [96] have also studied the phenomenon of cross resistance among capreomycin, kanamycin, amikacin, and viomycin. Mutations in the 16S rRNA gene (rrs) have been found to be associated with resistance to each of the drugs. Three rrs mutations (A1401G, C1402T, and G1484T) have been found to be linked with a particular pattern of cross resistance.

Akbergenov et al. [97] investigated the mechanisms of action of capreomycin, a cyclic peptide antibiotic, and viomycin, the first reported tuberactinomycin. Both of these antibiotics bind across the ribosomal interface involving 23S rRNA helix 69 (H69) and 16S rRNA helix 44 (h44). The binding site of tuberactinomycins in h44 partially overlaps with that of aminoglycosides, and they share with these drugs the side effect of irreversible hearing loss also. Using the approach of site-directed mutagenesis, [97] identified rRNA residues in h44 as the main determinants of phylogenetic selectivity of proposed mechanisms involved in tuberactinomycin ototoxicity. These leads are important from the point of view of future developments of better and less toxic antimycobacterial compounds.

Clofazimine (CLF)

Clofazimine (CLF), a riminophenazine compound, has a wide-spectrum antimycobacterial activity [8]. It was originally discovered as an anti-TB drug in experimental animals [98]. Due to its comparative less efficacy in human TB probably due to its distribution pattern in various tissues and availability of more potent antituberculosis agents subsequently, its use in TB did not gain wider acceptance till emergence of drug-resistant TB. It has been observed that CLF has affinity to lipid-rich tissues and also concentrates in cells of reticuloendothelial system. Over the years, CLF became more important as anti-leprosy drug not only for its anti-*M. leprae* activity, but it is a strong anti-reaction/anti-inflammatory drug and is a key component of management of leprosy [3]. While initially CLF was recommended to be part of multibacillary (MB) leprosy regimens [99], it was also observed to reduce morbidity in paucibacillary (PB) leprosy, thus useful across the spectrum of disease [100]. Latest WHO guidelines recommend a common CLF containing regimen with different durations, 6 months for PB and 12 months for MB leprosy [4]. CLF has also shown good activity against various NTMs including MAC and *M. kansasii* [2]. It has also been found to be useful to treat drug-resistant TB [6].

For a long time, there was very little understanding about its mode of action of CLF except for the information of its binding to guanine residues of DNA [101]. Only very recently important information has started becoming available. Zhang et al. [103] have observed that 97% of CLF-resistant mutants had a mutation in rv0678 encoding a transcription repressor for efflux pump MmpL5. Earlier reported A202G mutation (S68G) in rv0678 occurred less frequently in this study. The remaining 34 mutations were found to be scattered along the entire rv0678 gene. Hartskkorn et al. [102] have observed mutations in Rv0678 with concomitant upregulation of MmpL5 are responsible for CLF and bedaquiline resistance. Zhang et al. [103] have also described two new genes (rv1979c and rv2535c) associated with CLF resistance in strains without rv0678 mutations. Further, [104] have reported that mutations in MAB 2299c, MAB 1483, and MAB 0540 are the major mechanisms of CLF resistance in *M. abscessus* [104]. As whole genome sequencing has become less expensive and more widely used, better understanding of the molecular basis of CLF susceptibility/resistance in pathogenic mycobacteria can be expected in the future.

Bedaquiline (Bdq)

Since its discovery reported by Andries et al. [105], bedaquiline (Bdq) has emerged as an important drug to treat MDR/XDR-TB. It is a diarylquinoline (earlier identification TMC 207/R207910) which acts on mycobacterial ATP synthase. This drug is active against *M. tuberculosis* and several other mycobacteria. It also inhibits ATP synthase of non tuberculous rapid growing mycobacteria, the D29V or A64P substitutions in *atpE* conferred high resistance, thus resolving the target of BDQ in *M. abscessus* [106]. Bedaquiline acts by binding to the c-subunit in the membrane-bound F_O portion of the F_1F_O-adenosine triphosphate (ATP) synthase. Recent observations related to the bactericidal effects of bedaquiline show that it is a potent uncoupler of respiration-driven ATP synthesis [107]. A bedaquiline-containing regimen eradicated persistent TB infections and completely prevented disease relapse in mice [108]. This activity against persisters is very promising as this will reduce the relapses. In the case of *Mycobacterium xenopi*, if the highly conserved residue Ala63 is replaced by Met, this modification makes it resistant to R207910 [109].

Various studies on *M. tuberculosis* isolates from Australia [110], India [111], China [112], and Russia [113] showed that mutations in atpE were commonly associated with Bdq resistance. Mutations in mmpR gene were also found to be linked to high MIC to Bdq in Russian isolates [113]. Cross resistance between CLF and Bdq has been observed to be a common problem as even strains from patients not earlier exposed to these drugs have been observed to this cross resistance ([114, 115], Vilellas et al. 2018). Mutations at locus Rv0608 have been frequently shown to be associated with CLF and Bdq cross resistance [102, 116–118]. Hartskoorn et al. [102] observed that cross resistance between CLF and Bdq occurs through upregulation of MmpL5 in Mtb along with mutations in transcription regulator Rv0678. Ismail et al. [118] reported that while the Rv0678 RAVs were not the dominant mechanism of CFZ resistance, the cross resistance was limited to isolates with an Rv0678 mutation. In another study, point mutations or deletion of *MAB2299c* was associated with the concomitant upregulation of the *mmpS* and *mmpL* transcripts and accounted for this cross resistance [116]. Presence of mutations in these target genes of a significant proportion of isolates even without prior exposure to these drugs [114, 115] shows the need to search for alternate mechanisms of developing such resistance/cross resistance involving Bdq.

Delaminid

Delaminid, earlier known as OPC 67683, is a derivate of nitro-hydro-imidazoooxazole and has been demonstrated to be an inhibitor of mycolic acid biosynthesis [14]. This compound was initially reported by Matsumoto et al. [119] to be highly active against Mtb (sensitive as well as resistant isolates) in mouse model. While this drug has also become a preferred drug to treat drug-resistant TB, robust breakpoints still do not exist for bedaquiline and delamanid several years after their approval for human use [120]. There are few studies on molecular basis of action/resistance to delaminid. Yang et al. [121] have observed that the Gly81Ser and Gly81Asp mutations were associated with resistance to delamanid. On the other hand, Fujiwara et al. [122] have reported delamanid-resistant bacilli have mutations in one of the five genes in the F_{420}-dependent bio-activation pathway with distinct F_{420}HPLC elution patterns.

Linezolid (LZD)

Linezolid (LZD) has become an important drug for the treatment of multidrug-resistant tuberculosis (MDR-TB), infections due to NTMs, as well as other pathogens. While significant information has become available about its inhibitory effects on several organisms including Mtb, mechanisms of resistance to LZD are not well understood. During the recent years, several important studies on molecular mechanisms of resistance to LZD in *Mycobacterium tuberculosis* have been published. Zhang et al. [123] have reported byrrl and rplC mutations in Mtb, later being associated with high MICs. Ismail et al. [118] have reported mutations in the *rplC* and *rrl* genes of LZD-resistant Mtb isolates. Zimenkov et al. [113] have observed that mutations in 23S rRNA gene (g2294a and g2814t) and the C154R substitution in ribosomal protein L3 are associated with LZD resistance. Wasserman et al. [124] have observed that all isolates with phenotypic resistance studied by them were associated with known resistance mutations, most frequently due to the T460C substitution in rplC and rrl mutations such as G2814T, G2270C/T, and A2810C.

There are very few studies evaluating the mechanisms driving LZD resistance in nontuberculous mycobacteria. Kim et al. [125] have reported that novel mutations G2599A and A2137T in the 23S rRNA gene and mutations A439G and G443A in the rplD gene are associated with LZD resistance in *M. avium* complex isolates. Ye et al. [126] have reported higher transcriptional levels of efflux pumps *lmrS* and *mmpL9* in LZD-resistant isolates of *M. abscessus* and have suggested these as potential target genes for further studies.

Cycloserine

D-cycloserine is a useful second-line drug against tuberculosis. This drug competitively inhibits two enzymes which act consecutively: alanine racemase and D-alanine synthetase [127]. It has also been found to inhibit the synthesis of D-alanyl-D-alanine and of peptidoglycan in *M. tuberculosis* [128]. Till recently, there was very little information about molecular basis of resistance to D-cycloserine. Chen et al. [129] have identified mutations in 16 genes that are associated with D-cycloserine resistance. Interestingly, these mutations were found only in alr (rv3423c) encoding alanine racemase, but not in other known D-cycloserine resistance-associated genes such as ddl, cycA, or ald. Significantly, [129] have identified 13 new genes [rv0059, betP (rv0917), rv0221, rv1403c, rv1683, rv1726, gabD2 (rv1731), rv2749, sugI (rv3331), hisC2 (rv3772), the 5′ intergenic region of rv3345c and rv1435c, and the 3′ region of rv0759c] that had solo mutations associated with D-cycloserine resistance. Desjardins et al. [130] have also observed that loss-of-function mutations in ald (Rv2780), encoding L-alanine dehydrogenase, are associated with drug resistance. Chen et al. [129] also conclude that the mechanisms of D-cycloserine resistance are more complex than previously thought and involve genes participating in different cellular functions such as lipid metabolism, methyltransferase, the stress response, and transport systems.

Terizidone (TRD)

Terizidone (TRD), a structural analogue of cycloserine, was reported to be useful in treatment of pulmonary TB long back [131]. A meta-analysis published by Hwang et al. [132] concludes that terizidone (TRD) may be better tolerated than cycloserine. There is very few published literature regarding its mechanism(s) of action/ resistance.

Fluoroquinolones (FQs)

During the recent years, fluoroquinolones have emerged as important second-line anti-TB agents, also important in the treatment of infections caused by NTMs and also leprosy especially nonresponsive/resistant diseases. Choice of preferred FQs has also rapidly evolved from Cipro, Oflo, Peflo, to moxifloxacin and levofloxacin. Mode of action of FQs is by inhibition of topoisomerase II (DNA gyrase) and topoisomerase IV which are important for viability and are encoded by gyrA, gyrB, parC, and parE [14].

FQ Resistance in *Mycobacterium tuberculosis*

Over the years several mutations in gyrA such as Ala90-Val, Asp94-Gly/Tyr/Ala, etc. and gyrB have been identified in Mtb that are commonly associated with FQ resistance [133, 134]. It needs to remembered that several mutations like those at position 95 of gyrA are mainly polymorphisms found in both sensitive and resistant isolates [135]. Besides mutations, efflux pumps have also been observed to be contributing significantly to FQ resistance in Mtb [25].

Pantel et al. [136] have demonstrated that the eight substitutions in GyrB (D473N, P478A, R485H, S486F, A506G, A547V, G551R, and G559A), identified in FQ-resistant clinical strains of *M. tuberculosis* isolated in France, could not be implicated in FQ resistance. These results also underline that, as opposed to phenotypic FQ susceptibility testing, the DNA gyrase inhibition assay is the only way to prove the role of a DNA gyrase mutation in FQ resistance. Therefore, the use of FQ in the treatment of tuberculosis (TB) patients should not be ruled out only on the basis of the presence of mutations in gyrB.

FQ Resistance in *M. leprae*

There are several reports regarding the mutations in gyrA and gyrB loci of *M. leprae* isolated from human leprosy cases [37, 38, 42, 43, 137–139]. While mutations in these loci have been identified and thought to be responsible for FQ resistance, validation in mouse foot (MFP) assay has not been done in all cases. Veziris et al. [139] observed that DNA gyrase mutation is not always synonymous of lack of in vivo fluoroquinolone activity in *M. leprae*. These authors have also suggested potential contribution of an amino acid substitution of Asp to Gly or Asn at position 95 to fluoroquinolone resistance. Maeda et al. [137] studied mutations in the quinolone resistance-determining region of *gyrA* which were also reported for quinolone-resistant mycobacteria. Only one case of quinolone resistance in leprosy was reported, from Mali. One mutation (Ala→Val at position 91) was detected in five isolates of *M. leprae*, and another mutation (Gly→Cys at position 89) was found as in quinolone-resistant *M. tuberculosis*. Cambau et al. [38] have reported gyrA mutation in a MFP proven ofloxacin-resistant case. Nisha and Shanthi [140] have recently reported that binding energy measurements obtained from molecular docking studies suggest that hydrogen bond-mediated efficient binding of ofloxacin to Asp47 in the native GyrA-DNA complex in comparison with that of the mutant GyrA-DNA complex would be useful in explaining the mechanism(s) of FQ resistance in *Mycobacterium leprae*.

Macrolides

Macrolides like azithromycin and clarithromycin are considered as important antimycobacterial agents [2, 7]. Their activity is limited against *M. tuberculosis* because of an intrinsic resistance due to the presence of a sequence encoding a putative rRNA methyltransferase. The deduced protein is similar to Erm methyltransferases, which confer macrolide-lincosamide-streptogramin (MLS) resistance by methylation of 23S rRNA (named ErmMT). This resistance in *Mycobacterium tuberculosis* complex is a good example of a gene conferring resistance by target modification [141]. Nash and Inderlied [142] have reported a link between mutations within the 23S rRNA gene and clinically significant macrolide resistance in *M. avium*. Jamal et al. [143] have suggested that most of the resistance in *M. avium-intracellulare* complex (MAC) arose from the mutation in domain V of the 23S rRNA gene; however, other unknown mechanisms evidently exist in mycobacteria. Pfister et al. [144] have also observed that 23S rRNA A2058G alteration mediates macrolide resistance. Nash et al. [145] have reported that erm (39) is linked to macrolide resistance in *M. fortuitum*. Mougari et al. [146] have also observed that acquisition of clarithromycin resistance in *M. abscessus* strains in their series was 100% mediated by structural changes in 50S subunit mutations. Hemerythrin-like protein MSMEG3312 has also been observed to be involved in macrolide resistance [147].

Sulphones

Dapsone (4,4-diaminodiphenylsulphone, DDS), a synthetic chemotherapeutic agent, has been used in the treatment of leprosy for the last nearly seven decades. Another sulphone acedapsone was tried earlier as chemoprophylactic against leprosy. Its antibacterial activity is inhibited by para-aminobenzoate (PABA), because of this the mechanism(s) of action of DDS and other sulphonamides have been interpreted to be similar through folic acid synthesis [148]. These investigators have studied the role of folP1 and folP2 genes encoding dihydropteroate synthase (DHPS) and concluded that folP1 is functional in *M. leprae*, and mutations in this gene are associated with dapsone resistance. On the other hand, folP2 does not have role in DDS susceptibility/resistance [148]. Such conclusions can also be drawn from the data published by others as well [42, 43, 137, 149]. Some association between levels of resistance and mutations at codons 53 or 55 has also been described [149]. Chaitanya et al. [150] have investigated the underlying mechanism by in silico experiments. An increase in volume of the binding cavities of mutant structures was noted when compared to native form indicating a weakening in interaction leading to lower activity of enzyme.

Para-amino Salicylic Acid (PAS)

PAS was among the first group of chemotherapeutic agents found to be active against *Mycobacterium tuberculosis* and is still in the group C list of the WHO [6] to treat drug-resistant TB. Despite this long history of its use, the understanding of mechanisms for susceptibility/resistance to PAS is still incomplete. Mutations in *thy*A gene [151] and in *folC* [152] are associated with resistance to PAS. Zhao et al. [152] have reported that various missense mutations within the coding sequence of the dihydropteroate (H2Pte) binding pocket of dihydrofolate synthase (*Fol*C) confer PAS resistance in laboratory isolates of *M. tuberculosis* and *Mycobacterium bovis* [152].

Beta-lactam Antibiotics

Beta-lactam antibiotics inhibit the bacterial D,D-transpeptidases that are involved in cell wall biosynthesis. Their use has been extremely limited in the treatment of *Mycobacterium tuberculosis* infections because of this organism's resistance to beta-lactams. Clavulanate reacts with the enzyme quickly to form hydrolytically stable, inactive forms of the enzyme and thus has potential to be used in combination with beta-lactam antibiotics to treat multidrug resistant (MDR) and extremely drug-resistant (XDR) strains of *M. tuberculosis* [153]. Gupta et al. [154] have suggested that a combination of L,D-transpeptidase and beta-lactamase inhibitors could effectively target persisting bacilli during the chronic phase of tuberculosis. Cordilott et al. [155] have observed that the carbapenems, an important β-lactam class, inactivate L,D-transpeptidase (LDTs) of *Mycobacterium tuberculosis*, thereby affecting the cross linking of peptidoglycan. Imipenem inactivated LDTs more rapidly than ertapenem, and both drugs were more efficient than meropenem and doripenem, indicating that modification of the carbapenem side chain could be used to optimize their antimycobacterial activity [155]. Tiberi et al. [156] have also observed that meropenem/clavulanate are more effective than imipenem/clavulanate in treating MDR/XDR-TB patients. The single crystal X-ray structure of the extracellular portion of the L,D-transpeptidase (ex-LdtMt2 – residues 120–408) enzyme has been recently published and used to understand the mechanism of action of imipenem and meropenem on the peptidoglycan layer of *Mycobacterium tuberculosis* [157].

Inderlied et al. [158] have reported that meropenem is effective against GNB, GPB, and *M. avium*. Several beta-lactam antibiotics are recommended for the treatment of NTM infections [2, 7]. Imipenem remains the most active carbapenem against rapidly growing mycobacteria (RGM), including *Mycobacterium abscessus* subsp. *abscessus* [159]. Kumar et al. [160] have also demonstrated that inhibition of these enzymes by the carbapenem class of β-lactams determines their activity against *Mycobacterium tuberculosis*. Interestingly both the L,D-transpeptidases in *M. abscessus*, namely, Ldt_{Mab1} and Ldt_{Mab2}, were found to be inhibited by both the carbapenem and cephalosporin, but not penicillin. Contrary to the commonly held belief that combination therapy with β-lactams is redundant, doripenem and cefdinir exhibit synergy against both pan-susceptible *M. abscessus* and clinical isolates that are resistant to most antibiotics, which suggests that dual-β-lactam therapy has potential for the treatment of *M. abscessus* infections.

Tetracyclines (TET)

Tetracyclines are old antibiotics reported to be useful in various infectious and apparently some noninfectious disease conditions. Doxycycline and minocycline are among the tetracyclines found to be useful in treating infections due to rapidly growing mycobacteria (RGM) especially *M. fortuitum* and *M. chelonae* [2] as well as leprosy [161]. Compared to doxycycline, minocycline has been reported to be more effective/active against *M. marinum* [162], and against RGM [163].

Mechanism of action of tetracyclines appears to be complex and ill understood even after several decades of their discovery and clinical use. Robert [164] has focused on tetracycline resistance genes which are common among different genera and has concluded that Tet M determinant which codes for a ribosomal protection protein and Tet B efflux determinant is responsible for resistance to tetracycline, doxycycline, and minocycline. Ramon-Garcia et al. [165] have described the role of

Tap protein-mediated efflux in extruding tetracycline from *M. fortuitum* cells. Kyselková et al. [166] have also concluded that intrinsic efflux pumps may be more important for TET resistance than horizontally transferred genes in clinical RGM as well as in environment. Rudra et al. [167] have shown that high level of resistance to tetracycline and doxycycline in *M. abscessus* is conferred by a WhiB7-independent tetracycline-inactivating monooxygenase, MabTetX (MAB1496c). Despite these important advances, there is no consensus on the mechanisms of action of various tetracyclines on mycobacteria. Chukwudi [168] has commented that the 16S rRNA-binding mechanism currently held for the antibacterial action of the tetracyclines does not explain their activity against viruses, protozoa that lack mitochondria, and noninfectious conditions. In light of recent evidence that the tetracyclines bind to various synthetic double-stranded RNAs (dsRNAs) of random base sequences, [168] has suggested to consider possible alternative binding modes or sites that could help explain the mechanisms of action of the tetracyclines against various pathogens and disease conditions.

Future Perspective

This chapter has reviewed important mechanisms involved in susceptibility/resistance to important antimycobacterial drugs used for treatment of common mycobacterial diseases like tuberculosis, leprosy, and important nontuberculous mycobacteria. Thus there are inherent limitations that not all the antimycobacterial drugs (e.g., thiacetazone) mentioned in earlier reviews [8, 14] are discussed in this chapter. Further out of a pool of several hundred good-quality publications, only selected references could be quoted in this chapter. Considering the dynamics of ever-changing drug susceptibility profiles, the priorities of clinicians and public health professionals will also change. Wide gaps in understanding the molecular basis of action of several drugs are apparent. Further evidence available should be carefully analyzed in terms of genotype-phenotype correlations as such results may be affected by the proportion of mixed populations due to heteroresistance [169] and many unknown confounding factors. Appropriate use of high-throughput sequencing methods, structural genomic tools, and functional genomic studies with different combinations of drugs/environmental stresses will help in resolving many mysteries surrounding the action of antimycobacterial drugs.

References

1. An approved list of bacterial names with standing in nomenclature – Genus Mycobacterium. Available at URL: http://www.bacterio.net/mycobacterium.html. Last accessed on 28th May, 2019.
2. Katoch, V. M., Kumar, M., & Babbanjee. (2019). Non tuberculous mycobacterial infections. In S. K. Sharma & A. Mohan (Eds.), *Textbook of tuberculosis and non tuberculous mycobacterial diseases* (pp. 506–542). New Delhi: Jaypee Publishers.
3. National Leprosy Eradication Programme. (2013). Training Manual for Medical Officer, Central Leprosy Division, DGHS, MoHFW, Govt of India, pp 16–21.

4. World Health Organization. (2018). Guidelines for the Diagnosis, treatment and prevention of Leprosy.
5. Revised Tuberculosis Control Programme. (2016). Technical and operational guidelines for tuberculosis control in India. Central TB Division, DGHS, MoHFW, Govt of India, pp 30–45.
6. WHO. (2019). *WHO consolidated guidelines on drug resistant tuberculosis treatment*. Geneva: WHO.
7. Griffith, D. E., Aksamit, T., Brown-Elliott, B. A., Catanzaro, A., Daley, C., Gordin, F., et al. (2007). An official ATS/IDSA statement: Diagnosis, treatment, and prevention of nontuberculous mycobacterial diseases. *American Journal of Respiratory and Critical Care Medicine, 175*, 367–416.
8. Winder, F. G. (1982). Mode of action of the antimycobacterial agents and associated aspects of molecular biology of the mycobacteria. In J. L. Stanford & C. Ratledge (Eds.), *The biology of mycobacteria* (Vol. 1, pp. 383–438). London: Academic.
9. Cole, S. T. (1994). *Mycobacterium tuberculosis*: Drug-resistance mechanisms. *Trends in Microbiology, 2*, 411–415.
10. Ramaswamy, S., & Musser, J. M. (1998). Molecular genetic basis of antimicrobial agent resistance in *Mycobacterium tuberculosis*: 1998 update. *Tubercle and Lung Disease, 79*, 3–29.
11. Somoskovi, A., Parsons, L. M., & Salfinger, M. (2001). The molecular basis of resistance to isoniazid, rifampin, and pyrazinamide in *Mycobacterium tuberculosis*. *Respiratory Research, 2*, 164–168.
12. Almeida Da Silva, P. E., & Palomino, J. C. (2011). Molecular basis and mechanisms of drug resistance in *Mycobacterium tuberculosis*: Classical and new drugs. *The Journal of Antimicrobial Chemotherapy, 66*, 1417–1430.
13. Cohen, K. A., Bishai, W. R., & Pym, A. S. (2014).Molecular basis of drug resistance in *Mycobacterium tuberculosis*. *Microbiology Spectrum, 2*. https://doi.org/10.1128/microbiolspec.MGM2-0036-2013.
14. Palomino, J. C., & Martin, A. (2014). Drug resistance mechanisms in *Mycobacterium tuberculosis*. *Antibiotics (Basel), 3*, 317–340.
15. Flandrois, J. P., Lina, G., & Dumitrescu, O. (2014). MUBII-TB-DB: A database of mutations associated with antibiotic resistance in Mycobacterium tuberculosis. *BMC Bioinformatics, 15*, 107. https://doi.org/10.1186/1471-2105-15-107.
16. Cirillo, D. M., Miotto, P., Tagliani, E., & ReSeqTB Consortium. (2016). Reaching consensus on drug resistance conferring mutations. *International Journal of Mycobacteriology, 5* (Suppl1), S31–S32. https://doi.org/10.1016/j.ijmyco.2016.09.062. Epub 2016 Nov 15.
17. Cohen, K. A., Abeel, T., Manson McGuire, A., Desjardins, C. A., Munsamy, V., Shea, T. P., et al. (2015). Evolution of extensively drug-resistant tuberculosis over four decades: Whole genome sequencing and dating analysis of *Mycobacterium tuberculosis* isolates from KwaZulu-Natal. *PLoS Medicine, 12*(9), e1001880. https://doi.org/10.1371/journal.pmed.1001880. eCollection 2015 Sep.
18. Brynildsrud, O. B., Pepperell, C. S., Suffys, P., Grandjean, L., Monteserin, J., Debech, N., et al. (2018). Global expansion of *Mycobacterium tuberculosis* lineage 4 shaped by colonial migration and local adaptation. *Science Advances, 4*(10), eaat5869. https://doi.org/10.1126/sciadv.aat5869. eCollection 2018 Oct.
19. Vats, S., & Shanker, A. (2018). Groups of coevolving positions provide drug resistance to *Mycobacterium tuberculosis*: A study using targets of first-line antituberculosis drugs. *International Journal of Antimicrobial Agents*. . pii: S0924-8579(18)30325-X. https://doi.org/10.1016/j.ijantimicag.2018.10.027.
20. Miotto, P., Tessema, B., Tagliani, E., Chindelevitch, L., Starks, A.M., Emerson, C., et al. (2017). A standardised method for interpreting the association between mutations and phenotypic drug resistance in *Mycobacterium tuberculosis*. *European Respiratory Journal, 50*(6). pii: 1701354. https://doi.org/10.1183/13993003.01354-2017. Print 2017 Dec.
21. Kashyap, A., Singh, P. K., & Silakari, O. (2018). Mechanistic investigation of resistance via drug-inactivating enzymes in *Mycobacterium tuberculosis*. *Drug Metabolism Reviews, 50*, 448–465.

22. Nasiri, M. J., Haelll, M., Ghazi, M., Goudarzi, H., Pormohammad, A., Abbas, A., et al. (2017). New insights into intrinsic and acquired drug resistance mechanisms in mycobacteria. *Frontiers in Microbiology, 8*, 1–19.
23. Gupta, A. K., Katoch, V. M., Chauhan, D. S., Sharma, R., Singh, M., Venkatesan, K., & Sharma, V. D. (2010). Microarray analysis of efflux pump genes in multidrug-resistant *Mycobacterium tuberculosis* during stress induced by common anti-tuberculous drugs. *Microbial Drug Resistance, 16*(1), 21–28.
24. Gupta, A. K., Reddy, V. P., Lavania, M., Chauhan, D. S., Venkatesan, K., Sharma, V. D., et al. (2010). jefA (Rv2459), a drug efflux gene in *Mycobacterium tuberculosis* confers resistance to isoniazid & ethambutol. *The Indian Journal of Medical Research, 132*, 176–188.
25. Singh, M., Jadaun, G. P., Ramdas, Srivastava, K., Chauhan, V., Mishra, R., et al. (2011). Effect of efflux pump inhibitors on drug susceptibility of ofloxacin resistant *Mycobacterium tuberculosis* isolates. *The Indian Journal of Medical Research, 133*, 535–540.
26. Coelho, T., Machado, D., Couto, I., Maschmann, R., Ramos, D., von Groll, A., et al. (2015). Enhancement of antibiotic activity by efflux inhibitors against multidrug resistant *Mycobacterium tuberculosis* clinical isolates from Brazil. *Frontiers in Microbiology, 6*, 330. https://doi.org/10.3389/fmicb.2015.00330. eCollection 2015.
27. Sandhu, P., & Akhter, Y. (2016). The drug binding sites and transport mechanism of the RND pumps from *Mycobacterium tuberculosis*: Insights from molecular dynamics simulations. *Archives of Biochemistry and Biophysics, 592*, 38–49.
28. Blanchard, J. S. (1996). Molecular mechanisms of drug resistance in *Mycobacterium tuberculosis*. *Annual Review of Biochemistry, 65*, 215–239.
29. Siddiqi, N., Shamim, M., Jain, N. K., Rattan, A., Amin, A., Katoch, V. M., et al. (1998). Molecular genetic analysis of multi-drug resistance in Indian isolates of *Mycobacterium tuberculosis*. *Memórias do Instituto Oswaldo Cruz, 93*, 589–594.
30. Srivastava, K., Das, R., Jakhmola, P., Gupta, P., Chauhan, D. S., Sharma, V. D., et al. (2004). Correlation of mutations detected by INNO-LiPA with levels of rifampicin resistance in *Mycobacterium tuberculosis*. *The Indian Journal of Medical Research, 120*, 100–105.
31. Yang, B., Koga, H., Ohno, H., Ogawa, K., Fukuda, M., Hirakata, Y., et al. (1998). Relationship between antimycobacterial activities of rifampicin, rifabutin and KRM-1648 and *rpoB* mutations of *Mycobacterium tuberculosis*. *The Journal of Antimicrobial Chemotherapy, 42*, 621–628.
32. Cavusoglu, C., Karaca-Derici, Y., & Bilgic, A. (2004). In-vitro activity of rifabutin against rifampicin-resistant *Mycobacterium tuberculosis* isolates with known *rpoB* mutations. *Clinical Microbiology and Infection, 10*, 662–665.
33. Prakash, R., Gupta, R., Katoch, V. M., & Tiwari, P. K. (2017). Molecular modelling and docking analysis of katG and rpoB gene in MDR TB isolates from North Central Indian population. *Journal of Infection and Public Health, 10*, 593–599.
34. Molodtsov, V., Scharf, N. T., Stefan, M. A., Garcia, G. A., & Murakami, K. S. (2017). Structural basis for rifamycin resistance of bacterial RNA polymerase by the three most clinically important rpoB mutations found in *Mycobacterium tuberculosis*. *Molecular Microbiology, 103*, 1034–1045.
35. Comas, I., Borrell, S., Roetzer, A., Rose, G., Malla, B., Kato-Maeda, M., et al. (2011). Whole-genome sequencing of rifampicin-resistant *Mycobacterium tuberculosis* strains identifies compensatory mutations in RNA polymerase genes. *Nature Genetics, 44*, 106–110.
36. Honore, N., & Cole, S. T. (1993). Molecular basis of rifampin resistance in Mycobacterium leprae. *Antimicrobial Agents and Chemotherapy, 37*, 414–418.
37. Williams, D. L., Waguespack, C., Eisenach, K., Crawford, J. T., Portaels, F., Salfinger, M., et al. (1994). Characterization of rifampin-resistance in pathogenic mycobacteria. *Antimicrobial Agents and Chemotherapy, 38*, 2380–2386.
38. Cambau, E., Bonnafous, P., Perani, E., Sougakoff, W., Ji, B., & Jarlier, V. (2002). Molecular detection of rifampin and ofloxacin resistance for patients who experience relapse of multibacillary leprosy. *Clinical Infectious Diseases, 34*, 39–45.

39. Zhang, L., Namisato, M., & Matsuoka, M. (2004). A mutation at codon 516 in the rpoB gene of Mycobacterium leprae confers resistance to rifampin. *International Journal of Leprosy and Other Mycobacterial Diseases, 72*, 468–472.
40. Vedithi, S. C., Lavania, M., Kumar, M., Kaur, P., Turankar, R. P., Singh, I., et al. (2015). A report of rifampin-resistant leprosy from northern and eastern India: Identification and in silico analysis of molecular interactions. *Medical Microbiology and Immunology, 204*, 193–203.
41. HasanoorReja, A. H., Biswas, N., Biswas, S., Lavania, M., Chaitanya, V. S., Banerjee, S., et al. (2015). Report of rpoB mutation in clinically suspected cases of drug resistant leprosy: A study from Eastern India. *The Indian Journal of Dermatology Venereology and Leprology*. 2015 Mar-Apr, *81*(2), 155–161. https://doi.org/10.4103/0378-6323.152185.
42. Lavania, M., Hena, A., Reja, H., Nigam, A., Biswas, N. K., Singh, I., et al. (2016). Mutation at codon 442 in the rpoB gene of Mycobacterium leprae does not confer resistance to rifampicin. *Leprosy Review, 87*, 93–100.
43. Cambau, E., Saunderson, P., Matsuoka, M., Cole, S. T., Kai, M., Suffys, P., WHO surveillance network of antimicrobial resistance in leprosy, et al. (2018). Antimicrobial resistance in leprosy: Results of the first prospective open survey conducted by a WHO surveillance network for the period 2009–15. *Clinical Microbiology and Infection, 24*, 1305–1310.
44. Klein, J. L., Brown, T. J., & French, G. L. (2001). Rifampin resistance in Mycobacterium kansasii is associated with rpoB mutations. *Antimicrobial Agents and Chemotherapy, 45*, 3056–3058.
45. Peek, J., Lilic, M., Montiel, D., Milshteyn, A., Woodworth, I., Biggins, J. B., et al. (2018). Rifamycin congeners kanglemycins are active against rifampicin-resistant bacteria via a distinct mechanism. *Nature Communications, 9*(1), 4147. https://doi.org/10.1038/s41467-018-06587-2.
46. Middlebrook, G. (1954). Isoniazid resistance and catalase activity of tubercle bacilli. *American Review of Tuberculosis, 69*, 471–472.
47. Wayne, L. G., Diaz, G. A., & Doubek, J. R. (1968). Acquired isoniazid resistance and catalase activity of mycobacteria. *The American Review of Respiratory Disease, 97*, 909–913.
48. Ramaswamy, S. V., Reich, R., Dou, S. J., Jasperse, L., Pan, X., Wanger, A., et al. (2003). Single nucleotide polymorphisms in genes associated with isoniazid resistance in *Mycobacterium tuberculosis. Antimicrobial Agents and Chemotherapy, 47*, 1241–1250.
49. Purkan, P., Ihsanawati, I., Natalia, D., Syah, Y. M., Retnoningrum, D. S., & Siswanto, I. (2018). Molecular analysis of *katG* encoding catalase-peroxidase from clinical isolate of isoniazid-resistant *Mycobacterium tuberculosis. Journal of Medicine and Life, 11*, 160–167.
50. Ramasubban, G., Therese, K. L., Vetrivel, U., Sivashanmugam, M., Rajan, P., Sridhar, R., et al. (2011). Detection of novel coupled mutations in the katG gene (His276Met, Gln295His and Ser315Thr) in a multidrug-resistant *Mycobacterium tuberculosis* strain from Chennai, India, and insight into the molecular mechanism of isoniazid resistance using structural bioinformatics approaches. *International Journal of Antimicrobial Agents, 37*, 368–372.
51. Vilchèze, C., & Jacobs, W. R., Jr. (2014). Resistance to isoniazid and Ethionamide in *Mycobacterium tuberculosis*: Genes, mutations, and causalities. *Microbiology Spectrum, 2* (4), MGM2-0014-2013. https://doi.org/10.1128/microbiolspec.MGM2-0014-2013.
52. Viswanathan, G., Yadav, S., & Raghunand, T. R. (2016). Identification of novel loci associated with mycobacterial isoniazid resistance. *Tuberculosis (Edinburgh, Scotland), 96*, 21–26.
53. Argyrou, A., Vetting, M. W., Aladegbami, B., & Blanchard, J. S. (2006). *Mycobacterium tuberculosis* dihydrofolate reductase is a target for isoniazid. *Nature Structural & Molecular Biology, 13*, 408–413.
54. Wang, F., Jain, P., Gulten, G., Liu, Z., Feng, Y., Ganesula, K., et al. (2010). *Mycobacterium tuberculosis* dihydrofolate reductase is not a target relevant to the antitubercular activity of isoniazid. *Antimicrobial Agents and Chemotherapy, 54*, 3776–3782.
55. Ando, H., Miyoshi-Akiyama, T., Watanabe, S., & Kirikae, T. (2014). A silent mutation in *mabA* confers isoniazid resistance on *Mycobacterium tuberculosis. Molecular Microbiology, 91*, 538–547.

56. Singh, A., Singh, A., Grover, S., Pandey, B., Kumari, A., & Grover, A. (2018). Wild-type catalase peroxidase vs G279D mutant type: Molecular basis of isoniazid drug resistance in *Mycobacterium tuberculosis*. *Gene, 641*, 226–234.
57. Scardigli, A., Caminero, J. A., Sotgiu, G., Centis, R., D'Ambrosio, L., & Migliori, G. B. (2016). Efficacy and tolerability of ethionamide versus prothionamide: A systematic review. *The European Respiratory Journal, 48*, 946–951.
58. Schaefer, W. B. (1960). Effect of isoniazid on the dehydrogenase activity of *Mycobacterium tuberculosis*. *Journal of Bacteriology, 79*, 236–245.
59. Rist, N. (1960). L' activitieantituberculeuse de Ethionamide. *Ad Tubercle Research, 10*, 69–126.
60. Rueda, J., Realpe, T., Mejia, G. I., Zapata, E., Rozo, J. C., Ferro, B. E., & Robledo, J. (2015). Genotypic analysis of genes associated with independent resistance and cross-resistance to isoniazid and Ethionamide in *Mycobacterium tuberculosis* clinical isolates. *Antimicrobial Agents and Chemotherapy, 59*, 7805–7810.
61. Konno, K., Feldmann, F. M., & McDermott, W. (1967). Pyrazinamide susceptibility and amidase activity of tubercle bacilli. *The American Review of Respiratory Disease, 95*, 461–469.
62. Scorpio, A., & Zhang, Y. (1996). Mutations in pncA, a gene encoding pyrazinamidase/nicotinamidase, cause resistance to the antituberculous drug pyrazinamide in tubercle bacillus. *Nature Medicine, 2*, 662–667.
63. Singh, P., Mishra, A. K., Malonia, S. K., Chauhan, D. S., Sharma, V. D., Venkatesan, K., & Katoch, V. M. (2006). The paradox of pyrazinamide: An update on the molecular mechanisms of pyrazinamide resistance in mycobacteria. *The Journal of Communicable Diseases, 8*, 288–298.
64. Sheen, P., Lozano, K., Gilman, R. H., Valencia, H. J., Loli, S., Fuentes, P., et al. (2013). pncA gene expression and prediction factors on pyrazinamide resistance in *Mycobacterium tuberculosis*. *Tuberculosis (Edinburgh, Scotland), 93*, 515–522.
65. Unissa, A. N., Selvakumar, N., & Hassan, S. (2009). Insight to pyrazinamide resistance in *Mycobacterium tuberculosis* by molecular docking. *Bioinformation, 4*, 24–29.
66. Huy, N. Q., Lucie, C., Hoa, T. T. T., Hung, N. V., Lan, N. T. N., et al. (2017). Molecular analysis of pyrazinamide resistance in *Mycobacterium tuberculosis* in Vietnam highlights the high rate of pyrazinamide resistance-associated mutations in clinical isolates. *Emerging Microbes & Infections*.. Oct 11, *6*(10), e86. https://doi.org/10.1038/emi.2017.73.
67. Yang, J., Liu, Y., Bi, J., Cai, Q., Liao, X., Li, W., Guo, C., et al. (2015). Structural basis for targeting the ribosomal protein S1 of *Mycobacterium tuberculosis* by pyrazinamide. *Molecular Microbiology, 95*, 791–803.
68. Zimic, M., Fuentes, P., Gilman, R. H., Gutierrez, A. H., Kirwan, D., & Sheen, P. (2012). Pyrazinoic acid efflux rate in *Mycobacterium tuberculosis* is a better proxy of pyrazinamide resistance. *Tuberculosis (Edinburgh, Scotland), 92*, 84–91.
69. Zhang, Y., Zhang, J., Cui, P., Zhang, Y., & Zhang, W. (2017). Identification of novel efflux proteins Rv0191, Rv3756C, Rv3008 and Rv 1667C involved in pyrazinamide resistance in *Mycobacterium tuberculosis*. *Antimicrobial Agents and Chemotherapy*. 2017. July 25; *61*(8). pii e00940–17. https://doi.org/10.1128/AAC.00940-17. Print 2017 Aug.
70. Kilburn, J. O., & Greenberg, J. (1977). Effect of ethambutol on the viable cell count in Mycobacterium smegmatis. *Antimicrobial Agents and Chemotherapy, 11*, 534–540.
71. Winder, F., Collins, P., & Whelan, D. (1971). Effect of ethionamide and Isoxyl on mycolic acid synthesis of *Mycobacterium tuberculosis* BCG. *Journal of General Microbiology, 66*, 379–380.
72. Takayama, K., Armstrong, E. L., Kunugi, K. A., & Kilburn, J. O. (1979). Inhibition by ethambutol of mycolic acid transfer into cell wall of Mycobacterium smegmatis. *Antimicrobial Agents and Chemotherapy, 16*, 240–242.
73. Takayama, K., & Kilburn, J. O. (1989). Inhibition of arabinogalactan by ethambutol in Mycobacterium smegmatis. *Antimicrobial Agents and Chemotherapy, 33*, 1493–1499.

74. Mikusová, K., Slayden, R. A., Besra, G., & Brennan, P. J. (1995). Biogenesis of the mycobacterial cell wall and the site of action of ethambutol. *Antimicrobial Agents and Chemotherapy, 39*, 2484–2489.
75. Ramaswamy, S. V., Amin, A. G., Göksel, S., Stager, C. E., Dou, S. J., El Sahly, H., et al. (2000). Molecular genetic analysis of nucleotide polymorphisms associated with ethambutol resistance in human isolates of *Mycobacterium tuberculosis*. *Antimicrobial Agents and Chemotherapy, 44*, 326–336.
76. Sreevatsan, S., Stockbauer, K. E., Pan, X., Kreiswirth, B. N., Moghazeh, S. L., Jacobs, W. R., Jr., et al. (1997). Ethambutol resistance in *Mycobacterium tuberculosis*: Critical role of *embB* mutations. *Antimicrobial Agents and Chemotherapy, 41*, 1677–1681.
77. Jadaun, G. P., Das, R., Upadhyay, P., Chauhan, D. S., Sharma, V. D., & Katoch, V. M. (2009). Role of embCAB gene mutations in ethambutol resistance in *Mycobacterium tuberculosis* isolates from India. *International Journal of Antimicrobial Agents, 33*, 483–486.
78. Andres, S., Gröschel, M. I., Hillemann, D., Merker, M., Niemann, S., & Kranzer, K. (2019). A diagnostic algorithm to investigate pyrazinamide and ethambutol resistance in Rifampin-Resistant *Mycobacterium tuberculosis* Isolates in a Low-Incidence Setting. Antimicrobial Agents and Chemotherapy. 63(2). pii: e01798–18. https://doi.org/10.1128/AAC.01798-18. Print 2019 Feb
79. Waksman, S. A. (1949). *Streptomycin – Nature and practical applications*. London: Tindl and Cox.
80. Shaila, M. S., Gopinathan, K. P., & Ramakrishnan, T. (1973). Protein synthesis in *Mycobacterium tuberculosis* H37Rv and the effect of streptomycin in streptomycin susceptible and resistant strains. *Antimicrobial Agents and Chemotherapy, 4*, 205–213.
81. Shaila, M. S., Gopinathan, K. P., & Ramakrishnan, T. (1975). Mechanism of streptomycin action in *Mycobacterium tuberculosis* H37Rv. *Journal of the Indian Institute of Science, 57*, 301–316.
82. Yamada, T., Masuda, K., Mizuguchi, Y., & Suga, K. (1976). Altered ribosomes in resistant mutants of Mycobacterium smegmatis. *Antimicrobial Agents and Chemotherapy, 9*, 817–823.
83. Masuda, K., & Yamada, T. (1978). Genetic and biochemical studies on drug resistant Mycobacterium smegmatis III. Ribosomal binding sites of antibiotics. In E. Frrerksen, I. Tarnok, & J. H. Thumin (Eds.), *Genetics of Actinomycetales* (pp. 91–98). Stutgard: Gustav Verlag.
84. Finken, M., Kirschner, P., Meier, A., Wrede, A., & Böttger, E. C. (1993). Molecular basis of streptomycin resistance in *Mycobacterium tuberculosis*: Alterations of the ribosomal protein S12 gene and point mutations within a functional 16S ribosomal RNA pseudoknot. *Molecular Microbiology, 9*, 1239–1246.
85. Das, R., Gupta, P., Singh, P., Chauhan, D. S., Katoch, K., & Katoch, V. M. (2009). Association of mutations in rpsL gene with high degree of streptomycin resistance in clinical isolates of *Mycobacterium tuberculosis* in India. *The Indian Journal of Medical Research, 129*, 108–110.
86. Okamoto, S., Tamaru, A., Nakajima, C., Nishimura, K., Tanaka, Y., Tokuyama, S., et al. (2007). Loss of a conserved 7-methylguanosine modification in 16S rRNA confers low-level streptomycin resistance in bacteria. *Molecular Microbiology, 63*, 1096–1106.
87. Wong, S. Y., Lee, J. S., Kwak, H. K., Via, L. E., Boshoff, H. I., & Barry, C. E., 3rd. (2011). Mutations in gidB confer low-level streptomycin resistance in *Mycobacterium tuberculosis*. *Antimicrobial Agents and Chemotherapy, 55*, 2515–2522.
88. Verma, J. S., Gupta, Y., Nair, D., Manzoor, N., Rautela, R. S., Rai, A., & Katoch, V. M (2014). Evaluation of gidB alterations responsible for streptomycin resistance in *Mycobacterium tuberculosis*. *The Journal of Antimicrobial Chemotherapy, 69*, 2935–2941.
89. Rominski, A., Selchow, P., Becker, K., Brülle, J. K., Dal Molin, M., & Sander, P. (2017). Elucidation of Mycobacterium abscessus aminoglycoside and capreomycin resistance by targeted deletion of three putative resistance genes. *The Journal of Antimicrobial Chemotherapy, 72*, 2191–2200.

90. Viljoen, A., Dubois, V., Girard-Misguich, F., Blaise, M., Herrmann, J. L., & Kremer, L. (2017). The diverse family of MmpL transporters in mycobacteria: From regulation to antimicrobial developments. *Molecular Microbiology, 104*, 889–904.
91. Sharma, P., Kumar, B., Gupta, Y., Singhal, N., Katoch, V. M., Venkatesan, K., & Bisht, D. (2010). Proteomic analysis of streptomycin resistant and sensitive clinical isolates of *Mycobacterium tuberculosis. Proteome Science.*. 2010 Nov 18, *8*, 59. https://doi.org/10.1186/1477-5956-8-59.
92. Kumar, B., Sharma, D., Sharma, P., Katoch, V. M., Venkatesan, K., & Bisht, D. (2013). Proteomic analysis of *Mycobacterium tuberculosis* isolates resistant to kanamycin and amikacin. *Journal of Proteomics, 94*, 68–77.
93. Ahn, J. W., & Kim, K. J. (2013). Rv3168 phosphotransferase activity mediates kanamycin resistance in *Mycobacterium tuberculosis. Journal of Microbiology and Biotechnology, 23*, 1529–1535.
94. Reeves, A. Z., Campbell, P. J., Sultana, R., Malik, S., Murray, M., Plikaytis, B. B., et al. (2013). Aminoglycoside cross-resistance in *Mycobacterium tuberculosis* due to mutations in the 5′ untranslated region of whiB7. *Antimicrobial Agents and Chemotherapy, 57*, 1857–1865.
95. Maus, C. E., Plikaytis, B. B., & Shinnick, T. M. (2005). Mutation of tlyA confers capreomycin resistance in *Mycobacterium tuberculosis. Antimicrobial Agents and Chemotherapy, 49*, 571–577.
96. Maus, C. E., Plikaytis, B. B., & Shinnick, T. M. (2005). Molecular analysis of cross-resistance to capreomycin, kanamycin, amikacin, and viomycin in *Mycobacterium tuberculosis. Antimicrobial Agents and Chemotherapy, 49*, 3192–3197.
97. Akbergenov, R., Shcherbakov, D., Matt, T., Duscha, S., Meyer, M., Wilson, D. N., & Böttger, E. C. (2011). Molecular basis for the selectivity of antituberculosis compounds capreomycin and viomycin. *Antimicrobial Agents and Chemotherapy, 55*, 4712–4717.
98. Barry, V. C., Belton, J. G., Conalty, M. L., Denneny, J. M., Edward, D. W., O'Sullivan, J. F., et al. (1957). A new series of phenazines (rhimino compounds) with high antituberculosis activity. *Nature (London), 179*, 1013–1015.
99. World Health Organization. (1982). *Chemotherapy of leprosy for control programmes: Report of a WHO study group* (Technical rep series 675). Geneva: WHO.
100. Katoch, K., Natrajan, M., Katoch, V. M., Singh, H. B., & Bhatia, A. S. (1999). Chemotherapy trials in paucibacillary leprosy using clofazimine. *Indian Journal of Leprosy, 71*, 211–224.
101. Morrison, N. E., & Morley, G. M. (1976). Clofazimine binding studies with deoxyribonucleic acid. *International Journal of Leprosy, 44*, 475–481.
102. Hartskkorn, R. C., Uplekar, S., & Cole, S. T. (2014). Cross resistance between clofazimine and bedaquiline through upregulation of MmpL5 in *Mycobacterium tuberculosis. Antimicrobial Agents and Chemotherapy, 58*, 2979–2981.
103. Zhang, S., Chen, J., Cui, P., Shi, W., Zhang, W., & Zhang, Y. (2015). Identification of novel mutations associated with clofazimine resistance in *Mycobacterium tuberculosis. The Journal of Antimicrobial Chemotherapy, 70*, 2507–2510.
104. Chen, Y., Chen, J., Zhang, S., Shi, W., Zhang, W., Zhu, M., & Zhang, Y. (2018). Novel mutations associated with clofazimine resistance in Mycobacterium abscessus. *Antimicrobial Agents and Chemotherapy. 62*(7). pii: e00544–18. https://doi.org/10.1128/AAC.00544-18. Print 2018 Jul.
105. Andries, K., Verhasselt, P., Guillemont, J., Gohlmann, H. W., Neefs, J. M., Winkler, H., et al. (2005). A diarylquinolone drug active on the ATP synthase of *Mycobacterium tuberculosis. Science, 307*, 223–227.
106. Dupont, C., Viljoen, A., Thomas, S., Roquet-Banères, F., Herrmann, J.L., Pethe, K., & Kremer, L. (2017). Bedaquiline inhibits the ATP synthase in Mycobacterium abscessus and is effective in infected Zebrafish. *Antimicrobial Agents and Chemotherapy, 61*(11). pii: e01225–17. https://doi.org/10.1128/AAC.01225-17. Print 2017 Nov.

107. Nath, S. (2018). Interpretation of the mechanism of action of antituberculosis drug bedaquiline based on a novel two-ion theory of energy coupling in ATP synthesis. *Bioengineering & Translational Medicine, 4*, 164–170.
108. Hu, Y., Pertinez, H., Liu, Y., Davies, G., & Coates, A. (2019). Bedaquiline kills persistent *Mycobacterium tuberculosis* with no disease relapse: An in vivo model of a potential cure. *The Journal of Antimicrobial Chemotherapy, 74*, 1627–1633.
109. Petrella, S., Cambau, E., Chauffour, A., Andries, K., Jarlier, V., & Sougakoff, W. (2006). Genetic basis for natural and acquired resistance to the diarylquinoline R207910 in mycobacteria. *Antimicrobial Agents and Chemotherapy, 50*, 2853–2856.
110. Martinez, E., Hennessy, D., Jelfs, P., Crighton, T., Chen, S. C., & Sintchenko, V. (2018). Mutations associated with in vitro resistance to bedaquiline in *Mycobacterium tuberculosis* isolates in Australia. *Tuberculosis (Edinburgh, Scotland), 111*, 31–34.
111. Chawla, K., Martinez, E., Kumar, A., Shenoy, V. P., & Sintchenko, V. (2018). Whole-genome sequencing reveals genetic signature of bedaquiline resistance in a clinical isolate of *Mycobacterium tuberculosis*. *The Journal of Global Antimicrobial Resistance, 15*, 103–104.
112. Pang, Y., Zheng, H., Tan, Y., Song, Y., & Zhao, Y. (2017). In vitro activity of Bedaquiline against Nontuberculous Mycobacteria in China. *Antimicrobial Agents and Chemotherapy. 61* (5). pii: e02627–16. https://doi.org/10.1128/AAC.02627-16. Print 2017 May.
113. Zimenkov, D. V., Nosova, E. Y., Kulagina, E. V., Antonova, O. V., Arslanbaeva, L. R., Isakova, A. I., et al. (2017). Examination of bedaquiline- and linezolid-resistant *Mycobacterium tuberculosis* isolates from the Moscow region. *The Journal of Antimicrobial Chemotherapy, 72*, 1901–1906.
114. Xu, J., Wang, B., Hu, M., Huo, F., Guo, S., Jing, W., Nuermberger, E., & Lu, Y. (2017). Primary Clofazimine and Bedaquiline resistance among isolates from patients with multidrug-resistant tuberculosis. *Antimicrobial Agents and Chemotherapy*. 2017 May 24;*61*(6). pii: e00239–17. https://doi.org/10.1128/AAC.00239-17. Print 2017 Jun.
115. Villellas, C., Coeck, N., Meehan, C. J., Lounis, N., de Jong, B., Rigouts, L., & Andries, K. (2017). Unexpected high prevalence of resistance-associated Rv0678 variants in MDR-TB patients without documented prior use of clofazimine or bedaquiline. *The Journal of Antimicrobial Chemotherapy, 72*, 684–690.
116. Richard, M., Gutiérrez, A. V., Viljoen, A., Rodriguez-Rincon, D., Roquet-Baneres, F., Blaise, M., et al. (2018). Mutations in the MAB2299c TetR regulator confer cross-resistance to Clofazimine and Bedaquiline in *Mycobacterium abscessus*. *Antimicrobial Agents and Chemotherapy*. 2018 Dec 21;*63*(1). pii: e01316–18. https://doi.org/10.1128/AAC.01316-18. Print 2019 Jan.
117. Ismail, N. A., Omar, S. V., Joseph, L., Govender, N., Blows, L., et al. (2018). Defining Bedaquiline susceptibility, resistance, cross-resistance and associated genetic determinants: A retrospective cohort study. *eBioMedicine, 28*, 136–142.
118. Ismail, N., Omar, S. V., Ismail, N. A., & Peters, R. P. H. (2018). Collated data of mutation frequencies and associated genetic variants of bedaquiline, clofazimine and linezolid resistance in *Mycobacterium tuberculosis*. *Data in Brief, 20*, 1975–1983.
119. Matsumoto, M., Hashizume, H., Tomishige, T., Kawasaki, M., Tsubouchi, H., Sasaki, H., et al. (2006). OPC-67683, a nitro- Dihydro imidazooxazole derivative with promising action tuberculosis in vitro and in mice. *PLoS Medicine, 3*(11), e466.
120. Köser, C. U., Maurer, F. P., & Kranzer, K. (2019). Those who cannot remember the past are condemned to repeat it': Drug-susceptibility testing for bedaquiline and delamanid. *International Journal of Infectious Diseases, 80S*, S32–S35.
121. Yang, J. S., Kim, K. J., Choi, H., & Lee, S. H. (2018). Delamanid, Bedaquiline, and linezolid minimum inhibitory concentration distributions and resistance-related gene mutations in multidrug-resistant and extensively drug-resistant tuberculosis in Korea. *Annals of Laboratory Medicine, 38*, 563–568.

122. Fujiwara, M., Kawasaki, M., Hariguchi, N., Liu, Y., & Matsumoto, M. (2018). Mechanisms of resistance to delamanid, a drug for *Mycobacterium tuberculosis*. *Tuberculosis (Edinburgh, Scotland), 108*, 186–194.
123. Zhang, S., Chen, J., Cui, P., Shi, W., Shi, X., Niu, H., et al. (2016). *Mycobacterium tuberculosis* mutations associated with reduced susceptibility to linezolid. *Antimicrobial Agents and Chemotherapy, 60*, 2542–2544.
124. Wasserman, S., Louw, G., Ramangoaela, L., Barber, G., Hayes, C., Omar, S.V., et al. (2019). Linezolid resistance in patients with drug-resistant TB and treatment failure in South Africa. *Journal of Antimicrobial Chemotherapy*. 2019 May 12. pii: dkz206. https://doi.org/10.1093/jac/dkz206.
125. Kim, S. Y., Jhun, B. W., Moon, S. M., Jeon, K., Kwon, O. J., Huh, H. J., et al. (2019). Genetic mutations in linezolid-resistant *Mycobacterium avium* complex and Mycobacterium abscessus clinical isolates. *Diagnostic Microbiology and Infectious Disease, 94*, 38–40.
126. Ye, M., Xu, L., Zou, Y., Li, B., Guo, Q., Zhang, Y., et al. (2019) Molecular analysis of linezolid-resistant clinical isolates of *Mycobacterium abscessus*. *Antimicrobial Agents and Chemotherapy*. 2019 Jan 29;*63*(2). pii: e01842–18. https://doi.org/10.1128/AAC.01842-18. Print 2019 Feb.
127. Strominger, J. L., Eto, E., & Threnn, R. H. (1960). Competitive inhibition of enzymatic reactions by oxamycin. *Journal of the American Chemical Society, 82*, 998–999.
128. Takayama, K., David, H. L., Wang, L., & Goldman, D. S. (1970). Unpublished data quoted by David HL. *The American Review of Respiratory Disease, 102*, 68–74.
129. Chen, J., Zhang, S., Cui, P., Shi, W., Zhang, W., & Zhang, Y. (2017). Identification of novel mutations associated with cycloserine resistance in *Mycobacterium tuberculosis*. *The Journal of Antimicrobial Chemotherapy, 72*, 3272–3276.
130. Desjardins, C. A., Cohen, K. A., Munsamy, V., Abeel, T., Maharaj, K., Walker, B. J., et al. (2016). Genomic and functional analyses of *Mycobacterium tuberculosis* strains implicate ald in D-cycloserine resistance. *Nature Genetics, 48*, 544–551.
131. Fossati, C. (1971). Terizidone in the therapy of chronic pulmonary tuberculosis. *Giornale Italiano delle Malattie del Torace, 25*, 195–204.
132. Hwang, T. J., Wares, D. F., Jafarov, A., Jakubowiak, W., Nunn, P., & Keshavjee, S. (2013). Safety of cycloserine and terizidone for the treatment of drug-resistant tuberculosis: A meta-analysis. *The International Journal of Tuberculosis and Lung Disease, 17*, 1257–1266.
133. Von Groll, A., Martin, A., Jureen, P., Hoffner, S., Vandamme, P., Portaels, F., et al. (2009). Fluoroquinolone resistance in Mycobacterium tuberculosis and mutations in gyrA and gyrB. *Antimicrobial Agents and Chemotherapy, 53*, 4498–4500.
134. Maruri, F., Sterling, T. R., Kaiga, A. W., Blackman, A., van der Heijden, Y. F., Mayer, C., et al. (2012). A systematic review of gyrase mutations associated with fluroquinole-resistant Mycobacterium tuberculosis and a proposed gyrase numbering system. *The Journal of Antimicrobial Chemotherapy, 67*, 819–831.
135. Musser, J. M. (1995). Antimicrobial agent resistance in mycobacteria: Molecular insights. *Clinical Molecular Microbiology Review, 8*, 496–514.
136. Pantel, A., Petrella, S., Matrat, S., Brossier, F., Bastian, S., Reitter, D., et al. (2011). DNA gyrase inhibition assays are necessary to demonstrate fluoroquinolone resistance secondary to gyrB mutations in *Mycobacterium tuberculosis*. *Antimicrobial Agents and Chemotherapy, 55*, 4524–4529.
137. Maeda, S., Matsuoka, M., Nakata, N., Kai, M., Maeda, Y., Ken Hashimoto, K., et al. (2001). Multidrug resistant *Mycobacterium leprae* from patients with leprosy. *Antimicrobial Agents and Chemotherapy, 45*, 3635–3639.
138. Yokoyama, K., Kim, H., Mukai, T., Matsuoka, M., Nakajima, C., & Suzuki, Y. (2012). Amino acid substitutions at position 95 in GyrA can add fluoroquinolone resistance to Mycobacterium leprae. *Antimicrobial Agents and Chemotherapy, 56*, 697–702.
139. Veziris, N., Chauffour, A., Escolano, S., Henquet, S., Matsuoka, M., Jarlier, V., & Aubry, A. (2013). Resistance of M. leprae to quinolones: A question of relativity? *PLoS Neglected Tropical Diseases, 7*(11), e2559. https://doi.org/10.1371/journal.pntd.0002559. eCollection 2013 Nov.

140. Nisha, J., & Shanthi, V. (2018). Characterization of Ofloxacin interaction with mutated (A91V) quinolone resistance determining region of DNA gyrase in Mycobacterium Leprae through computational simulation. *Cell Biochemistry and Biophysics, 76*, 125–134.
141. Buriánková, K., Doucet-Populaire, F., Dorson, O., Gondran, A., Ghnassia, J. C., Weiser, J., & Pernodet, J. L. (2004). Molecular basis of intrinsic macrolide resistance in the *Mycobacterium tuberculosis* complex. *Antimicrobial Agents and Chemotherapy, 48*, 143–150.
142. Nash, K. A., & Inderlied, C. B. (1995). Genetic basis of macrolide resistance in *Mycobacterium avium* isolated from patients with disseminated disease. *Antimicrobial Agents and Chemotherapy, 39*, 2625–2630.
143. Jamal, M. A., Maeda, S., Nakata, N., Kai, M., Fukuchi, K., & Kashiwabara, Y. (2000). Molecular basis of clarithromycin-resistance in *Mycobacterium avium* intracellulare complex. *Tubercle and Lung Disease, 80*, 1–4.
144. Pfister, P., Corti, N., Hobbie, S., Bruell, C., Zarivach, R., Yonath, A., & Böttger, E. C. (2005). 23S rRNA base pair 2057-2611 determines ketolide susceptibility and fitness cost of the macrolide resistance mutation 2058A-->G. *Proceedings of the National Academy of Sciences of the United States of America, 102*, 5180–5185.
145. Nash, K. A., Zhang, Y., Brown-Elliott, B. A., & Wallace, R. J., Jr. (2005). Molecular basis of intrinsic macrolide resistance in clinical isolates of Mycobacterium fortuitum. *The Journal of Antimicrobial Chemotherapy, 55*, 170–177.
146. Mougari, F., Bouziane, F., Crockett, F., Nessar, R., Chau, F., Veziris, N. et al. (2016). Selection of resistance to clarithromycin in mycobacterium abscessus subspecies. *Antimicrobial Agents and Chemotherapy.* 2016 Dec 27;*61*(1). pii: e00943–16. https://doi.org/10.1128/AAC.00943-16. Print 2017 Jan.
147. Huang, L., Hu, X., Tao, J., & Mi, K. (2014). A hemerythrin-like protein MSMEG3312 influences erythromycin resistance in mycobacteria. *Wei Sheng Wu Xue Bao, 54*, 1279–1288.
148. Williams, D. L., Spring, L., Harris, E., Roche, P., & Gillis, T. P. (2000). Dihydropteroate synthase of *Mycobacterium leprae* and dapsone resistance. *Antimicrobial Agents and Chemotherapy, 44*, 1530–1537.
149. Cambau, E., Carthagena, L., Chauffour, A., Ji, B., & Jarlier, V. (2006). Dihydropteroate synthase mutations in the folP1 gene predict dapsone resistance in relapsed cases of leprosy. *Clinical Infectious Diseases, 42*(2), 238–241.
150. Chaitanya, V. S., Das, M., Bhat, P., & Ebenezer, M. (2015). Computational modelling of Dapsone interaction with dihydropteroate synthase in Mycobacterium leprae; insights into molecular basis of Dapsone resistance in leprosy. *Journal of Cellular Biochemistry, 116*, 2293–2303.
151. Mathys, V., Wintjens, R., Lefevre, P., Bertout, J., Singhal, A., Kiass, M., et al. (2009). Molecular genetics of para amino salicylic acid resistance in clinical isolates and spontaneous mutants of *Mycobacterium tuberculosis. Antimicrobial Agents and Chemotherapy, 53*, 2100–2109.
152. Zhao, F., Wang, X. D., Erber, L. N., Luo, M., Guo, A. Z., Yang, S. S., et al. (2014). Binding pocket alterations in dihydrofolate synthase confer resistance to para-aminosalicylic acid in clinical isolates of *Mycobacterium tuberculosis. Antimicrobial Agents and Chemotherapy, 58*, 1479–1487.
153. Hugonnet, J. E., & Blanchard, J. S. (2007). Irreversible inhibition of the *Mycobacterium tuberculosis* beta-lactamase by clavulanate. *Biochemistry, 46*(43), 11998–12004.
154. Gupta, R., Lavollay, M., Mainardi, J. L., Arthur, M., Bishai, W. R., & Lamichhane, G. (2010). The *Mycobacterium tuberculosis* protein LdtMt2 is a nonclassical transpeptidase required for virulence and resistance to amoxicillin. *Nature Medicine, 16*, 466–469.
155. Cordillot, M., Dubée, V., Triboulet, S., Dubost, L., Marie, A., Hugonnet, J. E., et al. (2013). In vitro cross-linking of *Mycobacterium tuberculosis* peptidoglycan by L,D-transpeptidases and inactivation of these enzymes by carbapenems. *Antimicrobial Agents and Chemotherapy, 57*, 5940–5945.
156. Tiberi, S., Sotgiu, G., D'Ambrosio, L., Centis, R., Abdo Arbex, M., Alarcon Arrascue, E., et al. (2016). Comparison of effectiveness and safety of imipenem/clavulanate- versus

meropenem/clavulanate containing regimens in the treatment of MDR and XDR-TB. *The European Respiratory Journal, 47*, 1758–1766.
157. Silva, J. R., Bishai, W. R., Govender, T., Lamichhane, G., Maguire, G. E., Kruger, H. G., et al. (2016). Targeting the cell wall of *Mycobacterium tuberculosis*: A molecular modeling investigation of the interaction of imipenem and meropenem with L,D-transpeptidase 2. *Journal of Biomolecular Structure & Dynamics, 34*, 304–317.
158. Inderlied, C. B., Lancero, M. G., & Young, L. S. (1989). Bacteriostatic and bactericidal in-vitro activity of meropenem against clinical isolates, including *Mycobacterium avium* complex. *The Journal of Antimicrobial Chemotherapy, 24*(Suppl A), 85–99.
159. Brown-Elliott, B. A., Killingley, J., Vasireddy, S., Bridge, L., & Wallace, R. J., Jr. (2016). In vitro comparison of ertapenem, meropenem, and imipenem against isolates of rapidly growing mycobacteria and Nocardia by use of broth microdilution and Etest. *Journal of Clinical Microbiology, 54*, 1586–1592.
160. Kumar, P., Chauhan, V., Silva, J. R. A., Lameira, J., d'Andrea, F. B., & Li, S. G., et al. (2017). Mycobacterium abscessus l,d-Transpeptidases are susceptible to inactivation by carbapenems and Cephalosporins but not penicillins. *Antimicrobial Agents and Chemotherapy*. Sep 22;*61* (10). pii: e00866–17. https://doi.org/10.1128/AAC.00866-17. Print 2017 Oct.
161. Katoch, K., Katoch, V. M., Natarajan, M., Gupta, U. D., Sharma, V. D., & Singh, H. B. (2008). Long term follow-up results of 1 year MDT in MB leprosy patients treated with standard MDT + once a month Minocycline and Ofloxacin. *Indian Journal of Leprosy, 80*, 331–344.
162. Cummins, D. L., Delacerda, D., & Tausk, F. A. (2005). *Mycobacterium marinum* with different responses to second-generation tetracyclines. *International Journal of Dermatology, 44*, 518–520.
163. Huang, T. S., Lee, S. S., Hsueh, P. R., Tsai, H. C., Chen, Y. S., & Wann, S. R. (2008). Antimicrobial resistance of rapidly growing mycobacteria in Western Taiwan: SMART program 2002. *Journal of the Formosan Medical Association, 107*, 281–287.
164. Roberts, M. C. (1997). Genetic mobility and distribution of tetracycline resistance determinants. *Ciba Foundation Symposium, 207*, 206–218. discussion 219–22.
165. Ramón-García, S., Martín, C., Aínsa, J. A., & De Rossi, E. (2006). Characterization of tetracycline resistance mediated by the efflux pump tap from Mycobacterium fortuitum. *The Journal of Antimicrobial Chemotherapy, 57*, 252–259.
166. Kyselková, M., Chroňáková, A., Volná, L., Němec, J., Ulmann, V., Scharfen, J., & Elhottová, D. (2012). Tetracycline resistance and presence of tetracycline resistance determinants tet (V) and tap in rapidly growing mycobacteria from agricultural soils and clinical isolates. *Microbes and Environments, 27*(4), 413–422. Epub 2012 May 17.
167. Rudra, P., Hurst-Hess, K., Lappierre, P., & Ghosh, P. (2018). High levels of intrinsic tetracycline resistance in mycobacterium abscessus are conferred by a tetracycline-modifying monooxygenase. *Antimicrobial Agents and Chemotherapy. 62*(6). pii: e00119–18. https://doi.org/10.1128/AAC.00119-18. Print 2018 Jun.
168. Chukwudi, C. U. (2016). rRNA binding sites and the molecular mechanism of action of the Tetracyclines. *Antimicrobial Agents and Chemotherapy, 60*, 4433–4441.
169. Rinder, H., Mieskes, K. T., & Loscher, T. (2001). Heteroresistance in *Mycobacterium tuberculosis*. *The International Journal of Tuberculosis and Lung Disease, 5*, 339–345.

The Challenge of Drug-Resistant Tuberculosis: An Update

Chanchal Kumar, Kamal Shrivastava, Shraddha Gupta, and Mandira Varma-Basil

Abstract

Tuberculosis (TB) is one of the leading causes of death from a single infectious agent, and millions of people fall sick due to this disease every year. In spite of effective treatment, drug-resistant tuberculosis is a major problem worldwide. Globally, 3.5% of new cases and 18% of previously treated cases were estimated to have MDR-TB in 2017. Among these, 8.5% cases were estimated to have XDR-TB. Multidrug-resistant (MDR) and extensively drug-resistant (XDR) TB are, thus, responsible for significant obstruction to global TB control. Drug resistance in *M. tuberculosis* primarily occurs through chromosomal mutations in genes encoding for drug target or drug-activating enzymes. However, alternative mechanisms, such as membrane impermeability, drug inactivation or modification by enzymes, and efflux pumps, may also be responsible for drug resistance. Insight into the mechanisms of drug resistance would help mankind in limiting the disease through the use and development of better diagnostic and therapeutic tools. Here we describe the prevalence of drug resistance in *M. tuberculosis*, the mechanisms of drug resistance, and the rapid diagnostic assays currently available to detect drug-resistant *M. tuberculosis*.

Keywords

Tuberculosis · Drug resistance · Multidrug resistance · Mechanisms of drug resistance · Diagnosis

C. Kumar · K. Shrivastava · S. Gupta · M. Varma-Basil (✉)
Department of Microbiology, Vallabhbhai Patel Chest Institute, University of Delhi, New Delhi, India
e-mail: mandirav@rediffmail.com

S. Hameed, Z. Fatima (eds.), *Pathogenicity and Drug Resistance of Human Pathogens*,
https://doi.org/10.1007/978-981-32-9449-3_2

Introduction

Tuberculosis (TB) is a global disease, found in every country in the world. Globally, TB is one of the leading causes of death from a single infectious agent, and millions of people fall sick due to this disease every year.

Despite the availability of an effective treatment regimen, 9.0–11.0 million individuals contracted active tuberculosis, and 1.3 million died from the disease in 2017 [100]. Fortunately, the rate of infection has declined globally with the number of TB deaths among HIV-negative people reducing from 1.8 million in 2000 to 1.3 million in 2017. The number of TB deaths among HIV-positive people has fallen from 534,000 in 2000 to 300,000 in 2017 [100]. However, the threat due to TB is unabated due to emergence of drug resistance, in particular, multidrug-resistant (MDR) and extensively drug-resistant (XDR) TB. In 2017, 558,000 people developed TB that was resistant to rifampicin (RIF), the most effective first-line antituberculosis (anti-TB) drug. Of these, 82% people had MDR-TB. These statistics underscore the urgent need for better diagnostic and therapeutic tools for drug-resistant TB.

It is very important to understand the epidemiology and mechanisms of drug resistance in TB in order to limit the spread of this disease. This chapter focuses on the prevalence of drug-resistant TB, the mechanisms leading to drug resistance, and the method of detection of drug-resistant *M. tuberculosis*.

Drug-Resistant TB: Magnitude of the Problem

Tuberculosis mortality rate has reduced by 42% since 2000 [100]. High TB burden countries such as the Russian Federation and Ethiopia have also shown a decline of 13% and 12% per year, respectively, during the time period 2013–2017. During this period, the rates of decline have also been high in Sierra Leone (10% per year), Kenya (8% per year), and Vietnam (8% per year) [100]. The European region showed a decline of 11% per year, and in Southeast Asia, the decline in mortality due to TB was 4% [100]. However, the current public health issues arise as a result of emergence of MDR-TB and XDR-TB and the increase in HIV-associated tuberculosis. Treatment of MDR-TB, though improved over the years, is extremely expensive and toxic.

The Global Burden of MDR-TB

Global surveillance has shown that drug-resistant TB is widespread and is a threat to TB control programs in many countries. Multidrug-resistant tuberculosis (MDR-TB), defined as tuberculosis caused by strains which are resistant to at least isoniazid (INH) and RIF, emerged as a threat to TB control in the 1990s [9]. However, of the WHO estimates of newly emerging MDR-TB or RIF-resistant TB, only one-fourth are detected or notified [105]. The geographical distribution of MDR-TB is extremely variable, and several regions with high prevalence exist in low- and middle-income countries, mainly in East Europe and Asia [44].

Globally, 3.5% of new cases and 18% of previously treated cases were estimated to have MDR-TB [100]. Among these, 8.5% of cases were estimated to have XDR-TB [100], defined as resistance to at least RIF and INH from among the first-line anti-TB drugs and, in addition, resistance to any fluoroquinolone and to at least one of the three injectable second-line anti-TB drugs used in the treatment of TB (capreomycin, kanamycin, amikacin) [51]. In fact, mathematical modeling predicts an increase in MDR-TB among incident TB patients and XDR-TB among incident MDR-TB patients in high-burden countries in the fourth coming decades [75]. In Europe, more than 50% of recurrent TB cases were either MDR or XDR-TB [44]. Moreover, the alarming fact is that only 55% of MDR-TB patients completed treatment in 2015. Of these, 8% failed treatment, 15% died, and 14% were lost to follow-up. There was no outcome information in 7% of patients [100]. What is encouraging is that globally treatment success has improved marginally in recent years. In 2015, successful treatment was observed mostly in Eastern Mediterranean region (62%) and least in Southeast Asia (50%) [100].

Another encouraging fact is that, globally in 2017, of the 6.7 million new and previously treated TB cases notified, 2.0 million (30%) were tested for RIF resistance (RR-TB). Of these, 24% of new TB patients and 70% of previously treated TB patients were tested for RR-TB. In contrast, in 2015, only 12% of new and 53% of previously treated TB cases had been tested for RIF resistance. The improved coverage highlighted more vigorous TB control programs worldwide. Drug susceptibility testing (DST) coverage also increased between 2016 and 2017, with the highest in the European region (57%). However, the DST coverage varied enormously between different countries and among the 30 high MDR-TB burden countries [100]. This resulted in an improvement in notification, and the number of reported MDR/RR-TB cases increased by more than 30% in some of the high MDR-TB burden countries, i.e., Angola, Democratic People's Republic of Korea, Indonesia, Nigeria, Somalia, and Thailand. In 2017, among the MDR/RR-TB patients notified, 50% were tested for resistance to both fluoroquinolones and second-line injectable drugs. In addition, 10,800 cases of XDR-TB were reported in 77 countries with 88% of cases being from Europe and Southeast Asia [100].

Risk Factors for Development of Drug-Resistant TB

Today MDR-TB spreads unchecked in large parts of the world. It is fueled by poverty, which at the individual level, limits access to effective treatment; and at the national level, under-resourced governments lack the capacity to tackle this disease.

Risk factors influence the probability of infection, disease, or outcome and operate on many scales (physiological, genetic, environmental, and behavioral).

The use of inferior TB drug regimens, high HIV infection rates, and previous exposure to anti-TB treatment are well-established risk factors for drug-resistant TB. The duration of RIF treatment beyond 4 months has been associated with increased risk of acquiring drug resistance in initially drug-sensitive strains [48].

An association has been seen between MDR-TB and age of patients, mainly in the age group 45–65 years. MDR-TB patients have been more likely to be males in Europe. However, in regions with a higher transmission of TB, male sex may not be a risk factor. Immigration has also been suggested as a factor leading to the increased prevalence of MDR-TB in Europe [23].

Moreover, social determinants, such as industrialization, urbanization, poverty, and overcrowding, factors associated with transmission of TB, are also associated with several other risk factors such as lack of accessibility or adherence to proper treatment that eventually give rise to drug resistance. Selection pressure from inappropriate use of anti-TB drugs accounts for initial emergence of resistance.

Mechanisms of Antituberculosis Drug Resistance and Factors Associated with Its Emergence

The MDR phenotype is caused by sequential accumulation of mutations in different genes involved in individual drug resistance, due to inappropriate treatment or poor adherence to treatment. The probability of acquiring resistance varies and is 1 in 10^8 bacilli for RIF and 1 in 10^6 bacilli for INH, EMB, and SM [17]. Hence, a bacillary load of 10^9 will contain several mutants resistant to any one anti-TB drug [28]. Also, simultaneous resistance of *M. tuberculosis* strains to two or more drugs depends on individual probabilities of mutations. Resistance to a drug does not benefit the bacterium unless it is exposed to that drug, when the sensitive strains are killed and the drug-resistant mutants continue to multiply [94].

Mechanism of Resistance to First-Line Anti-TB Drugs (Table 1)

Rifampicin (RIF)

RIF, introduced as anti-TB drug in 1963, is a semisynthetic drug extracted from culture filtrates of *Streptomyces mediterranei* in the form of rifamycin. It is a lipophilic drug and highly bactericidal against dividing and nondividing intracellular and extracellular bacilli due to its rapid diffusion property across the hydrophobic cell envelope. In comparison to other anti-TB drugs, RIF has highest rate of sputum conversion to negativity. RIF inhibits the activity of DNA-dependent RNA polymerase. It targets β-subunit of RNA polymerase, which serves as a catalytic subunit and helps in initiation and elongation of messenger RNA. RIF forms a stable complex by binding to the β-subunit of RNA polymerase, thus leading to the inhibition of transcription and resulting in cell death.

Table 1 *M. tuberculosis* targeted drug and associated drug resistance mutations

Drug	Mechanism of action	Genes associated with drug resistance	Frequency of mutations (%)
Rifampicin	Inhibition of RNA synthesis	*rpoB*	95
Isoniazid	Inhibition of mycolic acid Synthesis	*katG*	70
		inhA	10
		kasA	10
Ethambutol	Inhibition of arabinogalactan biosynthesis	*embB*	70
		embA	–
		ubiA	–
Pyrazinamide	Targets the energy production, trans-translation and perhaps pantothenate / coenzyme A synthesis	*pncA*	99
		rpsA	–
		panD	–
Streptomycin	Inhibition of protein synthesis	*rpsl*	70–85
		rrs	70–85
		gidB	–
Fluoroquinolones	Inhibition of DNA synthesis	*gyrA*	90
		gyrB	<5
Capreomycin, amikacin, kanamycin	Inhibition of protein synthesis	*rrs*	62–70
		eis	<80
		tlyA	3
Para-aminosalicylic acid (PAS)	Inhibition of folic acid and thymine nucleotide metabolism	*thyA*	40
		folC	–
		ribD	90
Ethionamide	Inhibition of mycolic acid synthesis	*ethA*	
		mshA	
		ndh	
		inhA	
		inhA promoter	
Clofazimine	Target membrane destabilization, production of ROS	*Rv0678*	Not defined
		pepQ	
Bedaquiline	Block mycobacterial ATP synthesis	*Rv0678*	Not defined
		atpE	
		pepQ	
Delamanid	Inhibition of mycolic acid biosynthesis, target cell wall	*fgd1*	*fgd1* determined in clinical resistance
		fbiC	
		fbiA	
		fbiB	

Studies conducted on RIF-resistant isolates of *Mycobacterium tuberculosis* have recognized various mutations and short deletions in the targeted *rpoB* gene [90]. In 95–99% cases, mutations in *rpoB* gene of clinical isolates of *Mycobacterium tuberculosis* were the reason for resistance to RIF. Most of the mutations are located in an 81 bp hotspot region of *rpoB* gene, commonly known as RIF resistance-determining region (RRDR) covering codons 507–533 [66]. The most common mutations found in approximately 90% RIF-resistant isolates are Ser531Leu, His526Tyr, and Asp516Val of RRDR [32, 66, 68]. In recent studies, various groups of researchers have also documented mutations outside the hotspot RRDR region such as Val146Phe and Ile572Phe [79]. Previous research has also reported mutations outside the hotspot region: at codons 504, 534, 535, 541, and 553 [9, 112].

Isoniazid (INH)

INH or isonicotinic acid hydrazide is a synthetic drug derived from nicotinic acid. Its anti-TB properties were first detected in 1951 [33, 68, 80]. It is still one of the most effective drugs against the TB bacilli. Combined with RIF, it plays a major role in antituberculosis therapy. However, it is active only against the rapidly growing bacilli with a bactericidal effect [113]. The primary action of INH is inhibition of synthesis of mycobacterial cell wall. INH is a prodrug which needs to be activated by the *M. tuberculosis* catalase peroxidase enzyme *KatG*. The activated INH then produces oxygen-derived and organic free radicals [114]. Being toxic in nature, these radicals affect biosynthesis of mycolic acids, by inhibiting NADH-dependent enoyl-ACP reductase encoded by *inhA*. Lack of mycolic acid synthesis eventually results in loss of cellular integrity which leads to death of TB bacilli [3].

Resistance to INH has become a cause for concern. Jenkins et al. produced the first analysis of global INH resistance data reported to the WHO [37]. Data was submitted from only 56% of the world's population from 1994 to 2009. Of this, the former Soviet Union countries reported the highest number of cases of INH resistance (44.9%), including mono-resistance and MDR-TB. Of these, 16.1% did not have concurrent RIF resistance. Among the rest of the world, 7.5% had INH resistance without RIF resistance. Later, by 2013, it was estimated that 9.5% of the global TB cases had INH resistance without RIF resistance [103].

Phenotypic resistance to INH involves mutations in various genes. The TB Drug Resistance Mutation Database has reported 22 mutations associated with INH resistance such as *katG*, *ahpC*, *inhA*, *kasA*, and *ndh* [71]. The most important cause of INH resistance is mutations in *katG* and *inhA* gene or in promoter region of *inhA* gene [30, 67, 78]. In fact, when Gardner Middlebrook isolated the first isolates of *M. tuberculosis* that were resistant to INH, he noticed that the strains had lost catalase peroxidase activity [56]. The most prevalent mutation responsible for INH resistance is missense mutation at codon 315 of *katG* gene, i.e., S315T, which accounts for 60–95% of INH resistance [30, 45]. The second most common INH resistance-associated mutation is C15T, which occurs in promoter region of fabG1-inhA. In contrast to *katG*, which is mostly associated with high-level INH resistance (minimum inhibitory values (MIC) $\geq$1 μg/ml), the *inhA* mutation is associated with low-level INH resistance (MIC <1 μg/ml). Although less frequent, mutations at the active sites such as S94A and I194T were also reported [49]. Some studies have also shown strong association of mutations in gene *ahpC* with INH resistance [30, 45].

The complete mechanism of INH resistance is still not known. The most important reason for this is knowledge gaps about the exact association between mutations associated with resistance, phenotypic resistance, and outcome of treatment. The high number of mutations associated with INH resistance makes the creation of a minimal predictive mutation set difficult.

Pyrazinamide (PZA)

PZA, an analog of nicotinamide, was included in anti-TB therapy in 1972. Its introduction has reduced the length of therapy. Similar to RIF and INH, PZA has bactericidal activity. It has the ability to kill semi-dormant bacilli living in acidic environment [58, 81]. PZA is also a prodrug, activated by pyrazinamidase, which converts it into pyrazinoic acid. This pyrazinoic acid under acidic environment recruits protons to the cell which results in enhanced acidification of cytoplasm and inhibits the transport of vital enzymes across the membrane [74], thus resulting in cellular damage. It is also a fact that PZA is highly specific for *Mycobacterium tuberculosis*. *M. bovis* is naturally resistant to PZA because of a unique C to G point mutation at codon 169 of the *pncA* gene.

Resistance to PZA has been found to be associated with mutations in *pncA* gene. Several mutations have been recognized in *pncA* gene in more than 70% PZA-resistant isolates. No hotspot region in *pncA* has been identified till date [73, 74, 82]. However, the most common mutations identified in *pncA* gene are at codons 68, 138, 141, and 162, present in the 561 bp region of the open reading frame. Other than this, mutations in an 82 bp portion of its putative region have also been reported [39, 74]. An Indian study has reported mutations in *pncA* gene in 78% of PZA-resistant isolates [61]. It has also been observed that there are some PZA-resistant isolates which do not have *pncA* mutations, suggesting the association of other genes and mechanisms with PZA resistance [11]. In this context, several other targets such as ribosomal protein S1 (*rpsA*) and aspartate decarboxylase (*panD*), which are targets of pyrazinoic acid, have also been investigated for PZA resistance.

Ethambutol (EMB)

EMB is a synthetic compound, namely, 2,2′-(1,2-ethanediyldiimino)bis-1-butanol. EMB was first used in 1966 as anti-TB drug. It is bactericidal for replicating bacilli, but for slow-growing and non-replicating bacteria, it demonstrates limited activity [43]. In mouse models, EMB mostly shows bacteriostatic activity which results in 2-log difference in bacterial burden among treated and untreated animals following 28 days of daily therapy at a dose equivalent to human dosage [38]. It is active against multiplying bacilli, where it interferes in the biosynthesis of cell wall arabinogalactan [87]. It inhibits arabinosyltransferases involved in cell wall biosynthesis [87].

Resistance toward EMB is usually caused by missense mutations at *embCAB* operon, especially at codons 306, 406, and 497 of *embB* gene. The most common mutation associated with resistance to EMB is *embB* M306 V [57]. Also, missense mutations in *Rv3806c* (*ubiA*) V188A, A237V, R240C, and A249G as well as overexpression of the gene have been shown to cause increased MIC of EMB [70].

Fluoroquinolones (Group A)

FQs are used as second-line drugs for MDR-TB treatment. FQs include ciprofloxacin (CIP) and ofloxacin (OFX) that are synthetic derivatives of nalidixic acid. These are bactericidal antibiotics. FQs target type II topoisomerase (DNA gyrase), which catalyzes the supercoiling of DNA [2]. Type II topoisomerase is a tetramer with two subunits, A and B, encoded by the *gyrA* and *gyrB*, respectively [12, 18, 88]. Resistance mechanism to FQs in *M. tuberculosis* includes mutation in quinolone resistance-determining region (QRDR) of *gyrA* (320 bp) and *gyrB* (375 bp) [27, 88]. The most frequent mutation in *gyrA* gene is Ala90 and Asp94; however several mutations have also been reported at Ala74, Gly88, and Ser91 [12, 85]. A mutation at codon 95 has also been reported, though not related to FQ resistance as it occurs in both FQ-susceptible and FQ-resistant strains [60]. In several studies performed on Indian FQ-resistant clinical isolates, the most common mutation was at codon 94 followed by mutation at codon 90 and 95 [77, 84]. Other studies have demonstrated that same *gyrA* mutations gave different MICs of OFX for laboratory-selected or clinical isolates [85], which indicates the involvement of other resistance mechanisms such as alterations in membrane permeability and increased expression of efflux pumps.

Second-Line Injectable Drugs (Group B)

Streptomycin (SM)

Streptomycin was introduced as an anti-TB drug in 1944. It is an aminoglycoside derived from actinobacterium *Streptomyces griseus*. It is used in retreatment cases of TB along with other first-line drugs, i.e., RIF, INH, EMB, and PZA. It is also a bactericidal drug. SM inhibits protein synthesis by interacting with the 16s rRNA (*rrs*) and S12 ribosomal proteins (*rpsl*) [90]. Interaction of SM with *rrs* and *rpsl* leads to ribosomal changes which result in misreading of mRNA and cause inhibition of protein synthesis [22, 26].

Resistance to SM involves mutations in both the targeted genes (*rrs* and *rpsl*), which results in modifications in SM binding site. These genes account for 65–67% of SM-resistant cases [66]. In *rpsl* gene, mutations occur in two hotspot regions at codon 43 and codon 88 with amino acid changes Lys-Arg/Thr and Lys Arg/Gln, respectively. In an Indian study by Das et al. [14], 55% of SM-resistant isolates had Lys-Arg mutation at codon 43, while codon 88 contributed only to 4% of SM resistance with Lys-Arg amino acid change. In the *rrs* gene, mutations have been observed in a 530-loop region, a part of binding site of aminoacyl-tRNA and involved in decoding process [8]. The mutations occur at positions 491, 512, and 516 with a C-T conversion and at position 513 with A-C/T transversion. However,

C-T transition at codon 491 has been observed in both SM-susceptible and SM-resistant isolates. Hence this mutation has not been associated with SM resistance [95]. Other mutations, in the 915 loop, have also been reported to have an association with SM resistance, namely, at codons 903 (C-A/G) and 904 (A-G) [8]. Mutations in *gidB* gene, involved in methylation of 16s rRNA, have been reported to confer low-level SM resistance [62].

Kanamycin (KAN) and Amikacin (AMK)

Both the drugs belong to aminoglycosides and inhibit protein synthesis. They inhibit the elongation of peptide chain by binding to the ribosomes. Resistance mechanisms to these drugs involve mutations in *rrs* gene which encodes 16s rRNA. It has been observed that KAN resistance is specifically associated with loci 1400, 1401, and 1483 of *rrs* gene [86]. A change of alanine to glycine at 1400 position in *rrs* leads to MIC of more than 200 μg/ml [86, 89]. Kaur et al. [40] have shown that G1484T mutation in *rrs* gene was the most prevalent (17.2%) followed by A1401G (10.3%) in KAN-resistant Indian isolates. Zaunbrecher et al. reported mutation in promotor region of *eis* gene, which led to low-level resistance to KAN, but not to AMK. They also reported that 80% of their clinical isolates with low-level resistance had a mutation in the promoter region of *eis* gene [111].

Other Core Second-Line Drugs (Group C)

Clofazimine (CFZ)

CFZ, a riminophenazine drug, was originally described in 1957 as having antimycobacterial activity. CFZ has been conventionally used for the treatment of leprosy. Interest in this drug has been renewed for its potential use in treatment of MDR-TB and XDR-TB and to shorten tuberculosis treatment [16]. The exact mechanism of action of CFZ remains ambiguous; it seems to have multiple effects on the organism, such as production of reactive oxygen species [110] and membrane destabilization and dysfunction [93]. The mechanism of resistance to CFZ is not yet clear. Recently, some studies have shown mutations in *Rv0678* gene, which is associated with efflux pump MmpS5-MmpL5. Overexpression of efflux pumps due to this mutation led to cross-resistance to CFZ and bedaquiline [29]. According to an in vitro study, the tentative breakpoint for CFZ resistance by the microplate alamar blue assay (MABA) was 1.2 μg/ml [107].

Recently, mutations in the putative proline amino peptidase gene *pepQ* (*Rv3525c*) have also been described as additional mechanisms associated with low-level cross-resistance between bedaquiline and CFZ in vitro and in mice [1].

Ethionamide (ETH)

Structurally, ETH is very similar to INH, involved in inhibition of mycolic acid synthesis. ETH is a prodrug that is activated by *ethA* gene. Co-resistance to INH and ETH can occur due to mutations that alter the *inhA* target or cause its overexpression. It can also occur due to mutations in *ndh* that increase the intracellular concentration of NADH [97]. Recently it has been suggested that a gene encoding a glycosyltransferase and involved in mycothiol biosynthesis (*mshA*) may also be a target for ETH [96].

D-Cycloserine (DCS)

DCS is a bacteriostatic drug. It is a cyclic analog of D-alanine, which is responsible for cross-linking during peptidoglycan synthesis [24, 66]. It acts by inhibiting cell wall synthesis by competing with D-alanine for the enzymes D-alanyl-D-alanine synthetase (*ddl*) and D-alanine racemase (*alr*). It also inhibits the synthesis of these proteins along with L-alanine dehydrogenase (*ald*). Conversion of G→T in *alr* promoter region leads to overexpression of *alr*, and this overexpression results in DCS resistance [24, 66]. Recently, researchers have demonstrated relationship between mutations in *ald* (Rv2780), leading to loss of function and DCS resistance. Since mutations in *ald* can occur anywhere along the length of the 1116 bp gene, it is difficult to develop molecular diagnostics with only a small number of SNPs for detecting DCS resistance [15].

Para-aminosalicylic Acid (PAS)

The exact mechanism of action of PAS is not yet known. It has been suggested that PAS may compete with para-aminobenzoic acid for dihydropteroate synthase, which is needed for folate biosynthesis. More recently, a study found that PAS-resistant clinical isolates of *M. tuberculosis* carried mutations in the gene *thyA* leading to decreased enzyme activity [69]. Other studies assessing the role of enzymes in the folate pathway determined that PAS was a prodrug and its activation required *thyA* [53]. However, only 37% of the *M. tuberculosis* isolates studied had mutations in *thyA*, suggesting the presence of additional mechanisms for resistance to PAS. These studies reported Thr202Ala as the most common mutation associated with PAS resistance, although some susceptible isolates have been found to harbor the same mutation [47]. Recently Feuerriegel et al. reported that Thr202Ala mutation of *thyA* was found exclusively in strains of the Latin American Mediterranean (LAM) lineage irrespective of PAS resistance [25].

Recently Introduced Anti-TB Drugs

Bedaquiline (BDQ)

The US Food and Drug Administration (FDA) granted approval to Johnson and Johnson's drug BDQ to treat drug-resistant TB, on 28 December 2012 [50]. BDQ is a diarylquinoline. It has a central heterocyclic nucleus with alcohol and amine side chains, responsible for its antimycobacterial activity. Bedaquiline is given in combination therapy to MDR and XDR patients and is administrated orally. It acts as an inhibitor of mycobacterium ATP synthase and is metabolized by cytochrome P450 isoenzyme 3AH (CYP3A4) to a less active N-monodesmethyl metabolite. In vitro studies show MIC of <0.0125 to 0.25 μg/mL [109]. According to available studies, mutations in *atpE* gene lead to high level of MIC to BDQ [36]. Additionally, mutations in gene *Rv0678*, coding for a drug efflux pump regulator, and in *Rv2535c* (pepQ) also confer resistance to BDQ [36].

Delamanid

Delamanid is also a unique anti-TB drug belonging to the nitroimidazole class of antibiotics and derivatives of dihydro-nitroimidazo oxazole. Delamanid is a prodrug which requires enzyme deazaflavin-dependent nitroreductase to get activated. It acts by inhibiting the synthesis of components of the mycobacterial cell wall, methoxymycolic and ketomycolic acid [106]. Delamanid has been reported to be highly active against *M. tuberculosis* with a minimum inhibitory concentration of 0.006–0.024 μg/mL in vitro [54]. Resistance to delamanid has been associated with prodrug activation genes such as *fgd1* and *ddn* or genes associated with deazaflavin biosynthesis pathway such as *fbiA*, *fbiB*, and *fbiC* [52].

Alternative Mechanisms of Drug Resistance

It is well established that resistance is not necessarily associated with genetic mutations in a proportion of clinical *M. tuberculosis* isolates [66, 67]. Hence, it is proposed that alternative mechanisms might be responsible in conferring resistance to anti-TB drugs. Evidence suggests that various other strategies exist, due to which bacteria become drug resistant. These alternative mechanisms include:

Membrane Impermeability

This is a strategy by which bacteria prevent entrance of the drug. Mycobacteria achieve this impermeability with the help of its cell wall that is very rich in mycolic acids [5]. This reduced membrane permeability results in reduction in the influx of drugs that further leads to a decrement of intracellular drug accumulation [63].

Drug Inactivation or Modification by Enzymes

Mycobacterial species, e.g., *M. smegmatis*, is found to be naturally resistant to RIF, in spite of the fact that it has no mutation in *rpoB* gene [64]. It demonstrates resistance by inactivation of RIF via ribosylation.

Target Alteration

It takes place because of mutation in the target gene which results in either reduction in binding capacity of drug or overexpression of the drug target. Sequence analysis of drug-resistant clinical isolates of *M. tuberculosis* has shown that this phenomenon specifically occurs in anti-TB drugs such as INH, PZA, and ETH [33]. Mutations occurring in DNA repair systems lead to inefficient repair of damaged DNA, thus further increasing mutation rates and providing selective advantage to bacteria that carry the mutations. In fact, whole genome sequence studies (WGS) have indicated increased variability in the genes encoding DNA repair proteins in Beijing strains [65, 72, 83]. Exposure to suboptimal drug concentrations is also associated with an increase in mutation rates.

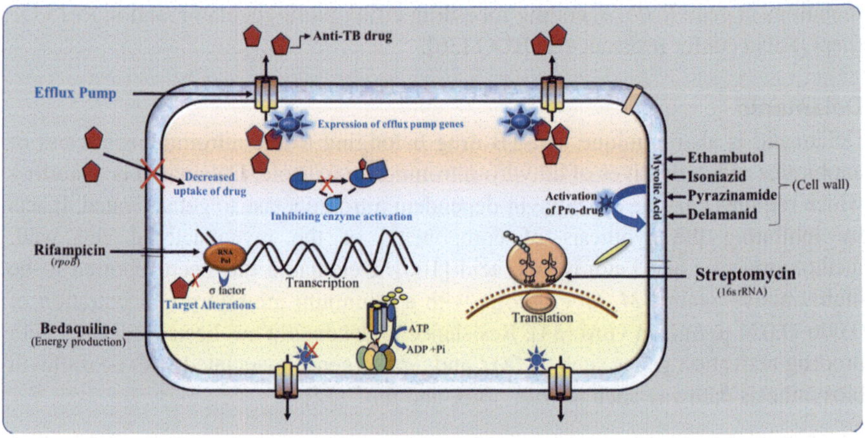

Fig. 1 Representation of common mechanisms of drug resistance to **(a)** first-line anti-TB drugs and recently added new drugs for TB treatment, i.e., bedaquiline and delamanid (in black color text; text in brackets mentions the drug target), **(b)** alternative mechanisms of drug resistance (blue color text)

Figure 1 summarizes the various mechanisms of drug resistance that have been observed in *M. tuberculosis*.

Drug Resistance in *M. tuberculosis* Is Very Complex and Some Other Complicating Factors Should Also Be Considered

Drug Tolerance and Persistence

Differential response of bacteria to drug treatment can arise from drug resistance, drug tolerance, or persistence. All these mechanisms are also observed in TB. Brauner et al. [4] differentiated between these processes on the basis of MIC and "minimum duration of killing" (MDK) values. MIC99 represented the minimum concentration of drug needed to kill 99% of the bacterial population, while MDK99 represented the minimum time needed to kill 99% of a population. Drug resistance, as mentioned above, usually occurs due to mutations in the gene encoding the drug target and is heritable. It results in an increase in the MIC99. Drug tolerance leads to transient survival of bacterial cells exposed to concentrations of antibiotic that would normally be lethal. Tolerance to a drug is observed in all the bacterial cells of a culture. Though the MIC99 in these cells may be similar to phenotypically drug-susceptible isolates, the MDK99 value is significantly higher. Persistence leads to transient survival in the presence of inhibitory concentrations of an antibiotic but is observed in only a small percentage of the population. Persisters are killed more slowly than a susceptible population of bacteria. In this case, though the MIC99 and MDK99 values are similar to the susceptible population, the MDK99.99 is much higher [4, 42].

Cross-Resistance

Cross-resistance can occur to anti-TB drugs within the same or different classes of drugs: viz., mutations in *gyrA* and *gyrB* can lead to cross-resistance to multiple fluoroquinolones [98]; and mutations in *rpoB* can confer cross-resistance to RIF and other rifamycins. Recently, it has been demonstrated that mutation in *Rv0678*, a transcriptional regulator, results in cross-resistance of *M. tuberculosis* to CFZ, an anti-leprosy drug, and the recently approved anti-TB drug BDQ, through upregulation of an efflux pump [29, 34]. Thus, *M. tuberculosis* can evade the effect of drugs through a number of mechanisms, further affecting the development of new anti-TB drugs.

Hetero-resistance

Hetero-resistance is another phenomenon that complicates the diagnosis and management of drug-resistant TB. Hetero-resistance is the coexistence of susceptible and resistant variants of *M. tuberculosis*, or of multiple resistant strains with different mutations in a single sample. Hetero-resistance can occur because of an infection with different strains of *M. tuberculosis* or due to mutations within a clonal *M. tuberculosis* population and occurs in 5.38% drug-resistant TB cases. Next-generation sequencing has revealed significant heterogeneity at loci associated with drug resistance [21, 92]. It has also been shown that minor variants (<1–5% of the population) may differ in frequency through the course of infection. This has implications for patient management and also underscores the importance of new diagnostic techniques that can identify a resistant phenotype even if it is a minority variant, as minority variants may be missed by phenotypic DST [42].

Detection of Drug-Resistant *M. tuberculosis*

Phenotypic Drug Susceptibility Testing (PDST) Methods

PDST is the gold standard for detection of drug-resistant *M. tuberculosis*. It is performed to test the in vitro capacity of a particular drug to inhibit bacterial growth. Inhibition of bacterial growth also predicts the success of therapy. Phenotypic DST is generally considered technically demanding; in addition, proper biosafety precautions should be taken for handling live *M. tuberculosis* isolates. Reliability of the results may vary on the basis of concentration of drugs, medium, handling, and incubation time. PDST can be performed on solid (e.g., Löwenstein-Jensen/LJ) or liquid media. Liquid media show faster results as compared to solid media [7]. Traditionally, LJ culture has been used for drug sensitivity testing by the following methods: (i) absolute concentration, (ii) resistance ratio, and (iii) proportion method [76, 94].

Absolute Concentration Method

In this method, the MIC of the drug is determined by inoculating *M. tuberculosis* culture into a drug-containing medium and a drug-free control. Several sequential twofold dilutions of the drugs are used in the medium. Resistance is determined by the lowest concentration of the drug which inhibits growth (<20 colonies) [94]. Broth dilution method can also be used. Variation in the inoculum size is an important source of error in this method of drug susceptibility testing [19].

Resistance Ratio Method

In resistance ratio method, the resistance of clinical strain of tubercle bacilli is compared with that of a standard susceptible strain of *M. tuberculosis*. The susceptible strains used can be H37Rv, or a recently isolated drug susceptible wild-type strain [19]. A ratio of the MIC of the test strain and the susceptible isolate is considered, thus avoiding inter and intra-laboratory variations. If the ratio is 2 or less, the strain is considered to be susceptible, while a ratio of 8 or more indicates high-level resistance [19].

Proportion Method of Drug Susceptibility Testing (PDST)

In the proportion method, the ratio of the number of colonies growing on drug-containing medium to the number of colonies growing on drug-free medium is measured as a proportion of drug-resistant bacilli present in the bacterial population. Below a critical proportion (1%), the strain is classified as susceptible and above that as resistant [76, 94]. The drug concentrations used as per WHO recommendations are 4 mg/l for SM, 0.2 mg/l for INH, 40 mg/l for RIF, and 2 mg/l for EMB [59]. Inoculated LJ slants are incubated at 37 °C and the reading recorded after 21–28 days.

Nitrate Reductase Method

The phenotypic assays mentioned above are extremely slow. Hence, alternative rapid methods such as the nitrate reductase assay have been suggested. This assay is a low-cost colorimetric assay performed on LJ medium. It is simple to perform and has been successfully implemented in low-resources countries [13, 46]. The assay is based on the ability of *M. tuberculosis* to reduce nitrate to nitrite. A color change is observed in the medium using Griess reagent. Results from culture isolates are obtained within 14 days. A few studies have also analyzed the assay directly on clinical specimens. Although standardization of readings is required, the assay has given good results with INH, RIF, and ofloxacin [35]. The pooled sensitivity for RIF and INH has been reported to be 99% and 94%, respectively. The specificity for both the drugs is reported to be 100% [6]

Microscopic Observation Drug Susceptibility (MODS)

This is a low-technology liquid culture-based rapid DST assay. Drugs are added to liquid medium, and resistance is detected with the observation of characteristic cord-like structures of *M. tuberculosis* under an inverted microscope. Growth is compared

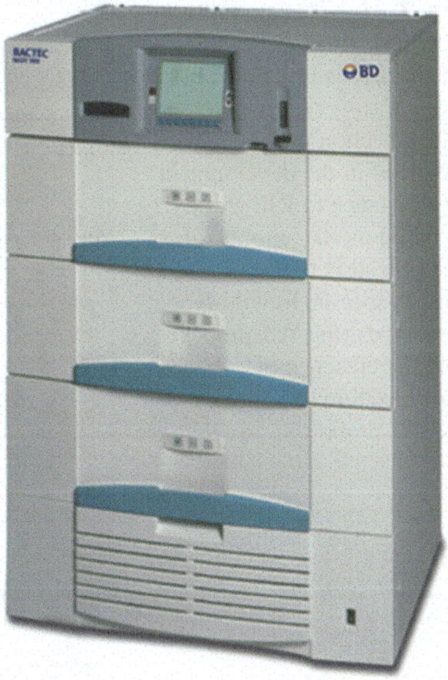

Fig. 2 The MGIT 960 instrument

to a drug-free control. The reported pooled sensitivity and specificity for detection of RIF resistance were 96% and 96% with MODS and that for INH was 92% and 96% [6].

Mycobacteria Growth Indicator Tube (MGIT)

The MGIT system (Becton-Dickinson) is a rapid, nonradioactive method for detection and susceptibility testing of *M. tuberculosis* (Adjers-Koskela and Katila 2003) (Fig. 2). MGIT contains an oxygen-quenched fluorochrome, and growth is detected by the consumption of oxygen by the organisms and the production of fluorescence. The fluorescence produced is detected by using an ultraviolet transilluminator. Scrutiny of the fluorescence is done to determine if the tube is growth positive. Studies carried out show comparable results with the proportion method of DST [20].

Molecular and Genotypic Methods

M. tuberculosis acquires drug resistance by random chromosomal mutations. Nucleotide changes such as point mutations, deletions, and/or insertions confer resistance to single drugs. The stepwise evolution of these mutations leads to the development of MDR strains of *M. tuberculosis*. Drug-resistant strains appear when chemotherapy is intermittent or inadequate, highlighting the importance of early detection of drug resistance. Here, we describe the most relevant and current assays for molecular detection of drug resistance, especially the assays endorsed by WHO.

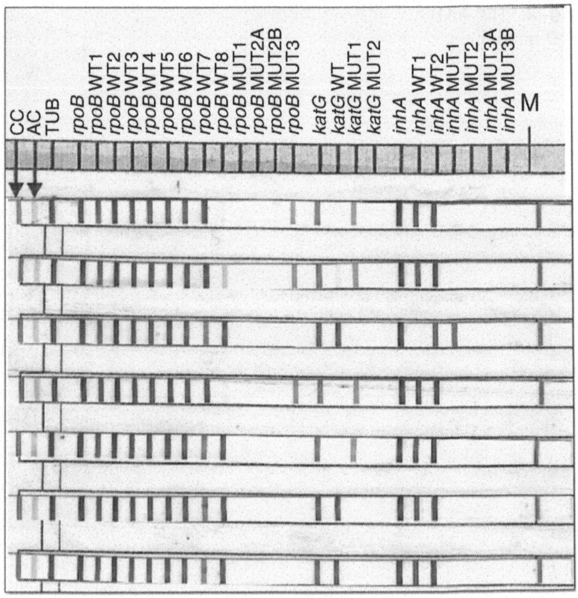

Fig. 3 Hybridization results obtained with the MTBDRplus line probe assay from *M. tuberculosis* isolates. The detection of mutations involved in the development of resistance to RIF and INH is based on the absence of hybridization to one or several probes specific to wild-type genotypes (the WT probes) and the presence of hybridization to a mutant (MUT) probe

Line Probe Assay

Molecular line probe assay (LPA) technology was endorsed by WHO for rapid detection of MDR-TB, in 2008. LPA (Fig. 3) involves amplification of the requisite gene associated with drug resistance by using biotinylated primers, from culture isolates (indirect testing) or from clinical specimens (direct testing). The amplicons are hybridized with specific oligonucleotide probes that have been immobilized on nitrocellulose membrane. Captured labeled hybrids are detected by colorimetric development. The assay detects the presence of *M. tuberculosis* complex along with wild-type and mutant targets associated with drug resistance. If a mutation is present in one of the target regions, there will be absence of hybridization with the relevant probe. The reaction leads to the development of colored bands on the strip at the site of probe binding. The GenoType MTBDR*sl* test has been introduced for the rapid determination of genetic mutations associated with resistance to fluoroquinolones, aminoglycosides (kanamycin, amikacin, SM), cyclic peptides (capreomycin), and EMB. Resistance to fluoroquinolones is detected by mutations at the *gyrA* gene (encoding DNA gyrase). Resistance to EMB is detected by mutations at the *embB* gene. The assay format allows reporting the results within 24 h [101, 102].

LPA has sensitivity and specificity of 98% and 99%, respectively, for detection of RIF resistance; 92% and 99%, respectively, for detection of INH resistance; and 97% and 100%, respectively, for detection of MDR-TB [108].

GeneXpert

Xpert MTB/RIF (Fig. 4) is an automated, cartridge-based nucleic acid amplification. It is a real-time PCR-based assay for detection of *M. tuberculosis* DNA and

Fig. 4 GeneXpert instrument

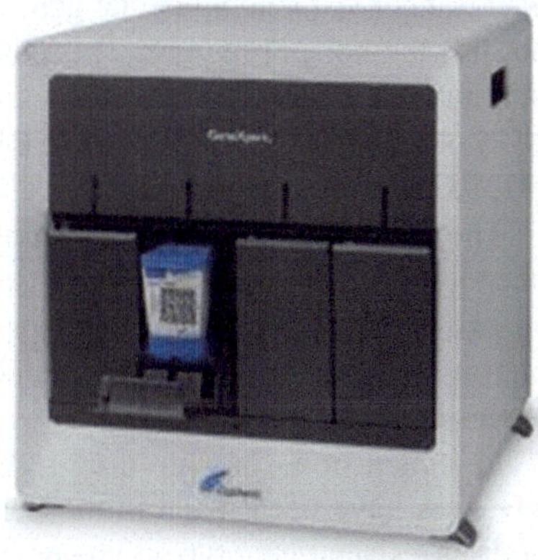

mutations in the *rpoB* gene, associated with RIF resistance. Xpert platform was developed as a partnership between the Foundation for Innovative New Diagnostics (FIND), Cepheid, USA, and the University of Medicine and Dentistry, New Jersey, USA. The technology was supported by the National Institute of Health, USA. The Xpert MTB/RIF platform is a fully automated DNA testing equipment that has been designed for direct testing on clinical samples and was recommended for use by WHO in December 2010. The assay is based on molecular beacons technology and uses three specific primers and five unique molecular probes with a high degree of specificity. The molecular beacon probes are complementary to an 81 bp core region of the *rpoB* gene. Results are provided within 2 h, directly from clinical specimens. The limit of detection (LOD) is 131 CFU/ml of sputum [31, 104], making it more sensitive than smear microscopy. When used as an initial diagnostic test in place of smear microscopy, the Xpert MTB/RIF achieved an overall pooled sensitivity and specificity of 88% and 99%, respectively [102]. The WHO recommends the use of Xpert MTB/ RIF as an initial diagnostic test for diagnosing pulmonary TB in adults and children suspected of having MDR-TB or human immunodeficiency virus (HIV)-associated TB. The evidence for pediatric TB and extrapulmonary TB is currently very low. However, the WHO recommends that Xpert MTB/RIF be used as an initial diagnostic test on cerebrospinal fluid (CSF) for TB meningitis. It may also be used for certain extrapulmonary specimens such as lymph node aspirates. A next-generation cartridge, Xpert Ultra, was launched in 2017 [10]. Xpert Ultra is more sensitive and is able to detect synonymous mutations and silent mutations in the targeted region. The LOD for Ultra is 15.6 CFU/ml in sputum samples.

Whole Genome Sequencing (WGS)

WGS offers new opportunities in the clinical management of drug-resistant TB. WGS can identify various genetic polymorphisms, including single nucleotide polymorphisms (SNPs) and small insertions and deletions (indels) that can be associated with drug resistance. Currently WGS is used to identify mutations from culture isolates. However, direct testing on clinical samples is also being tried and can bring about a paradigm shift in management of TB. The advent of next-generation sequencing (NGS) has hastened this process. The advantage of WGS over Sanger sequencing is that genes can be sequenced across the entire length and not just the hotspot regions. This can help us identify a number of novel mutations that may be responsible for drug resistance. The Illumina (San Diego, CA, USA) NGS platform was used later for epidemiological analysis and rapid molecular DST [41, 91].

The accuracy of WGS in routine drug susceptibility testing, as compared to phenotypic DST, was found to be 96%. In addition, it was rapid (72 h for WGS as against 28 days for DST). Another approach being used is targeted NGS which improves the sensitivity and can reliably estimate proportion of resistant to wild-type alleles in mixtures [55].

In conclusion, though drug-resistant TB is a challenging problem, several advances have been made in the field of molecular diagnostics for rapid detection of drug-resistant genotypes. The potential to use WGS in routine diagnostics is very promising and will not only hasten the time to diagnosis but will also provide information on novel loci associated with drug resistance. Further innovations may also consider the complexities of drug resistance in *M. tuberculosis*, to curb this disease.

References

1. Almeida, D., Ioerger, T., Tyagi, S., et al. (2016). Mutations in *pepQ* confer low-level resistance to bedaquiline and clofazimine in *Mycobacterium tuberculosis. Antimicrobial Agents and Chemotherapy, 60*, 4590–4599.
2. Aubry, A., Pan, X. S., Fisher, L. M., et al. (2004). *Mycobacterium tuberculosis* DNA gyrase: Interaction with quinolones and correlation with antimycobacterial drug activity. *Antimicrobial Agents and Chemotherapy, 48*, 1281–1288.
3. Barry, C. E., 3rd, Lee, R. E., Mdluli, K., et al. (1998). Mycolic acids: Structure, biosynthesis and physiological functions. *Progress in Lipid Research, 37*, 143.
4. Brauner, A., Fridman, O., Gefen, O., et al. (2016). Distinguishing between resistance, tolerance and persistence to antibiotic treatment. *Nature Reviews. Microbiology, 14*, 320–330.
5. Brennan, P. J., & Nikaido, H. (1995). The envelope of mycobacteria. *Annual Review of Biochemistry, 64*, 29–63.
6. Bwanga, F., Hoffner, S., Haile, M., et al. (2009). Direct susceptibility testing for multi drug resistant tuberculosis: A meta-analysis. *BMC Infectious Diseases, 9*, 67.
7. Canetti, G., Fox, W., Khomenko, A. A., et al. (1969). Advances in techniques of testing mycobacterial drug sensitivity, and the use of sensitivity tests in tuberculosis control programmes. *Bulletin of the World Health Organization, 41*, 21.
8. Carter, A. P., Clemons, W. M., Brodersen, D. E., et al. (2000). Functional insights from the structure of the 30S ribosomal subunit and its interactions with antibiotics. *Nature, 407*, 340–348.
9. Cavusoglu, C., Hilmioglu, S., Guneri, S., et al. (2002). Characterization of *rpoB* mutations in rifampin-resistant clinical isolates of *Mycobacterium tuberculosis* from Turkey by DNA sequencing and line probe assay. *Journal of Clinical Microbiology, 40*, 4435–4438.
10. Chakravorty, S., Simmons, A. M., & Parmar, R. M. (2017). The new Xpert MTB/RIF ultra: Improving detection of *Mycobacterium tuberculosis* and resistance to rifampin in an assay suitable for point-of-care testing. *MBio, 8*, e00812–e00817.
11. Cheng, S. J., Thibert, L., Sanchez, T., et al. (2000). *pncA* mutations as a major mechanism of pyrazinamide resistance in *Mycobacterium tuberculosis*: Spread of a monoresistant strain in Quebec. *Canada. Antimicrob Agents Chemother, 44*, 528–532.

12. Cheng, A. F., Yew, W. W., Chan, E. W., et al. (2004). Multiplex PCR amplimer conformation analysis for rapid detection of *gyrA* mutations in fluoroquinolone-resistant *Mycobacterium tuberculosis* clinical isolates. *Antimicrobial Agents and Chemotherapy, 48*, 596–601.
13. Coban, A. Y., Birinci, A., Ekinci, B., et al. (2004). Drug susceptibility testing of *Mycobacterium tuberculosis* with nitrate reductase assay. *International Journal of Antimicrobial Agents, 24*, 304–306.
14. Das, R., Gupta, P., Singh, P., et al. (2009). Association of mutations in *rpsL* gene with high degree of streptomycin resistance in clinical isolates of *Mycobacterium tuberculosis* in India. *The Indian Journal of Medical Research, 129*, 108–111.
15. Desjardins, C. A., Cohen, K. A., Munsamy, V., et al. (2016). Genomic and functional analyses of *Mycobacterium tuberculosis* strains implicate ald in D-cycloserine resistance. *Nature Genetics, 48*, 544.
16. Dey, T., Brigden, G., Cox, H., et al. (2013). Outcomes of clofazimine for the treatment of drug-resistant tuberculosis: A systematic review and meta-analysis. *The Journal of Antimicrobial Chemotherapy, 68*, 284–293.
17. Dookie, N., Rambaran, S., Padayatchi, N., Mahomed, S., & Naidoo, K. (2018). Evolution of drug resistance in *Mycobacterium tuberculosis*. *The Journal of Antimicrobial Chemotherapy, 73*, 1138–1151.
18. Drlica, K. (1999). Mechanism of fluoroquinolone action. *Current Opinion in Microbiology, 2*, 504–508.
19. Drobniewski, F., Rüsch-Gerdes, S., & Hoffner, S. (2007). Antimicrobial susceptibility testing of *Mycobacterium tuberculosis* (EUCAST document E. DEF 8.1)–report of the subcommittee on antimicrobial susceptibility testing of *Mycobacterium tuberculosis* of the European Committee for Antimicrobial Susceptibility Testing (EUCAST) of the European Society of Clinical Microbiology and Infectious Diseases (ESCMID). *Clinical Microbiology and Infection, 13*, 1144–1156.
20. Duque, A., Lin, S. Y. G., Desmond, E., et al. (2013). Evaluation of the BD Bactec MGIT 320 system for detection of mycobacteria and drug susceptibility testing of *Mycobacterium tuberculosis*. *Journal of Clinical Microbiology, 51*, 3403–3405.
21. Eldholm, V., Norheim, G., von der Lippe, B., et al. (2014). Evolution of extensively drug-resistant *Mycobacterium tuberculosis* from a susceptible ancestor in a single patient. *Genome Biology, 15*, 490.
22. Escalante, P., Ramaswamy, S., Sanabria, H., et al. (1998). Genotypic characterization of drug-resistant *Mycobacterium tuberculosis* isolates from Peru. *Tubercle and Lung Disease, 79*, 111–118.
23. Faustini, A., Hall, A. J., & Perucci, C. A. (2006). Risk factors for multidrug resistant tuberculosis in Europe: A systematic review. *Thorax, 61*, 158–163.
24. Feng, Z., & Barletta, R. G. (2003). Roles of *Mycobacterium smegmatis* D-alanine: D-alanine ligase and D-alanine racemase in the mechanisms of action of and resistance to the peptidoglycan inhibitor D-cycloserine. *Antimicrobial Agents and Chemotherapy, 47*, 283–291.
25. Feuerriegel, S., Köser, C., Trübe, L., et al. (2010). Thr202Ala in thyA is a marker for the Latin American Mediterranean lineage of the *Mycobacterium tuberculosis* complex rather than para-aminosalicylic acid resistance. *Antimicrobial Agents and Chemotherapy, 54*, 4794–4798.
26. Finken, M., Kirschner, P., Meier, A., et al. (1993). Molecular basis of streptomycin resistance in *Mycobacterium tuberculosis*: Alterations of the ribosomal protein S12 gene and point mutations within a functional 16S ribosomal RNA pseudoknot. *Molecular Microbiology, 9*, 1239–1246.
27. Ginsburg, A. S., Grosset, J. H., & Bishai, W. R. (2003). Fluoroquinolones, tuberculosis, and resistance. *Lancet Infectious Diseases, 3*, 432–442.
28. Grange, J. M. (1990). Drug resistance and tuberculosis elimination. *Bulletin of the International Union against Tuberculosis and Lung Disease, 65*, 57.
29. Hartkoorn, R. C., Uplekar, S., & Cole, S. T. (2014). Cross-resistance between clofazimine and bedaquiline through upregulation of *MmpL5* in *Mycobacterium tuberculosis*. *Antimicrobial Agents and Chemotherapy, 58*, 2979–2981.

30. Hazbon, M. H., Brimacombe, M., Bobadilla del Valle, M., et al. (2006). Population genetics study of isoniazid resistance mutations and evolution of multidrug-resistant *Mycobacterium tuberculosis*. *Antimicrobial Agents and Chemotherapy, 50*, 2640–2649.
31. Helb, D., Jones, M., Story, E., et al. (2010). Rapid detection of *Mycobacterium tuberculosis* and rifampin resistance by use of on-demand, near-patient technology. *Journal of Clinical Microbiology, 48*, 229–237.
32. Herrera, L., Jiménez, S., Valverde, A., et al. (2003). Molecular analysis of rifampicin-resistant *Mycobacterium tuberculosis* isolated in Spain (1996–2001). Description of new mutations in the *rpoB* gene and review of the literature. *International Journal of Antimicrobial Agents, 21*, 403–408.
33. Heym, B., & Cole, S. T. (1997). Multidrug resistance in *Mycobacterium tuberculosis*. *International Journal of Antimicrobial Agents, 8*, 61–70.
34. Hoagland, D. T., Liu, J., Lee, R. B., et al. (2016). New agents for the treatment of drug-resistant *Mycobacterium tuberculosis*. *Advanced Drug Delivery Reviews, 102*, 55–72.
35. Horne, D. J., Pinto, L. M., Arentz, M., et al. (2013). Diagnostic accuracy and reproducibility of WHO-endorsed phenotypic drug susceptibility testing methods for first-line and second-line anti-TB drugs. *Journal of Clinical Microbiology, 51*, 393–401.
36. Ismail, N. A., Omar, S. V., Joseph, L., et al. (2018). Defining bedaquiline susceptibility, resistance, cross-resistance and associated genetic determinants: A retrospective cohort study. *E BioMedicine, 28*, 136–142.
37. Jenkins, H. E., Zignol, M., & Cohen, T. (2011). Quantifying the burden and trends of isoniazid resistant tuberculosis, 1994–2009. *PLoS One, 6*, e22927.
38. Jia, L., Tomaszewski, J. E., Hanrahan, C., et al. (2005). Pharmacodynamics and pharmacokinetics of SQ109, a new diamine-based antitubercular drug. *British Journal of Pharmacology, 144*, 80–87.
39. Juréen, P., Werngren, J., Toro, J.-C., et al. (2008). Pyrazinamide resistance and *pncA* gene mutations in *Mycobacterium tuberculosis*. *Antimicrobial Agents and Chemotherapy, 52*, 1852–1854.
40. Kaur, S., Rana, V., Singh, P., et al. (2016). Novel mutations conferring resistance to kanamycin in *Mycobacterium tuberculosis* clinical isolates from northern India. *Tuberculosis, 96*, 96–101.
41. Köser, C. U., Bryant, J. M., Becq, J., et al. (2013). Whole-genome sequencing for rapid susceptibility testing of *M. tuberculosis*. *The New England Journal of Medicine, 369*, 290–292.
42. Koch, A., Cox, H., & Mizrahi, V. (2018). Drug-resistant tuberculosis: Challenges and opportunities for diagnosis and treatment. *Current Opinion in Pharmacology, 42*, 7–15.
43. Lakshminarayana, S. B., Huat, T. B., Ho, P. C., et al. (2014). Comprehensive physicochemical, pharmacokinetic and activity profiling of anti-TB agents. *The Journal of Antimicrobial Chemotherapy, 70*, 857–867.
44. Lange, C., Chesov, D., Heyckendorf, J., et al. (2018). Drug-resistant tuberculosis: An update on disease burden, diagnosis and treatment. *Respirology, 23*, 656–673.
45. Larsen, M. H., Vilcheze, C., Kremer, L., et al. (2002). Overexpression of *inhA*, but not *kasA*, confers resistance to isoniazid and ethionamide in *Mycobacterium smegmatis*, *M. bovis* BCG and *M. tuberculosis*. *Molecular Microbiology, 46*, 453–466.
46. Lemus, D., Montoro, E., Echemendía, M., et al. (2006). Nitrate reductase assay for detection of drug resistance in *Mycobacterium tuberculosis:* Simple and inexpensive method for low-resource laboratories. *Journal of Medical Microbiology, 55*, 861–863.
47. Leung, K. L., Yip, C. W., Yeung, Y. L., et al. (2010). Usefulness of resistant gene markers for predicting treatment outcome on second-line anti-tuberculosis drugs. *Journal of Applied Microbiology, 109*, 2087–2094.
48. Lew, W., Pai, M., Oxlade, O., et al. (2008). Initial drug resistance and tuberculosis treatment outcomes: Systematic review and meta-analysis. *Annals of Internal Medicine, 149*, 123–134.
49. Machado, D., Couto, I., Perdigão, J., et al. (2013). Contribution of efflux to the emergence of isoniazid and multidrug resistance in *Mycobacterium tuberculosis*. *PLoS One, 7*, e34538.

50. Mahajan, R. (2013). Bedaquiline: First FDA-approved tuberculosis drug in 40 years. *International Journal of Applied & Basic Medical Research, 3*(1).
51. Manissero, D., & Fernandez de la Hoz, K. (2006). Surveillance methods and case definition for extensively drug resistant TB (XDR-TB) and relevance to Europe: Summary update. *Euro Surveillance, 11*, e061103.
52. Manjunatha, U. H., Boshoff, H., Dowd, C. S., et al. (2006). Identification of a nitroimidazo-oxazine-specific protein involved in PA-824 resistance in *Mycobacterium tuberculosis*. *Proceedings of the National Academy of Sciences of the United States of America, 103*, 431–436.
53. Mathys, V., Wintjens, R., Lefevre, P., et al. (2009). Molecular genetics of Para-aminosalicylic acid resistance in clinical isolates and spontaneous mutants of *Mycobacterium tuberculosis*. *Antimicrobial Agents and Chemotherapy, 53*, 2100–2109.
54. Matsumoto, M., Hashizume, H., Tomishige, T., et al. (2006). OPC-67683, a nitro-dihydroimidazooxazole derivative with promising action against tuberculosis in vitro and in mice. *PLoS medic, 3*, e466.
55. Metcalfe, J. Z., Streicher, E., Theron, G., et al. (2017). Cryptic microheteroresistance explains *Mycobacterium tuberculosis* phenotypic resistance. *American Journal of Respiratory and Critical Care Medicine, 196*, 1191–1201.
56. Middlebrook, G. (1954). Isoniazid-resistance and catalase activity of tubercle bacilli. *Am Rev Tuberc, 69*, 471–472.
57. Miotto, P., Zhang, Y., Cirillo, D. M., & Yam, W. C. (2018). Drug resistance mechanisms and drug susceptibility testing for tuberculosis. *Respirology, 23*, 1098–1113.
58. Mitchison, D. A. (1985). The action of antituberculosis drugs in short-course chemotherapy. *Tubercle, 66*, 219–225.
59. Mondal, R., & Jain, A. (2007). Extensively drug-resistant *Mycobacterium tuberculosis*, India. *Emerging Infectious Diseases, 13*, 1429.
60. Musser, J. M. (1995). Antimicrobial agent resistance in mycobacteria: Molecular genetic insights. *Clinical Microbiology Reviews, 8*, 496–514.
61. Muthaiah, M., Jagadeesan, S., Ayalusamy, N., et al. (2010). Molecular epidemiological study of pyrazinamide-resistance in clinical isolates of *Mycobacterium tuberculosis* from South India. *International Journal of Molecular Sciences, 11*, 2670–2680.
62. Pandey, B., Grover, S., Goyal, S., et al. (2018). Novel missense mutations in *gidB* gene associated with streptomycin resistance in *Mycobacterium tuberculosis*: Insights from molecular dynamics. *Journal of Biomolecular Structure & Dynamics*, 1–16.
63. Pasca, M. R., Guglierame, P., De Rossi, E., et al. (2005). *mmpL7* gene of *Mycobacterium tuberculosis* is responsible for isoniazid efflux in *Mycobacterium smegmatis*. *Antimicrobial Agents and Chemotherapy, 49*, 4775–4777.
64. Quan, S., Venter, H., & Dabbs, E. R. (1997). Ribosylative inactivation of rifampin by *Mycobacterium smegmatis* is a principal contributor to its low susceptibility to this antibiotic. *Antimicrobial Agents and Chemotherapy, 41*, 2456–2460.
65. Rad, M. E., Bifani, P., Martin, C., et al. (2003). Mutations in putative mutator genes of *Mycobacterium tuberculosis* strains of the W-Beijing family. *Emerging Infectious Diseases, 9*, 838.
66. Ramaswamy, S., & Musser, J. M. (1998). Molecular genetic basis of antimicrobial agent resistance in *Mycobacterium tuberculosis*: 1998 update. *Tubercle and Lung Disease, 79*, 3–29.
67. Ramaswamy, S. V., Reich, R., Dou, S. J., et al. (2003). Single nucleotide polymorphisms in genes associated with isoniazid resistance in *Mycobacterium tuberculosis*. *Antimicrobial Agents and Chemotherapy, 47*, 1241–1250.
68. Rattan, A., Kalia, A., & Ahmad, N. (1998). Multidrug-resistant *Mycobacterium tuberculosis*: molecular perspectives. *Emerging Infectious Diseases, 4*, 195.
69. Rengarajan, J., Sassetti, C. M., Naroditskaya, V., et al. (2004). The folate pathway is a target for resistance to the drug para-aminosalicylic acid (PAS) in mycobacteria. *Molecular Microbiology, 53*, 275–282.

70. Safi, H., Lingaraju, S., Amin, A., et al. (2013). Evolution of high-level ethambutol-resistant tuberculosis through interacting mutations in decaprenylphosphoryl-β-D-arabinose biosynthetic and utilization pathway genes. *Nature Genetics, 45*, 1190.
71. Sandgren, A., Strong, M., Muthu krishnan, P., et al. (2009). Tuberculosis drug resistance mutation database. *PLoS Medicine, 6*, e1000002.
72. Saunders, N. J., Trivedi, U. H., Thomson, M. L., et al. (2011). Deep resequencing of serial sputum isolates of *Mycobacterium tuberculosis* during therapeutic failure due to poor compliance reveals stepwise mutation of key resistance genes on an otherwise stable genetic background. *The Journal of Infection, 62*, 212–217.
73. Scorpio, A., Lindholm-Levy, P., Heifets, L., et al. (1997). Characterization of *pncA* mutations in pyrazinamide-resistant *Mycobacterium tuberculosis. Antimicrobial Agents and Chemotherapy, 41*, 540–543.
74. Scorpio, A., & Zhang, Y. (1996). Mutations in pncA, a gene encoding pyrazinamidase/nicotinamidase, cause resistance to the antituberculous drug pyrazinamide in tubercle bacillus. *Nature Medicine, 2*, 662–667.
75. Sharma, A., Hill, A., Kurbatova, E., et al. (2017). Estimating the future burden of multidrug-resistant and extensively drug-resistant tuberculosis in India, the Philippines, Russia, and South Africa: A mathematical modelling study. *The Lancet Infectious Diseases, 17*, 707–715.
76. Sharma, S. K., & Mohan, A. (2004). Multidrug-resistant tuberculosis. *The Indian Journal of Medical Research, 120*, 354–376.
77. Siddiqi, N., Shamim, M., Hussain, S., et al. (2002). Molecular characterization of multidrug-resistant isolates of *Mycobacterium tuberculosis* from patients in North India. *Antimicrobial Agents and Chemotherapy, 46*, 443–450.
78. Silva, M. S. N., Senna, S. G., Ribeiro, M. O., et al. (2003). Mutations in *katG, inhA,* and *ahpC* genes of Brazilian isoniazid-resistant isolates of *Mycobacterium tuberculosis. Journal of Clinical Microbiology, 41*, 4471–4474.
79. Siu, G. K. H., Zhang, Y., Lau, T. C., et al. (2011). Mutations outside the rifampicin resistance-determining region associated with rifampicin resistance in *Mycobacterium tuberculosis. Antimicrob Chemother, 66*, 730–733.
80. Slayden, R. A., & Barry, C. E., III. (2000). The genetics and biochemistry of isoniazid resistance in *Mycobacterium tuberculosis. Microbes and Infection, 2*, 659–669.
81. Somoskovi, A., Parsons, L. M., & Salfinger, M. (2001). The molecular basis of resistance to isoniazid, rifampin, and pyrazinamide in *Mycobacterium tuberculosis. Respiratory Research, 2*, 164.
82. Sreevatsan, S., Pan, X., Zhang, Y., et al. (1997). Mutations associated with pyrazinamide resistance in *pncA* of *Mycobacterium tuberculosis* complex organisms. *Antimicrobial Agents and Chemotherapy, 41*, 636–640.
83. Stagg, H. R., Cohen, T., Becerra, M. C., et al. (2014). A tale of two settings: The role of the Beijing genotype in the epidemiology of multidrug-resistant tuberculosis. *The European Respiratory Journal, 43*, 632–635.
84. Sulochana, S., Narayanan, S., Paramasivan, C. N., et al. (2007). Analysis of fluoroquinolone resistance in clinical isolates of *Mycobacterium tuberculosis* from India. *Journal of Chemotherapy, 19*, 166–171.
85. Sun, Z., Zhang, J., Zhang, X., et al. (2008). Comparison of *gyrA* gene mutations between laboratory-selected ofloxacin-resistant *Mycobacterium tuberculosis* strains and clinical isolates. *International Journal of Antimicrobial Agents, 31*, 115–121.
86. Suzuki, Y., Katsukawa, C., Tamaru, A. K. I., et al. (1998). Detection of kanamycin-resistant *Mycobacterium tuberculosis* by identifying mutations in the 16S rRNA gene. *Journal of Clinical Microbiology, 36*, 1220–1225.
87. Takayama, K., & Kilburn, J. O. (1989). Inhibition of synthesis of arabinogalactan by ethambutol in *Mycobacterium smegmatis. Antimicrobial Agents and Chemotherapy, 33*, 1493–1499.
88. Takiff, H. E., Salazar, L., Guerrer, C., et al. (1994). Cloning and nucleotide sequence of *Mycobacterium tuberculosis* gyrA and gyrB genes and detection of quinolone resistance mutations. *Antimicrobial Agents and Chemotherapy, 38*, 773–780.

89. Taniguchi, H., Chang, B., Abe, C., et al. (1997). Molecular analysis of kanamycin and viomycin resistance in *Mycobacterium smegmatis* by use of the conjugation system. *Journal of Bacteriology, 179*, 4795–4801.
90. Telenti, A., Imboden, P., Marchesi, F., et al. (1993). Detection of rifampicin-resistance mutations in *Mycobacterium tuberculosis. The Lancet, 341*, 647–651.
91. Török, M. E., Reuter, S., Bryant, J., et al. (2013). Rapid whole-genome sequencing for investigation of a suspected tuberculosis outbreak. *Journal of Clinical Microbiology, 51*, 611–614.
92. Trauner, A., Liu, Q., Via, L. E. R., et al. (2017). The within-host population dynamics of *Mycobacterium tuberculosis* vary with treatment efficacy. *Genome Biology, 18*, 71.
93. van Rensburg, C. E., Joone, G. K., O'sullivan, J. F., et al. (1992). Antimicrobial activities of clofazimine and B669 are mediated by lysophospholipids. *Antimicrobial Agents and Chemotherapy, 36*, 2729–2735.
94. Vareldzis, B. P., Grosset, J., de Kantor, I., et al. (1994). Drug-resistant tuberculosis: Laboratory issues. World Health Organization recommendations. *Tubercle and Lung Disease, 75*, 1–7.
95. Victor, T. C., van Rie, A., Jordaan, A. M., et al. (2001). Sequence polymorphism in the *rrs* gene of *Mycobacterium tuberculosis* is deeply rooted within an evolutionary clade and is not associated with streptomycin resistance. *Journal of Clinical Microbiolgy, 39*, 4184–4186.
96. Vilchèze, C., Av-Gay, Y., Attarian, R., et al. (2008). Mycothiol biosynthesis is essential for ethionamide susceptibility in *Mycobacterium tuberculosis. Molecular Microbiology, 69*, 1316–1329.
97. Vilcheze, C., Weisbrod, T. R., Chen, B., et al. (2005). Altered NADH/NAD+ ratio mediates coresistance to isoniazid and ethionamide in mycobacteria. *Antimicrobial Agents and Chemotherapy, 49*, 708–720.
98. Willby, M., Sikes, R. D., Malik, S., et al. (2015). Correlation between *GyrA* substitutions and ofloxacin, levofloxacin, and moxifloxacin cross-resistance in *Mycobacterium tuberculosis. Antimicrobial Agents and Chemotherapy, 59*, 5427–5434.
99. World Health Organization. (2004). *Global tuberculosis report*. Geneva: WHO.
100. World Health Organization. (2012). *Global tuberculosis report*. Switzerland: WHO, Geneva.
101. World Health Organization. (2013). *Global tuberculosis report*. Switzerland: WHO, Geneva.
102. World Health Organization. (2013). Using the Xpert MTB/RIF assay to detect pulmonary and extrapulmonary tuberculosis and rifampicin resistance in adults and children: Expert group meeting report. http://who.int/tb/publications/xpert-mtb-rif-assay-diagnosis-meeting-report/eng/pdf
103. World Health Organization. (2014). Global tuberculosis report. http://apps.who.int/iris/bitstream/10665/137094/1/9789241564809_eng.pdf
104. World Health Organization. (2014). *Xpert MTB/RIF implementation manual: Technical and operational 'How-to': Practical considerations*. Geneva: WHO.
105. World Health Organization. (2017). *Global tuberculosis report*. Geneva: WHO.
106. Xavier, A. S., & Lakshmanan, M. (2014). Delamanid: A new armor in combating drug-resistant tuberculosis. *Journal of Pharmacology and Pharmacotherapeutics, 5*, 222.
107. Xu, J., Wang, B., Hu, M., et al. (2017). Primary clofazimine and bedaquiline resistance among isolates from patients with multidrug-resistant tuberculosis. *Antimicrobial Agents and Chemotherapy, 61*, e00239–e00217.
108. Yadav, R. N., Singh, B. K., Sharma, S. K., et al. (2013). Comparative evaluation of GenoType MTBDRplus line probe assay with solid culture method in early diagnosis of multidrug resistant tuberculosis (MDR-TB) at a tertiary care Centre in India. *PLoS One, 8*, e72036.
109. Yang, J. S., Kim, K. J., Choi, H., et al. (2018). Delamanid, bedaquiline, and linezolid minimum inhibitory concentration distributions and resistance-related gene mutations in multidrug-resistant and extensively drug-resistant tuberculosis in Korea. *Annals of Laboratory Medicine, 38*, 563–568.

110. Yano, T., Kassovska-Bratinova, S., Teh, J. S., et al. (2011). Reduction of clofazimine by mycobacterial type 2 NADH: Quinone oxidoreductase a pathway for the generation of bactericidal levels of reactive oxygen species. *The Journal of Biological Chemistry, 286*, 10276–10287.
111. Zaunbrecher, M. A., Sikes, R. D., Metchock, B., et al. (2009). Overexpression of the chromosomally encoded aminoglycoside acetyltransferase *eis* confers kanamycin resistance in *Mycobacterium tuberculosis. Proceedings of the National Academy of Sciences of the United States of America, 106*, 20004–20009.
112. Zaw, M. T., Emran, N. A., & Lin, Z. (2018). Mutations inside rifampicin-resistance determining region of *rpoB* gene associated with rifampicin-resistance in *Mycobacterium tuberculosis. Journal of Infection and Public Health, 11*, 605–610.
113. Zhang, M., Yue, J., Yang, Y. P., et al. (2005). Detection of mutations associated with isoniazid resistance in *Mycobacterium tuberculosis* isolates from China. *Journal of Clinical Microbiology, 43*, 5477–5482.
114. Zhang, Y., & Young, D. B. (1993). Molecular mechanisms of isoniazid: A drug at the front line of tuberculosis control. *Trends in Microbiology, 1*, 109–113.

Eliminating Mycobacterial Persistence: Novel Targets for Anti-TB Therapy

Ashima Bhaskar, Ved Prakash Dwivedi, and Vinay Kumar Nandicoori

Abstract

Despite enormous efforts toward eradication of tuberculosis (TB), it remains one of the leading causes of deaths from a single infectious agent. *Mycobacterium tuberculosis* (*Mtb*), the causative agent of TB, has successfully persisted in the human host for centuries. Current anti-TB therapy is effective in most of the cases but due to a very long duration of treatment and adverse side effects of medication, it leads to non-compliance resulting in the emergence of drug resistance. Moreover, host never gets completely sterilized of the bug and thus remains susceptible to disease relapse. Thus in order to completely eliminate tuberculosis in the near future, one must attempt to understand the dynamics of dormant, persistent, and latent bacilli which are resistant to majority of anti-TB drugs and are a major cause for recurrence incidents. Here, we briefly describe dormancy, persistence, and latency in terms of tuberculosis disease. We attempt to elaborate on various models used to study these phenomena which have led to a better understanding of the mechanisms adopted by *Mtb* in order to survive in the hostile environment of the host. These studies are critical for developing newer and effective strategies to target TB.

Keywords

Mycobacteria · Dormancy · Latency · Persisters · Wayne model · Bedaquiline

A. Bhaskar (✉) · V. K. Nandicoori
Signal Transduction Laboratory, National Institute of Immunology, New Delhi, India
e-mail: ashimabhaskar@gmail.com

V. P. Dwivedi
International Centre for Genetic Engineering and Biotechnology, New Delhi, India

S. Hameed, Z. Fatima (eds.), *Pathogenicity and Drug Resistance of Human Pathogens*,
https://doi.org/10.1007/978-981-32-9449-3_3

Introduction

Globally, tuberculosis (TB) is one of the top ten causes of death and the leading cause from a single infectious agent [45]. *Mycobacterium tuberculosis* (*Mtb*), the causative agent of TB, remains one of the most successful human pathogen, responsible for 1.3 million deaths in the year 2017. The tubercle bacilli are unique in terms of possessing a remarkable ability of surviving in vivo in a dormant state for years only to reactivate. Approximately 1.7 billion people, which accounts for 23% of the global population, are infected with latent TB and thus provide a massive reservoir for this deadly pathogen. The existing TB treatment, which includes 60-day treatment of four frontline anti-TB drugs, isoniazid (INH), rifampin (RIF), ethambutol (EMB), and pyrazinamide (PZA), followed by a 120-day treatment with INH and RIF, is effective and has been successful in averting millions of deaths each year, but it fails to completely sterilize the individual. This lengthy anti-TB therapy of 6 months or more leads to non-compliance, thus resulting in the emergence of drug-resistant variants of *Mtb*: MDR and XDR TB. It takes 9 months or even more to treat MDR and XDR TB, with the success rate being below 50% [79]. The current situation is further worsened by HIV-TB coinfections wherein immunocompromised state of HIV-positive individuals put them at high risk of relapse. The reason for this long treatment and failure to achieve 100% sterility is believed to be the presence of dormant, latent, or persistent bacilli, which can revert back to active form and cause reinfection [92]. Understanding the dynamics of relationship between latent bacilli and host is critical for the development of shorter and more effective treatments for TB. In order to achieve the goal of ending the TB epidemic, it becomes imperative for us to study in detail the various mechanisms of mycobacterial dormancy, latency, and persistence.

TB pathogenesis is a complex phenomenon wherein infection by *Mtb* can result in multiple outcomes depending upon the load, infectiousness of the bug, and the immune state of the host [87]. Bacteria can either be cleared by the host or can cause active infection. Third possibility is the emergence of dormant bacilli, making the host asymptomatic but a carrier. Finally, reactivation of the bacteria due to multiple host factors leads to reemergence of infection [136].

Mycobacterial dormancy, latency, and persistence are three very different yet overlapping phenomena. This review attempts at distinctly defining mycobacterial dormancy, latency, and persistence all of which are used interchangeably leading to confusion in the field. We also review various models for studying these phenotypes and factors responsible for their existence. Finally an attempt is made to understand the various strategies being employed to control latent TB.

Dormancy

Dormancy is defined as a phase of life where an organism shuts its metabolic activity to a minimal level leading to its growth arrest.

(continued)

Latency
Latency is a clinical term referring to the in vivo state of an infectious organism where the host remains asymptomatic.

Persistence
Persistence is the phenomenon where phenotypic heterogeneity gives rise to a subpopulation of transiently antibiotic-tolerant bacterial cells that may or may not be slow-growing or growth-arrested.

Dormant Bacilli

Dormancy phenomenon occurs in multiple organisms ranging from bacteria to plants and animals. Dormancy is a metabolically inactive state induced in order to survive in adverse conditions [116]. Dormant bacteria are characterized by slow or no growth, inability to form colony-forming units, resistance to antibiotics and multiple stresses, and ability to resuscitate under favorable conditions [46] (Fig. 1). In mycobacterial physiology, dormancy refers to a stable but reversible growth arrest in response to multiple host stresses [46]. Dormant mycobacteria have been shown to be present in aged liquid cultures, macrophages, and mice tissue [14, 15, 40, 46, 54]. Non-growing but metabolically active bacteria have also been shown in the lungs of chronically infected and isoniazid-treated mice suggesting a role in post-chemotherapeutic relapses [86]. Majority of currently available anti-TB

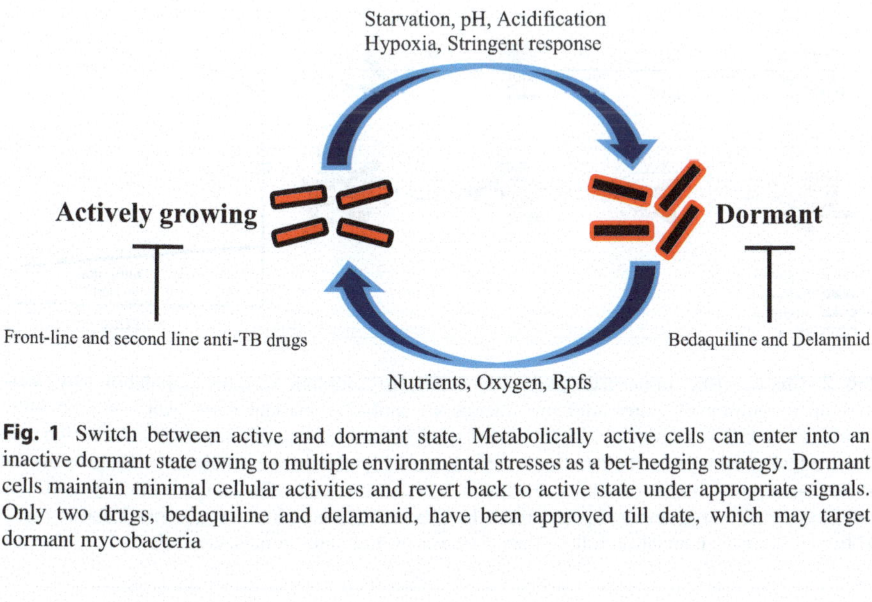

Fig. 1 Switch between active and dormant state. Metabolically active cells can enter into an inactive dormant state owing to multiple environmental stresses as a bet-hedging strategy. Dormant cells maintain minimal cellular activities and revert back to active state under appropriate signals. Only two drugs, bedaquiline and delamanid, have been approved till date, which may target dormant mycobacteria

drugs target processes critical for an actively growing bacteria such as DNA replication, translation, or cell wall synthesis. Non-growing or slow-growing state of dormant bacilli make them resistant toward these drugs. This allows for long-term survival in the host, which may result in resuscitation and relapse. Therefore, complete prevention and cure of TB require targeting dormant *Mtb* and evolving novel strategies and therapeutics to kill these quiescent bugs.

One of the earliest and most studied models of dormancy is the Wayne model where bacteria when subjected to gradual depletion of atmospheric oxygen enter a non-replicating state [142]. Since then, many models have been designed to study the mechanism of dormancy and factors responsible for resuscitation. This chapter will give a brief overview of the various models used to study dormancy, the bacterial and host factors responsible for the entry and exit from dormant state, and lastly the mechanisms of exit from mycobacterial dormancy.

Persisters

Persisters are multidrug-tolerant cells, which are able to survive lethal doses of antibiotics. They represent a small subpopulation of otherwise antibiotic-sensitive cells, which become transiently refractory to antibiotics. They are different from antibiotic-resistant cells in terms of nonheritable phenotypic resistance, which arises due to stochastic or deterministic epigenetic factors that are not passed onto the next generation (Fig. 2a). Joseph Bigger first coined the term persisters in 1944 when he discovered the inability of penicillin to completely sterilize a *Staphylococcus* culture [9, 51]. Since their discovery, persisters have been identified not only in bacteria but also in eukaryotic systems including major human pathogens and diseases [12, 47,

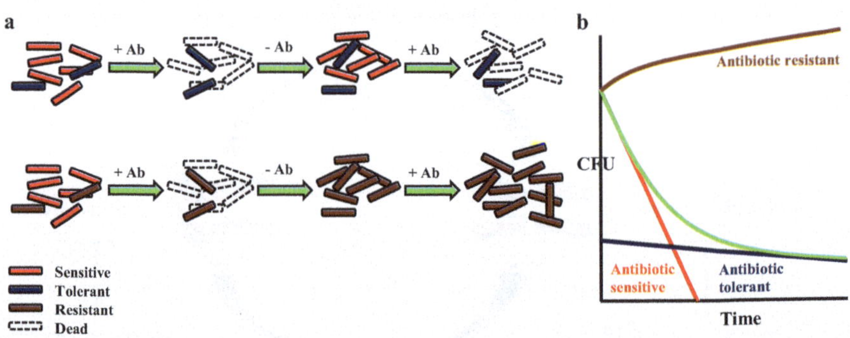

Fig. 2 Characteristics of persister and resistant cells. (**a**) Antibiotic-sensitive population may carry a small proportion of either antibiotic-tolerant or antibiotic-resistant cells. Antibiotic pressure results in the killing of sensitive population leaving behind tolerant and resistant cells, which can be distinguished based on the fact that upon removal of antibiotic stress, tolerant cells give rise to same parent population with similar sensitivity, whereas resistant cells being genetically modified produce a totally drug-resistant population. (**b**) Classic biphasic kill curve generated due to rapid killing of sensitive population followed by a slower or no killing of persister cells

61, 63, 69, 97]. Experimentally, persisters are identified and characterized by a biphasic kill curve wherein, upon addition of a lethal dose of antibiotics, bacterial survival follows a two-phase trend over time (Fig. 2b). The initial phase is associated with a rapid decline in bacterial survival followed by slower killing rate in the second phase representing persister fraction. Persisters can be differentiated from resistant population where the regrowth after antibiotic removal yields a similar antibiotic-sensitive population. Also, persisters fail to divide in the presence of antibiotics and randomly switch back to normal-growing phenotype.

Since most of the antibiotics target pathways essential for actively growing cells, persisters were initially thought to be non-growing dormant cells making them refractory to the killing action of these antibiotics [72]. This hypothesis was first supported by the transcriptome analysis of persister fraction showing downregulation of metabolic and biosynthetic pathways with a small set of upregulated genes common to various dormancy models, suggesting dormant state of persisters [61, 121]. However, many other mechanisms have now been shown to be involved in the generation of this transient tolerance [1, 101, 141].

Walsh McDermott first demonstrated mycobacterial persisters way back in 1956 [89, 91]. In the context of tuberculosis, persistence generally refers to the unusual capacity of mycobacteria to persist in the host for decades. It is able to withstand the insults conferred by macrophages and the much developed host immune system. Moreover, extremely long drug therapy reflects another facet of mycobacterial persistence. Several human pathogens cause chronic and persistent infections; however, clinical link between persisters and chronic infections came from a study where antibiotic-treated cystic fibrosis patients showed the presence of high persister mutants [31, 100]. Recently, persisters have also been shown to mediate the emergence of multidrug resistance due to the existence of overlapping survival strategies employed by these cells such as efflux pumps, stringent response, biofilm formation, etc. [1, 18, 71]. Persisters also draw homology with drug-tolerant human cancerous cells responsible for treatment failure and relapse years after chemotherapy [35, 42, 44]. Together, these studies emphasize the importance of novel anti-persister strategies in order to prevent recalcitrant chronic infections, reduce the risk of drug resistance during treatment and to achieve eradication of drug-tolerant cancer cells. In this review, we focus on different bacterial and host mechanisms leading to persister formation and various attempts being made to combat mycobacterial persisters.

Latent Bacteria

Dormants or persisters are the terms used to describe the bacterial physiology, whereas latency refers to a state of equilibrium between the host and the pathogen where pathogen resides in the host without causing any apparent symptoms, a classic case of latent tuberculosis infection (LTBI) [103]. Entry of aerosolized *Mtb* pathogen into the host airways can result in multiple possible outcomes. Close exposure to *Mtb* for several hours can lead to a 30–40% chance of getting infected with the pathogen.

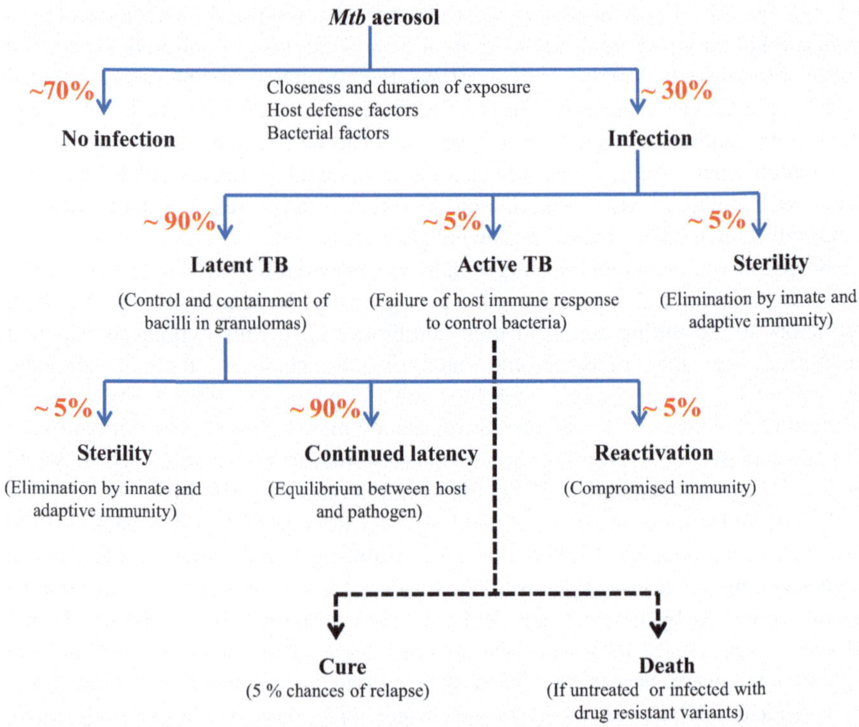

Fig. 3 Possible outcomes after an individual is exposed to *Mycobacterium tuberculosis*

In around 90% of infected individuals, innate and adaptive host immunity is able to control and confine the bacteria in specialized cellular structures called granulomas leading to latent infection. These individuals are at a 5–10% risk of developing active disease later in life. However this risk increases multiple folds during immunocompromised conditions such as HIV infection. The individuals with active TB can be cured effectively; however, there exists a risk of relapse in these individuals [68] (Fig. 3).

It is estimated that approximately one-third of the global population is latently infected with TB. LTBI is detected by various tests such as tuberculin skin test (TST) and interferon gamma release assays (IGRA), which detect the previous exposure of an individual to *Mtb* [5]. The most effective treatment for LTBI till date is administration of isoniazid (INH) for 6–9 months. Since INH blocks cell wall synthesis, which is essential for actively growing bacilli, it has been shown to be ineffective against dormant tubercles [150]. The proven effectiveness of INH in prophylaxis of LTBI indicates that in a patient suffering from LTBI, bacteria can either exist in a metabolically active and growing state where it is controlled by the host to achieve equilibrium or it could be dormant and hiding only to get reactivated upon appropriate signals. There are contradictory reports in the field wherein few studies indicate the presence of replicating bacilli during LTBI [36], while others show that restriction fragment length polymorphism remains unchanged even decades after initial exposure to TB bug [74]. Therefore, according to a recent and most accepted theory, individuals with LTBI contain heterogeneous bacterial population with different physiologies, including actively growing bacteria, which are susceptible to INH and metabolically quiescent drug-tolerant bacilli.

LTBI individuals represent a massive reservoir of hidden TB bacilli waiting to get ruptured to cause new infections. Therefore global eradication of TB must emphasize on efforts to detect and effectively treat LTBI. For this, an improved understanding of the host and bacterial mechanisms underlying LTBI is of utmost importance in order to develop improved diagnostics of LTBI detection and shorter and novel therapeutic strategies [104, 138].

LTBI is an outcome of complex interactions that occur between the host and the pathogen. Beginning from its entry into the human body, *Mtb* faces multiple insults coming from phagocytic cells and immune cells. Host has developed multiple mechanisms to control infection and reactivation, whereas *Mtb* employs various strategies to avoid elimination and cause persistent infection.

Models for Studying Dormancy, Persistence, and Latency

In order to achieve a complete global eradication of tuberculosis, one must focus on mechanisms adopted by the tubercle bacilli to maintain decade-long interactions with the human host. Persistent efforts are required to develop newer in vitro and in vivo models of *Mtb* latency for understanding, in detail, the various aspects of TB pathogenesis leading to discovery of novel therapeutics. This section will focus on in vitro and in vivo models being currently used for simulating latent tuberculosis.

In Vitro Models

The Wayne Model

Very early research in the field of TB pathology suggested a strong link between oxygen levels and bacterial growth rate in multiple animal models [50, 94, 114] with ample evidence of human granulomas being hypoxic in nature [59]. One of the best characterized and extensively studied models of TB dormancy was developed by Wayne and Hayes in 1996 [143], where TB cultures are exposed to gradual oxygen depletion in order to mimic the low oxygen environment faced by the bacilli inside macrophages or granulomas. TB bacilli are grown in airtight tubes with defined air to culture ratio leading to a two-stage progression into non-replicating state (NRP). One percent oxygen level results in NRP-1, characterized by a shift to the glyoxylate cycle, thickening of the outer cell wall, and caseation of growth. NRP-2 is achieved at oxygen levels below 0.06% and is characterized by altered susceptibility to antimicrobial drugs [145, 146]. The model suffers from various limitations such as lack of standardization, generation of fully viable colony forming bacilli, and sensitivity to metronidazole, which fails to show any effect in the mouse model of latency [66]. Nevertheless, being a relatively easy system, the Wayne model and the variants like defined hypoxia model [148] and rapid anaerobic model [70] have led to many important discoveries in the field of bacterial dormancy.

Nutrient Starvation Model

This model was developed based on the hypothesis that, inside the host, *Mtb* is deprived of nutrients and constantly faces starvation stress. Here, bacterial cultures grown in nutrient-rich media are washed, transferred, and cultured in buffered saline for several days which results in a gradual shutdown of respiration and a global shift in metabolism with downregulation of multiple pathways and upregulation of rescue pathways [7]. The establishment of latency is characterized by resistance to most frontline anti-TB drugs while sensitive to pyrazinamide [7, 55]. It is a simple model, which takes into account only one aspect of LTBI. However, it is widely accepted and extensively used to screen drugs against latent tubercle bacilli.

Single Stress Models

Upon entry into alveolar macrophages, *Mtb* is encountered by a plethora of stresses such as oxido-reductive stress, NO stress, pH stress, nutrient stress, etc. Several authors have used in vitro conditions to mimic these stress conditions in order to study the mechanisms employed by *Mtb* to establish persistent infections in the host. Some of these stresses result in growth arrest and hence provide a tool to study dormancy. In acidic pH model, exposure of *Mtb* to an acidic pH <5.5 leads to growth arrest and resistance toward isoniazid [102]. Similarly, inorganic phosphate restriction results in a quiescent inactive state with an increase in drug tolerance [115]. Same as hypoxia, exposure to NO inhibits *Mtb* respiration, replication, transcription, and translation, thus initiating persistence response [140].

Multiple Stress Model

Single stress models are easy to adopt and are sufficient to induce a metabolic switch in mycobacteria leading to dormancy. However, these models face a major limitation of not being able to completely simulate the environment faced by bacilli during human LTBI. To overcome this limitation and to closely represent the conditions inside a granuloma, a multiple stress model was developed where the bacteria are exposed to low oxygen (5%), nutrient starvation, acidic pH (pH 5), and high CO_2 (10%) [22]. When exposed to 5% O_2 + 10% CO_2 + 85% N under low nutrient conditions and acidic pH for 18 days, *Mtb* lose acid-fastness, become quiescent, accumulate wax esters and lipids, and develop phenotypic drug tolerance. It is accompanied by a change in global gene expression with downregulation of genes involved in energy generation, transcription, and translation along with induction of stress-responsive genes [22].

Puissegur et al. in 2004 developed an interesting in vitro granuloma model which uses either mycobacterial antigen-coated beads or live mycobacteria to induce formation of granulomas in human PBMCs. These granulomas are characterized by recruitment of macrophages around live bacilli or beads, their differentiation into multinucleated giant cells and epithelioid cells, and the final recruitment of a ring of activated lymphocytes [110].

Persister Model

The easiest and widely accepted assay to isolate persister fraction of an isogenic bacterial population is to expose the mycobacterial culture to lethal doses of antibiotics for an extended period followed by enumeration of surviving fraction by plating CFU [8]. Multiple factors influence the number of persisters obtained by the end of this assay such as nature of the antibiotics, treatment duration, culture stage, aeration, etc. The simplest model of *Mtb* persistence is the stationary phase culture. These cultures being subjected to stresses like low oxygen and nutrient starvation show upregulation of stringent response and hence become refractory to antibiotic killing [81]. With advancements in the field of live cell imaging, single cell microfluidics is being extensively used in order to demonstrate drug tolerance at a single cell level [6], leading to the discovery of novel mechanisms underlying bacterial persistence [141].

In Vivo Models

Cornell Mouse Model

The most widely accepted in vivo model of LTBI was developed by McCune et al. at Cornell University in 1956. In this model, mice are infected intravenously with a high dose of virulent *Mtb* followed by treatment with anti-TB drugs like isoniazid and pyrazinamide for 3 months in order to achieve a state of pseudo-sterilization where no detectable bacterial burden is observed. Spontaneous reactivation occurs in about one-third of the mice after 3 months of termination of chemotherapy [91], whereas 100% relapse is achieved when immunosuppressant therapy is given [90]. In this model, paucibacillary infection is not achieved by the host immune response as happens in humans, and an external intervention like drug treatment is required to achieve this stage. Therefore, this model is more close to a model of mycobacterial persistence rather than mycobacterial latency. To overcome this limitation, a low-dose model was developed which results in a steady bacterial burden leading to disease reactivation after 18 months [13]. Immunosuppression leads to faster and more frequent relapse cases. This model solely depends on the host immune system to develop LTBI [34]. Till date, these models have been subjected to many variations such as dose of infection, route, duration of treatment, combination of drugs, etc. [27, 60, 119, 149].

Guinea Pig/Rabbit Model

The murine TB granulomas differ significantly from human granulomas in several aspects. They are small, poorly differentiated, less hypoxic, and excessively cellular and lack multinucleated giant cells and caseous necrotic centers [33]. Granulomas in guinea pigs and rabbits resemble human granulomas more closely by virtue of being similar in cellular composition and architecture [33]. Both guinea pigs and rabbits show tissue necrosis, a hallmark trait of human TB pathology [139]. In addition, these animals harbor a heterogeneous population TB lesions facilitating differentiation between primary granulomas and secondary lesions [82]. However, guinea pigs are highly susceptible to *Mtb* infection and succumb to the disease failing to establish LTBI with few exceptions [2, 132]. Rabbits on the other hand show extreme resistance to *Mtb* infection leading to a paucibacillary infection making them a better option to study LTBI. These animals are less prone to spontaneous reactivation making them more close to human TB. Reactivation can be achieved by administration of immunosuppressant treatment [131]. Since many parameters can be manipulated in order to achieve a spectrum of TB pathology in the rabbit, an ideal model of latency can be established in these animals. However, same as guinea pigs, rabbits suffer from lack of immunological tools available to study them.

Nonhuman Primate Model

Being closest to humans, primates most accurately reproduce the TB pathology of humans [16] making them the most preferred model. High-dose *Mtb* infection results in an acute infection, whereas low-dose infection leads to establishment of latency in 40% of monkeys for at least 6 months [16]. Monkeys with active disease show a wide spectrum of TB lesions, whereas latently infected monkeys harbor a limited number of small granulomas containing few culturable bacteria [76]. Infection with simian immunodeficiency virus or tumor necrosis factor alpha (TNF-α) neutralization has been used to reactivate the disease [25, 75]. Despite unlimited advantages, NHP model suffers from being resource intensive, with high cost involved in handling these animals, containment of biohazards, and ethical issues.

Mechanisms of Mycobacterial Persistence

With the aid of abovementioned models, many bacterial and host factors have been implicated in the long-term persistence of mycobacteria in the host. This section aims at highlighting various bacterial genes and pathways identified using multiple approaches responsible for the entry and maintenance of TB bacilli in a dormant or persistent state inside the host. A detailed account of mechanisms involved in mycobacterial latency is discussed below.

Metabolic Adaptations

Glyoxylate Shunt and Gluconeogenesis

Inside the host, *Mtb* faces two major stresses of oxygen depletion and nutrient starvation leading to the induction of NRP stage. Under hypoxic conditions, macrophages accumulate triglycerides. Furthermore, intercellular space of granulomas is rich in cholesterol and fatty acids [64]. *Mtb* utilizes these host lipids to accumulate lipid inclusion bodies containing triacylglycerides. These lipid body-positive bacilli isolated from sputum smear samples of patients are shown to be involved in long term survival in the host and show phenotypic tolerance to antibiotics [39, 49]. *Mtb* utilizes fatty acids, cholesterol, and lipids as a carbon and energy source in the murine model of tuberculosis [11] resulting in the accumulation of acetyl coenzyme A (AcCoA) and propionyl coenzyme A (PropCoA), which are assimilated by glyoxylate shunt and gluconeogenesis. This is accompanied by a massive metabolic shunt wherein TCA cycle is downregulated and glyoxylate cycle is upregulated [123]. Isocitrate lyase (ICL) is the initial enzyme in the glyoxylate cycle, which catalyzes the conversion of isocitrate to glyoxylate and succinate. It has been strongly implicated in mycobacterial persistence. ICL is significantly upregulated in infected macrophages and non-necrotic lesions in human lungs and during the chronic infection in mice [32, 52, 93].

Nitrate Respiration

Inside the hypoxic host granulomas, *Mtb* is thought to use nitrate respiration to provide energy for sustained survival. During oxygen depletion, *Mtb* reduces nitrate to nitrite to maintain the proton motive gradient and generate ATP for maintaining redox homeostasis and energy production [144]. Mycobacterium carries functional nitrate reductase encoded by four genes, viz., *narG*, *narH*, *narJ*, and *narI*, present in an operon *narGHJI.* The nitrate reductase activity is significantly upregulated during oxygen depletion [129]. This increase is due to the induction of the nitrate transporter encoded by *narK2* [128]. *M. bovis* BCG mutant lacking narGHJI fails to persist in the lungs of infected mice [37]. Nitrite reductase (NirBD) reduces nitrite to ammonia by utilizing NADH pool as electron donor. The NirBD mutant shows survival defects in Wayne's and human macrophage based models of dormancy [3].

Regulatory Mechanisms

Stringent Response

Stringent response is a conserved global stress response mediated by the accumulation of an alarmone hyperphosphorylated guanine ribonucleotide (ppGpp), which controls bacterial gene expression, thus enabling the bacterium to adjust under stress conditions within the host [109]. Under various growth-limiting conditions, poly(P), a linear polymer of inorganic phosphate accumulates via an increase in the activity of polyphosphate kinase (PPK) and a decrease in exopolyphosphatase (PPX) activity [67]. Poly(P) acts as a phosphate donor and activates two-component sensor histidine kinase MprB which in turn activates SigE along with many other stress-responsive genes. SigE facilitates increased transcription of relA leading to synthesis of ppGpp [133]. Stringent response as a whole as well as individual genes involved in this response has been linked to mycobacterial dormancy, virulence gene expression, persistence, latency, and drug tolerance. For instance, Δrel_{Mtb} mutant fails to maintain a sustained persistent infection in mice [21, 60] and guinea pig lungs [65].

Dos Regulon

The transcriptional response where around 48 genes got upregulated during defined hypoxia was termed as Dos (dormancy survival) regulon (DosR) [122]. The Dos two-component system consists of two histidine kinases DosS and DosT and the response regulator DosR [107]. Upon sensing changes in the environment such as low O_2, NO, CO, H_2O_2, ascorbic acid, residence in macrophages and mice, etc. [28], DosS and DosT autophosphorylate and transduce the signal to DosR which subsequently binds to the DNA and influence the transcription of many genes. DosR is thought to be a key regulator of mycobacterial dormancy. However, varied and contrasting results have been obtained from animal studies involving DosS-DosT-DosR regulon. DosR deletion leads to hypervirulence in SCID mice or immunocompetent DBA mice [83, 106], whereas Converse et al. showed reduced survival of DosR mutant in rabbits, mice, and guinea pigs [20]. Furthermore, being constitutively overexpressed in the Beijing strains of *Mtb*, DosR is believed to be responsible for the global epidemic spread and increased drug-resistant variants of these strains [30]. However, many genes regulated by DosR have been shown to be indispensable during establishment of dormancy in vitro and in vivo.

Toxin-Antitoxin Modules

Toxin-antitoxin (TA) system is very strongly implicated in bacterial drug tolerance and persistence. It comprises of a set of two closely linked genes that encode for a toxin whose activity is controlled by a corresponding antitoxin. Upon encountering stress, antitoxin is degraded releasing the toxin to exert its effect of promoting cell dormancy. TA modules control the switch between active and dormant state either by sensing external stimuli or by random stochastic factors [80]. Among the six TA families characterized till date, *Mtb* majorly possess type II TA modules where the toxin inhibits protein translation. For an intracellular pathogen, *Mtb* possess the largest number of TA modules (around 80) which are thought to serve as an *armor* for various stresses encountered in the host, making the bacteria refractory to these forces and enabling it to establish lifelong infections in the host [111]. Extensively studied in *E. coli*, many TA modules have been shown to produce drug-tolerant persisters [62, 137]. However, functional redundancy makes it difficult to elucidate the role of individual TA modules in establishing chronic infections in animal models. Two major TA families in *Mtb*, VapBC, and MazEF have been found to be upregulated during nutrient stress, hypoxia, and during infection ex vivo and in vivo [61, 135]. Some knockout studies have tried to establish the requirement of these systems during chronic infections and antibiotic tolerance [135]. Around ten TA modules were overexpressed in *Mtb* persisters isolated after treatment with D cycloserine [1].

Transcription Factors

Most bacteria carry an array of sigma (σ) factors which upon specific environmental cues associate with core RNA polymerase leading to promoter recognition and modulation of gene expression. In *Mtb*, varied in vitro stresses are known to induce the expression of sigma factors, which ultimately aid the pathogen to survive in multiple phases of infection [84]. For instance, expression of SigE is increased inside macrophages, while the mutant displays growth defect in the host ex vivo [85]. SigE overexpression leads to increased expression of many genes involved in bacterial dormancy such as mprA, mprB, and sigB. Furthermore, SigE mutant shows downregulation of genes involved in protein translation, electron transport, acyl-CoA biosynthesis, etc. [85]. SigF and SigH are also induced upon entry into macrophages and various in vitro stresses that replicate the environment within the host [23, 112]. While both the mutants fail to show any survival defect in mice model of TB infection, SigH-deleted *Mtb* strain exhibit reduced growth in the lungs of rhesus macaques [84, 95].

Mtb possess Fe-S cluster containing redox-sensitive transcription factors of the WhiB-like (Wbl) family known to play a role in bacterial cell cycle, stress response, redox balancing, antibiotic resistance, and pathogenesis. Among the seven WhiB proteins in *Mtb*, WhiB3 is established as an intracellular redox sensor finding a role in maintaining redox balance, lipid anabolism, and virulence [105, 125]. It interacts with σ^A to regulate the expression of pathogenicity genes [130]. It controls the biosynthesis of virulence polyketides and storage lipids and modulates host innate immunity [124]. However, there are contrasting reports on the role of WhiB3 in in vivo growth of *Mtb* [96, 130].

Other Mechanisms

Efflux Pumps

Mtb genome codes for many putative efflux pump implicated in extruding antituberculous drugs during TB treatment, thereby contributing to multiple drug resistance and phenotypic drug tolerance. Several of these transporters are shown to be highly upregulated in MDR strains of *Mtb* as compared to sensitive clinical isolates [73]. Out of around 20 efflux pumps that are transcriptionally induced in macrophages [120], 7 were shown to be essential for intramacrophage survival [113]. Macrophage residence leads to induction of bacterial efflux pump Rv1258c, which promotes tolerance toward antimicrobials [1]. Use of verapamil as an adjunct to the current anti-TB therapy led to a significant reduction in the bacterial load in murine model of infection [48].

Cell Wall Components

Mycobacteria exhibit high degree of intrinsic resistance to many antibiotics owing to its highly complex and well-structured cell wall, which is composed of a plasma membrane, the mycolyl-arabinogalactan-peptidoglycan complex, and outer capsule-like layer containing glycol (lipids) and proteins [57]. Due to their importance in bacterial survival and virulence, many cell wall components are the targets of current anti-TB drugs such as INH and ethionamide (ETH). Many reports suggest the role of several cell wall components in establishing chronic infections in the host such as mycolic acids [43], lipomannan (LM) and lipoarabinomannan (LAM) [38, 108], trehalose dimycolates (TDM) [56], PDIMs [26], etc. A complete account of cell wall determinants implicated in *Mtb* pathogenesis has been reviewed in detail elsewhere [57].

Strategies to Target Latent TB

Targeting Resuscitation-Promoting Factors (Rpf)

Rpf are needed to revive and activate dormant bacilli into actively growing form. *Mtb* possess five Rpf genes, rpfA–E, which are known to induce replication and resuscitation of dormant *Mtb* possibly via cell wall remodeling or muropeptide-mediated signaling [58]. *Mtb* strains deleted for more than one Rpf gene show reduced virulence in both acute and chronic phases of infection in mice [10, 118]. Being highly immunogenic and protective, Rpfs have been proposed as novel subunit vaccines against TB [117]. Many studies have focused on the discovery of compounds with potent anti-Rpf activities, rendering the bacilli incapable of reactivation. 2-Nitrophenylthiocyanate (NPT) compounds represent a new class of drugs which demonstrate inhibitory effects on the activity of Rpfs, making them potential candidates for inhibiting reactivation of latent organisms [24]. On the other hand, Rpfs or Rpf-activating factors can be used for reactivating and sensitizing dormant bacilli which may lead to shortening of anti-TB therapy [98].

Anti-persister Strategies

Even metabolically dormant persister bacteria require basal activity of multiple pathways for successful persistence such as energy metabolism, maintaining membrane potential, membrane integrity, etc. AM-0016, a novel xanthone-based antibacterial, rapidly kills mycobacterial persisters by disrupting membrane potential and causing structural damage to the cell envelope [99]. Pyrazinamide, a prototype persister-targeting drug against *Mtb*, works by disrupting membrane energy [151]. New TB drugs, bedaquiline and delamanid show promising activity against both latent and active TB. Both target enzymes essential for survival of even dormant quiescent bacilli where bedaquiline targets a key respiratory chain enzyme F1/F0-ATPase [17] and delamanid inhibits the synthesis of mycobacterial cell wall components, methoxy mycolic acid and ketomycolic acid [17]. Other drugs under clinical trials include lassomycin [41] and ADEP4 [19], both of which target ClpP and teixobactin [77] and CPZEN-45 [127] targeting bacterial cell wall. Maintenance of optimal redox balance is critical for persister cell viability. In this regard, clofazimine was successful in eradicating persister fraction by stimulating ROS production in *Mycobacterium smegmatis* [17, 147]. With the increasing literature on the mechanisms controlling persistence, a range of potential targets have been identified as novel anti-persister therapeutics such as TA modules, PPK [126], relA, etc. Metabolic enzymes such as ICL and PcaA are actively pursued as novel drug targets against dormant bacilli [78].

Alternative Treatment Strategies

Intermittent drug treatment initially proposed by J. Bigger [9] is based on the hypothesis that upon withdrawal of antibiotics, dormant bacilli may resuscitate and become susceptible. However this approach suffers from many disadvantages such as risk of drug resistance, adverse reactions in the host, and increased chances of relapse [152]. Another approach uses high doses of antibiotics in order to fasten the process of bacterial killing which may lead to the generation of fewer persister cells [53]. Adverse toxic effects of drugs are one of the major concerns of this treatment method. Recently, host-directed immunotherapy as an adjunct treatment has gained increased attention. Many reports suggest increased efficacy of current anti-TB drugs when given in combination with natural or chemical immunomodulators [29, 88].

Concluding Remarks

Despite all the efforts directed toward the eradication of TB, it remains one of the major causes of mortality worldwide. The long duration and toxic effects of chemotherapy involved in the treatment of TB frequently leads to non-compliance contributing toward the resistant and recurrent forms of infection. The scenario is further worsened by the existence of latent TB infections (LTBI), which affects approximately two billion people worldwide. This results from the tremendous ability of tubercle bacilli to persist for decades inside the host.

In this chapter, we try to make clear distinction between the terms dormancy, persistence, and latency which are often used interchangeably leading to confusion. Further, we describe various models used to study these phenomena. We have briefly described the various mechanisms by which *Mtb* is able to survive in the hostile environment of the host. These studies are critical for developing newer and effective strategies to target latent TB. However it is still not clear whether targeting the factors responsible for *Mtb* latency with newer drugs will lead to sterilization in vivo, nor it is essential that results obtained in mouse model will replicate in humans. Therefore continuous efforts need to be put in developing more effective models of latency to study the efficacy of novel therapeutics.

References

1. Adams, K. N., Takaki, K., Connolly, L. E., et al. (2011). Drug tolerance in replicating mycobacteria mediated by a macrophage-induced efflux mechanism. *Cell, 145*(1), 39–53.
2. Ahmad, Z., Klinkenberg, L. G., Pinn, M. L., et al. (2009). Biphasic kill curve of isoniazid reveals the presence of drug-tolerant, not drug resistant, Mycobacterium tuberculosis in the guinea pig. *The Journal of Infectious Diseases, 200*, 1136–1143.
3. Akhtar, S., Khan, A., Sohaskey, C. D., et al. (2013). Nitrite reductase NirBD is induced and plays an important role during in vitro dormancy of Mycobacterium tuberculosis. *Journal of Bacteriology, 195*, 4592–4599.
4. Albrethsen, J., Agner, J., Piersma, S. R., et al. (2013). Proteomic profiling of Mycobacterium tuberculosis identifies nutrient-starvation-responsive toxin-antitoxin systems. *Molecular & Cellular Proteomics, 12*, 1180–1191.
5. Al-Orainey, I. O. (2009). Diagnosis of latent tuberculosis: Can we do better? *Ann Thorac Med, 4*(1), 5–9.
6. Balaban, N. Q., Merrin, J., & Chait, R. (2004). Bacterial persistence as a phenotypic switch. *Science, 305*, 1622–1625.
7. Betts, J. C., Lukey, P. T., Robb, L. C., et al. (2002). Evaluation of a nutrient starvation model of Mycobacterium tuberculosis persistence by gene and protein expression profiling. *Molecular Microbiology, 43*, 717–731.
8. Bhaskar, A., De Piano, C., Gelman, E., McKinney, J. D., & Dhar, N. (2018). Elucidating the role of (p)ppGpp in mycobacterial persistence against antibiotics. *IUBMB Life, 70*(9), 836–844.
9. Bigger, J. W. (1944). Treatment of staphylococcal infections with penicillin. *Lancet, 244*, 497–500.
10. Biketov, S., Potapov, V., Ganina, E., et al. (2007). The role of resuscitation promoting factors in pathogenesis and reactivation of Mycobacterium tuberculosis during intra-peritoneal infection in mice. *BMC Infectious Diseases, 7*, 146.
11. Bloch, H., & Segal, W. (1956). Biochemical differentiation of Mycobacterium tuberculosis grown in vivo and in vitro. *Journal of Bacteriology, 72*, 132–141.
12. Bojsen, R., Regenberg, B., & Gresham, D. (2016). A common mechanism involving the TORC1 pathway can lead to amphotericin B-persistence in biofilm and planktonic Saccharomyces cerevisiae populations. *Scientific Reports, 6*, 21874.
13. Brown, D. H., Miles, B. A., & Zwilling, B. S. (1995). Growth of Mycobacterium tuberculosis in BCG-resistant and - susceptible mice: Establishment of latency and reactivation. *Infection and Immunity, 63*, 2243–2247.
14. Burdon, K. L. (1946a). Disparity in appearance of true Hansen's bacilli and cultured 'leprosy bacilli' when stained for fat. *Journal of Bacteriology, 52*, 679–680.
15. Burdon, K. L. (1946b). Fatty material in bacteria and fungi revealed by staining dried, fixed slide preparations. *Journal of Bacteriology, 52*, 665–678.
16. Capuano, S. V., III, Croix, D. A., Pawar, S., et al. (2003). Experimental Mycobacterium tuberculosis infection of cynomolgus macaques closely resembles the various manifestations of human M. tuberculosis infection. *Infection and Immunity, 71*, 5831–5844.
17. Cholo, M. C., Mothiba, M. T., Fourie, B., et al. (2016). Mechanisms of action and therapeutic efficacies of the lipophilic antimycobacterial agents clofazimine and bedaquiline. *Journal of Antimicrobial Chemotherapy, 72*(2), 338–353.
18. Cohen, N. R., Lobritz, M. A., & Collins, J. J. (2013). Microbial persistence and the road to drug resistance. *Cell Host & Microbe, 13*(6), 632–642.
19. Conlon, B. P., Nakayasu, E. S., Fleck, L. E., et al. (2013). Activated ClpP kills persisters and eradicates a chronic biofilm infection. *Nature, 503*, 365–370.
20. Converse, P. J., Karakousis, P. C., Klinkenberg, L. G., et al. (2009). Role of the dosR-dosS two-component regulatory system in Mycobacterium tuberculosis virulence in three animal models. *Infection and Immunity, 77*, 1230–1237.

21. Dahl, J. L., Kraus, C. N., Boshoff, H. I., et al. (2003). The role of RelMtb-mediated adaptation to stationary phase in long-term persistence of Mycobacterium tuberculosis in mice. *Proceedings of the National Academy of Sciences of the United States of America, 100*, 10026–10031.
22. Deb, C., Lee, C. M., Dubey, V., et al. (2009). A novel in vitro multiple-stress dormancy model for Mycobacterium tuberculosis generates a lipid-loaded, drug-tolerant, dormant pathogen. *PLoS One, 4*, e6077.
23. DeMaio, J., Zhang, Y., Ko, C., et al. (1996). A stationary-phase stress-response sigma factor from Mycobacterium tuberculosis. *Proceedings of the National Academy of Sciences of the United States of America, 93*, 2790–2794.
24. Demina, G. R., Makarov, V. A., Nikitushkin, V. D., et al. (2009). Finding of the low molecular weight inhibitors of resuscitation promoting factor enzymatic and resuscitation activity. *PLoS One, 4*, e8174.
25. Diedrich, C. R., Mattila, J. T., Klein, E., et al. (2010). Reactivation of latent tuberculosis in cynomolgus macaques infected with SIV is associated with early peripheral T cell depletion and not virus load. *PLoS One, 5*, e9611.
26. Domenech, P., & Reed, M. B. (2009). Rapid and spontaneous loss of phthiocerol dimycocerosate (PDIM) from Mycobacterium tuberculosis grown in vitro: Implications for virulence studies. *Microbiology, 155*, 3532–3543.
27. Dutta, N. K., Illei, P. B., Jain, S. K., et al. (2014). Characterization of a novel necrotic granuloma model of latent tuberculosis infection and reactivation in mice. *The American Journal of Pathology, 184*, 2045–2055.
28. Dutta, N. K., & Karakousis, P. C. (2014). Latent tuberculosis infection: Myths, models, and molecular mechanisms. *Microbiology and Molecular Biology Reviews, 78*(3), 343–371.
29. Dwivedi, V. P., Bhattacharya, D., Singh, M., et al. (2018). Allicin enhances antimicrobial activity of macrophages during Mycobacterium tuberculosis infection. *Journal of Ethnopharmacology, 2018*. Dec 8.pii: S0378–8741(18)31763-X. https://doi.org/10.1016/j.jep.2018.12.008.
30. Fallow, A., Domenech, P., & Reed, M. B. (2010). Strains of the east Asian (W/Beijing) lineage of Mycobacterium tuberculosis are DosS/DosT-DosR two-component regulatory system natural mutants. *Journal of Bacteriology, 192*(8), 2228–2238.
31. Fauvart, M., et al. (2011). Role of persister cells in chronic infections: Clinical relevance and perspectives on anti-persister therapies. *Journal of Medical Microbiology, 60*, 699–709.
32. Fenhalls, G. L., Stevens, L., Moses, J., et al. (2002). In situ detection of Mycobacterium tuberculosis transcripts in human lung granulomas reveals differential gene expression in necrotic lesions. *Infection and Immunity, 70*, 6330–6338.
33. Flynn, J., & Chan, J. (2004). Animal models of tuberculosis. In W. Rom & S. Garay (Eds.), *Tuberculosis* (2nd ed., pp. 237–250). Philadelphia: Lippincott Williams & Wilkins.
34. Flynn, J. L., Scanga, C. A., Tanaka, K. E., et al. (1998). Effects of aminoguanidine on latent murine tuberculosis. *Journal of Immunology, 160*, 1796–1803.
35. Ford, A. M., Mansur, M. B., & Furness, C. L. (2015). Protracted dormancy of pre-leukemic stem cells. *Leukemia, 29*, 1–24.
36. Ford, C. B., Lin, P. L., Chase, M. R., et al. (2011). Use of whole genome sequencing to estimate the mutation rate of Mycobacterium tuberculosis during latent infection. *Nature Genetics, 43*, 482–486.
37. Fritz, C., Maass, S., Kreft, A., et al. (2002). Dependence of Mycobacterium bovis BCG on anaerobic nitrate reductase for persistence is tissue specific. *Infection and Immunity, 70*, 286–291.
38. Fukuda, T., Matsumura, T., Ato, M., et al. (2013). Critical roles for lipomannan and lipoarabinomannan in cell wall integrity of mycobacteria and pathogenesis of tuberculosis. *MBio, 4*(1), e00472–e00412.
39. Garton, N., Waddell, S., Sherratt, A., et al. (2008). Cytological and transcript analyses reveal fat and lazy persister-like bacilli in tuberculous sputum. *PLoS Medicine, 5*, 0634–0645.

40. Garton, N. J., Christensen, H., Minnikin, D. E., et al. (2002). Intracellular lipophilic inclusions of mycobacteria in vitro and in sputum. *Microbiology, 148*, 2951–2958.
41. Gavrish, E., Sit, C. S., Cao, S., et al. (2014). Lassomycin, a ribosomally synthesized cyclic peptide, kills Mycobacterium tuberculosis by targeting the ATP-dependent protease ClpC1P1P2. *Chemistry & Biology, 21*, 509–518.
42. Giancotti, F. G. (2013). Mechanisms governing metastatic dormancy and reactivation. *Cell, 155*(4), 750–764.
43. Glickman, M. S., Cox, J. S., & Jacobs, W. R., Jr. (2000). A novel mycolic acid cyclopropane synthetase is required for cording, persistence, and virulence of Mycobacterium tuberculosis. *Molecular Cell, 5*(4), 717–727.
44. Glickman, M. S., & Sawyers, C. L. (2012). Converting cancer therapies into cures: Lessons from infectious diseases. *Cell, 148*(6), 1089–1098.
45. Global tuberculosis report (2018). Geneva: World Health Organization Licence: CC BY-NC-SA 3.0 IGO.
46. Gomez, J. E., & McKinney, J. D. (2003). M. tuberculosis persistence, latency, and drug tolerance. *Tuberculosis, 84*, 29–44.
47. Groote, D., Verstraeten, V. N., Fauvart, N., et al. (2009). Novel persistence genes in Pseudomonas aeruginosa identified by high-throughput screening. *FEMS Microbiology Letters, 297*, 73–79.
48. Gupta, S., Tyagi, S., Almeida, D. V., et al. (2013). Acceleration of tuberculosis treatment by adjunctive therapy with verapamil as an efflux inhibitor. *American Journal of Respiratory and Critical Care Medicine, 188*, 600–607.
49. Hammond, R., Baron, V., Oravcova, K., et al. (2015). Phenotypic resistance in mycobacteria: Is it because I am old or fat that I resist you? *The Journal of Antimicrobial Chemotherapy, 70*, 2823–2827.
50. Hanlon, C. R., Scott, J. H. W., & Olson, B. J. (1950). Experimental tuberculosis. I. Effects of anastomosis between systemic and pulmonary arteries on tuberculosis in monkeys. *Surgery, 28*, 209–224.
51. Hobby, G. L., Meyer, K., & Chaffee, E. (1942). Observations on the mechanism of action of penicillin. *Proceedings of the Society for Experimental Biology and Medicine, 50*, 281–285.
52. HönerZuBentrup, K., Miczak, A., Swenson, D. L., et al. (1999). Characterization of activity and expression of isocitratelyase in Mycobacterium avium and Mycobacterium tuberculosis. *Journal of Bacteriology, 181*, 7161–7167.
53. Hu, Y., Liu, A., Ortega-Muro, F., et al. (2015). High-dose rifampicin kills persisters, shortens treatment duration, and reduces relapse rate in vitro and in vivo. *Frontiers in Microbiology, 6*, 641.
54. Hu, Y., Mangan, J. A., Dhillon, J., et al. (2000). Detection of mRNA transcripts and active transcription in persistent Mycobacterium tuberculosis induced by exposure to rifampin or pyrazinamide. *Journal of Bacteriology, 182*, 6358–6365.
55. Huang, Q., Chen, Z. F., Li, Y. Y., et al. (2007). Nutrient-starved incubation conditions enhance pyrazinamide activity against Mycobacterium tuberculosis. *Chemotherapy, 53*, 338–343.
56. Hunter, R. L., Olsen, M., Jagannath, C., et al. (2006). Trehalose 6,6′-dimycolate and lipid in the pathogenesis of caseating granulomas of tuberculosis in mice. *The American Journal of Pathology, 168*, 1249–1261.
57. Jackson, M. (2014). The mycobacterial cell envelope-lipids. *Cold Spring Harbor Perspectives in Medicine, 4*(10), a021105.
58. Kana, B. D., & Mizrahi, V. (2010). Resuscitation-promoting factors as lytic enzymes for bacterial growth and signaling. *FEMS Immunology and Medical Microbiology, 58*, 39–50.
59. Kaplan, G., Post, F. A., Moreira, A. L., et al. (2003). Mycobacterium tuberculosis growth at the cavity surface: A microenvironment with failed immunity. *Infection and Immunity, 71*, 7099–7108.

60. Karakousis, P. C., Yoshimatsu, T., Lamichhane, G., et al. (2004). Dormancy phenotype displayed by extracellular Mycobacterium tuberculosis within artificial granulomas in mice. *The Journal of Experimental Medicine, 200*, 647–657.
61. Keren, I., Minami, S., Rubin, E., et al. (2011). Characterization and transcriptome analysis of Mycobacterium tuberculosis persisters. *MBio, 2*, e00100–e111es.
62. Keren, I., Shah, D., Spoering, A., et al. (2004). Specialized persister cells and the mechanism of multidrug tolerance in Escherichia coli. *Journal of Bacteriology, 186*(24), 8172–8180.
63. Kim, J. J., & Tannock, I. F. (2005). Repopulation of cancer cells during therapy: An important cause of treatment failure. *Nature Reviews. Cancer, 5*, 516–525.
64. Kim, M., Wainwright, H., Locketz, M., et al. (2010). Caseation of human tuberculosis granulomas correlates with elevated host lipid metabolism. *EMBO Molecular Medicine, 2*, 258–274.
65. Klinkenberg, L., Lee, J., Bishai, W., & Karakousis, P. C. (2010). The stringent response is required for full virulence of Mycobacterium tuberculosis in Guinea pigs. *The Journal of Infectious Diseases, 202*, 1397–1404.
66. Klinkenberg, L. G., Sutherland, L. A., Bishai, W. R., et al. (2008). Metronidazole lacks activity against Mycobacterium tuberculosis in an in vivo hypoxic granuloma model of latency. *The Journal of Infectious Diseases, 198*, 275–283.
67. Kornberg, A., Rao, N. N., & Ault-Riche, D. (1999). Inorganic polyphosphate: A molecule of many functions. *Annual Review of Biochemistry, 68*, 89–125.
68. Koul, A., Arnoult, E., Lounis, N., et al. (2011). The challenge of new drug discovery for tuberculosis. *Nature, 469*, 483–490.
69. LaFleur, M. D., Kumamoto, C. A., & Lewis, K. (2006). Candida albicans biofilms produce antifungal-tolerant persister cells. *Antimicrobial Agents and Chemotherapy, 50*, 3839–3846.
70. Leistikow, R. L., Morton, R. A., Bartek, I. L., et al. (2010). The Mycobacterium tuberculosis DosR regulon assists in metabolic homeostasis and enables rapid recovery from nonrespiring dormancy. *Journal of Bacteriology, 192*, 1662–1670.
71. Levin-Reisman, I., Ronin, I., Gefen, O., et al. (2017). Antibiotic tolerance facilitates the evolution of resistance. *Science, 355*(6327), 826–830.
72. Lewis, K. (2010). Persister cells. *Annual Review of Microbiology, 64*, 357–372.
73. Li, G., Zhang, J., Guo, Q., et al. (2015). Efflux pump gene expression in multidrug-resistant Mycobacterium tuberculosis clinical isolates. *PLoS One, 10*, e0119013.
74. Lillebaek, T., Dirksen, A., Vynnycky, E., et al. (2003). Stability of DNA patterns and evidence of Mycobacterium tuberculosis reactivation occurring decades after the initial infection. *The Journal of Infectious Diseases, 188*, 1032–1039.
75. Lin, P. L., Myers, A., Smith, L., et al. (2010). Tumor necrosis factor neutralization results in disseminated disease in acute and latent Mycobacterium tuberculosis infection with normal granuloma structure in a cynomolgus macaque model. *Arthritis and Rheumatism, 62*, 340–350.
76. Lin, P. L., Pawar, S., Myers, A., et al. (2006). Early events in Mycobacterium tuberculosis infection in cynomolgus macaques. *Infection and Immunity, 74*, 3790–3803.
77. Ling, L. L., Schneider, T., Peoples, A. J., et al. (2015). A new antibiotic kills pathogens without detectable resistance. *Nature, 517*, 455–459.
78. Liu, Y., Zhou, S., Deng, Q., et al. (2016). Identification of a novel inhibitor of isocitratelyase as a potent antitubercular agent against both active and non-replicating Mycobacterium tuberculosis. *Tuberculosis, 97*, 38–46.
79. Longo, D. L., Horsburgh, C. R., Barry, C. E., et al. (2015). Treatment of tuberculosis. *The New England Journal of Medicine, 373*, 2149–2160.
80. Loris, R., & Garcia-Pino, A. (2014). Disorder- and dynamics-based regulatory mechanisms in toxin-antitoxin modules. *Chemical Reviews, 114*, 6933–6947.
81. Luidalepp, H., Joers, A., Kaldalu, N., et al. (2011). Age of inoculum strongly influences persister frequency and can mask effects of mutations implicated in altered persistence. *Journal of Bacteriology, 193*, 3598–3605.

82. Ly, L. H., Russell, M. I., & McMurray, D. N. (2008). Cytokine profiles in primary and secondary pulmonary granulomas of guinea pigs with tuberculosis. *American Journal of Respiratory Cell and Molecular Biology, 38*, 455–462.
83. Malhotra, V., Sharma, D., Ramanathan, V. D., et al. (2004). Disruption of response regulator gene, devR, leads to attenuation in virulence of Mycobacterium tuberculosis. *FEMS Microbiology Letters, 231*, 237–245.
84. Manganelli, R., Provvedi, R., Rodrigue, S., et al. (2004). Sigma factors and global gene regulation in Mycobacterium tuberculosis. *Journal of Bacteriology, 186*, 895–902.
85. Manganelli, R., Voskuil, M. I., Schoolnik, G. K., et al. (2001). The Mycobacterium tuberculosis ECF sigma factor sigmaE: Role in global gene expression and survival in macrophages. *Molecular Microbiology, 41*, 423–437.
86. Manina, G., Dhar, N., & McKinney, J. D. (2015). Stress and host immunity amplify Mycobacterium tuberculosis phenotypic heterogeneity and induce nongrowing metabolically active forms. *Cell Host & Microbe, 17*, 32–46.
87. Mathema, B., Kurepina, N., Fallows, D., et al. (2008). Lessons from molecular epidemiology and comparative genomics. *Seminars in Respiratory and Critical Care Medicine, 29*, 467–480.
88. Mayanja-Kizza, H., Jones-Lopez, E., Okwera, A., et al. (2005). Immunoadjuvant prednisolone therapy for HIV-associated tuberculosis: A phase 2 clinical trial in Uganda. *The Journal of Infectious Diseases, 191*, 856–865.
89. McCune, R., Lee, S. H., Deuschle, K., et al. (1957). Ineffectiveness of isoniazid in modifying the phenomenon of microbial persistence. *American Review of Tuberculosis, 76*, 1106–1109.
90. McCune, R. M., Feldmann, F. M., Lambert, H. P., et al. (1966). Microbial persistence I. the capacity of tubercle bacilli to survive sterilization in mouse tissues. *The Journal of Experimental Medicine, 123*, 445–468.
91. McCune, R. M., Jr., Tompsett, R., & McDermott, W. (1956). The fate of Mycobacterium tuberculosis in mouse tissues as determined by the microbial enumeration technique II. The conversion of tuberculous infection to the latent state by the administration of pyrazinamide and a companion drug. *The Journal of Experimental Medicine, 104*, 763–802.
92. McDermott, W., McCune, R. M., & Tompsett, R. (1974). Dynamics of antituberculous chemotherapy. *American Review of Tuberculosis, 74*, 100–108.
93. McKinney, J. D., HonerzuBentrup, K., Munoz-Elias, E. J., et al. (2000). Persistence of Mycobacterium tuberculosis in macrophages and mice requires the glyoxylate shunt enzyme isocitratelyase. *Nature, 406*, 735–738.
94. Medlar, E. M. (1948). The pathogenesis of minimal pulmonary tuberculosis: A study of 1225 necropsies in cases of sudden and unexpected death. *American Review of Tuberculosis, 58*, 583–611.
95. Mehra, S., Golden, N. A., Stuckey, K., et al. (2012). The Mycobacterium tuberculosis stress response factor SigH is required for bacterial burden as well as immunopathology in primate lungs. *The Journal of Infectious Diseases, 205*, 1203–1213.
96. Mehta, M., Rajmani, R. S., & Singh, A. (2016). Mycobacterium tuberculosis WhiB3 responds to vacuolar pH-induced changes in mycothiol redox potential to modulate phagosomal maturation and virulence. *The Journal of Biological Chemistry, 291*, 2888–2903.
97. Menchón, S. A., & Condat, C. A. (2011). Quiescent cells: A natural way to resist chemotherapy. *Physica A, 390*, 3354–3361.
98. Mukamolova, G. V., Turapov, O., Malkin, J., et al. (2010). Resuscitation-promoting factors reveal an occult population of tubercle bacilli in sputum. *American Journal of Respiratory and Critical Care Medicine, 181*, 174–180.
99. Mukherjee, D., Zou, H., Liu, S., et al. (2016). Membrane-targeting AM-0016 kills mycobacterial persisters and shows low propensity for resistance development. *Future Microbiology, 11*(5), 643–650.

100. Mulcahy, L. R., et al. (2010). Emergence of Pseudomonas aeruginosa strains producing high levels of persister cells in patients with cystic fibrosis. *Journal of Bacteriology, 192*, 6191–6199.
101. Nguyen, D., Joshi-Datar, A., & Lepine, F. (2011). Active starvation responses mediate antibiotic tolerance in biofilms and nutrient-limited bacteria. *Science, 334*, 982–986.
102. Niki, M., Tateishi, Y., Ozeki, Y., et al. (2012). A novel mechanism of growth phase-dependent tolerance to isoniazid in mycobacteria. *The Journal of Biological Chemistry, 287*, 22743–22752.
103. Orme, I. M. (2001). The latent tubercle bacillus (I'll let you know if I ever meet one). *The International Journal of Tuberculosis and Lung Disease, 5*, 589–593.
104. Pai, M. (2010). Spectrum of latent tuberculosis-existing tests cannot resolve the underlying phenotypes. *Nature Reviews. Microbiology, 8*, 242.
105. Pang, X., Vu, P., Byrd, T. F., et al. (2007). Evidence for complex interactions of stress associated regulons in an mprAB deletion mutant of Mycobacterium tuberculosis. *Microbiology, 153*, 1229–1242.
106. Parish, T., Smith, D. A., Kendall, S., et al. (2003). Deletion of two-component regulatory systems increases the virulence of Mycobacterium tuberculosis. *Infection and Immunity, 71*, 1134–1140.
107. Park, H. D., Guinn, K. M., Harrell, M. I., et al. (2003). Rv3133c/dosR is a transcription factor that mediates the hypoxic response of Mycobacterium tuberculosis. *Molecular Microbiology, 48*, 833–843.
108. Pitarque, S., Larrouy-Maumus, G., Payré, B., et al. (2008). The immunomodulatory lipoglycans, lipoarabinomannan and lipomannan, are exposed at the mycobacterial cell surface. *Tuberculosis, 88*(6), 560–565.
109. Primm, T. P., Andersen, S. J., Mizrahi, V., et al. (2000). The stringent response of Mycobacterium tuberculosis is required for long-term survival. *Journal of Bacteriology, 182*, 4889–4898.
110. Puissegur, M. P., Botanch, C., Duteyrat, J. L., et al. (2004). An in vitro dual model of mycobacterial granulomas to investigate the molecular interactions between mycobacteria and human host cells. *Cellular Microbiology, 6*, 423–433.
111. Ramage, H. R., Connolly, L. E., & Cox, J. S. (2009). Comprehensive functional analysis of Mycobacterium tuberculosis toxin-antitoxin systems: Implications for pathogenesis, stress responses, and evolution. *PLoS Genetics, 5*, e1000767.
112. Raman, S., Song, T., Puyang, X., et al. (2001). The alternative sigma factor SigH regulates major components of oxidative and heat stress responses in Mycobacterium tuberculosis. *Journal of Bacteriology, 183*, 6119–6125.
113. Rengarajan, J., Bloom, B. R., & Rubin, E. J. (2005). Genome-wide requirements for Mycobacterium tuberculosis adaptation and survival in macrophages. *Proceedings of the National Academy of Sciences of the United States of America, 102*, 8327–8332.
114. Rich, A. R., & Follis, R. H., Jr. (1942). The effect of low oxygen tension upon the development of experimental tuberculosis. *Bulletin of the Johns Hopkins Hospital, 71*, 345–363.
115. Rifat, D., Bishai, W. R., & Karakousis, P. C. (2009). Phosphate depletion: A novel trigger for Mycobacterium tuberculosis persistence. *The Journal of Infectious Diseases, 200*, 1126–1135.
116. Rittershaus, E. S. C., Baek, S. H., & Sassetti, C. M. (2013). The normalcy of dormancy: Common themes in microbial quiescence. *Cell Host & Microbe, 13*, 643–651.
117. Romano, M., Aryan, E., Korf, H., et al. (2012). Potential of Mycobacterium tuberculosis resuscitation-promoting factors as antigens in novel tuberculosis sub-unit vaccines. *Microbes and Infection, 14*(1), 86–95.
118. Russell-Goldman, E., Xu, J., Wang, X., et al. (2008). A Mycobacterium tuberculosis Rpf double-knockout strain exhibits profound defects in reactivation from chronic tuberculosis and innate immunity phenotypes. *Infection and Immunity, 76*, 4269–4281.
119. Scanga, C. A., Mohan, V., Joseph, H., et al. (1999). Reactivation of latent tuberculosis: Variations on the Cornell murine model. *Infection and Immunity, 67*, 4531–4538.

120. Schnappinger, D., Ehrt, S., Voskuil, M. I., et al. (2003). Transcriptional adaptation of Mycobacterium tuberculosis within macrophages: Insights into the Phagosomal environment. *The Journal of Experimental Medicine, 198*(5), 693–704.
121. Shah, D., et al. (2006). Persisters: A distinct physiological state of E. coli. *BMC Microbiology, 6*, 53.
122. Sherman, D. R., Voskuil, M., Schnappinger, D., et al. (2001). Regulation of the Mycobacterium tuberculosis hypoxic response gene encoding alpha-crystallin. *Proceedings of the National Academy of Sciences of the United States of America, 98*, 7534–7539.
123. Shi, L., Sohaskey, C. D., Pfeiffer, C., et al. (2010). Carbon flux rerouting during Mycobacterium tuberculosis growth arrest. *Molecular Microbiology, 78*, 1199–1215.
124. Singh, A., Crossman, D. K., Mai, D., et al. (2009). Mycobacterium tuberculosis WhiB3 maintains redox homeostasis by regulating virulence lipid anabolism to modulate macrophage response. *PLoS Pathogens, 5*, e1000545.
125. Singh, A., Guidry, L., Narasimhulu, K. V., et al. (2007). Mycobacterium tuberculosis WhiB3 responds to O2 and nitric oxide via its [4Fe-4S] cluster and is essential for nutrient starvation survival. *Proceedings of the National Academy of Sciences of the United States of America, 104*, 11562–11567.
126. Singh, M., Tiwari, P., Arora, G., et al. (2016). Establishing virulence associated polyphosphate kinase 2 as a drug target for Mycobacterium tuberculosis. *Scientific Reports, 6*, 26900.
127. Siricilla, S., Mitachi, K., Wan, B., et al. (2014). Discovery of a capuramycin analog that kills nonreplicating Mycobacterium tuberculosis and its synergistic effects with translocase I inhibitors. *The Journal of Antibiotics, 68*, 271–278.
128. Sohaskey, C. D. (2005). Regulation of nitrate reductase activity in Mycobacterium tuberculosis by oxygen and nitric oxide. *Microbiology, 151*, 3803–3810.
129. Sohaskey, C. D., & Modesti, L. (2009). Differences in nitrate reduction between Mycobacterium tuberculosis and Mycobacterium bovis are due to differential expression of both narGHJI and narK2. *FEMS Microbiology Letters, 290*, 129–134.
130. Steyn, A. J., Collins, D. M., Hondalus, M. K., et al. (2002). Mycobacterium tuberculosis WhiB3 interacts with RpoV to affect host survival but is dispensable for in vivo growth. *Proceedings of the National Academy of Sciences of the United States of America, 99*, 3147–3152.
131. Subbian, S., O'Brien, P., Kushner, N. L., et al. (2013). Molecular immunologic correlates of spontaneous latency in a rabbit model of pulmonary tuberculosis. *Cell Communication and Signaling: CCS, 11*, 16.
132. Sugawara, I., Udagawa, T., Aoki, T., et al. (2009). Establishment of a Guinea pig model of latent tuberculosis with GFP-introduced Mycobacterium tuberculosis. *The Tohoku Journal of Experimental Medicine, 219*, 257–262.
133. Sureka, K., Ghosh, B., Dasgupta, A., et al. (2008). Positive feedback and noise activate the stringent response regulator rel in mycobacteria. *PLoS One, 3*, e1771.
134. Targeted tuberculin testing and treatment of latent tuberculosis infection. American Thoracic Society. MMWR Recomm Rep 49:1–51.
135. Tiwari, P., Arora, G., Singh, M., et al. (2015). MazF ribonucleases promote Mycobacterium tuberculosis drug tolerance and virulence in guinea pigs. *Nature Communications, 6*, 6059.
136. Trauner, A., Lougheed, K. E., Bennett, M. H., et al. (2012). The dormancy regulator DosR controls ribosome stability in hypoxic mycobacteria. *The Journal of Biological Chemistry, 287*, 24053–24063.
137. Vazquez-Laslop, N., Lee, H., & Neyfakh, A. A. (2006). Increased persistence in Escherichia coli caused by controlled expression of toxins or other unrelated proteins. *Journal of Bacteriology, 188*(10), 3494–3497.
138. Vernon, A. (2013). Treatment of latent tuberculosis infection. *Seminars in Respiratory and Critical Care Medicine, 34*, 67–86.
139. Via, L. E., Lin, P. L., Ray, S. M., III, et al. (2008). Tuberculous granulomas are hypoxic in guinea pigs, rabbits, and nonhuman primates. *Infection and Immunity, 76*, 2333–2340.

140. Voskuil, M. I., Schnappinger, D., Visconti, K. C., et al. (2003). Inhibition of respiration by nitric oxide induces a Mycobacterium tuberculosis dormancy program. *The Journal of Experimental Medicine, 198*, 705–713.
141. Wakamoto, Y., Dhar, N., Chait, R., et al. (2013). Dynamic persistence of antibiotic-stressed mycobacteria. *Science, 339*, 91–95.
142. Wayne, L. G. (1994). Dormancy of Mycobacterium tuberculosis and latency of disease. *European Journal of Clinical Microbiology & Infectious Diseases, 13*, 908–914.
143. Wayne, L. G., & Hayes, L. G. (1996). An in vitro model for sequential study of shift down of Mycobacterium tuberculosis through two stages of nonreplicating persistence. *Infection and Immunity, 64*, 2062–2069.
144. Wayne, L. G., & Hayes, L. G. (1998). Nitrate reduction as a marker for hypoxic shiftdown of Mycobacterium tuberculosis. *Tubercle and Lung Disease, 79*, 127–132.
145. Wayne, L. G., & Lin, K. Y. (1982). Glyoxylate metabolism and adaptation of Mycobacterium tuberculosis to survival under anaerobic conditions. *Infection and Immunity, 37*, 1042–1049.
146. Wayne, L. G., & Sramek, H. A. (1994). Metronidazole is bactericidal to dormant cells of Mycobacterium tuberculosis. *Antimicrobial Agents and Chemotherapy, 38*, 2054–2058.
147. Xavier, A. S., & Lakshmanan, M. (2014). Delamanid: A new armor in combating drug-resistant tuberculosis. *Journal of Pharmacology and Pharmacotherapeutics, 5*(3), 222–224.
148. Yuan, Y., Crane, D. D., Simpson, R. M., et al. (1998). The 16-kDa – crystallin (Acr) protein of Mycobacterium tuberculosis is required for growth in macrophages. *Proceedings of the National Academy of Sciences of the United States of America, 95*, 9578–9583.
149. Zhang, T., Zhang, M., Rosenthal, I. M., et al. (2009). Short-course therapy with daily rifapentine in a murine model of latent tuberculosis infection. *American Journal of Respiratory and Critical Care Medicine, 180*, 1151–1157.
150. Zhang, Y. (2003). Chapter 49: Isoniazid. In W. N. Rom & S. Garay (Eds.), *Tuberculosis* (2nd ed., pp. 739–758). New York: Lippincott Williams & Wilkins, A Wolters Kluwer Company.
151. Zhang, Y., Wade, M. M., Scorpio, A., et al. (2003). Mode of action of pyrazinamide: Disruption of Mycobacterium tuberculosis membrane transport and energetics by pyrazinoic acid. *Journal of Antimicrobial Chemotherapy, 52*, 790–795.
152. Zierski, M. (1979). Intermittent treatment regimens in pulmonary tuberculosis. *Lung, 156*, 17–32.

Response of *Mycobacterium tuberculosis* to pH Stress: Promising Approach to Control Tuberculosis

Saif Hameed and Zeeshan Fatima

Abstract

Widespread and prolonged usage of antitubercular drugs in treating Tuberculosis (TB) caused by *Mycobacterium tuberculosis* (MTB) has led to the emergence of drug resistance by a phenomenon termed as Multi-Drug Resistance (MDR-TB). Despite reasonable advances in understanding of major players which contribute to drug resistance, it appears unavoidable to consider novel mechanisms combating MDR. The ability of pathogenic MTB, to acclimatize and become accustomed to changes in the host environment is essential for its survival that confers the basis for success of MTB as dreadful pathogen. One such significant environmental factor that MTB must surmount is pH adaptation, since they encounter diverse anatomical sites during the establishment of infection within the host. Considering the importance of MTB, being the second most common cause of mortality, this chapter focuses on gaining insights of various pH dependent mechanisms in MTB and how they can be exploited pharmacologically as efficient anti-mycobacterial therapeutic strategy.

Keywords

pH · Mycobacterium · MDR · Phagosome · Drug target

Introduction

Mycobacterium tuberculosis (MTB), the causative agent of tuberculosis (TB) has been responsible for deaths worldwide claiming about 1.8 million lives annually [1]. The pathogen actively co-infects during immunocompromised conditions such as HIV infection or organ transplantation. Several first line drugs have been

S. Hameed · Z. Fatima (✉)
Amity Institute of Biotechnology, Amity University Haryana, Manesar, Gurugram, India
e-mail: drzeeshanfatima@gmail.com

S. Hameed, Z. Fatima (eds.), *Pathogenicity and Drug Resistance of Human Pathogens*,
https://doi.org/10.1007/978-981-32-9449-3_4

identified and are being used now-a-days viz. isoniazid, rifampicin, ethambutol, pyrazinamide etc. but *Mycobacterium* has become resistant to these first line drugs resulting in species that are Multidrug Resistant (MDR) and in some cases Extensively Drug Resistant (XDR) [2]. Thus elucidation of novel strategies to combat MDR-TB is the need of hour. In recent years, emerging evidences has demonstrated that there exist such novel mechanisms which can be targeted to combat MDR which may also facilitate the development of novel therapies to overcome this deadly infection. For instance targeting significant micronutrient such as iron acquisition of MTB has been demonstrated to be an efficient strategy that could be exploited against MTB [3–5]. Similarly usage of natural compounds that can overcome MDR in mycobacteria is a useful therapeutic option [6]. Likewise maintaining pH homeostasis is another crucial condition that MTB must surmount to establish successful infection. The current chapter deals with the progress that has been accomplished to identify putative drug targets which may be exploited as effective therapeutic strategy.

Significance of pH in MTB Pathogenesis

When any organism including MTB infects, it encounters a variety of sites from oral cavity to gastrointestinal to urogenital tract with varied pH ranges. Therefore the microorganism needs to acclimatize its physiology within the hostile niche in such a manner that the optimal activity of the surface proteins, proton gradient and nutrient availability is not compromised. pH is also suppose to have profound effect on drug therapy, since many, if not all anti-TB drugs are often hydrophobic molecules which are also typically charged. Thus, changes in the electrochemical parameters of MTB cell membrane due to diverse pH environment have profound effects on the transmembranous diffusion and cellular retention of these drugs. Furthermore, fluctuations in these parameters can also modulate the functioning of immunological factors which could affect the signaling cascade associated with induction of apoptosis. Thus it deserves special mention that for establishment of the infection, understanding MTB response to the environmental cues intrinsic to the host niche such as pH, is crucial.

MTB is an aerobe which primarily invades the mammalian respiratory tract and is taken up by the macrophages of alveoli in the lungs. The cell wall of the MTB prevents the fusion of phagosomes and lysosomes and hence terminates its maturation. Therefore, mycobacterium resides in a slightly acidic medium inside the host body (i.e. ~6.2) and even during acidic cytoplasmic conditions, the bacteria is able to maintain homeostasis. Through various researches it has been observed that maintaining homeostasis inside host cell is one of the important factors for the survivability of mycobacterium. The pH of the phagolysosomes of the macrophages varies throughout depending upon the immunological state of the host body. Phagosomal acidification is one of the important factors for the survival of the microorganism inside the host body for which various factors are responsible. Many antibiotics have been used extensively, which work even during the acidic conditions, among which pyrazinamide is the most effective; but the mechanism of action of such antibiotics still remains unclear [7].

Factors Governing pH Responses in MTB

As the mycobacterium enters the host body, it encounters environments of varying pH and the pH at which the organism can survive successfully within the host body is ~6.2. Once the microorganism enters the host, it is taken up by macrophages by phagocytosis, and resides in the phagosomes. These phagosomes are prevented to further fuse with lysosomes hence restricting it from maturation which helps in maintaining the homeostasis. Therefore in order to cause virulence, maintaining a pH homeostasis becomes important once it has entered the host macrophage, as the pH in the various body parts of the host can vary. The pH of the lungs usually ranges from 7.38 to 7.42 during normal conditions. Even if the pH conditions change to acidic or alkaline it is important to maintain a neutral pH for its survival and virulence. The intracellular lethal pH is <6.0 and cell viability is affected greatly of the mycobacterium. When the external pH changes i.e. from neutral to either acidic or alkaline, F_1-F_0 ATPase plays a key role in maintaining the intracellular pH to near neutral [8]. The following sections will deal with the factors that govern pH homeostasis in MTB (Fig. 1).

Cell Envelope

The reason behind survival of MTB in acidic compartments of macrophages during the course of infection is due its ability to arrest the fusion of acidic lysosomes with

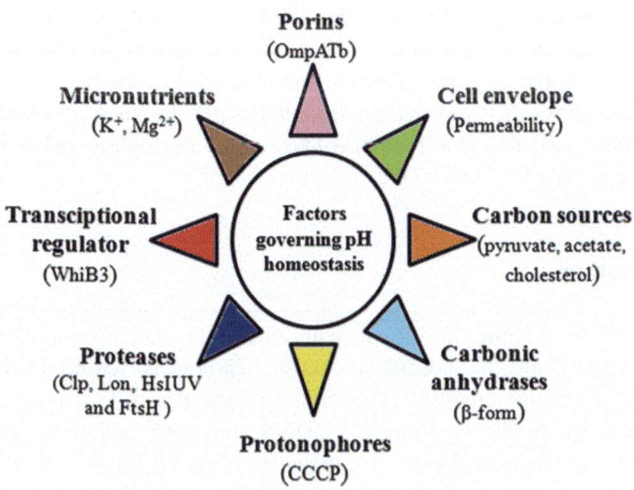

Fig. 1 Factors governing pH homeostasis in MTB

phagosomes [9]. It acquires resistance against this acidity by maintaining its intracellular pH; and this is because of a complex but efficient permeability barrier, the lipid rich cell envelope, which restricts entry of protons thereby regulating the intrabacterial pH. This is made possible with the contribution of a group of mechanisms including, cell envelope modification, proton pumps, production of ammonia, etc. [10]. Studies of acid sensitive mutants have revealed an altered regulation of genes involved in cell wall or lipid biosynthesis under acidic conditions [11–13]. The permeability of the cell envelope is influenced by the presence of small pore forming proteins, porins, which allow the movement of small hydrophilic molecules across the membrane [14]. A common condition known to be responsible for the increase tolerance to antibiotics is the decreased translocation of protons [15].

Porins

The complex mycobacterial cell wall controls the membrane permeability with the help of certain porin forming proteins, present on the surface, which helps in transportation of molecules across the membrane. OmpATb is an operon which regulates the functioning of four porin proteins MspA, MspC, MspD. The chief permeability hurdle of mycobacteria is recognised by an unusual outer membrane. Porins like MspA the channel-forming proteins aid transportation of minute hydrophilic solutes across the outer membrane of *Mycobacterium smegmatis* but MspA-like porins do not exist in MTB. Similarly like porin protein OmpA from Gram-negative bacteria, OmpATb of MTB forms channels in vitro [16]. Raynaud et al. [17] demonstrated that loss of the ompATb gene would not have any profound effect on the growth under normal conditions but this ability was lost when the pH was lowered because the permeability of the solutes was lost [17]. It was also observed through infection experiments that the *ompATb* operon is not essential for complete virulence in mice indicative of the fact that MTB has multiple other mechanisms for avoiding acidification of phagosomes. There are data indicating that OmpATb has mainly two functions: as a pore-forming protein with properties of a porin, and enabling response towards reduced environmental pH. It is not yet known whether the second function is linked to the porin-like activity during low pH or involves an entirely separate role for OmpATb [16].

Carbon Sources

All living cells have some nutritional requirements in order to complete their life processes. Among all the necessary factors, carbon is required to build the components for cellular structures. As acidic pH can disrupt biochemical reactions and cause damage to the mycobacterial DNA, proteins, and lipids. It has been observed that host-associated carbon sources are required for growth at acidic pH, such as pyruvate, acetate, oxaloacetate and cholesterol [18]. Few acidic pH induced and PhoPR-regulated genes which are related with carbon metabolism are as follows: *PKS2*, *PKS3*, and *PKS4* genes, involved in the production of cell envelope lipids and the aprABC locus is associated with the regulation of carbon and propionate metabolism genes. Although availability of carbon sources in vivo is not known but studies showed that isocitrate lyase is required to cause infection, hence suggesting that MTB metabolizes acetyl-CoA to form long chain fatty acids through glyoxylate shunt or propionyl-CoA from cholesterol through the methyl citrate cycle [19–21].

β-Carbonic Anhydrase

Carbonic anhydrases are Zn-dependent enzymes which catalyse reversible hydration reaction, that converts the membrane permeable CO^{2-} ions into ionic bicarbonate ions. These reactions are necessary for other biochemical processes like fatty acid biosynthesis, carbon assimilation and pH homeostasis. These enzymes are further divided into three classes α, β and γ. This reaction mechanism works in the presence of a reactive species, Zn-coordinated hydroxide. It was previously demonstrated that the Rv3588c gene of MTB which codes for carbonic anhydrase in a pH dependent manner, was found to be active at pH 8.4 but not at pH 7.5 or below. The structure of this dimeric protein with a blocked active site was thoroughly studied, a structure of the thiocynate complexed protein was presented in an altered crystal form. It was observed that the protein forms distinct tetramers showing enormous structural changes including the carboxylic shift yielding an available active site. It was further demonstrated through this structure that carbonic anhydrase is capable of switching between two states. This suggested that a carboxylate shift on/off switch for the enzyme might be controlled by a dimer to tetramer equilibrium which could be demonstrated by dynamic light scattering measurements [22].

Protonophores

While there are various factors that help the microorganism in maintaining homeostasis once it encounters change in pH, there are few factors which suppress the intrinsic acid resistance. Carbonyl cyanide m-chlorophenylhydrazone (CCCP) is an electrogenic protonophore which has been reported to be used as a suppressor for the acid resistance in *M. smegmatis,* mainly due to restricted permeability of the cell membrane to protons. It helps to suppress the intrinsic resistance to acid stress by equalising the external and internal pH. Various studies have been carried out in order to study the relation between cellular growth and CCCP in acidic pH, all of which reinforces that this protonophore is responsible for proton exchange, thus helps to maintain the pH homeostasis [8].

Proteases

Proteases and their associated chaperones have a crucial role in maintaining homeostasis in all living cells They also play dynamic roles in controlling the amount of cellular proteins involved in basic life processes, including transcription, replication, metabolism as well as virulence. There are different proteases in bacteria but only few of them take part in significant protein degradation reactions, such as Clp, Lon, HslUV and FtsH [23]. Previous studies have revealed that Clp is present in all bacteria which controls the transition from exponential to stationary growth phase, and manages 50% of the total cellular protein turnover. It was also observed that ClgR acts as gene regulator that has a role in transcriptional activation of these systems and virulence of the bacterial pathogen, so that they can replicate in the macrophages. ClgR activates the transcription of ten genes, among which four code for protease systems (ClpP1/C, ClpP2/C, PtrB and HtrA-like protease Rv1043c) while three code for chaperones (Acr2, ClpB and the chaperonin Rv3269) [24, 25].

Transcriptional Regulator

Recent studies show that MTB is able to neutralize its intracellular pH both in resting as well as activated macrophages. This indicates that in response to the acidic phagosomal environment, the intrabacterial pH acts as a signal that triggers the genetic alterations in MTB. A putative transcriptional factor, WhiB3 has been determined as a known cytoplasmic redox sensor, which along with mycothiol is essential in maintaining the homeostasis under acidic environments. Studies have revealed that in response to acidic pH, WhiB3 also regulates the expression of various genes related to lipid anabolism, secretion, and redox metabolism [26].

Micronutrients

Potassium

Micronutrients are essential elements for human life and equally for the growth and propagation of microbes. Alterations of micronutrients in the host environment can trigger virulence attributes of the organism. Even if rifampicin is known to be an important first-line antibiotic for the treatment of mycobacterial infections, the bacteria has acquired resistance against this antibiotic. While mutations in the rpoB gene give rise to most of the rifampicin-resistant strains, some of them show no rpoB mutations. This suggested that this resistance may be the result of some alternative mechanisms. Just like other micronutrients, K^+ is also essential for the maintenance of the intracellular ionic balance. Disruption of a putative regulator of K^+ uptake, trkA, have been shown to increase rifampicin resistance and studies indicate that TrkA-mediated K^+ uptake is necessary for maintaining the *M. smegmatis* growth rate, ionic balance and membrane potential. Inactivation of trkA also results in resistance to other hydrophobic agents, such as novobiocin, while increases sensitivity to isoniazid and positively charged aminoglycosides [27].

Magnesium

Mycobacterial cells encounter a relatively hostile environment during the course of infection because it resides inside the host macrophages. These macrophages undergo a maturation phase via the fusion of phagosome and lysosome. This maturation leads to an acidic as well as a nutrient limited condition for the invaded pathogen. Certain studies have been conducted to determine the role of various divalent cations like Mg^{2+}, Ca^{2+}, Zn^{2+} etc. in the survivability of the mycobacteria during mildly acidic conditions. Among all, Mg^{2+} has come out to be the most significant one, but the mechanism for indispensability of Mg^{2+} still remains unknown. Similarly these studies also revealed that no other divalent cations can be used as a substituent to meet the requirement of Mg^{2+}. Higher levels of Mg^{2+} were required for the growth of mycobacteria in acidic medium as compared to the growth in neutral medium [28].

Targeting pH Homeostasis as Promising Approach

Targeting Cell Envelope

A screen of acid-susceptible MTB transposon mutants showed that the disruption of 21 genes led to hypersensitivity of the mutants to low pH [29]. This motivated Vandal et al. [9] to study the effect of other forms of stress on the acid sensitive mutants. Their study indicated that protection against other forms of stress can be associated with the emergence of acid tolerance because the transposon insertions mutants were hypersensitive to antibiotics, reactive oxygen and nitrogen intermediates, sodium dodecyl sulfate and heat shock [9]. As it is a well known fact that proton translocation appears to have a critical role in the intracelllular pH homeostasis; therefore, any impairment in the proton homeostasis would result in decrease of the intrabacterial concentration of protons. This translocation is possible through ATP synthase enzyme, which catalyses ATP synthesis [30, 31]. Antibiotics targeting the ATP-driven biochemical pathways will eventually decrease the ATP turnover, hence decreasing the proton influx. This will further cause an increase in the intracelular pH, which could result in membrane disruption of the pathogen. Bartek et al. 2016 proposed a model which would result in reduction of intracellular proton concentration due to an imbalance in the proton homeostasis, further resensitizing the antibiotic-induced cell death [15].

Targeting Porins

It has been observed by Raynaud et al. 2002 that transcription of ompATb operon is increased at lower pH which closes the porin channel frequently in an acidic medium and reduces the membrane permeability of the bacteria [17]. Further chemical analysis by Song et al. 2011 revealed that ompATb operon encoded proteins are responsible for generation of a rapid ammonia burst, when MTB encounters an acidic environment. This addition of ammonia helps in neutralising the pH by counterbalancing the acidic medium and hence allows an exponential growth of MTB [16].

Targeting Carbon Sources

Jacob et al. 2014 have demonstrated the linkage of carbon availability and acidic environments that help regulate the physiology of MTB. They showed growth arrest of MTB at low pH indicating requirement of host associated carbon sources, involved in glycolysis and TCA cycle as well other carbon sources. Addition of pyruvate also restored this arrest suggesting that this growth arrest was the result of a pH-dependent checkpoint on the metabolism. It was also demonstrated by their work that the *phoPR* two-component regulatory system is essential for slow growth of MTB at acidic pH and also helps maintain redox homeostasis [18].

Targeting β-Carbonic Anhydrase

Covarrubias et al. 2005 studied the structure and function of the two identified β-class carbonic anhydrases in MTB, Rv1284 and Rv3588c and found the active site of the former to be smaller and shielded whereas larger and open for the latter. While Rv3588c was found to be a completely functional carbonic anhydrase, there was a lack in the activity of Rv1284. This was assumed to be due to a significant depletion of zinc at the active site of Rv1284. This was further supported by the Rv1284 structure itself which revealed that the electron density of the metal is only half as compared to Rv3588c. Moreover, zinc supplementation caused the protein to precipitate. The carbonic anhydrases in human belong to α-class which has a complete different structure than the bacterial β-Carbonic anhydrase. This difference in both the organisms can be exploited further in order to carry out detailed analysis of the enzyme to consider it as a potential target against the fatal pathogen [32].

Targeting Protonophores

Studies executed by Sieu L. Tran et. al. [8] showed that an electrogenic protonophore CCCP could be used to cause intracellular acid stress to cells at external pH values that are ordinarily not harmful to the cell. They observed that at acidic pH *M. smegmatis* cells, with a high degree of acid resistance, were unable to grow at different concentrations of CCCP. They identified few genes, disruption of which may respond to CCCP at acidic pH, either by making the pathogen sensitive or resistant to it. Therefore, through their observations they were able to conclude that acid sensitive mutants were unable to grow in the absence of a desired concentration of CCCP, thereby concluding that it helps in maintaining homeostasis by proton exchange during acidic environment [8].

Targeting Proteases

Since membrane acts as a physical barrier for the movement of molecules in and out of the cell, therefore membrane associated macromolecules present on the surface play crucial roles in the pH homeostasis. It was observed in MTB that mutant lacking a membrane-associated protease; Rv3671c was susceptible towards acidic environments. Its inability to maintain the intrabacterial pH in vitro and also in activated macrophages indicated that the membrane protease Rv3671c, crucially is essential in causing virulence and can be a possible cause of the MTB resistance to these acidic environments [29].

Targeting Transcriptional Regulator

The fact that MTB is able to stably maintain its intracellular pH in the acidic phagosomes during the course of infection, was essential to study the genetic alterations which the pathogen undergoes when it encounters the hostile environment. Since it was found that WhiB3 was the only transcription factor, the expression of which was pH-responsive, Mehta et al. [26] used a genetic biosensor of mycothiol redox potential in order to demonstrate the role of WhiB3 in homeostasis. They found that even a modest decrease in phagosomal pH was enough to create redox heterogeneity in mycothiol redox potential of the MTB population which was found to be in a WhiB3 dependent manner. This data also indicated that low external pH acts as a signal and hence there is an alteration in the cytoplasmic mycothiol redox potential which further activates WhiB3 mediated gene expression and helps in acid resistance. WhiB3 was also found to down regulate the expression of innate immune genes and therefore blocks the phagosomal maturation. This indicated that the WhiB3-dependent production of polyketide lipids was partly responsible for blockage in the phagosomal maturation. With the help of a WhiB3 mutant MtbΔwhiB3, it was confirmed that the intramacrophage survival defect of the mutant could be rescued by inhibiting the acidification of the phagosome. These results indicated a crucial link between vacuolar acidification, redox physiology, and virulence MTB and confirmed that WhiB3 can act as a mediator for phagosomal maturation arrest and hence acid resistance in MTB [26].

Targeting Micronutrients

Since a nutrient limited environment enhances the stress caused by low pH, acid sensitive mutants of MTB have been studied, which revealed the association of various ions in the intrabacterial pH homeostasis during the course of infection. K^+ uptake is essential for the maintenance of the ionic balance inside the cell, which suggests that loss in this activity can result in hypersensitivity to acidic environment. The growth of a trkA knockout mutant, lacking the K^+ transporter, was observed over a range of various pH (pH 5–8) and observed an impaired growth of the mutant at acidic pH as compared to wild type; but no significant difference in the growth at alkaline pH for both wild type and the mutant were observed. Furthermore, this growth defect was restored on supplementation of K^+ at various concentrations during acidic pH. These results suggested that K^+ uptake by TrkA might increase the intracellular pH when cells are exposed to acidic medium via membrane hyperpolarization [27].

Magnesium dependent growth of a MTB mutant lacking a putative magnesium transporter, MgtC, at low pH in vitro indicated that survival of the bacteria in the acidic phagosomes requires a sufficient level of Mg^{2+} [33]. Similarly whether other divalent cations possess the ability to meet this requirement of Mg^{2+} or not was examined by Piddington et al. [28]. It was observed that no other divalent cations, except Ca^{2+} could partially substitute and was able to rescue the mycobacteria under acidic environment unlike Mg^{2+} [28]. Several other studies have also suggested that Mg^{2+} is required for proton translocation catalysed by enzyme ATP synthase, and also for maintaining the integrity of the cell membrane, both mechanisms known to be associated with acid tolerance of MTB [34].

Conclusión

Although the strategy of targeting MTB pH homeostasis is at rudimentary level, it holds promising potential. The significance of elucidating the pH regulation mechanisms in MTB will help to elucidate how they are adapted to the various hostile niches. Therefore, disruption of any pH dependent mechanism in MTB certainly merits a closer look to develop novel antimycobacterial pharmacological approach.

Conflict of Interests None to declare.

References

1. Banaei-Esfahani, A., Nicod, C., Aebersold, R., & Collins, B. C. (2017). Systems proteomics approaches to study bacterial pathogens: Application to *Mycobacterium tuberculosis*. *Current Opinion in Microbiology, 39*, 64–72.
2. Jain, A., & Dixit, P. (2008). Multidrug-resistant to extensively drug resistant tuberculosis: What is next? *Journal of Biosciences, 33*, 605–616.
3. Hameed, S., Pal, R., & Fatima, Z. (2015). Iron acquisition mechanisms: Promising target against *Mycobacterium tuberculosis*. *The Open Microbiology Journal, 9*, 91–97.
4. Pal, R., Hameed, S., & Fatima, Z. (2015). Iron deprivation affects drug susceptibilities of mycobacteria targeting membrane integrity. *Journal of Pathogens, 2015*, 1–10.
5. Pal, R., Hameed, S., Sharma, S., & Fatima, Z. (2016). Influence of iron deprivation on virulence traits of mycobacteria. *The Brazilian Journal of Infectious Diseases, 20*, 585–591.
6. Sharma, S., Hameed, S., & Fatima, Z. (2016). Natural compounds for overcoming multidrug resistance in mycobacteria. *Recent Patents on Biotechnology, 10*, 167–174.

7. Zhang, Y., Scorpio, A., Nikaido, H., & Sun, Z. (1999). Role of acid pH and deficient efflux of Pyrazinoic acid in unique susceptibility of Mycobacterium tuberculosis to pyrazinamide. *Journal of Bacteriology, 181*, 2044–2049.
8. Tran, S. L., Rao, M., Simmers, C., et al. (2005). Mutants of *Mycobacterium smegmatis* unable to grow at acidic pH in the presence of the protonophore carbonyl cyanide m-chlorophenylhydrazone. *Microbiology, 151*, 665–672.
9. Vandal, O. H., Roberts, J. A., & Odaira, T. (2009a). Acid-susceptible mutants of Mycobacterium tuberculosis share hypersusceptibility to Cell Wall and oxidative stress and to the host environment. *Journal of Bacteriology, 191*, 625–631.
10. Vandal, O. H., Nathan, C. F., & Ehrt, S. (2009b). Acid resistance in *Mycobacterium tuberculosis*. *Journal of Bacteriology, 191*, 4714–4721.
11. Fisher, M. A., Plikaytis, B. B., & Shinnick, T. M. (2002). Microarray analysis of the *Mycobacterium tuberculosis* transcriptional response to the acidic conditions found in phagosomes. *Journal of Bacteriology, 184*, 4025–4032.
12. Rohde, K. H., Abramovitch, R. B., & Russell, D. G. (2007). *Mycobacterium tuberculosis* invasion of macrophages: Linking bacterial gene expression to environmental cues. *Cell Host & Microbe, 2*, 352–364.
13. Saviola, B., Woolwine, S. C., & Bishai, W. R. (2003). Isolation of acid inducible genes of *Mycobacterium tuberculosis* with the use of recombinase based in vivo expression technology. *Infection and Immunity, 71*, 1379–1388.
14. Kartmann, B., Stenger, S., & Weis, M. N. (1999). Porins in the Cell Wall of *Mycobacterium tuberculosis*. *Journal of Bacteriology, 181*, 6543–6546.
15. Bartek, I. L., Reichlen, M. J., & Honaker, R. W. (2016). Antibiotic bactericidal activity is countered by maintaining pH homeostasis in *Mycobacterium smegmatis*. *American Society of Microbiology, 1*, e00176–e00116.
16. Song, H., Huff, J., & Janik, K. (2011). Expression of the ompATb operon accelerates ammonia secretion and adaptation of *Mycobacterium tuberculosis* to acidic environments. *Molecular Microbiology, 80*, 900–918.
17. Raynaud, C., Papavinasasundaram, K. G., Speight, R. A., et al. (2002). The functions of OmpATb, a pore-forming protein of *Mycobacterium tuberculosis*. *Molecular Microbiology, 46*, 191–201.
18. Baker, J. J., Johnson, B. K., & Abramovitch, R. B. (2014). Slow growth of Mycobacterium tuberculosis at acidic pH is regulated by phoPR and host-associated carbon sources. *Molecular Microbiology, 94*, 56–69.
19. Upton, A. M., & McKinney, J. D. (2007). Role of the methylcitrate cycle in propionate metabolism and detoxification in *Mycobacterium smegmatis*. *Microbiology, 153*, 3973–3982.
20. Savvi, S., Warner, D. F., Kana, B. D., et al. (2008). Functional characterization of a vitamin B-12-dependent methylmalonyl pathway in *Mycobacterium tuberculosis*: Implications for propionate metabolism during growth on fatty acids. *Journal of Bacteriology, 190*, 3886–3895.
21. Eoh, H., & Rhee, K. Y. (2014). Methylcitrate cycle defines the bactericidal essentiality of isocitrate lyase for survival of *Mycobacterium tuberculosis* on fatty acids. *Proceedings of the National Academy of Sciences, 111*, 4976–4981.
22. Covarrubias, A. S., Bergfors, T., Jones, T. A., & Högbom, M. (2006). Structural mechanics of the pH-dependent activity of β-carbonic anhydrase from *Mycobacterium tuberculosis*. *The Journal of Biological Chemistry, 281*, 4993–4999.
23. Gottesman, S., Wickner, S., & Maurizi, M. R. (1997). Protein quality control: Triage by chaperones and proteases. *Genes & Development, 11*, 815–823.
24. Schnappinger, D., Ehrt, S., Voskuil, M. I., et al. (2003). Transcriptional adaptation of *Mycobacterium tuberculosis* within macrophages: Insights into the phagosomal environment. *The Journal of Experimental Medicine, 198*, 693–704.
25. Estorninho, M., Smith, H., Thole, J., et al. (2010). ClgR regulation of chaperone and protease systems is essential for *Mycobacterium tuberculosis* parasitism of the macrophage. *Microbiology, 156*, 3445–3455.

26. Mehta, M., Rajmani, R. S., & Singh, A. (2016). Mycobacterium tuberculosis WhiB3 responds to vacuolar pH-induced changes in Mycothiol redox potential to modulate Phagosomal maturation and virulence. *The Journal of Biological Chemistry, 291*, 2888–2903.
27. Castañeda-García, A., Thuy Do, T., & Blázquez, J. (2011). The K1 uptake regulator TrkA controls membrane potential, pH homeostasis and multidrug susceptibility in *Mycobacterium smegmatis*. *Journal of Antimicrobial Chemotherapy, 66*, 1489–1498.
28. Piddington, D. L., Kashkouli, A., & Buchmeier, N. A. (2000). Growth of Mycobacterium tuberculosis in a defined medium is very restricted by acid pH and Mg^{2+} levels. *Infection and Immunity, 68*, 4518–4522.
29. Vandal, O. H., Pierini, L. M., Schnappinger, D., et al. (2008). A membrane protein preserves intrabacterial pH in intraphagosomal *Mycobacterium tuberculosis*. *Nature Medicine, 14*, 849–854.
30. Mitchell, P. (1961). Coupling of phosphorylation to electron and hydrogen transfer by a chemiosmotic type of mechanism. *Nature, 191*, 144–148.
31. Hattori, T., Watanabe, K., Uechi, Y., Yoshioka, H., & Ohta, Y. (2005). Repetitive transient depolarizations of the inner mitochondrial membrane induced by proton pumping. *Biophysical Journal, 88*, 2340–2349.
32. Covarrubias, A. S., Larsson, A. M., Ho¨gbom, M., et al. (2005). Structure and function of carbonic anhydrases from *Mycobacterium tuberculosis*. *The Journal of Biological Chemistry, 280*, 18782–18789.
33. Buchmeier, N., Blanc-Potard, A., Ehrt, S., et al. (2000). A parallel intraphagosomal survival strategy shared by *Mycobacterium tuberculosis* and *Salmonella enteric*. *Molecular Microbiology, 35*, 1375–1382.
34. Cotter, P. D., & Hill, C. (2003). Surviving the acid test: Responses of gram-positive Bacteria to low pH. *Microbiology and Molecular Biology Reviews, 67*, 429–453.

Roadmap for the Eradication of Multidrug Resistant Tuberculosis

Mohit Agarwal and Ashok Rattan

Abstract

Tuberculosis is an ancient disease which has become rampant in recent times due to it's multidrug resistant nature. Global community is concerned about it and measures to control it are being taken in the form of Millennium Development Goals and Stop TB Strategy. These goals can be achieved by early case finding and better diagnosis. Detection of latent TB infection and it's proper treatment is also necessary to eliminate disease. A newer vaccine which could either replace or accentuate the current BCG vaccine is also demand of the time. And the last step in the direction of elimination of TB will be judicious use of currently available drugs. It is also necessary that we come out with newer anti-TB drugs and regimens which could handle the issues like cost and toxicity. In India Revised National TB Program (RNTCP) has also implemented National Strategic Plan to eliminate TB by 2030.

Keywords

End TB strategy · FAST strategy · Front loaded microscopy · QFT-Plus · TB vaccine · Nix-TB

Introduction

Tuberculosis (TB) is probably as old as mankind. The earliest archaeological evidences of human TB in the form of spinal TB (Pott's disease) are found in Egyptian art and mummies as early as 9000 years ago [1, 2]. But in spite of having

M. Agarwal
Mahatma Gandhi Medical College and Hospital, Jaipur, India

A. Rattan (✉)
Mahatma Gandhi Medical College and Hospital, Jaipur, India

Pathkind Labs, Udyog Vihar, Gurugram, India
e-mail: drashokrattan@gmail.com

S. Hameed, Z. Fatima (eds.), *Pathogenicity and Drug Resistance of Human Pathogens*,
https://doi.org/10.1007/978-981-32-9449-3_5

such a long history it continues to haunt the mankind even in the presence of several spectacular advances in the diagnosis and management.

Worldwide, TB is one of the top 10 causes of death and the leading cause from a single infectious agent (above HIV/AIDS) [3].

In 2017, TB caused an estimated 1.3 million deaths among HIV-negative people and there were an additional 300,000 deaths from TB among HIV-positive people. Globally, the best estimate is that ten million people developed TB disease in 2017. India tops the list with 27% of these new cases [3].

Drug-resistant TB continues to be a public health crisis. The best estimate is that, worldwide in 2017, 558000 people developed TB that was resistant to rifampicin (RR-TB), and of these, 82% had multidrug-resistant TB (MDR-TB). India once again tops the list with 24% of these cases [3].

Among cases of MDR-TB in 2017, 8.5% were estimated to have extensively drug-resistant TB (XDR-TB) [3].

Treatment of TB is long and difficult and as resistance rises it becomes even more difficult. There have been efforts at global level to reduce the incidence and prevalence of TB and how to stop emergence of resistance in TB bacilli.

Global Efforts to Control and Eliminate TB

Upto 2015 Millennium Development Goals (MDGs) of United Nations were driving factor to reduce the burden of TB. Target 6c of MDG 6 was to "halt and reverse" TB incidence [4].

The Stop TB Partnership which was established in2001 adopted this target and set two additional targets: to halve TB prevalence and TB mortality rates by 2015 compared with their levels in 1990 [5].

WHO also developed it's Stop TB Strategy for the decade of 2006–2015 with adoption of all these three goals [3].

As per WHO's Global Tuberculosis Report, the MDG target to halt and reverse TB incidence was achieved on a worldwide basis and the TB incidence rate was 18% lower than the level of 2000 [6].

TB prevalence in 2015 was 42% lower than in 1990. Though the target of halving the rate compared with 1990 could not be achieved worldwide, it was achieved in three WHO regions and in nine high-burden countries including India. Target of halving the mortality rate was achieved in four WHO regions and in 11 high-burden countries including India [6].

In 2016 Sustainable Development Goals (SDGs) succeeded the MDGs for a period of 2015–2030. WHO also endorsed it's End TB Strategy for 2016–2035. These two together provides a framework for all efforts at national or international level to end TB epidemic [7].

Sustainable Development Goal 3 deals with health. It states 'Ensure healthy lives and promote well-being for all at all ages. There are 13 Targets for this goal and Target 3.3 explicitly mentions TB. Target 3.3 is "By 2030, end the epidemics of AIDS, tuberculosis, malaria and neglected tropical diseases and combat hepatitis, water-borne diseases and other communicable diseases". TB incidence rate has been chosen as TB indicator for this Target [7].

The overall Goal of End TB Strategy is "End the global TB epidemic" [8]. To achieve this Goal three indicators are chosen. These three indicators are-

1. The number of TB deaths per year
2. The TB incidence rate
3. The percentage of TB-affected households that experience catastrophic costs as a result of TB disease.

Tb incidence rate has been defined as new cases per 100,000 population per year. For each indicator there are Milestones and Targets [8].

	Milestones		Targets	
Indicators	2020	2025	2030	2035
1. Percentage reduction in the absolute number of TB deaths	35	75	90	95
2. Percentage reduction in the TB incidence rate	20	50	80	90
3. Percentage of TB-affected households experiencing catastrophic costs due to TB	0	0	0	0

For Indicator one and two baseline year is 2015. Targets set for 2030 are 90% reduction in absolute number of TB deaths and 80% reduction in TB incidence rate as compared to 2015. Targets are further heightened and for 2035, 95% reduction in absolute number of TB deaths and 90% reduction in TB incidence rate is targeted as compared to 2015. For the third Indicator a zero percent milestone is to achieved by 2020 and will be maintained further [8].

For the achievement of these Milestones and Targets there are four underlying principles and three pillars.

Four underlying principles broadly define the roles of global community, governments, organizations and also talk about human rights and ethics. These principles are

1. Government stewardship and accountability, with monitoring and evaluation
2. Strong coalition with civil society organizations and communities
3. Protection and promotion of human rights, ethics and equity
4. Adaptation of the strategy and targets at country level, with global collaboration

For three Pillars ten components are also defined.

Pillars	Components
Integrated, patient-centred care and prevention	Early diagnosis of tuberculosis including universal drug-susceptibility testing, and systematic screening of contacts and high-risk groups
	Treatment of all people with tuberculosis including drug-resistant tuberculosis, and patient support

(continued)

	Collaborative TB/HIV activities, and management of comorbidities
	Preventive treatment of persons at high risk, and vaccination against TB
Bold policies and supportive systems	Political commitment with adequate resources for TB care and prevention
	Engagement of communities, civil society organizations, and public and private care providers
	Universal health coverage policy, and regulatory frameworks for case notification, vital registration, quality and rational use of medicines, and infection control
	Social protection, poverty alleviation and actions on other determinants of TB
Intensified research and innovation	Discovery, development and rapid uptake of new tools, interventions and strategies
	Research to optimize implementation and impact, and promote innovations

WHO has also identified ten priority indicators for monitoring of progress in implementing the End TB Strategy [8].

Priority indicator	Recommended target level (2025)
TB treatment coverage	≥90%
TB treatment success rate	≥90%
Percentage of TB-affected households that experience catastrophic costs due to TB	0%
Percentage of new and relapse TB patients tested using a WHO-recommended rapid diagnostic (WRD) at the time of diagnosis	≥90%
Latent TB infection (LTBI) treatment coverage	≥90%
Contact investigation coverage	≥90%
Drug-susceptibility testing (DST) coverage for TB patients	100%
Treatment coverage, new TB drugs	≥90%
Documentation of HIV status among TB patients	100%
Case fatality ratio (CFR)	≤5%

Control, Elimination and Eradication

Health expert agree that Control is reduction of disease incidence, prevalence, morbidity or mortality to a locally acceptable level as a result of deliberate efforts. Elimination is reduction to zero of the incidence of disease or infection in a defined geographical area. Eradication is permanent reduction to zero of the worldwide incidence of infection [9]. It is to be noted that to control or eliminate a disease or infection continued interventions are required. These interventions prevent the re-emergence and re-establishment of transmission. But once rates of incidence and prevalence start to decline, countries start to cut budgets and this neglect may cause the disease to re-emerge [10].

TB control is a strategy which aims at diagnosing and rapidly rendering infectious cases non-infectious, so that chain of transmission can be broken.

TB elimination, defined as less than one TB case per million population, a rate so low that it no longer constitutes a public health problem [11]. Complete TB eradication is not considered possible due to various reasons [12].

Strong support for the elimination hypothesis is derived from mathematical modelling. It is suggested that to achieve TB elimination, the incidence rate must be cut by a factor of more than thousand [13]. To achieve this rate cut it is expected that we immediately optimize use of current and new emerging tools. Simultaneously we should introduce new tools as more effective vaccines, newer drugs, shorter drug regimens and improved diagnostic tests to fight active as well as LTBI.

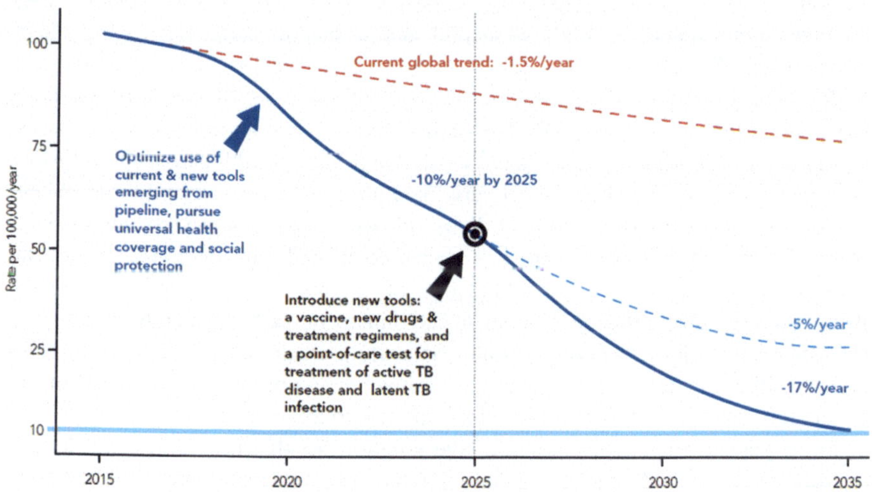

Case Finding

Stopping an epidemic requires stopping transmission. WHO estimates that three million people with tuberculosis are missed each year by health systems [14]. These missed cases lead to the persistence of infection and transmission in families and society. Targeted active case finding is a fundamental strategy for disease control though most tuberculosis programmes in countries with high burdens of tuberculosis have adopted policies that rely on passive case-finding. It is important for each tuberculosis programme to identify what combination of targeted active case-finding activities is the most effective in the local context. One high risk group which is present in all settings is contacts of patients with tuberculosis. First component of Pillar one emphasizes the importance of contact screening. Health facilities are a unique setting in which targeted active case-finding can be done. 'Finding TB cases Actively, Separating safely, and Treating effectively'(FAST) strategy has helped facilities reduce the risk and duration of exposure to tuberculosis for both patients and health-care workers [15].

Diagnosis of TB and Future Prospects

Sputum smear microscopy plays a major role in the diagnosis of TB in low and middle income countries (LMIC). In LMIC, it is the only cost-effective tool for diagnosing infectious patients, monitor their progress in treatment and confirm cure. The most regular practice is acid-fast staining using carbol fuchsin. Direct microscopic examination of sputum for AFB is inexpensive, rapid, and easy to perform. Although specific, it lacks sensitivity. Almost 85.8% of TB cases are detected with the first sputum specimen. With the second sputum specimen, the average incremental yield is 11.9% [16].

As suggested by mathematical modelling, optimum use of currently available tools is required to eliminate TB. The optimisation of sputum microscopy services, often the only TB diagnostic services possible at primary health care level in LMIC, is urgently needed. Need to collect serial sputum specimens over multiple patient visits results in patient drop-out [17]. Tuberculosis is disease of poverty and the high costs of transportation, food, and lost wages associated with diagnostic visits can consume major part of the household income of patients with suspected tuberculosis. It leads a large proportion to dropout of the diagnostic pathway before completing sputum examination, receiving results, or starting treatment [17–20]. A dropout patient will come again to healthcare system when symptoms are worsened but in meanwhile duration it has transmitted the disease to others as well.

Front loaded microscopy provides the solution for this problem. Front-loaded microscopy is a new diagnostic strategy in which two smears are prepared from two spot sputum specimens obtained on the first day a patient is assessed [21]. When all samples are collected and the results reported on 1 day, the strategy is termed same-day microscopy [22]. Studies have proven that front loaded microscopy has same yield as standard strategy where two sputum specimens are collected as spot and morning [23]. Using this strategy may lead to prevent the dropout of patients. It also helps to achieve the first component of first pillar of End TB Strategy which mentions about early diagnosis of TB.

Apart from microscopy there have been a formidable progress in the field of TB diagnostics. Several assays, such as Xpert MTB/RIF (Xpert), Xpert MTB/RIFUltra (Ultra), urine lateral flow lipoarabinomannan (LF-LAM) or loop-mediated isothermal amplification (TB-LAMP) havebeen WHO endorsed [24].

Xpert is a real-time quantitative PCR assay that detects TB and rifampicin resistance simultaneously. It is approved by WHO as a frontline test for pulmonary, extrapulmonary, and paediatric TB. It is a step forward in the direction of universal drug susceptibility testing.

Xpert MTB/RIF Ultra is a successor technology to Xpert that uses the same test hardware. Overall, sensitivity of the Xpert Ultra is 5% higher than that of Xpert but specificity is 3.2%lower [25].

Point of care (POC) testing can be a boon for TB patients as well as for state health programs because they can be deployed at the most decentralised level by health care workers with minimal training. WHO has released Ideal characteristics of a POC assay for diagnosing TB [26]. An ideal POC test should be able to detect infection in all populations like children and adults, HIV-positive or HIV-negative TB presumptive cases. It should detect pulmonary or extrapulmonary tuberculosis (TB). It should deliver results in less than 20 min. It should be cheap also [26].

Point-of-care diagnostic development should combine the most innovative technologies. Biosensors and combination of nanotechnology and biosensing technology has great potential in the medical diagnostics field [27].

Latent Tuberculosis Infection

Tuberculosis is a disease where pathogenesis involves a period of asymptomatic subclinical infection that might last for weeks to decades. Different people mount different immune responses after the initial infection and it affects the risk of tuberculosis infection progressing to active disease. An effective immune response may eradicate bacilli from body while an intermediate response may lead to containment of infection but still harbouring the organism in the body. People with no effective immunity against tuberculosis progress rapidly from tuberculosis infection to disease [28]. People with latent tuberculosis infection serve as seedbeds of infection and to break the chain of infection it is necessary to find these people and treat them [29]. Interferon-gamma release assay (IGRA) or tuberculin skin test (TST) can be used to identify the people with latent TB infection. IGRAs have better role to play in LTBI detection in comparison to TST. Though theoretically it has been suggested that interferon gamma values go down with treatment, current IGRAs are not recommended to determine treatment efficacy or for treatment monitoring. The new generation IGRA, QuantiFERON-TB Gold Plus (QFT-Plus) utilizes antigens designed specifically to stimulate both $CD8^{+}$T-cells and$CD4^{+}$T-cells. It has been demonstrated that there is increased differential activity of $CD8^{+}$T-cells inactive TB, and a functional decline in $CD8^{+}$T-cell activity that correlates with bacterial clearance during treatment. So addition of $CD8^{+}$T-cells in newer IGRA may be helpful in differentiating latent and early re-active TB [30]. Treatment regimens could be either isoniazid or rifampicin alone or in combination.

TB Vaccine

To interrupt the TB pandemic role of vaccination is extremely important. Component 1D and 2A of End TB Strategy mention about vaccination.

Bacille Calmette-Guérin (BCG) is currently the only licensed vaccine available against tuberculosis (TB). The vaccine has been given to over four billion people yet, TB still poses a major public health threat globally [31].

BCG is a live attenuated strain of *M. bovis*, developed by French scientists Albert Calmette and Camille Guerin. The vaccine was first given in 1921 by oral route and later on changed to intradermal route in 1927. From 1974 it was included in WHO Expanded Programme on Immunization. Currently more than 100 million infants receive vaccine annually. After the first BCG creation by Calmette and Guerin multiple substrains generated due to diverse culture methods [32]. These BCG substrains differ in various characteristics including immunogenicity, and virulence in animals. This leads to substantial variation in efficacy [33].

Though BCG is widely used certain countries like United States, Canada, Italy, Belgium, and The Netherlands never had universal BCG vaccination programs. These countries recommend BCG vaccination in certain cases only [34]. Countries like Ecuador, Australia and New Zealand had universal immunization in past but later on they also adopted BCG vaccination in selected population only.

BCG vaccination is contraindicated in HIV infected persons including infants. This becomes important considering the fact that HIV infection promotes the chances of acquiring TB infection [35, 36].

Countries which adopt universal BCG vaccination there it is traditionally administered in newborns, and in that population, it has a protective effect. However, this protective effect wanes over the time, and the general consensus is now that the vaccine provides little protection in adult individuals [37]. This may be related to insufficient immunological memory resulting in waning of immunity during childhood.

Mathematical modelling shows that, compared with an infant vaccine, an adult vaccine would have considerably greater early impact on the epidemic and would be cost-effective, even with a relatively low efficacy and short protection duration.

This has prompted a concerted effort over the past two decades to develop new candidate vaccines, to improve present BCG, to boost it, or to replace it altogether [38, 39].

Currently there are 16 TB vaccine candidates in different stages of clinical development globally. The candidates include whole, live mycobacteria such as recombinant BCGs (e.g. *M. vaccae,* DAR-901) and recombinant *M. tuberculosis*, lysates of whole mycobacteria, adjuvanted recombinant protein vaccines (H1 + IC31, H4 + IC31) and viral-vectored candidates (M72 + AS01E, MVA85A/AERAS-485) [38, 40].

These vaccine candidates work across spectrum of TB infection and disease. There are many approaches suggested to develop vaccine. First idea is the development of a vaccine candidate that could act rapidly to prevent the actual establishment of a site of infection in the first place. Though idea is fascinating, it is a at the level of concept only. No candidate has been developed yet for clinical trial [39].

Next idea is to improve current BCG vaccine. This can be done by either using recombinant strain or by fusion protein constructs.

Another idea is to boost immunity engendered by neonatal BCG vaccine because it is worldwide coverage is good. Leading candidates in this field are MVA85A which is based on vaccinia virus, and Aeras-402 uses adenovirus type 35 [41].

Postexposure or therapeutic vaccines are the next candidates. These vaccines could be given to individuals after they have been actively infected or exposed. RUTI vaccine is the most advanced candidate from this field [42].

Therapeutic vaccines are those which can facilitate the activity of the drug regimens used. ID93 is such a candidate [43].

Reaching the goals of End TB strategic will be influenced by the availability of new vaccines to contribute to the global fight against tuberculosis. These are required to complement available and newer drugs and diagnostic technologies. Development of new TB vaccines would be a critical step in halting the spread of both drug-sensitive and drug-resistant-TB.

TB Drugs and Regimens

Anti-TB drugs used in first line of treatment are more than 40 years old. Though these drugs are effective, treatment duration is long that is minimum of 6 months. Simultaneously if patient defaults in this duration then drug resistance may also develop. Current treatment regimens for MDR-TB are far from satisfactory due to lengthy duration (upto 20 months), lower cure rates, more toxicity and high cost [44].

To remove these difficulties there is an urgent need of new anti-TB drugs. These new drugs would be helpful in making better, safer, less toxic, shorter and cheaper regimen so that treatment default can be stopped. Considering the dearth of new anti-TB drugs and emergence of drug resistant TB it appears that tuberculosis has evolved faster than our medicines.

Development of new drugs is very essential for elimination of TB. Many drugs are in pipeline of development.

Drug	Class	Phase
Bedaquiline	Diarylquinoline	Approved
Delamanid	Nitroimidazole	Approved
Pretomanid	Nitroimidazole	Approved
Sutezolid	Oxazolidinone	II
Telacebec	Imidazopyridine amide	II
Delpazolid	Oxazolidinone	II
Macozinone	Benzothiazinone	I

Bedaquiline got accelerated approval from U.S. Food and Drug Administration (FDA) in December 2012. It was indicated to be used as part of combination therapy in adults for pulmonary MDR-TB [45]. It is a bactericidal drug. It's unique and specific anti-mycobacterial activity derives from inhibition of the proton pump of mycobacterial ATP synthase. Binding of bedaquiline to the oligomeric and proteolipic subunit-c of mycobacterial ATP synthase leads to inhibition of ATP synthesis, which subsequently results in bacterial death. Recently bedaquiline has been classified by WHO as Group A drug for use in Longer MDR-TB regimens [46].

Delamanid was granted a conditional marketing authorisation by the European Medicines Agency (EMA) in April 2014. It was also indicated to be used as part of combination therapy for pulmonary MDR-TB. It is a bactericidal drug. It acts by inhibiting the synthesis of mycobacterial cell wall components, methoxy mycolic acid and ketomycolic acid. Delamanid may be included in the treatment of MDR-TB patients aged 3 years or more on longer regimens [47]. Recently delamanid has been classified by WHO as Group C drug for use in Longer MDR-TB regimens [46].

Drugs licensed for other conditions but used to treatMDR/XDR-TB are called repurposed drugs [48]. Fluoroquinolones, kanamycin, amikacin, clofazimine, linezolid, carbapenems, amoxicillin/clavulanic acid are all repurposed drugs.

Current treatment for drug-susceptible TB consists of the standard four-drug regimen comprising, isoniazid (H), rifampicin (R), ethambutol (E) and pyrazinamide (Z) for 6 months, (2HREZ/4HR). Though this regimen achieves high cure rates of 90–95%, patient has to take medicine for 6 months. Problem with compliance may lead to resistance.

Drug resistance in TB can be classified as RR-TB (rifampicin resistant TB), MDR-TB (rifampicin and isoniazid resistant TB), Pre XDR TB (MDR- TB associated with resistance to at least one fluoroquinolone or a second-line injectable like amikacin, kanamycin, or capreomycin) and XDR TB (MDR- TB associated with resistance to at least one fluoroquinolone and a second-line injectable like amikacin, kanamycin, or capreomycin).

The World Health Organisation (WHO) has recently updated the classification of anti-TB drugs other than first line for the treatment of RR-TB and MDR-TB [49].

Group A	Fluoroquinolones	Levofloxacin
		Moxifloxacin
		Gatifloxacin
Group B	Second-line injectable agents	Amikacin
		Capreomycin
		Kanamycin
		(Streptomycin)
Group C	Other core second-line agents	Ethionamide/prothionamide
		Cycloserine/terizidone
		Linezolid
		Clofazimine
Group D	Add-on agents (not part of the core MDR-TB regimen)	D1
		Pyrazinamide
		Ethambutol
		High-dose isoniazid.

(continued)

		D2
		Bedaquiline
		Delamanid
		D3
		Para-aminosalicylic acid
		Imipenem plus cilastatin (requires clavulanate)
		Meropenem (requires clavulanate)
		Amoxicillin plus clavulanate
		(Thioacetazone)

Current guidelines recommend that patients with RR or MDR- TB, should receive at least five active TB medicines during the intensive phase, including pyrazinamide and four core second-line TB medicines -one is chosen from group A, one from group B, and at least two from group C. If the minimum of effective TB medicines cannot be composed as above, an agent from group D2 and other agents from D3 may be added to bring the total to five.

There is no regulatory-approved regimen for curing XDR-TB. Instead, healthcare providers try to individualize treatment, often using repurposed antibiotics. This treatment lasts for more than 2 years and patient has to take thousands of pills plus injections and face horrible side effects. It is extraordinarily costly as well. Despite the length, cost, and intensity of the treatment, outcomes are extremely poor. Most XDR-TB is not treated at all because the cost and complexity of such programs are out of reach for many health systems in TB-endemic countries.

Novel regimens based on three or more oral agents with little or nopre-existing resistance would provide simpler, more universally active regimens [50]. If such novel regimens are more effective than the current first-line regimen for drug-susceptible TB, they may shorten and simplify treatment for pulmonary TB irrespective of resistance to existing drugs.

One such combination of bedaquiline, pretomanid and sutezolid has shown some promising result in murine models. As the clinical development of sutezolidis stalled, linezolid is being tried in new experimental regimens [50].

Nix-TB trial was done in cases of XDR-TB. It used a three drugs regimen consisting of bedaquiline, pretomanid and linezolid (BPaL) [51]. ZeNix trial is a successor to NixTB using the same BPaL combination. Objectives of ZeNix were to evaluate whether the efficacy of the BPaL drug regimen can be maintained, while reducing toxicity by testing a lower dose and shorter duration of the drug linezolid [52]. Recently USFDA approved the use of pretomanid as part of combination regimen with bedaquiline and linezolid for the treatment of adults with pulmonary XDR, treatment-intolerant or nonresponsive multidrug-resistant (MDR) tuberculosis (TB) [54].

SimpliciTB is another trial evaluating the efficacy, safety and tolerability of a drug regimen (BPaMZ) for patients with drug-sensitive (DS) and drug-resistant (MDR) pulmonary tuberculosis. The BPaMZ regimen is comprised of four different antimicrobials: Bedaquiline (B), Pretomanid (Pa), Moxifloxacin (M) and Pyrazinamide (Z). This regimen is given for 4 months in case of drug susceptible TB and 6 months in case of MDR-TB [52].

TB Elimination in India

The National Strategic Plan (NSP) 2017–2025 has been prepared to eliminate TB in India by 2030. It is a framework to guide the activities of all stakeholders. It provides goals and strategies for the country's response to the disease. Although India has managed to scale up basic TB services in the public health system, the rate of decline is too slow to meet the 2030 Sustainable Development Goals (SDG) and 2035 End TB targets. The requirements for moving toward TB elimination have been integrated into the four strategic pillars of "Detect–Treat–Prevent–Build" (DTPB). By taking this approach the national programme can achieve significant positive change and make a real difference in the lives of the many people it serves [53].

Conclusion

Although epidemiological plausibility for TB elimination exists, a comprehensive effort is needed by national TB programmes. A significant progress has been made in the fight against TB over the last 25 years, significant challenges remain and much greater political and funder investment is still needed to achieve global elimination. More urgency, commitment, and funding than have been shown so far are needed if tuberculosis is really to be eliminated.

References

1. Cave, A. J. E. (1939). The evidence for the incidence oftuberculosis in ancient Egypt. *British Journal of Tuberculosis, 33*, 142–152.
2. Morse, D., Brothwell, D. R., & Ucko, P. J. (1964). Tuberculosis inancient Egypt. *The American Review of Respiratory Disease, 90*, 5224–5241.
3. World Health Organization. (2018). *Global tuberculosis report* (p. 2018). Geneva: World Health Organization.
4. Ortblad, K. F., Salomon, J. A., Bärnighausen, T., & Atun, R. (2015). Stopping tuberculosis: A biosocial model for sustainable development. *The Lancet, 386*(10010), 2354–2362.
5. Stop TB Partnership. The stop TB partnership factsheet. http://www.stoptb.org/assets/documents/resources/publications/acsm/FINAL%20UPDATED%20STOP%20TB%20BROCHURE.pdf. Accessed on 19.01.2019.
6. World Health Organization. (2015). *Global tuberculosis report* (p. 2015). Geneva: World Health Organization.
7. United Nations. Sustainable Development Goals. https://www.un.org/sustainabledevelopment/sustainable-development-goals. Accessed 21 Jan 2019.
8. World Health Organization. *Implementing the end TB strategy: The essentials.* Geneva: WHO. https://www.who.int/tb/publications/2015/The_Essentials_to_End_TB/en/. Accessed 21 Jan 2019.
9. Dowdle, W. R. (1998). The principles of disease elimination and eradication. *Bull WorldHealth Organ, 76*(Suppl 2), 23–25.
10. Schito, M., Hanna, D., & Zumla, A. (2017). Tuberculosis eradication versus control. *International Journal of Infectious Diseases, 56*, 10–13.

11. Lönnroth, K., Migliori, G. B., Abubakar, I., et al. (2015). Towards tuberculosis elimination: An action framework for low-incidence countries. *The European Respiratory Journal, 45*, 928–952.
12. Matteelli, A., Rendon, A., Tiberi, S., et al. (2018). Tuberculosis elimination: Where are we now? *European Respiratory Review, 27*, 180035.
13. Dye, C., Glaziou, P., Floyd, K., et al. (2013). Prospects for tuberculosis elimination. *Annual Review of Public Health, 34*, 271–286.
14. Yuen, C. M., Amanullah, F., Dharmadhikari, A., et al. (2015). Turning off the tap: Stopping tuberculosis transmission through active case-finding and prompt effective treatment. *Lancet, 386*(10010), 2334–2343.
15. Barrera, E., Livchits, V., & Nardell, E. (2015). F-A-S-T: A refocused, intensified, administrative tuberculosis transmission control strategy. *The International Journal of Tuberculosis and Lung Disease, 19*, 381–384.
16. Mase, S., Ramsay, A., Ng, N., Henry, M., Hopewell, P. C., Cunningham, J., et al. (2007). Yield of serial sputum specimen examinations in the diagnosis of pulmonary tuberculosis: A systematic review. *The International Journal of Tuberculosis and Lung Disease, 11*(5), 485–495.
17. Squire, S. B., Belaye, A. K., Kashoti, A., et al. (2005). 'Lost' smear positive pulmonary tuberculosis cases: Where are they and why did we lose them? *International Journal of Tuberculosis and Lung Disease, 9*(1), 25–31.
18. Chandrasekaran, V., Ramachandran, R., Cunningham, J., et al. (2005). Factors leading to tuberculosis diagnostic drop-out and delayed treatment initiation in Chennai, India. *The International Journal of Tuberculosis and Lung Disease, 9*, 172.
19. Nota, A., Ayles, H., Perkins, M., & Cunningham, J. A. (2005). Factors leading to tuberculosis diagnostic drop-out and delayed treatment initiation in urban Lusaka. *The International Journal of Tuberculosis and Lung Disease, 9*, 305.
20. Thongraung, W., Chongsuvivatwong, V., & Pungrassamee, P. (2008). Multilevel factors affecting tuberculosis diagnosis and initial treatment. *Journal of Evaluation in Clinical Practice, 14*, 378–384.
21. Ramsay, A., Yassin, M. A., Cambanis, A., et al. (2009). Front-loading sputum microscopy services: An opportunity to optimise smear-based case detection of tuberculosis in high prevalence countries. *Journal of Tropical Medicine, 2009*, 1–6.
22. WHO. (2011). *Same-day diagnosis of tuberculosis: Policy statement.* Geneva: World Health Organization.
23. Davis, J. L., Cattamanchi, A., Cuevas, L. E., et al. (2013). Diagnostic accuracy of same-day microscopy versus standard microscopy for pulmonary tuberculosis: A systematic review and meta-analysis. *The Lancet Infectious Diseases, 13*(2), 147–154.
24. García-Basteiro, A. L., DiNardo, A., Saavedra, B., et al. (2018). Point of care diagnostics for tuberculosis. *Pulmonology, 24*(2), 73–85.
25. World Health Organization. (2017). WHO meeting report of a technical expert consultation: non-inferiority analysis of Xpert MTF/RIF Ultra compared to Xpert MTB/RIF. Geneva: World Health Organization.
26. World Health Organization. (2014). High-priority target product profiles for new tuberculosis diagnostics: report of a consensus meeting. Geneva: WHO/HTM7TB/2014.18.
27. He, X., Zhou, L., He, D., et al. (2011). Biosensing technologies for mycobacterium tuberculosis detection: Status and new developments. *Clinical & Developmental Immunology, 2011*, 193963.
28. Ernst, J. D. (2012). The immunological life cycle of tuberculosis. *Nature Reviews. Immunology, 12*, 581–591.
29. Rangaka, M. X., Cavalcante, S. C., Marais, B. J., et al. (2015). Controlling the seedbeds of tuberculosis: Diagnosis and treatment of tuberculosis infection. *Lancet, 386*(10010), 2344–2353.

30. Kamada, A., & Amishima, M. (2017). QuantiFERON-TB® gold plus as a potential tuberculosis treatment monitoring tool. *The European Respiratory Journal, 49*, 1601976.
31. Hatherill, M. (2011). Prospects for elimination of childhood tuberculosis: The role of new vaccines. *Archives of Disease in Childhood, 96*, 851–856.
32. Zhang, Y., Wallace, R. J., Jr., & Mazurek, G. H. (1995). Genetic differences between BCG substrains. *Tubercle and Lung Disease, 76*, 43–50.
33. Osborn, T. W. (1983). Changes in BCG strains. *Tubercle, 64*, 1–13.
34. Zwerling, A., Behr, M. A., Verma, A., et al. (2011). The BCG world atlas: A database of global BCG vaccination policies and practices. *PLoSMed., 8*(3), e1001012.
35. WHO. (2007). Revised BCG vaccination guidelines for infants at risk for HIV infection. *Weekly Epidemiological Record, 21*, 193–196.
36. Hesseling, A. C., Cotton, M. F., Fordham von Reyn, C., et al. (2008). Consensus statement on the revised World Health Organization recommendations for BCG vaccination in HIV-infected infants. *The International Journal of Tuberculosis and Lung Disease, 12*(12), 1376–1379.
37. Smith, K. C., Orme, I. M., & Starke, J. (2012). The BCG vaccine. In S. Plotkin, W. Orenstein, & P. Offit (Eds.), *Vaccines* (6th ed., pp. 789–811). Philadelphia: W. B. Saunders.
38. Kaufmann, S. H., Lange, C., Rao, M., et al. (2014). Progress in tuberculosis vaccine development and host-directed therapies-a state of the art review. *The Lancet Respiratory Medicine, 2*, 301–320.
39. Orme, I. M. (2015). Tuberculosis vaccine types and timings. *Clinical and Vaccine Immunology, 22*, 249–257.
40. da Costa, C., Walker, B., & Bonavia, A. (2015). Tuberculosis vaccines—state of the art, and novel approaches to vaccine development. *International Journal of Infectious Diseases, 32*, 5–12.
41. Brennan, M. J., Clagett, B., Fitzgerald, H., et al. (2012). Preclinical evidence for implementing a prime-boost vaccine strategy for tuberculosis. *Vaccine, 30*, 2811–2823.
42. Vilaplana, C., Gil, O., Caceres, N., et al. (2011). Prophylactic effect of a therapeutic vaccine against TB based on fragments of Mycobacterium tuberculosis. *PLoS One, 6*, e20404.
43. Orr, M. T., Beebe, E. A., Hudson, T. E., et al. (2014). A dual TLR agonist adjuvant enhances the immunogenicity and protective efficacy of the tuberculosis vaccine antigen ID93. *PLoS One, 9*, e83884.
44. Tiberi, S., Mu˜noz-Torrico, M., Duarte, R., et al. (2018). New drugs and perspectives for new anti-tuberculosis regimens. *Pulmonology, 24*(2), 86–98.
45. Pontali, E., Sotgiu, G., D'Ambrosio, L., et al. (2016). Bedaquiline and multidrug-resistant tuberculosis: A systematic and critical analysis of the evidence. *The European Respiratory Journal, 47*, 394–402.
46. WHO treatment guidelines for multidrug- and rifampicin-resistant tuberculosis, 2018 update. Geneva.
47. World Health, Organization. (2016). *The use of delamanid in the treatment of multidrug-resistant tuberculosis in children and adolescents: interim policy, guidance*. Geneva: World Health Organization.
48. Lienhardt, C., Raviglione, M., Spigelman, M., et al. (2012). New drugs for the treatment of tuberculosis: Needs, challenges, promise, and prospects for the future. *The Journal of Infectious Diseases, 205*(Suppl. 2), S241–S249.
49. WHO treatment guidelines for drug-resistant tuberculosis, 2016 update. Geneva.
50. Tasneen, R., Betoudji, F., Tyagi, S., et al. (2015). Contribution of Oxazolidinones to the efficacy of novel regimens containing Bedaquiline and Pretomanid in a mouse model of tuberculosis. *Antimicrobial Agents and Chemotherapy, 60*(1), 270–277.
51. TB Alliance. Nix TB factsheet. www.tballiance.org/downloads/NixTB/NixTB_factsheet.pdf. Accessed on 25.01.2019.
52. Silva, D. R., Dalcolmo, M., Tiberi, S., et al. (2018). New and repurposed drugs to treat multidrug- and extensively drug-resistant tuberculosis. *Jornal Brasileiro de Pneumologia, 44* (2), 153–160.

53. Central TB Division; Ministry of Health & Family Welfare; Government of India. (2017). National strategic plan for tuberculosis elimination 2017–2025 (draft). New Delhi. http://tbcindia.gov.in/WriteReadData/NSP%20Draft%2020.02.2017%201.pdf. Accessed 25.01.2019.
54. U.S. Food and Drug Administration. Drug Approvals and Databases. (2019). Available from https://www.fda.gov/drugs/new-drugs-fda-cders-new-molecular-entities-and-new-therapeutic-biological-products/noveldrug-approvals-2019.

Part II

Candida Infections and Therapeutic Strategies

Fungal Diseases and Their Treatment: A Holistic Approach

Sushil Kumar, Tushar Jain, and Dibyendu Banerjee

Abstract

Fungal infections are one of the most common infections in humans and other animals. Fungal diseases are more common in tropical and subtropical countries. Worldwide more than a billion people suffer from fungal infections annually. Fungal pathogens (yeasts and molds) are diverse in their pathogenesis and they can cause mild infections to life-threatening fungal diseases. For the treatment of fungal infections, only a few classes of antifungal drugs are available because there are many similarities between fungal and human cells. Treatment of fungal infections is becoming more challenging because of the emergence of resistance to currently available drugs. Azole drug is the most common class of antifungal drug which is widely used for the treatment of superficial to systemic infections but unfortunately, many fungal pathogens (e.g., *Candida*, *Aspergillus*, *Histoplasma*, and *Paracoccidioides* etc.) have developed resistance. Most of the azole resistant fungi also develop resistance against the echinocandin class of drugs. The most common drug-resistance mechanisms include hyphal switching, alteration of drug targets, increased drug efflux by transporter proteins (ABC transporters or Facilitated diffusion superfamily transporter), and permeability barriers associated with biofilms. Thus, this area is in need of some outstanding work to control drug resistant fungal infections. Recently some new therapeutic approaches such as new formulations for antifungal agents, nanoparticle based drugs and immunotherapy (such as vaccines) are under trial to prevent and treat fungal diseases and inducing the production of host antimicrobial molecules. In this book chapter, we describe common fungal diseases, fungal pathogens, disease-causing factors, drug resistant fungal pathogens, mechanisms involved in the emergence of drug resistance, epidemiology of

S. Kumar · T. Jain · D. Banerjee (✉)
CSIR-Central Drug Research Institute, Lucknow, India
e-mail: d.banerjee@cdri.res.in; dibyenducdri@gmail.com

S. Hameed, Z. Fatima (eds.), *Pathogenicity and Drug Resistance of Human Pathogens*,
https://doi.org/10.1007/978-981-32-9449-3_6

fungal diseases, advanced diagnostic techniques, current and future treatment plans and discuss new approaches for the management of fungal diseases.

Keywords
Fungal infections · *Candida* · *Aspergillus* · Drug-resistance · Azoles · Echinocandins · Pathogenesis

Fungi are single celled or multicellular eukaryotic organisms existing in many different forms such as free living, symbiotic and parasitic forms [1]. *Fungi can have both advantageous as well as adverse effects on human health* [2]. *Fungi like Candida albicans, Cryptococcus spp. etc., which are a part of normal micro-flora, can be considered harmless but they are also a frequent cause of opportunistic infections in case of immune-compromised host. Fungal infections might be contributing substantially to human morbidity and mortality* [3] *but the impact of these diseases on human health is not widely appreciated. There are various reasons for fungal infections which include exposure to hot and humid environment, bad hygiene, or a weak host immune system. The improper and indiscriminate use of antifungals has led to an increase in incidences of drug-resistant fungal infections* [4]. *Hence, it becomes important for us to focus for the development of efficient diagnostic tests as well as safe and effective new anti-fungal drugs and vaccines. In this chapter, community based approaches for preventing and limiting the spread of fungal infections have also been discussed. Nosocomial (hospital acquired) infections leading to septicaemia and death also need to be controlled by spreading proper awareness and hygiene among patients and healthcare workers. A better understanding of the host-pathogen relationship has provided a clue for research into new anti-fungal drugs and vaccines, and fungal vaccines are on the anvil to become part of a new repertoire for prevention or treatment of fungal infections in the near future.*

Introduction

Gut microbiota refers to the group of microorganisms residing in the mammalian gastrointestinal tract. The human microbiota is a collection of bacteria and fungi that render several beneficial functions. An interesting fact about the bacteria residing in human gut is that they are not alone but with at least 100 distinct types of fungi. Gut of human is home to >50 genera of fungi with *Candida, Saccharomyces* and *Cladosporium* species being particularly common. The balance of power between bacteria and fungi might be the key to our gut health. Our gut microbiota helps in digestion and keeps the host healthy by protecting against infections. Therefore they may also be referred to as probiotics [5]. Hence, probiotics might be defined as live microorganisms which when administered in adequate amounts, confer a health benefit for the host [6] like helping to digest food and preventing diarrhoea and irritable bowel syndrome (IBS). Many bacterial strains such as *Lactobacillus*,

Bifidobacterium, *Enterococcus*, *Streptococcus*, *Pediococcus*, *Leuconostoc*, *Bacillus*, *Escherichia coli* genera are considered to be probiotic. However, in spite of the health benefits, some of the microbiota can cause mild to severe infections; especially in conditions of immune-suppression (for e.g. *Candida albicans*, haemolytic *Streptococcus*, pneumococcus *Haemophilus influenzae* type B).The symptoms of fungal infections depend on the type of infection and location within the body. Some types of fungal infections can be mild, such as a rash or a mild respiratory illness. However, other fungal infections can be severe, such as fungal pneumonia [7] or bloodstream infection, and can lead to serious complications such as meningitis or death. On a cellular level the human body is outnumbered by microbes by ratio of 1:10. The large number of gut microbiota is often considered as a human microbial organ. From an ecological point of view it could be argued that humans are a super-organism, a communal collective of human and microbial cells working as a single unit. The indiscriminate use of prescription antibiotics, compounded with non-compliance or improper usage by patients is leading to an increase in cases of drug resistance, causing problems in the treatment of various diseases including fungal infections. Multidrug Resistance (MDR) as a phenomenon is defined as resistance against a spectrum of drugs that share neither a common target nor a common structure. It was first described several years ago, when Biedler and Riehm (1970) noted that cell lines made resistant to Actinomycin D or Vinca alkaloids displayed cross-resistance to a wide range of other components. In medical terms, drug resistance is defined as the persistence or progression of an infection inspite of the appropriate drug treatment given to the disease. Therefore, to win the race against drug resistance, it is important to be one step ahead of the pathogen and to continuously come up with newer targets and strategies to beat the infection.

Fungi

Fungi are multicellular eukaryotic or unicellular [8], heterotrophic organisms which can play an important role in the nutrient cycling in an ecosystem. They can reproduce either through asexual mode or sexual mode of reproduction or both. They can form association with bacteria, plants and animals.

Type of Fungi

On the basis of morphology fungi can be classified into three groups

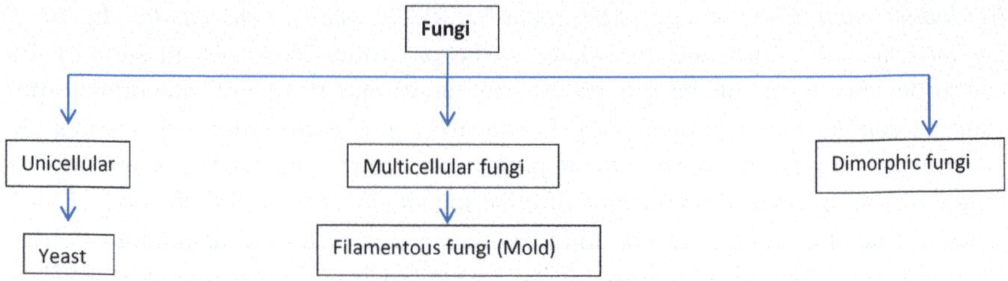

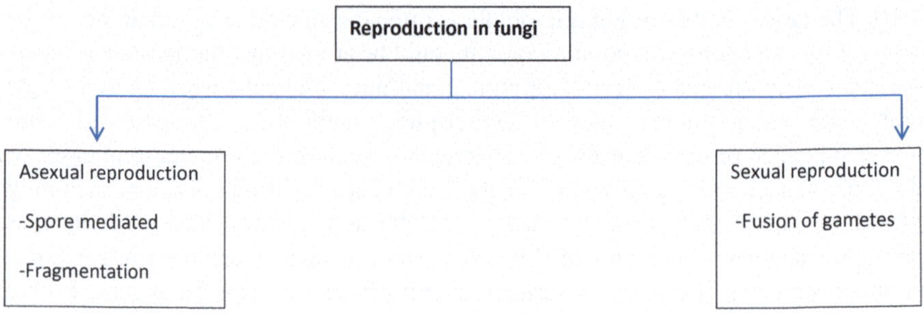

How Fungi Interact with Human?

The interactions between human and fungi can be both beneficial and harmful

- **Beneficial Effects of Fungi**
  - They carry out the process of decomposition which leading to nutrient and carbon recycling [9].
  - They act as biosynthetic factories that can be used for the production of drugs, antibiotics, alcohol, acids as well as food.
  - Agaricus and Marchella are two most common fungi which are usually used as a food supplement for human beings.
  - Moulds of the genus *Penicillium* ripen many cheeses.
  - *Saccharomyces cerevisiae*, also known as baker's yeast, is an important ingredient in bread [10].
  - Fungi naturally produce antibiotics (for e.g. penicillin and cephalosporins) that kill or inhibit the growth of other harmful pathogens [11].

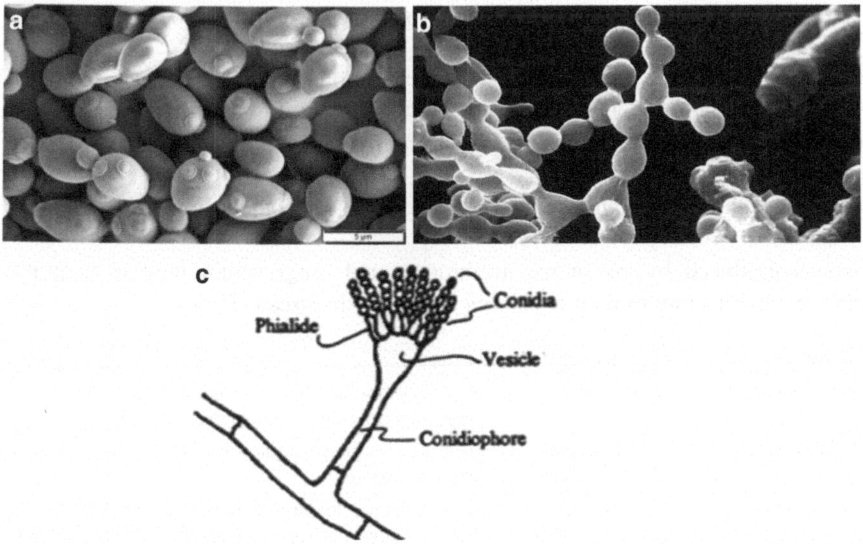

Fig. 1 (**a**) Scanning electron microscopic (SEM) image of yeast Saccharomyces cerevisiae (Mogana Das Murtey and Patchamuthu Ramasamy 2015). (**b**) Scanning electron microscopic (SEM) image of Neurospora crassa (Steven M. Carr 2014). (**c**) Light microscopic image of Aspergillus species (John W. Taylor 1993)

Model organisms such as *Neurospora*, *Saccharomyces* and *Aspergillus* can used for biochemical and genetic studies related to research [12, 13] (Fig. 1).

- **Harmful Effects of Fungi**
  - Fungal activities can cause spoiling of food, lumber, paper, and cloth.
  - Fungi cause many different kinds of diseases in animals as well as human, including allergies [14].

S.No.	Name of fungi	Fungal associated diseases
1.	*Candida albicans*	Candidiasis
2.	*Aspergillus fumigatus*	Aspergilosis
3.	*Histoplacapsulatum*	Histoplasmosis
4.	*Coccidiodesimmitis*	Coccidiomycosis
5.	*Blastomysisdermitidis*	Blastomycosis
6.	*Sporotricumscheneki*	Sporotrichosis
7.	*Dermatophytes*	Dermatomycosis

Pathogenesis of Fungi

- They can penetrate host barriers.
- Form capsule (mucopolysaccharide), that can inhibits phagocytosis of the yeast.

- Most of human pathogenic fungi well adopted to Grow at 37 °C temperature which is the human body temperature.
- Pathogenic fungi have morphological variability [15] (e.g., yeasts, hyphae, spherules, and sclerotic bodies) to survive in different tissue environment.
- They can produce harmful enzymes (Keratinase [16], phenyl oxidase) that damage host tissues.
- Some pathogenic fungi also inhibit phagosome-lysosome fusion.

Toxins produced by poisonous mushrooms and fungi within food (e.g., grain, cheese, etc.) leading to food poisoning and other diseases [17].

Toxin	Toxicity	Effects
Alpha-amanitin	Fatal	Causes often fatal liver damage 1–3 days after ingestion. Principal toxin in the death cap
Phallotoxin	Not fatal but infectious	Causes extreme gastrointestinal upset. It is found in various mushrooms
Orellanine	Fatal	Redox cycler similar to paraquat. Causes kidney failure within 3 weeks after ingestion. Principal toxin in genus *Cortinarius*
Muscarine	Potentially fatal	Causes SLUDGE syndrome. Found in various mushrooms. Antidote is atropine
Monomethylhydrazine (MMH)	Fatal	Causes brain damage, seizures, gastrointestinal upset, and hemolysis. Metabolic poison. Principal toxin in genus *Gyromitra*. Antidote is large doses of intravenous pyridoxine hydrochloride [22]
Coprine	Not fatal	Causes illness when consumed with alcohol. Principal toxin in genus *Coprinus*
Ibotenic acid	Potentially fatal	Excitotoxin. Principal toxin in *Amanita muscaria*, *A. pantherina*, and *A. gemmata*
Muscimol	Psychoactive	Causes CNS depression and hallucinations. Principal toxin in *Amanita muscaria*, *Amantiapantherina*, and *Amantiagemmata*
Psilocybin and psilocin	Psychoactive	Causes CNS arousal and hallucinations. Principal effects in psilocybin mushrooms, many of which belonging to the genus *Psilocybe* (often used recreationally)
Arabitol	Non-lethal	Causes diarrhea in some people
Bolesatine	Non-lethal	Causes gastrointestinal irritation, vomiting, nausea
Ergotamine	Deadly	Affects the vascular system and can lead to loss of limbs and/or cardiac arrest. Found in genus *Claviceps*

Sources of Fungal Infections

Fungi are omnipresent and can be found in soil, on plants, trees, and other vegetation as well as on our skin, mucous membranes, and intestinal tracts [18]. Most fungi are not dangerous, and some can be helpful for production of penicillin, bread, wine, and beer, etc.

The reproductive form of fungus is spores which are similar to the seeds of a plant. Spores extremely resistant to external stimulus such as heat, cold or medications and can survive in adverse conditions for longer periods. Furthermore, spores can be transmitted from one person to another and can spread fungal infection in this way [19]. Thus, treatment should also always aim to wipe out spores completely; otherwise the disease may flare up again or may transmit to the other people.

Type of Infections Caused by Fungal Pathogens

Mycoses are diseases caused by fungi and those affecting humans can be divided into four groups based on their penetrance level into the body tissues:

Fungal diseases on the basis of site of infection in the human body

- **Superficial mycoses**:- These are caused by fungi that grow on the surface of the skin or hair [20, 21].

S.No.	Fungal pathogen	Diseases conditions
1.	*Trichophyton rubrum*	Tinea corporis, tinea cruris, tinea pedis, tinea manuum, and onychomycosis and Majocchi granuloma
2.	*Trichophyton mentagrophytes varinterdigitale*	Tinea corporis, tinea cruris, tinea pedis (interdigital), tinea manuum, and onychomycosis
3.	*Epidermophyton floccosum*	Tinea cruris, tinea pedis, tinea manuum, and onychomycosis
4.	*Microsporum gypseum*	Inflammatory lesions
5.	*Trichophyton violaceum*	Tinea capitis—endothrix
6.	*Trichophyton verrucosum*	Tinea barbae, capitis, and kerion
7.	*Trichophyton nanum*	Tinea capitis, tinea corporis, and tinea cruris
8.	*Trichophyton concentricum*	Tinea imbricata
9.	*Trichophyton schoenleinii*	Favus
10.	*Trichophyton soudanense*	Tinea capitis
11.	*Microsporum audouinii*	Tinea capitis-more common in Europe

- **Cutaneous mycoses or dermatomycoses: –** Cutaneous mycosis is that include such infections as athlete's foot and ringworm, where growth occurs only in the superficial layers of skin, nails, or hair [22].

S.No.	Causative organisms	Disease
1.	Dermatophytes (*Arthroderma, Lophophyton, Microsporum, Nannizzia, Trichophyton, Epidermophyton*)	Dermatophytosis, ringworm of the scalp, glabrous skin and nails
2.	*Candida, Debaryomyces, Kluyveromyces, Meyerozyma, Pichia*, etc.	Candidiasis of skin, mucous membranes and nails
3.	Non-dermatophyte moulds	Dermatomycosis
	Neoscytalidium, Scopulariopsis	

- **Subcutaneous mycoses: –** Subcutaneous mycoses are those that penetrate below the skin to involve the subcutaneous, connective, and bone tissue. Subcutaneous mycoses can occur in healthy individuals whose common examples are sporotrichosis, chromoblastomycosis, phaeohyphomycosis, eumycoticmycetoma, and hyalohyphomycosis [23].

S.No.	Causative organisms	Disease
1.	*Sporothrix* spp.	Sporotrichosis
2.	*Fonsecaea, Phialophora, Cladophialophora* etc.	Chromoblastomycosis
3.	*Cladophialophora, Exophiala, Bipolaris, Exserohilum* etc	Phaeohyphomycosis
4.	*Scedosporium, Madurella, Trematosphaeria, Acremonium, Exophiala* etc.	Mycotic mycetoma
5.	*Basidiobolusranarum, Conidioboluscoronatus*	Subcutaneous zygomycosis (Entomophthoromycosis)
6.	*Rhizopus, Mucor, Rhizomucor,Lichtheimia, Saksenaea* etc.	Subcutaneous zygomycosis (Mucormycosis)
7.	*Loboaloboi*	Lobomycosis
8.	*Rhinosporidium seeberi*	Rhinosporidiosis

- **Systemic or deep mycoses: –** Systemic or deep mycoses can penetrate internal organs and become widely disseminated throughout the body. This type is often fatal [24].

S.No.	Causal organism	Disease
1	*Histoplasma capsulatum*	Histoplasmosis
2.	*Coccidioide simmitis*	Coccidioidomycosis
3.	*Blastomyces dermatitidis*	Blastomycosis
4.	*Paracoccidioides brasiliensis*	Paracoccidiodomycosis
5.	*Penicillium marneffei*	Penicilliosis

(continued)

6.	*Candida species*	Candidiasis
7.	*Aspergillus species*	Aspergillosis
8.	*Cryptococcus species*	Cryptococcosis
9.	*Zygomycetes species*	Zygomycosis

An estimate of the percentage of various fungal species causing invasive mycoses [25]

S.No.	Pathogens	% of invasive mycosis
1	*Candida*	70–90%
2	*Aspergillus*	10–20%
3	All others (*Cryptococcus, Mucorales, Fusarium, Scedosporium,* etc.)	~5%

How Common Are Fungal Diseases?

Fungal infections are extremely common and their treatment and caring for patients with is becoming an increasing economic burden. An estimated numbers of patients who will need expensive antifungal drugs for some of the life-threatening fungal illnesses are summarised here.

- Candidaemia infections are estimated to have incidence rate of 300,000 worldwide per year, with a mortality of 30–55% [25].
- Invasive aspergillosis can occur in different patient groups. It has been estimated that around 10% of new leukaemic cases might develop invasive aspergillosis (30,000 per year). Stem cell transplants are prone to aspergillus infections. It has been estimated that out of 54,000 of stem cell transplants done annually in USA, UK, Europe and Japan, 5400 are found to be infected and need urgent treatment [26].
- According to a research, it has been predicted that in chronic obstructive pulmonary disease 1.2% will need antifungals for aspergillosis (216,000 per year). Over 50% of invasive aspergillosis patients will die from their infection, even when they receive treatment [26].
- AIDS patients are prone to contract cryptococcal meningitis infections with an incidence rate of 600,000, 70% of which are in the sub-saharan Africa [27].
- Infections of low potential also occur which affect a large number of people worldwide. The examples of such infections include cutaneous fungal infections, nail infections and athletes' foot. These affect some 1.5 billion people, or 25% of the world's population.
- *Tinea capitis* is a kind of hair infection that is common in young children and is predicted to affect about 200 million population worldwide.

Infection on the Basis of Affected Organs of the Human Body

- **Fungal infections of the skin:** -Entry of fungal spores into our skin is an easy process, especially if the skin is broken [28]. These fungal infections produce spores which can be easily spread to the environment when the infected skin flakes off. If anyone comes into contact with these flakes, these spores may cause a new fungal infection through any cut on the skin of the new host. If yeast infects the skin on a baby's bottom area, it causes a diaper rash. One can pick up a Fungal infection on the foot that can be caused while walking bare foot in a humid environment. This is the reason why many types of skin infection have names like "athlete's foot" or "swimmers eczema". Athelete's foot occurs in sports persons (who sweat a lot) or in people living in damp and humid conditions. It is, in fact, a very common fungal infection which generally occurs between the toes and is extremely contagious. One can get athlete's foot just by using a communal bathing place or even sharing a towel. If left untreated the condition can spread to the soles and sides of the foot and to the toenails. Athlete's foot may be further complicated by a bacterial infection, accompanied by an unpleasant smell [29].

  Signs and symptoms of Athlete's foot often include:
  - Cracking and peeling of skin between the toes,
  - Inflammation,
  - Areas of white, dead skin,
  - Moistness or blistering,
  - Itching or soreness.

  Pets act as another source for skin infections. At first the fungal infections in animals might not be noticeable but can be manifested once they become infected through contact with their skin. Children often get infection by cuddling guinea pigs, dogs, cats or other pet animals. People who spend a lot of time around horses are prone to fungal infections on their faces while brushing the horses.

  The principle etiological agents for fungal infections are dermatophytic moulds belonging to the genera *Microsporum, Trichophyton* and *Epidermophyton* which cause ringworm or tinea of the scalp, glabrous skin and nails; *Malassezia fufur,* a lipophilic yeast responsible for pityriasis versicolor, follicular pityriasis, seborrhoeic dermatitis and dandruff; and *Candida albicans* and related species, causing candidiasis of skin, mucous membranes and nails [30].
- **Fungal infection of the mouth:-**The normal flora of the mouth comprises of fungal yeasts along with other microorganisms which means that under normal circumstances these yeasts grow silently, maintaining a balance with other microorganisms like bacteria, on the mucous membranes without doing any harm. Problems arise in case of immune compromised hosts [30]. This could occur after a bout of bad flu, chemotherapy, or treatment with immune-suppressant medication, e.g. after a transplant or AIDS. A fungal infection in the mouth may be regarded as the first sign of the onset of AIDS. Thus fungal infection is often an added problem for people already suffering health problems. People with false teeth are also susceptible to fungal infections because in people with false teeth, the mucous membrane of the mouth is often compromised, and

the fungus stands at a greater chance to inhabit the area. The fungus can even live in the false teeth itself and infect the mouth repeatedly. Oral thrush [31], a kind of fungal infection of the mouth is quite common among babies. Babies not only have a chance to get infected with *Candida* in the birth canal but also through contact with people's hands. Owing to the weak immune system of the baby, infection by *Candida* is able to take hold more easily and effectively.

- **Fungal infections of the upper respiratory tract (Throat infection)**
  Candida usually causes an infection in the throat that is called *Candida* esophagitis [32]. This type of infection commonly occurs in those people having immune-compromised or weak immune systems due to disease or disease treatments, such as HIV [33], leukaemia or chemotherapy.
- **Fungal infection in the lower respiratory tract (Lungs infection)**
  Fungi of the genus *Candida, Aspergillus* and *Cryptococcus* sometimes infect the lower respiratory tract especially in persons suffering from diseases such as cancers of the blood, bone marrow, and lymph nodes, or those with human immunodeficiency virus (HIV) infection [34]. Fungal infection is also common in those patients who take immunosuppressive drugs. Invasive pulmonary aspergillosis and systemic candidiasis are the most prevalent opportunistic fungal infections.
- **Fungal infection in the urogenital tract**
  Vulvo-vaginal candidiasis (VVC) is the most common cause of vaginal fungal infections [35]. It is characterized by itching, burning, soreness, pain during intercourse and/or urination, and vaginal discharge [36]. Men can get a yeast infection, too. This is more likely in men who are not circumcised. The symptoms of yeast infection in men include red rash on penis and itching and burning on the upper part of penis. The imbalance in the composition of the vaginal flora is one of the most common reasons for fungal infection of the vagina. VVC is caused more in the late phase of luteal cycle when estrogen and progesterone levels are high. The mucous membrane of the vagina normally self-protects by maintaining a slightly acidic pH. The reason for imbalance in the vagina may occur due to a sudden decrease in acidity, or because of using too much soap when taking a bath. The alkaline nature of the soap can sometimes make the vaginal environment less acidic and this may increase chances of fungal infections. Use of intra-uterine device or coil device for contraception may also sometimes increase the risk of bacterial and fungal infections. Vaginal candidiasis is rarely associated with serious underlying health problems in women of reproductive age. However, some women experience recurrent infections that are difficult to treat, decreasing their quality of life. It has been reported in research studies that recurrent infections sometimes contribute to clinical depression, thus decreasing the quality of life. However, most of the vaginal Candida can be treated with a single dose of fluconazole or with clotrimazolepessaries. Applying clotrimazole creamtwo to three times a day can be helpful to get rid of local irritation. Though probiotic preparations are sometimes helpful in treating vaginal thrush, but there is no clear evidence in support of their efficacy.

- **Infections of the internal organs**
  If the fungal hyphae penetrate deeper into the tissue, it leads to a serious condition known as systemic mycosis. This type of infection can be extensive and frequently life-threatening and is difficult to treat. *Aspergillus* which is a common fungus is one of the causes of lung and systemic fungal infections. Spores of *Aspergillus* is common in our environment and infections can be life-threatening and difficult to treat; particularly for people with a compromised immune system. These fungal spores can easily grow on construction and demolition sites and in cellars, stables, bird shelters and grain storage areas as well as in furniture covers, wallpaper, cane furniture, compost and household dust. *Aspergillus* infections are seen as a professional risk. In addition to pigeon flyers and farmers, the infection can also affect beer brewers (working with grain), furniture makers and construction workers. The disease can be controlled if hay is not left after cutting in the open air but stored in conditions. Another notorious source of infection is from air-conditions installed in hospitals. Many patients who have just been operated on, or have had chemotherapy to treat cancer end up having an *Aspergillus* infection. Systemic infections with *Candida* species are also common. Disseminated candidiasis (candidaemia and invasive candidiasis) is a life-threatening syndrome with an attributable mortality of 10–50%. Amphotericin B deoxycholate, a first line agent for treatment of disseminated candidiasis is an effective agent for this syndrome but its use is limited by toxicity that it poses while treatment [37].
- **Fungemia**
  Fungemia is a condition in which fungi is present in blood of human beings. *Candida* [38], *Saccharomyces*, *Aspergillus*, *Histoplasma* and *Cryptococcus* are major fungal pathogens that cause Fungemia [39].It is more common in immunocompromised patients.

Epidemiology of Pathogenic Fungi

Special human activities and changes in human behaviour are the major independent factors which are responsible for different level of susceptibilities in individual populations. Nosocomial infections [40] are transmitted in hospital through three main resion such as environmental routes-air, surface contact and water. According to current report of epidemiology of fungi, the mortality rate of fungal diseases are approximate 1.5 million people and the morbidity rate affect over a billion people [41]. Fungal infections are usually more severe in those patients who already have health issues such as asthma, AIDS, cancer, organ transplantation and corticosteroid therapies [42].

Global estimate of fungal diseases

S.No.	Name of disease	Global estimate (No. of cases reported annually)
1	Chronic pulmonary aspergillosis	3,000,000
2	Cryptococcal meningitis complicating HIV/AIDS	~223,100

(continued)

3	Invasive candidiasis	~700,000
4	Pneumocystis jirovecii pneumonia	~500,000
5	Invasive aspergillosis	~250,000
6	Histoplasmosis	~100,000
7	Fungal asthma	10,000,000
8	Fungal keratitis	~1000,000

Epidemiological studies suggested that patients who are in the intensive care unit they have greater risk of *Candida* infection but according to current report, the number of patients outside intensive care units are also infected through *Candida* with increasing rate.

- Invasive fungal diseases are the most common cause of large amount of mortality and morbidity especially in immunocompromised patients.
- Autopsy report is the best valuable tool for defining the correct epidemiology of diseases because it explains the pattern of fungal disease [43].
- From the last 20 years the occurrence of Aspergillosis significantly decreased where as Candiasis is increased.
- In tropical and subtropical countries superficial fungal infections are more common.
- According to WHO report, worldwide 20–25% superficial fungal infections are found.
- Among non Dermatophytic fungal infections candidiasis, aspergillosis, and zygomycosis are more common.
- Among *non-albicans candida, Candida tropicalis* is the most common yeast which can cause invasive candidiasis [44].
- According to Chakrabarty, India has only 71 well equipped laboratories that test fungal infections. He said India needs at least 1000 well equipped laboratories for management of such kind of infections.

S.No.	Name of hospital who recorded fungal infections	Chance of fungal infection in every 1000 patients
1.	Global Hospitals of Hyderabad	39.55 people
2.	Delhi's Safdarjung Hospital	32.75 people

Source: PGIMER Chandigarh

S.No.	Place where the fungal infections are more common	Total number of cases out of every 1000 patients
1.	India	1–12
2.	South Asia	
3.	US	0.8
4.	Europe	0.2
5.	Australia	0.9

Source: PGIMER Chandigarh

Name of the fungal pathogens and the place where they are abundant

S.No.	Name of fungal pathogens	Place where they are abundant
1.	*Fusarium infections*	France and Italy
2.	*Scedosporium prolificans*	Spain and Australia
3.	*Geotrichum capitatum*	Mediterranean countries

Epidemiology of fungal infections in India

S.No.	Type of fungal infection	% of occurrence of disease	Most common pathogen	Total number of patients and time of study
1	Superficial fungal infection	27.6% (82/297)	*Tinea corporis* (78%)	15,950 in 1 year of study
2	Dermatophytosis	75.6% (62/82)	*Trichophyton rubrum* (79%)	
3	Non-dermatophytosis	24.4% (20/82)	*Candida* (60%)	

Azole drug resistant biofilm forming albicans and non albicans species of *Candida*

S.No.	Candida species	Biofilm
1.	*Candida albicans*	+
2.	*Candida pseudotropicalis*	+
3.	*Candida parapsilosis*	+
4.	*Candida. Glabrata*	+

Diversity in the fungal pathogens that cause neurological disorders specially in immunocompromised patients

S.No.	Fungal pathogens	Disease subtypes
1.	*Cryptococcus neoformans*	Fungal meningitis
2.	*Candida albicans*	
3.	*Coccidioidesimmitis*	
4.	*Histoplasma capsulatum*	
5.	*Aspergillus* spp	Mass lesions in brain
6.	Zygomycetes	
7.	Melanized fungi	
8.	*C. neoformans*	True neurotropic fugi
9.	*Cladophialophorabantiana*	
10.	*Exophialadermatitidis*	
11.	*Ramichloridiummackenzie*	
12.	*Ochroconisgallopava*	

(Source:-ArunalokeChakrabarti, PGMIR Chandigarh)

Diversity in the specimens used for detection of fungal infections [45]

S.No.	Name of specimen	Chance of occurrence of fungal infections
1.	Sputum	29.6%
2.	Skin swabs	25.7%
3.	Urine	15.4%
4.	Blood	10.35%
5.	Broncho alveolar lavage (BAL)	9.5%

Diagnostic Techniques for Fungal Diseases

The diagnosis and treatment regimens for fungal diseases are very different.

Diagnosis of Superficial Mycoses

- Symptomatic based diagnosis
- Skin scrapings from the affected area of the body or fragments of infected hairs or hair follicles or nail should be collected as specimens for the laboratory microbiological examination.
- Microscopic examination of infected tissue or culture. Perhaps the most commonly used first line method for detection when samples are available for such examination. Requires trained person and can be done with unstained (wet mounts) or stained preparations.
- Culture technique:-Specimens are inoculated on agar culture plates and the organisms are often identified by the measuring typical size and shape of the cells by using appropriate tools and techniques.
- Fungi are slow growing organisms that may take up to 2–3 weeks of time to grow on culture medium; before this time, the clinician cannot provide any definitive treatments related to the pathogenic infection. Before the final diagnosis, doctors may prescribe a broad-spectrum antifungal medication to the patient.

Diagnosis of Systemic Fungal Infections

Systemic fungal infection is not easy to diagnose because symptoms frequently do not appear until the patient has developed severe illness. At present *Histoplasma*, *Cryptococcus*, *Candida* and *Aspergillus* are the major fungal pathogens responsible for systemic mycoses.

- ***Microscopy***
- ***Cultural techniques:*** **–** Blood and urine culture [46]
- ***Serological diagnosis***
- ***Molecular diagnosis***

- **Laser Microdissection:** Laser microdissection is the highly sensitive advanced microscopy based laser technology which is used for the study of specific cell types. This technology is highly useful for selective isolation of and detection of

selected cells of interest such as cells which carry antigens in the DNA or RNA or Proteins [47].
- Application of this technology is very common in the following area:-Generation of cDNA library, profiling of RNA transcript etc.
- Blood and urine culture. It is a relatively lengthy process but very useful to determine species and antibiotic resistance profile of the fungi under question.

Serological Diagnosis of Fungal Diseases

- **Immuno-Histochemistry(IHC):** IHC is the most common technique which is used for detection of fungal infection in the intact tissue sections by using antibodies to detect desired fungal antigen thus we can examine the morphology of the target as well as the surrounding tissues.
- **Galactomannan test:** Galactomannan is an essential component of the cell wall of most filamentous fungi such as *Aspergillus.* This component of fungi is release during growth of most of filamentous fungi. Now it is used as a diagnostic marker. For example examination of *Aspergillus* infection in human is detected by checking the level of galactomannan in blood [48]. This test is performed by using double-sandwich ELISA that was approved by the FDA in 2003. The test is more appropriate in patients who had hemopoetic cell transplants [49].
- **BDG(1, 3)-β-D-glucan test:** BDG is the essential component of most fungi except *Cryptococcus* and *Zygomycetes;* therefore this component is used for screening test for invasive fungal diseases [50].

Molecular Diagnosis of Fungal Diseases
The following are some of the more common laboratory techniques used for diagnosing fungal infections [51].

- **PCR-Based Methods:** Polymerase Chain Reaction is a highly sensitive molecular technique for the amplification of fungal DNA using DNA-primers for identifying the species causing the infection. The success rate of PCR is ~95%.
- **In Situ Hybridization:** In this technique probe based detection of fungal nucleic acid is carried out while preserving the tissue morphology.

Treatment of Fungal Diseases
Antifungal drugs are also called anti-mycotic drugs. They belong to various classes of compounds and are able to treat both superficial and deep-seated infections. The common antifungal groups of compounds that are used for treatment include:

- Polyenes, including amphotericin B and nystatin
- Immidazoles, which include fluconazole, ketoconazole, miconazole, clotrimazole and others,
- Anti-metabolites such as Flucytonsine (Ancobon)
- Combination therapy with more than one compatible antifungal drugs
- Antifungal vaccines are not yet available but are on the anvil [52].

List of currently available anti-fungal drugs which are available for the treatment of common fungal infections [53]

Antifungal class	Drugs in the class	Medical usage
Polyenes	Amphotericin B	Oral preparations of amphotericin B are used to treat thrush. The oral preparations are non-toxic in contrast to typical intravenous (IV) doses. Intravenously used for treating various systemic fungal infections (e.g., in critically ill, co-morbidly infected or immune-compromised patients), including Cryptococcal meningitis
	Nystatin	It is a topical antifungal drug
	Natamycin	It can be applied as a topical ophthalmic agent
Analogues of pyrimidine	5-Fluorocytosine (5FC)	It can be applied for the treatment of *Candida* and *Cryptococcus* fungal infection. Sometimes it can also use against moulds such as *Aspergillus* spp.
Imidazole's	Benzimidazoles	They act by binding to the fungal microtubules and stop hyphal growth. They also bind to the spindle microtubules and block nuclear division
	Clotrimazole	It is effective against fungal infections such as
		Oral candidiasis, vaginal yeast infections, and ringworm, treat athlete's foot and jock itch etc.
Tri-Azoles	Ketoconazole	It can give orally to the patients. It is highly effective treatment for oral candidiasis, coccidiomycosis, dermatophyte infections, endemic mycoses
	Fluconazole	Fluconazole is water-soluble;it is extensively used for the treatment of superficial and invasive candidiasis, cryptococosis
	Iitraconazole	It is first oral drug for aspergillosis. It has a broad spectrum antifungal activity against volvo-vaginal candidiasis and oral candidiasis, blastomycosis, sporotrichosis, histoplasmosis, and onychomycosis. Toxicity is the major issues with this drug
	Voriconazole	It is a broad spectrum antifungal agent which can be used against *Aspergillus* spp., *Fusarium* spp. and *Candida* spp. its side effects are more than those of fluconazole
	Posaconazole	Posaconazole is structurally similar to itraconazole. It has broad spectrum antifungal activity against *Aspergillus* spp. and *Candida* spp. Fatty food increase the Oral bioavailability
Allylamines	Terbinafine	It belongs from allylamine class which have clinical use. Terbinafine is the better option for the treatment of dermatomycosis
Echinocandins	Cilofungin	Cilofunginis one of the echinocandins which is used clinically as antifungal agents
	Caspofungin	Caspofunginis FDA approved antifungal drug which can block the synthesis of fungal cell wall by inhibiting β-1,3-glucan synthesis. Intravenous is the beast mode of injection of Cospofungin. e.g. A*spergillosis*

(continued)

	Micafungin	Intravenous mode of injection of micafungin is the best way to deliver drug in the human body. It is used for the treatment of candidiasis
	Anidulafungin	It is the derivative of echinocandins B. It is used for the treatment of oesophageal candidiasis, invasive *Candida* infection as well as in invasive *Aspergillus* infection

Caution: This table is for reference only. Medication should be taken only on the advice of a certified medical practitioner who can prescribe the correct medicine at the right dose for the optimum length of time

Current Trends of Available Drugs against Fungi

Treatments of fungal diseases is getting difficult day by day due to increase of multidrug resistance properties in the fungal pathogens. The discovery of novel drugs to treat fungal infections is slow because the chance of discovering potential drugs that would be highly effective against fungal pathogens and produce fewer side effects in patients is difficult due to the availability of few unique druggable targets.

Multidrug Resistance in Pathogenic Fungi

Multidrugresistant fungal infections are the major challenge to treat fungal associated diseases for all clinicians and healthcare workers. Worldwide, drug resistant fungi are responsible for high rates of morbidity and mortality specially in immunocompromised patients [54]. The resistance may be intrinsic, randomly acquired or nosocomial. Multidrug resistant pathogenic strains passes different mechanisms to develop drug resistance ranging from genetic mutations to metabolic adaptations to the action of several antifungal compounds available for treatment of fungal associated diseases. Major limitations of the currently available antifungal drugs are drug-drug interactions and toxicities thus we cannot use these drugs for prolong time. Currently Azole-drug resistance in *Candida albicans* and non-*albicans Candida* is the major problem to treat candidiasis. Resistance in *C. glabrata* is usually occurring against echinocandins. Azole resistance is also reported in *Aspergillus fumigatus.*

Mechanisms of Drug Resistance in Fungi

- Mutations in the genes that encode target proteins can play an important role in the pathogenesis of fungi.
- High rate of expression of multidrug efflux pumps altering the stoichiometry of the inhibitors [55].

- At the most, in one azole resistant fungus, four altered gene pathways have been identified.
- Pleiotropism is the major factor which induces Multidrugresistance in *Saccharomyces cerevisiae.*
- According to recent reports, molecular chaperone heat shock protein (Hsp90) can change the relationship between genotype and phenotype resulting in fungal cells that can develop drug resistance [56].
- Formation of aneuploidy is another factor which is responsible for drug resistance in fungi.

Ways to Overcome Drug Resistance of Fungi

- Discovery of novel bioactive molecules which can target ergosterol and β-glucan biosynthesis that will solve the drug resistance problem for the treating of fungal associated diseases.
- Innate and adaptive immunity can play an important role for the controlling fungal infections, therefore the vaccines are a very promising viable tool to prevent and treat fungal pathogens.
- The complete study of host pathogen interaction and pathogenic factors such as peptides, glycoproteins, glycolipids, and glycan's which is the component of cell wall and capsule of pathogenic fungi, these antigenic components can be used to selectively induce immunological bioactive molecules which can solve the drug resistance problem of many pathogenic fungi [57].

Future Directions of the Drugs Against Resistance Fungi

Now needs to look at novel therapeutic antifungal compounds which may be achieved from natural or synthetic sources. It is necessary to discover new classes of antifungal agents which can selectively target multi-drug resistant fungi, because treatment of these pathogenic fungal infections is the major challenge for all clinicians, healthcare persons and doctors [58]. Therefore need to find out some novel mode of action carrying molecules which can solve the multidrug resistance problem [59].

General Precautions to Prevent Fungal Infections

1. One should avoid living in damp and humid conditions to prevent fungal infections.

2. One should maintain personal hygiene and should always wear clean clothes and make sure to wash dirty clothes at 60 °C to destroy fungal spores.
3. Healthcare workers should wash their hands frequently before touching patients to avoid the spread of infections including fungal infections.
4. Scratching should be avoided to prevent the spread of spores through nails and air.
5. Fingernails should be cut short.

Community Based Infections May Be Controlled by the Following Measures

- Avoiding the sharing of items which is already used by someone such as clothes, combs, foot wears, gloves, etc.
- Frequent hand washing with soap and water before cooking and serving food,
- Treating infections with proper medication before it spreads to other people,
- Using antifungal sprays, creams and powders on external body surfaces such as socks and shoes which can help prevent infectious agents.

Conclusions

- Fungal diseases are more common in tropical countries than temperate counties. They are very common in a developing country especially in those people who belongs from poor background. Each year millions of cases of fungal diseases are diagnosed which may be fatal or non-fatal for diseased person.
- Approximate 600 different fungi have been reported that can cause infection in humans. The range of infections varies from the common to the fatal. Commonly, fungi infect the mucosa, skin, hair, and nails and sometimes also cause allergies.
- In the area of fungal aetiology, pathology, epidemiology, disease biology and their economic impact are essential to be understood to raise scientific interest and increase global investments into antifungal research. For the betterment of the life of the infected people, it is essential to have some new therapeutic approaches for the treatment of fungal diseases.
- The development of vaccines for fungal diseases is the very interesting area remained open for all the researchers. The developed vaccines can play a great significance in the prevention of fungal diseases worldwide. Till now, no approved human vaccines are available for any fungal pathogen. However, the scientific community is currently testing several vaccine candidates and vaccines against some common fungal pathogens and we hope that vaccines will soon become available.

Acknowledgement The authors would like to acknowledge CSIR (Council for Scientific and Industrial Research) and CSIR-CDRI (Central Drug Research Institute, Lucknow) for funding and providing the logistics for writing this article. This manuscript will bear a CDRI communication number upon acceptance.

References

1. Thomas, H. (2017). Gut microbiota: Intestinal fungi fuel the inflammatory fire in alcoholic liver disease. *Nature Reviews. Gastroenterology & Hepatology, 14*, 385. https://doi.org/10.1038/nrgastro.2017.78.
2. Zhang, F., et al. (2018). The interactions between gut microbiota and entomopathogenic fungi: A potential approach for biological control of Blattella germanica (L.). *Pest Management Science, 74*, 438–447. https://doi.org/10.1002/ps.4726.
3. Grant, I. W. (1969). Human diseases attributable to fungi. *Health Bulletin, 27*, 26.
4. Tap, R. M., et al. (2016). First two cases of fungal infections associated with multi-drug resistant yeast, Fereydounia khargensis. *Mycopathologia, 181*, 531–537. https://doi.org/10.1007/s11046-016-0002-y.
5. Reese, A. T., & Carmody, R. N. (2018). Thinking outside the cereal box: Non-carbohydrate routes for dietary manipulation of the gut microbiota. *Applied and Environmental Microbiology*. https://doi.org/10.1128/AEM.02246-18.
6. Dargahi, N., Johnson, J., Donkor, O., Vasiljevic, T., & Apostolopoulos, V. (2019). Immunomodulatory effects of probiotics: Can they be used to treat allergies and autoimmune diseases? *Maturitas, 119*, 25–38. https://doi.org/10.1016/j.maturitas.2018.11.002.
7. Young, A. Y., Leiva Juarez, M. M., & Evans, S. E. (2017). Fungal pneumonia in patients with hematologic malignancy and hematopoietic stem cell transplantation. *Clinics in Chest Medicine, 38*, 479–491. https://doi.org/10.1016/j.ccm.2017.04.009.
8. Esposto, M. C., et al. (2018). Yeast-like filamentous fungi: Molecular identification and in vitro susceptibility study. *Medical Mycology*. https://doi.org/10.1093/mmy/myy133.
9. Jiang, T. T., et al. (2017). Commensal fungi recapitulate the protective benefits of intestinal Bacteria. *Cell Host & Microbe, 22*, 809–816 e804. https://doi.org/10.1016/j.chom.2017.10.013.
10. Fadhel, M., Patel, S., Liu, E., Levitt, M., & Asif, A. (2019). *Saccharomyces cerevisiae* fungemia in a critically ill patient with acute cholangitis and long term probiotic use. *Medical Mycology Case Reports, 23*, 23–25. https://doi.org/10.1016/j.mmcr.2018.11.003.
11. Clardy, J., Fischbach, M. A., & Currie, C. R. (2009). The natural history of antibiotics. Current biology : CB, 19(11), R437–R441. https://doi.org/10.1016/j.cub.2009.04.001.
12. McCluskey, K., & Baker, S. E. (2017). Diverse data supports the transition of filamentous fungal model organisms into the post-genomics era. Mycology, 8(2), 67–83. https://doi.org/10.1080/21501203.2017.1281849.
13. Riquelme, M., & Martinez-Nunez, L. (2016). Hyphal ontogeny in Neurospora crassa: A model organism for all seasons. F1000Research, 5, 2801. https://doi.org/10.12688/f1000research.9679.1.
14. Lopez Martinez, R., Ruiz Sanchez, D., Guadalupe Huerta, J., Esquenaze, A., & Alvarez, M. T. (1986). Seasonal variation of allergy-causing fungi in the southern part of Mexico City. Allergologia et Immunopathologia, 14, 43–48.
15. Dasgupta, A., Fuller, K. K., Dunlap, J. C., & Loros, J. J. (2016). Seeing the world differently: Variability in the photosensory mechanisms of two model fungi. Environmental Microbiology, 18, 5–20. https://doi.org/10.1111/1462-2920.13055.
16. Rossmann, S. N., Cernoch, P. L., & Davis, J. R. (1996). Dematiaceous fungi are an increasing cause of human disease. Clinical Infectious Diseases : An Official Publication of the Infectious Diseases Society of America, 22, 73–80.
17. Wright, D. E. (1968). Toxins produced by fungi. Annual Review of Microbiology, 22, 269–282. https://doi.org/10.1146/annurev.mi.22.100168.001413.
18. Dittrich, O., & Rieth, H. (1992). Fungal infections in humans. Part 1: Fungus classification and sources of fungal infection. Medizinische Monatsschrift fur Pharmazeuten, 15, 66–69.
19. Li, X., et al. (2018). Microfluidic system for rapid detection of airborne pathogenic fungal spores. ACS Sensors, 3, 2095–2103. https://doi.org/10.1021/acssensors.8b00615.

20. Veasey, J. V., Avila, R. B., Miguel, B. A. F., & Muramatu, L. H. (2017). White piedra, black piedra, tinea versicolor, and tinea nigra: Contribution to the diagnosis of superficial mycosis. Anais Brasileiros de Dermatologia, 92, 413–416. https://doi.org/10.1590/abd1806-4841.20176018.
21. Hata, Y., & Nishikawa, T. (2015). Superficial mycosis. Medical Mycology Journal, 56, J69–J72. https://doi.org/10.3314/mmj.56.J69.
22. Slawinska, M., et al. (2018). The role of dermoscopy in the diagnosis of deep mycoses and systemic mycoses with cutaneous involvement: Comment on 'Dermoscopy in disseminated sporotrichosis'. Journal of the European Academy of Dermatology and Venereology : JEADV. https://doi.org/10.1111/jdv.15155.
23. Carrasco-Zuber, J. E., et al. (2016). Cutaneous involvement in the deep mycoses: A literature review. Part I-subcutaneous mycoses. Actas Dermo-Sifiliograficas, 107, 806–815. https://doi.org/10.1016/j.ad.2016.05.017.
24. Van Schalkwyk E., et.al.(2019). Defining breakthrough invasive fungal infection-Position paper of the mycoses study group education and research consortium and the European Confederation of Medical Mycology. Mycoses, 62(9):716–729. https://doi.org/10.1111/myc.12960.
25. Mathur, P., et al. (2018). Five-year profile of candidaemia at an Indian trauma centre: High rates of Candida auris blood stream infections. Mycoses. https://doi.org/10.1111/myc.12790.
26. Azanza, J. R., Sadaba, B., & Gomez-Guiu, A. (2014). Pharmacology of the antifungals used in the treatment of aspergillosis. Revista Iberoamericana de Micologia, 31, 255–261. https://doi.org/10.1016/j.riam.2014.05.001.
27. Brizendine, K. D., & Pappas, P. G. (2010). Cryptococcal meningitis: Current approaches to management in patients with and without AIDS. Current Infectious Disease Reports, 12, 299–305. https://doi.org/10.1007/s11908-010-0113-4.
28. Gupta, A. K., MacLeod, M. A., Foley, K. A., Gupta, G., & Friedlander, S. F. (2017). Fungal skin infections. Pediatrics in Review, 38, 8–22. https://doi.org/10.1542/pir.2015-0140.
29. Skin disease. Fighting athlete's foot. Harvard health letter, 26, 5 (2001).
30. De Magalhaes Lima, K., et al. (2008). Non-dermatophytic moulds: Onychomycosis in four patients infected with the human immunodeficiency virus. Revista Iberoamericana de Micologia, 25, 45–49.
31. Awaad, A. S., Al-Mudhayyif, H. A., Al-Othman, M. R., Zain, M. E., & El-Meligy, R. M. (2017). Amhezole, a novel fungal secondary metabolite from Aspergillus terreus for treatment of microbial mouth infection. Phytotherapy Research : PTR, 31, 395–402. https://doi.org/10.1002/ptr.5760.
32. Kondo, T., & Terada, K. (2017). *Candida* esophagitis. *The New England Journal of Medicine, 376*, 1574. https://doi.org/10.1056/NEJMicm1614893.
33. Stuehler, C., et al. (2016). Immune recovery in HIV-infected patients after *Candida* esophagitis is impaired despite long-term antiretroviral therapy. *AIDS, 30*, 1923–1933. https://doi.org/10.1097/QAD.0000000000001126.
34. Shailaja, V. V., Pai, L. A., Mathur, D. R., & Lakshmi, V. (2004). Prevalence of bacterial and fungal agents causing lower respiratory tract infections in patients with human immunodeficiency virus infection. *Indian Journal of Medical Microbiology, 22*, 28–33.
35. Jaeger, M., Plantinga, T. S., Joosten, L. A., Kullberg, B. J., & Netea, M. G. (2013). Genetic basis for recurrent vulvo-vaginal candidiasis. *Current Infectious Disease Reports, 15*, 136–142. https://doi.org/10.1007/s11908-013-0319-3.
36. Sturkenboom, M. C., et al. (1995). Vulvo-vaginal candidiasis associated with acitretin. *Journal of Clinical Epidemiology, 48*, 991–997.
37. Janiga, J., Kentley, J., Nabhan, C., & Abdulla, F. (2018). Current systemic therapeutic options for advanced mycosis fungoides and Sezary syndrome. *Leukemia & Lymphoma, 59*, 562–577. https://doi.org/10.1080/10428194.2017.1347650.
38. Chen, Y., et al. (2018). Emergency of fungemia cases caused by fluconazole-resistant *Candida auris* in Beijing, China. *The Journal of Infection*. https://doi.org/10.1016/j.jinf.2018.09.002.

39. Noginskiy, I., Samra, A., Nielsen, K., & Kalavar, M. R. (2018). A case of multiple myeloma presenting as *Streptococcus pneumoniae* meningitis with *Candida auris* Fungemia. *Case Reports in Oncology, 11*, 705–710. https://doi.org/10.1159/000493852.
40. Suleyman, G., & Alangaden, G. J. (2016). Nosocomial fungal infections: Epidemiology, infection control, and prevention. *Infectious Disease Clinics of North America, 30*, 1023–1052. https://doi.org/10.1016/j.idc.2016.07.008.
41. Bongomin, F., Gago, S., Oladele, R. O., & Denning, D. W. (2017). Global and Multi-National Prevalence of Fungal Diseases-Estimate Precision. Journal of fungi (Basel, Switzerland), 3(4), 57. https://doi.org/10.3390/jof3040057.
42. Lelievre, L., & Lanternier, F. (2015). The challenge of nosocomial fungal infections. La Revue du Praticien, 65, 1324.
43. Du, B., Zhang, H., & Chen, D. (1996). Invasive fungal infection in 3447 autopsy cases. *Zhonghua Yi Xue Za Zhi, 76*, 352–354.
44. Makinen, A., Nawaz, A., Makitie, A., & Meurman, J. H. (2018). Role of non-albicans *Candida* and *Candida albicans* in Oral squamous cell Cancer patients. *Journal of Oral and Maxillofacial Surgery: Official Journal of the American Association of Oral and Maxillofacial Surgeons, 76*, 2564–2571. https://doi.org/10.1016/j.joms.2018.06.012.
45. Sangoi, A. R., et al. (2009). Challenges and pitfalls of morphologic identification of fungal infections in histologic and cytologic specimens: A ten-year retrospective review at a single institution. *American Journal of Clinical Pathology, 131*, 364–375. https://doi.org/10.1309/AJCP99OOOZSNISCZ.
46. Marques, M. B., Waites, K. B., Jaye, D. L., Kilby, J. M., & Reddy, V. V. (2000). Histologic examination of bone marrow core biopsy specimens has limited value in the diagnosis of mycobacterial and fungal infections in patients with the acquired immunodeficiency syndrome. *Annals of Diagnostic Pathology, 4*, 1–6. https://doi.org/10.1053/adpa.2000.0001.
47. Ishimaru, T., et al. (2018). Laser microdissection-based tissue specific transcriptome analyses reveals novel regulatory network of genes involved in heat-induced grain chalk in rice endosperm. *Plant & Cell Physiology*. https://doi.org/10.1093/pcp/pcy233.
48. Jenks, J. D., et al. (2018). Bronchoalveolar lavage *Aspergillus* galactomannan lateral flow assay versus *Aspergillus*-specific lateral flow device test for diagnosis of invasive pulmonary aspergillosis in patients with hematological malignancies. *The Journal of Infection*. https://doi.org/10.1016/j.jinf.2018.10.014.
49. Kumar, J., Singh, A., Seth, R., Xess, I., & Kabra, S. K. (2018). Galactomannan antigen test for early diagnosis of invasive *Aspergillus* infection in pediatric febrile neutropenia. *Indian Pediatrics, 55*, 257–258.
50. Theel, E. S., & Doern, C. D. (2013). Beta-D-glucan testing is important for diagnosis of invasive fungal infections. *Journal of Clinical Microbiology, 51*, 3478–3483. https://doi.org/10.1128/JCM.01737-13.
51. Huppler, A. R., Fisher, B. T., Lehrnbecher, T., Walsh, T. J., & Steinbach, W. J. (2017). Role of molecular biomarkers in the diagnosis of invasive fungal diseases in children. *Journal of the Pediatric Infectious Diseases Society, 6*, S32–S44. https://doi.org/10.1093/jpids/pix054.
52. Armstrong-James, D., et al. (2017). Immunotherapeutic approaches to treatment of fungal diseases. *The Lancet Infectious Diseases, 17*, e393–e402. https://doi.org/10.1016/S1473-3099(17)30442-5.
53. Cornely, O. A., et al. (2018). Isavuconazole for treatment of rare invasive fungal diseases. *Mycoses, 61*, 518–533. https://doi.org/10.1111/myc.12778.
54. Morschhauser, J. (2010). Regulation of multidrug resistance in pathogenic fungi. *Fungal Genetics and Biology : FG & B, 47*, 94–106. https://doi.org/10.1016/j.fgb.2009.08.002.
55. Cowen, L. E., Sanglard, D., Howard, S. J., Rogers, P. D., & Perlin, D. S. (2014). Mechanisms of antifungal drug resistance. *Cold Spring Harbor Perspectives in Medicine, 5*(7), a019752. https://doi.org/10.1101/cshperspect.a019752.
56. Cowen, L. E. (2009). Hsp90 orchestrates stress response Signaling governing fungal drug resistance. *PLoS Pathogens, 5*(8), e1000471. https://doi.org/10.1371/journal.ppat.1000471.

57. Srinivasan, A., Lopez-Ribot, J. L., & Ramasubramanian, A. K. (2014). Overcoming antifungal resistance. *Drug Discovery Today. Technologies, 11*, 65–71. https://doi.org/10.1016/j.ddtec.2014.02.005.
58. Wiederhold, N. P. (2017). Antifungal resistance: Current trends and future strategies to combat. *Infection and Drug Resistance, 10*, 249–259. https://doi.org/10.2147/IDR.S124918.
59. Wiederhold, N. P. (2018). The antifungal arsenal: Alternative drugs and future targets. *International Journal of Antimicrobial Agents, 51*(3), 333–339. https://doi.org/10.1016/j.ijantimicag.2017.09.002.

Candida Pathogenicity and Alternative Therapeutic Strategies

Nikhat Manzoor

Abstract

Human fungal infections are widespread and difficult to treat. *Candida albicans*, a common fungal pathogen, exhibits different strategies for growth, proliferation, and survival within the host, as it is armed with mechanisms to escape the host defense system. This opportunistic commensal expresses several virulence factors such as adhesins, yeast to hyphal transition, phenotypic switching, biofilm formation, and secretion of hydrolytic enzymes, mainly proteases and phospholipases. Depending upon the mode of action, several classes of antifungal drugs have been developed till date. However, most of them are toxic with side effects. Excessive use and abuse of these drugs have led to the evolution of multidrug-resistant strains at an alarming rate leading to treatment failures. Since a change has been observed in pathogenesis strategies of *Candida* species, the treatment strategies also need to be improved. Continuous chemotherapies in immunocompromised patients, especially in HIV patients with oropharyngeal *Candida* infections, have led to severe host tissue toxicity. Therefore, safer phytomedicines, which are more efficacious, nontoxic, easily available, and do not develop resistance in fungal strains, are required. Plant extracts, essential oils, and their constituents that show promising antifungal potential are better and safer alternatives. Alternate therapeutic strategies also include chemosensitizing the pathogenic fungi to conventional antifungals using natural plant products.

Keywords

Candida · Multidrug resistance · Antifungal drugs · Natural products · Synergy

N. Manzoor (✉)
Medical Mycology Lab, Department of Biosciences, Jamia Millia Islamia, New Delhi, India
e-mail: nikhatmanzoor@yahoo.co.in

S. Hameed, Z. Fatima (eds.), *Pathogenicity and Drug Resistance of Human Pathogens*,
https://doi.org/10.1007/978-981-32-9449-3_7

Introduction

Candida albicans is a common human fungal pathogen which although a commensal becomes pathogenic under immunocompromised conditions. It colonizes mucosal surfaces of patients with cancer and HIV, underweight new born babies, and patients requiring multiple catheters. Depending upon magnitude and location, candidiasis varies from superficial thrush, chronic mucocutaneous candidiasis to deep-seated *Candida* myocarditis, and *Candida* septicemia [13, 17, 21]. Inefficiency in diagnosis, unavailability of appropriate drug treatment, and drug resistance are some issues that cause hindrance in successful drug discovery and development against fungal infections. Azoles, such as fluconazole (FLC) and itraconazole (ITR) in particular, constitute the largest class of antifungal therapeutics against a wide variety of fungi, including yeasts, dermatophytes, and some molds. The target site for the azoles is *ERG11* gene product, cytochrome P450 lanosterol 14-α-demethylase, which is part of the ergosterol biosynthetic pathway. Among other antifungal targets, 1,3-β-D-glucan synthase enzyme complex, essential for the formation of glucan polymers of fungal cell wall, is noncompetitively inhibited by echinocandins and numerous other closely related pneumocandins. Hyphal invasion by *C. albicans* inhibits the expression of human β-defensins in oral candidiasis. Besides this, many studies unraveling the early stages of fungal infection have shown that yeast-hyphae morphogenesis induces major changes in the macrophage morphology [5], their division, and proliferation [36]. Studies indicate that subsequent development of extracellular polymeric substances, commonly referred to as biofilms, serves as an incessant source of pathogens, particularly in cases of oral candidiasis and catheter-associated infections [14, 47]. However, relevant elucidation of drug interactions at molecular level is still less illuminated in this area.

The efficacy of antifungal drugs may be predicted by susceptibility evaluation in terms of minimum inhibitory concentration (MIC) and minimum fungicidal concentration (MFC) against different *Candida* species. The azole class of antifungals, with excellent efficacy-toxicity profiles, has emerged as principal drugs used in the treatment of *Candida* infections in non-neutropenic patients. FLC is the most widely used drug in both treatment and prevention of candidiasis [18, 49]. However, in recent years, prolonged use of FLC has contributed to the development of drug resistance in *C. albicans* and other *Candida* species [58]. Major consequences that favor the development of azole resistance in fungi include upregulation of genes controlling drug efflux, decreased affinity of azoles for the cellular target, increased levels of the target, and alterations in sterol biosynthesis. Traditional system of medication provides safer and effective formulations and is the current impetus in clinical research [11]. Moreover, recent drug development approaches campaign for combinatorial studies (synergism) of conventional antifungals with phytoextracts and essential oils as a promising area of antifungal therapeutics [51]. The present work is an effort to understand *Candida* infection, its etiology, treatment options, and their failures. The work also throws some light on the development of alternate phytomedicinal and synergistic therapeutics.

C. albicans and Other *Candida* Species

Candida species are the most common cause of nosocomial bloodstream infections [44, 46]. Although there are several *Candida* species known today, those that are more common and clinically important include *C. albicans*, *C. glabrata*, *C. parapsilosis*, *C. tropicalis*, and *C. krusei* [52]. *C. auris* is another species that is emerging fast as a multidrug-resistant organism and hence becoming a serious threat internationally. In a brief span of only 8 years, this fungus has spread to several countries including India [12] where it is the second most prevalent species after *C. tropicalis* [40]. Analyses of *Candida* blood stream infections have shown trends to selection of non-*albicans* species, some of which are difficult to treat with first-generation azoles. *C. tropicalis* has greater genetic similarity *with C. albicans* in comparison to other *Candida* species. *C. glabrata* has a haploid genome and grows only as blastoconidia unlike *C. albicans* [6]. Due to the emergence of multidrug-resistant non-albicans *Candida* species like *C. auris*, their identification up to the species level is clinically relevant.

Virulence and Pathogenicity

Although *Candida* is a commensal, when opportunity arises, this harmless yeast may express certain factors that contribute to pathogenesis of the disease (candidiasis or candidosis). *Candida* virulence is mainly due to expression of adhesins, yeast to hyphal transition, secretion of hydrolytic enzymes (aspartyl proteases, phospholipases, and hemolysins), formation of biofilms, and phenotypic switching [8]. These factors help the fungus in evading the host defenses and helping in colonization and establishment of infection. The fact that this fungal pathogen can easily grow at a range close to the body temperature also helps in spreading the infection fast. Invasive candidiasis is still the fourth most commonly acquired among the several nosocomial infections, and mortality rates are high, up to 50% [45]. Identifying genes that express for adhesion, biofilm formation, and other virulence factors during infection will lead to unraveling of molecular mechanisms for the infection process and also help in designing mechanism-based drugs to combat both virulence and multidrug resistance in pathogenic fungi.

Extracellular hydrolytic enzymes play an important role in invasion of the host tissues by facilitating adherence and tissue penetration. Aspartyl proteinases (Saps) are encoded by 10 *SAP* genes that play different roles in *Candida* infection. It has been reported that SAP1-6 participates in adherence, tissue damage, and evasion of host immune responses. While the role of SAP7-10 is not well defined, there is evidence that Sap9 and Sap10 are not secretory but have regulatory functions [42, 66]. Phospholipases are enzymes that hydrolyze phospholipids into fatty acids and other lipophilic substances. Several genes have been identified, but only a few have been well characterized (PLB1, PLB2, PLC1, and PLD1) [56]. Depending on the type of hydrolytic activity, these enzymes are classified into four major classes (A, B, C, and D). Phospholipase B has been shown to be essential for virulence and is secreted by *C. albicans* during the infectious process [16]. Although the mechanism(s) through which phospholipase modulates fungal virulence is still under investigation, early data suggests that direct host cell damage and lysis are the main mechanisms contributing to fungal virulence. Since the importance of phospholipases in fungal virulence is already known, the challenge is to utilize these lytic enzymes as therapeutic and diagnostic targets. The hemolytic activity of *C. albicans* allows it to acquire iron from host erythrocytes [50] and hence is pivotal for the survival and pathogenicity of the fungus.

Adhesion is the first step in the infection process. In human host, this phenomenon is complex and multifactorial involving several molecules called adhesins. The *Candida* cell wall houses certain proteins which provide hydrophobic interactions to make a strong bond between the fungus and the host surface [35]. The adhesion profiles vary to a large extent among *Candida* species [53]. Fungal biofilms are microbial communities encased in a matrix of extracellular polymeric substances. They play a vital role in the infection process mainly due to high antibiotic resistance associated with them [10]. Their clinical relevance lies in the fact that drug resistance in biofilms is sometimes more than 1000-fold greater than in planktonic cells. *Candida* cells have a tendency to form robust biofilms on both animate and inanimate surfaces. Indwelling medical devices such as dental implants, catheters, heart valves, vascular bypass grafts, ocular lenses, artificial joints, and central nervous system shunts are vulnerable to the growth of fungal biofilms [10, 53].

Phenotypic switching is another virulence trait expressed during *Candida* pathogenesis. The usual white colonies of the cells switch to opaque colonies which are considerably larger and more elongated and not so smooth. Opaque cells are better colonizers of the skin than white cells [35, 63]. Finally, like many other fungi, *C. albicans* can form chlamydospores, thick-walled, large spherical cells that seem to represent a resting stage. The available data indicates that the walls of yeast cells, pseudohyphae, and hyphae possess a similar molecular organization. However, the walls of opaque cells and chlamydospores have been studied less extensively. *C. albicans* is polymorphic and undergoes reversible transitions between different morphological forms (budding, pseudohyphal, and hyphal growth forms) that promote virulence of this pathogenic fungus [65]. The regulatory networks that control morphogenesis are being elucidated; however, the primary signals that trigger morphogenesis are complex and still ambiguous [41].

Conventional Antifungal Drugs and Their Mode of Action

The currently available antifungal drugs can be classified depending upon their site of action. The azoles are a class of five-membered heterocyclic compounds containing a nitrogen atom and at least one other non-carbon atom (i.e., nitrogen, sulfur, or oxygen) as part of the ring. Ketoconazole (an imidazole) was the first to be discovered followed by FLC and ITR. The latter two are triazoles having improved safety profiles and broader spectrum of antifungal activity [39]. Azoles inhibit the enzyme 14α-demethylase leading to the depletion of ergosterol and accumulation of sterol precursors. Since ergosterol is important for fungal cell membrane integrity and activity of several membrane-bound enzymes, its depletion results in altered membrane structure and function [43]. The polyene class of antimycotics include drugs like amphotericin B, nystatin, and natamycin. These antibiotics, obtained from some species of *Streptomyces* bacteria, are macrolides with multiple conjugated double bonds. Due to their amphiphilic nature, these molecules bind to ergosterol in fungal membranes making it porous and causing leakage of cellular content [4, 49]. Similarly, allylamines, which include drugs like amorolfine, butenafine, naftifine, and terbinafine, inhibit squalene epoxidase, another enzyme required for ergosterol synthesis, while flucytosine is a pyrimidine analogue that blocks nucleic acid synthesis. A more novel class of antifungals, called echinocandins (caspofungin, micafungin, and anidulafungin), inhibit the enzyme 1, 3 β-glucan synthase and hence block the synthesis of a very vital structural component of fungal cell walls (β-glucans) [43, 48].

Many different types of mechanisms contribute to the development of resistance to antifungals. These mechanisms include alteration in drug target, alteration in sterol biosynthesis, reduction in the intercellular concentration of target enzyme, and overexpression of the antifungal drug target. The shortcoming of the currently available drugs is that most of them are fungistatic. As a result fungal pathogens develop tolerance to these drugs and with time become ineffective. Increasing tolerance to drugs or multidrug resistance (MDR) is multifactorial as it can be the result of several mechanisms [49]. Due to the ever-increasing problem of MDR in fungal pathogens, inadequate supply of effective antifungals, and toxicity to the hosts, there is a persistent requirement for the development of novel antifungal drugs that are broad-spectrum and more efficacious. The diverse approaches in this area of drug discovery and development include synthesis of new antifungal drugs, both synthetic and semisynthetic [37]. Another development is using natural products with antifungal properties, as such or after modification as these compounds prove to be nontoxic, and there are negligible chances of developing MDR in fungal species. The clinically resistant *Candida* strains are evolving very fast, and the universally prescribed antifungal azoles need to be modified or replaced to produce more effective and nontoxic drugs.

Phytomedicine and New Approaches

Plants synthesize a vast number of secondary metabolites having therapeutic properties. These natural plant products have been used since antiquity in traditional medical systems like Ayurveda, Unani, and Siddha [25]. But recently they are being explored, and tremendous efforts are being put in to isolate, characterize, and "encash" their therapeutic potential as they are easily available, are nontoxic, and do not lead to microbial resistance. Since they are cheaper, the treatment offered for chronic illnesses like fungal infections becomes affordable [9]. Around 80% of the population in developing countries depends on the traditional forms of medicine to meet their health-care needs [67]. Medicinal herbs have been the foundation of the traditional systems of treatment as plants synthesize certain molecules (phytoalexins) as defense against pathogens and other abiotic agents [25]. These bioactive compounds include terpenoids, glycosides, flavonoids, and polyphenols. They exhibit excellent therapeutic properties [26] and are also being used as scaffolds for several synthetic and semisynthetic drugs [20, 69].

Several studies have explored the antimicrobial potential of plant essential oils and tried to unravel their mode of action [11, 19]. Due to the different degree of lipophilicity and hydrophilicity, these compounds alter cell permeability by getting inserted between fatty acid chains of lipid bilayers and hence alter the properties and functions of the cell membranes [7, 57]. The most widely used quantitative susceptibility estimate is measurement of the MIC of an antimicrobial. For any inhibitor-microbe combination in a particular growth environment, the endpoint of a susceptibility measurement depends on the inoculum size, temperature and pH of the medium, and growth phase of the cell cycle [59]. The smaller the value of MIC of the antifungal, the greater is its efficacy. Figure 1 shows some natural compounds that have promising antifungal potential exhibiting an MIC $\leq$500 μg/ml against strains of *C. albicans*. The essential oils of medicinal herbs like *Ocimum sanctum* (OSEO), *Coriandrum sativum*, and *Mentha piperata* (MEO) were found to be very effective against the *Candida* spp. [28, 54, 55, 62]. A study conducted by Khan et al. showed OSEO to be fungicidal against both FLC-sensitive and FLC-resistant clinical *Candida* isolates [29, 32]. Studies showed that the antifungal activity of OSEO and its major constituents (methyl chavicol and linalool) was due to disruption of ergosterol biosynthesis and cell membrane integrity [30]. This essential oil inhibited

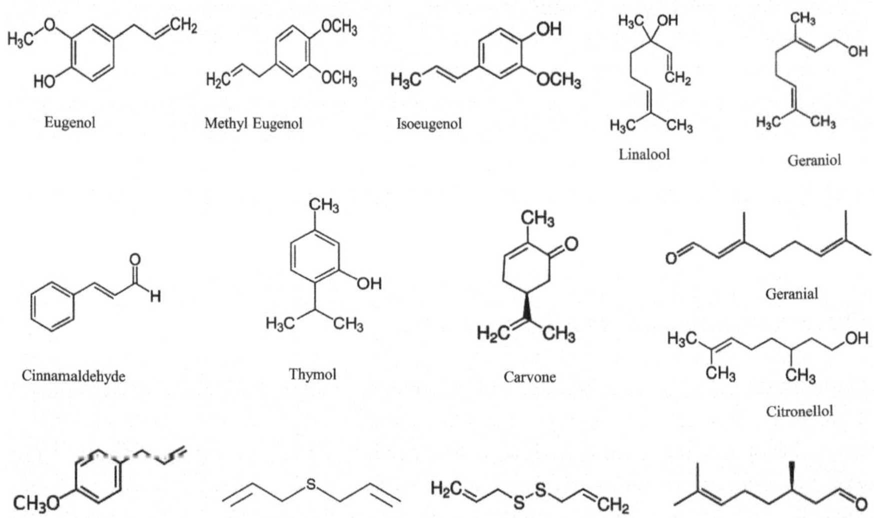

Fig. 1 Some natural compounds that have promising antifungal potential (MIC $\leq$500 μg/ml against *Candida albicans*)

morphogenesis and secretion of hydrolytic enzymes, as well as downregulated the associated genes (*HWP1*, *SAP1*, and *PLB2*) even at sub-inhibitory concentrations [32]. In another study it was shown that *Candida* cells exposed to OSEO, linalool, and methyl chavicol displayed apoptotic features at low concentrations and necrosis at higher concentrations [33].

The isolation of biomolecules from natural sources and their modifications encounter several difficulties, but the promising potential of natural products, along with the fact that they are low in toxicity, makes antifungal drug discovery less challenging [38]. The natural compounds studied have multiple modes of antifungal action due to varying hydrophilicity and hydrophobicity of the compounds. Fungal cell wall acts as a physical barrier and filter to the external environment. Changes in the cell wall architecture and cross-linking of cell wall polysaccharides to each other are likely to affect the porosity of the wall. Its porosity is a measure of its relative permeability to molecules and is normally reflective of the molecular size and charge on the permeant [64]. Natural compounds are being explored as scaffolds for the synthesis of more efficacious antifungal drugs [69]. Studies have been done on synthetic lead molecules [3, 22–24] and molecules derived from natural products [60]. α-methyl cinnamaldehyde showed greater anti-*Candida* activity than cinnamaldehyde suggesting toward the significance of the type and location of the functional groups [61]. Gold and silver nanoparticles having antifungal properties have been prepared using plant extracts [2, 34].

The antifungal drugs currently prescribed have high toxicity. By reducing the doses of antifungal drugs, the toxicity can be toned down, and hence combinatorial therapeutic strategies are now a better option for the management of chronic fungal infections like candidiasis. Plant extracts, essential oils or their major constituents when used in combination with conventional antifungal drugs mainly FLC and amphotericin B, have shown significant synergistic effects [51, 70]. Synergy produced maybe due to improved solubility and hence enhanced bioavailabilty of one or more biomolecules present in the extract/oil. The combination effects can be calculated in terms of fractional inhibitory concentration indices (FICI). Synergy is expressed when combinations of antifungal compounds are able to exert inhibitory effects that are greater than the sum of their effects alone (FICI <1.0), while antagonism is when the inhibitory effects are less than the sum of their effects alone (FICI >1.0). OSEO and MEO have both shown synergy when used in combination with FLC and ketoconazole against both sensitive and resistant strains [31, 55]. Antifungal synergy has been proposed to work by several mechanisms. The interaction may involve the inhibition of different stages of the same biochemical pathway. The penetration of antifungal agents through cell membranes and cell walls may increase as a result of the activity of the other agents. Also there is the possibility of simultaneous inhibitions of different fungal cell targets [27]. In another study, 12 natural phenolics were examined for fungicidal activity against 9 reference strains of *Candida*. Combinations of benzoic acid or thymol with ITR showed highest synergistic activity [15]. Yan et al. [68] reported potent synergistic antifungal activities of pure compounds isolated from traditional Chinese medicinal plant extracts with FLC against azole-resistant *C. albicans.* Thymol and carvacrol, the principal bioactive components of thyme oil, showed significant synergy with FLC, reversing the efflux-mediated resistance in clinical *Candida* isolates [1].

Conclusion

As both infection frequency and MDR are on the rise, there is an ardent need to search for novel antifungals and targets so that this chronic disease can be managed efficiently. Plant products that have antifungal potential can be used as scaffolds for the synthesis of novel antifungal drugs. These essential oils and compounds can also be used as chemosensitizing agents to enhance the efficacy of conventional antifungal drugs. This can further lower the effective dosage of drugs which is toxic and can lead to resistance on being prescribed repeatedly. In vitro studies have to be extended to animal models before the natural compounds can be used as drugs in invasive candidiasis.

References

1. Ahmad, A., Khan, A., & Manzoor, N. (2013). Reversal of efflux mediated antifungal resistance underlies synergistic activity of two monoterpenes with fluconazole. *European Journal of Pharmaceutical Sciences, 48*, 80–86.
2. Ahmad, T., Wani, I. A., Manzoor, N., Ahmed, J., Kalam, A., & Al-Shihri, A. (2014). Structural characterization, antifungal activity and optical properties of gold nanoparticles prepared by reverse micelles. *Advanced Science Letters, 20*, 1631–1636.
3. Aneja, B., Irfan, M., Kapil, C., Jairajpuri, M. A., Maguire, R., Kavanagh, K., Rizvi, M. M., Manzoor, N., Azam, A., & Abid, M. (2016). Effect of novel triazole-amino acid hybrids on growth and virulence of *Candida* species: In vitro and in vivo studies. *Organic & Biomolecular Chemistry, 14*(45), 10599–10619.
4. Baginski, M., & Czub, J. (2009). Amphotericin B and its new derivatives – mode of action. *Current Drug Metabolism, 10*(5), 459–469.
5. Boyce, K. J., & Andrianopoulos, A. (2015). Fungal dimorphism: The switch from hyphae to yeast is a specialized morphogenetic adaptation allowing colonization of a host. *FEMS Microbiology Reviews, 39*(6), 797–811.
6. Butler, G., Rasmussen, M. D., Lin, M. F., Santos, M. A., Sakthikumar, S., Munro, C. A., et al. (2009). Evolution of pathogenicity and sexual reproduction in eight *Candida* genomes. *Nature, 459*, 657–662.
7. Braga, P. C., Alfieri, M., Culci, M., & Dal Sasso, M. (2007). Inhibitory activity of thymol against the formation and viability of *Candida albicans* hyphae. *Mycoses, 50*, 502–506.
8. Calderone, R. A., & Fonzi, W. A. (2001). Virulence factors of *Candida albicans*. *Current Trends in Microbiology, 9*, 327–335.
9. Ciocan, I. D., & Bara, I. I. (2007). Plant products as antimicrobial agents. *Genetics and Biology Molecular, 8*, 151–156.
10. Chandra, J., Kuhn, D. M., Mukherjee, P. K., Hoyer, L. L., McCormick, T., & Ghannoum, M. A. (2001). Biofilm formation by the fungal pathogen *Candida albicans*: Development, architecture, and drug resistance. *Journal of Bacteriology, 183*(18), 5385–5394.
11. Chouhan, S., Sharma, K., & Guleria, S. (2017). Antimicrobial activity of some essential oils-present status and future perspectives. *Medicine, 4*(3), 58–69.

12. Chowdhary, A., Sharma, C., & Meis, J. F. (2017). *Candida auris*: A rapidly emerging cause of hospital-acquired multidrug-resistant fungal infections globally. *PLoS Pathogens, 13*(5), e1006290.
13. de Oliveira Santos, G. C., Vasconcelos, C. C., Lopes, A., de Sousa, C. M., Filho, A., do Nascimento, F., Ramos, R. M., Pires, E., de Andrade, M. S., Rocha, F., & de Andrade Monteiro, C. (2018). *Candida* infections and therapeutic strategies: Mechanisms of action for traditional and alternative agents. *Frontiers in Microbiology, 9*, 1351.
14. Donlan, R. M. (2001). Biofilms and device-associated infections. *Emerging Infectious Diseases, 7*, 277–281.
15. Faria, N. C. G., Kim, J. H., Goncalves, L. A. P., Martins, M. L., Chan, K. L., & Campbell, B. C. (2011). Enhanced activity of antifungal drugs using natural phenolics against yeast strains of *Candida* and *Cryptococcus*. *Letters in Applied Microbiology, 52*, 506–513.
16. Ghannoum, M. A. (2000). Potential role of phospholipases in virulence and fungal pathogenesis. *Clinical Microbiology Reviews, 13*(1), 122–143.
17. Ghannoum, M. A. (2001). *Candida*: A causative agent of an emerging infection. *Society for Investigative Dermatology, 6*, 1088–1096.
18. Ghannoum, M. A., & Rice, L. B. (1999). Antifungal agents, mode of action, mechanisms of resistance, and correlation of these mechanisms with bacterial resistance. *Clinical Microbiology Reviews, 12*, 501–517.
19. Griffin, S. G., Wyllie, S. G., Markham, J. L., & Leach, D. (1999). The role of structure and molecular properties of terpenoids in determining their antimicrobial activity. *Flavour and Fragrance Journal, 14*, 322–332.
20. Guo, Z. (2017). The modification of natural products for medical use. *Acta Pharmaceutica Sinica B, 7*(2), 119–136.
21. Gow, N. A. R., & Yadav, B. (2017). Microbe profile: *Candida albicans*: A shape-changing, opportunistic pathogenic fungus of humans. *Microbiology, 163*, 1145–1147.
22. Irfan, M., Aneja, B., Yadava, U., Khan, S. A., Manzoor, N., Daniliuc, C. G., & Abid, M. (2015). Synthesis, QSAR and anticandidal evaluation of 1,2,3-triazoles derived from naturally bioactive scaffolds. *European Journal of Medicinal Chemistry, 93*, 246–254.
23. Irfan, M., Alam, S., Manzoor, N., & Abid, M. (2017). Effect of quinoline based 1,2,3-triazole and its structural analogues on growth and virulence attributes of *Candida albicans*. *PLoS One, 12*(4), e0175710.
24. Irfan, M., Khan, S. I., Manzoor, N., & Abid, M. (2016). Biological activities and in silico physico-chemical properties of 1,2,3-triazoles derived from natural bioactive alcohols. *Anti-Infective Agents, 14*(2), 126–131.
25. Jaiswal, Y. S., & Williams, L. L. (2016). A glimpse of Ayurveda – The forgotten history and principles of Indian traditional medicine. *Journal of Traditional and Complementary Medicine, 7*(1), 50–53.
26. Jeandet, P. (2015). Phytoalexins: Current progress and future prospects. *Molecules, 20*(2), 2770–2774.
27. Johnson, M. D., MacDougall, C., Ostrosky-Zeichner, L., Perfect, J. R., & Rex, J. H. (2004). Combination antifungal therapy. *Antimicrobial Agents and Chemotherapy, 48*(3), 693–715.
28. Khan, A., Ahmad, A., Manzoor, N., & Khan, L. A. (2010). Antifungal activities of *Ocimum sanctum* essential oil and its lead molecules. *Natural Products Communications, 5*(2), 345–349.
29. Khan, A., Ahmad, A., Immaculata, X., Khan, L. A., & Manzoor, N. (2010). Anticandidal effect of *Ocimum sanctum* essential oil and its synergy with fluconazole and ketoconazole. *Phytomedicine, 17*, 921–925.
30. Khan, A., Ahmad, A., Akhtar, F., Yousuf, S., Xess, I., Khan, L. A., & Manzoor, N. (2010). *Ocimum sanctum* essential oil and its active principles exert their antifungal activity by disrupting ergosterol biosynthesis and membrane integrity. *Research Microbiology, 161*, 816–823.

31. Khan, A., Ahmad, A., Akhtar, F., Yousuf, S., Xess, I., Khan, L. A., & Manzoor, N. (2010). Anticandidal effect of *Ocimum sanctum* essential oil and its synergy with fluconazole and ketoconazole. *Phytomedicine, 17*, 921–925.
32. Khan, A., Ahmad, A., Xess, I., Khan, L. A., & Manzoor, N. (2014). *Ocimum sanctum* essential oil inhibits virulence attributes in *Candida albicans*. *Phytomedicine, 21*, 448–452.
33. Khan, A., Ahmad, A., Khan, L. A., & Manzoor, N. (2014). *Ocimum sanctum* (L.) essential oil and its lead molecules induce apoptosis in *Candida albicans*. *Research Microbiology, 165*, 411–419.
34. Khatoon, N., Mishra, A., Alam, H., Manzoor, N., & Sardar, M. (2015). Biosynthesis, characterization and antifungal activity of the Silver nanoparticles against pathogenic *Candida* species. *BioNanoScience, 5*, 65–74.
35. Klis, F. M., Sosinska, G. J., de Groot, P. W., & Brul, S. (2009). Covalently linked cell wall proteins of *Candida albicans* and their role in fitness and virulence. *FEMS Yeast Research, 9*(7), 1013–1028.
36. Lewis, L. E., Bain, J. M., Lowes, C., Gow, N. A., & Erwig, L. P. (2012). *Candida albicans* infection inhibits macrophage cell division and proliferation. *Fungal Genetics and Biology, 49* (9), 679–680.
37. McCarthy, M. W., Kontoyiannis, D. P., Cornely, O. A., Perfect, J. R., & Walsh, T. J. (2017). Novel agents and drug targets to meet the challenges of resistant fungi. *The Journal of Infectious Diseases, 216*(3), S474–S483.
38. McRae, J., Yang, Q., Crawford, R., & Palombo, E. (2007). Review of the methods used for isolating pharmaceutical lead compounds from traditional medicinal plants. *Environment, 27* (1), 165–174.
39. Maertens, J. A. (2004). History of the development of azole derivatives. *Clinical Microbiology and Infection, 10*(1), 1–10.
40. Mathur, P., Hasan, F., Singh, P. K., Malhotra, R., Walia, K., & Chowdhary, A. (2018). Five year profile of candidaemia at an Indian trauma centre: High rates of *Candida auris* blood stream infections. *Mycoses, 61*, 674–680.
41. Mayer, F. L., Wilson, D., & Hube, B. (2013). *Candida albicans* pathogenicity mechanisms. *Virulence, 4*(2), 119–128.
42. Naglik, J. R., Rodgers, C. A., Shirlaw, P. J., Dobbie, J. L., Fernandes-Naglik, L. L., Greenspan, D., Agabian, N., & Challacombe, S. J. (2003). Differential expression of *Candida albicans* secreted aspartyl proteinase and phospholipase B genes in humans correlates with active oral and vaginal infections. *The Journal of Infectious Diseases, 188*, 469–479.
43. Odds, F. C., Brown, A. J. P., & Gow, N. A. R. (2003). Antifungal agents: Mechanisms of action. *Trends in Microbiology, 11*, 272–279.
44. Perlroth, J., Choi, B., & Spellberg, B. (2007). Nosocomial fungal infections: Epidemiology, diagnosis, and treatment. *Medical Mycology, 45*(4), 321–346.
45. Pfaller, M. A. (2012). Antifungal drug resistance: Mechanisms, epidemiology, and consequences for treatment. *The American Journal of Medicine, 125*(1), S3–S13.
46. Pfaller, M. A., Moet, G. J., Messer, S. A., Jones, R. N., & Castanheira, M. (2011). *Candida* bloodstream infections: Comparison of species distributions and antifungal resistance patterns in community-onset and nosocomial isolates in the SENTRY antimicrobial surveillance program, 2008–2009. *Antimicrobial Agents and Chemotherapy, 55*(2), 561–566.
47. Polaquini, S. R. B., Svidzinski, T. I. E., Kemmelmeier, C., & Gasparetto, A. (2006). Effect of aqueous extract from neem (*Azadirachta indica A. Juss*) on hydrophobicity, biofilm formation and adhesion in composite resin by *Candida albicans*. *Archives of Oral Biology, 51*, 482–490.
48. Pierce, C. G., & Lopez-Ribot, J. L. (2013). Candidiasis drug discovery and development: New approaches targeting virulence for discovering and identifying new drugs. *Expert Opinion on Drug Discovery, 2013*(9), 1117–1126.
49. Prasad, R., Shah, A. H., & Rawal, M. K. (2016). Antifungals: Mechanism of action and drug resistance. *Advances in Experimental Medicine and Biology, 892*, 327–349.

50. Almeida, R. S., Wilson, D., & Hube, B. (2009). *Candida albicans* iron acquisition within the host. *FEMS Yeast Research, 9*(7), 1000–1012.
51. Salih, K. A. (2016). Synergistic effects of plant extracts and antifungal drugs on *C. albicans*. *Journal of Developing Drugs, 5*, 165.
52. Sardi, J. C., Scorzoni, L., Bernardi, T., Fusco-Almeida, A. M., & Mendes Giannini, M. J. (2013). *Candida* species: Current epidemiology, pathogenicity, biofilm formation, natural antifungal products and new therapeutic options. *Journal of Medical Microbiology, 62*, 10–24.
53. Silva-Dias, A., Miranda, I. M., Branco, J., Monteiro-Soares, M., Pina-Vaz, C., & Rodrigues, A. G. (2015). Adhesion, biofilm formation, cell surface hydrophobicity, and antifungal planktonic susceptibility: Relationship among *Candida* spp. *Frontiers in Microbiology, 6*, 205.
54. Samber, N., Varma, A., & Manzoor, N. (2014). Evaluation of *Mentha piperita* essential oil and its major constituents for antifungal activity in *Candida* spp. *IJIRSET, 3*(2), 9404–9411.
55. Samber, N., Amber, K., Varma, A., & Manzoor, N. (2015). Synergistic anti-candidal activity and mode of action of *Mentha piperita* essential oil and its major components. *Pharmaceutical Biology, 53*(10), 1496–1504.
56. Samaranayake, Y. H., Dassanayake, R. S., Cheung, B. P. K., Jayatilake, J. A., Yeung, K. W. S., Yau, J. Y. Y., & Samaranayake, L. P. (2006). Differential phospholipase gene expression by *Candida albicans* in artificial media and cultured human oral epithelium. *APMIS, 114*, 857–866.
57. Sanchez, M. E., Turina, A., Garcia, D. A., Veronica-Nolan, M., & Perillo, M. A. (2004). Surface activity of thymol: Implications for an eventful pharmacological activity. *Colloids and Surfaces B Biointerfaces, 34*, 77–86.
58. Sanglard, D., Ischer, F., Parkinson, T., Falconer, D., & Bille, J. (2003). *Candida albicans* mutations in the ergosterol biosynthetic pathway and resistance to several antifungal agents. *Antimicrobial Agents and Chemotherapy, 47*, 2404–2412.
59. Sanglard, D., & Odds, F. C. (2002). Resistance of *Candida* species to antifungal agents: Molecular mechanisms and clinical consequences. *The Lancet Infectious Disease, 2*, 73–85.
60. Shreaz, S., Rayees, A. S., Bhatia, R., Hashmi, A. A., Manzoor, N., & Khan, L. A. (2010). Anticandidal activity of cinnamaldehyde, its ligand and Ni(II) complex: Effect of increase in ring and side chain. *Microbial Pathogenesis, 49*, 75–82.
61. Shreaz, S., Rayees, A. S., Bhatia, R., Khan, N., Imran, S., Hashmi, A. A., Manzoor, N., Basir, S. F., & Khan, L. A. (2011). Antifungal activity of α-methyl trans cinnamaldehyde, its ligand and metal complexes: Promising growth and ergosterol inhibitors. *Biometals, 24*(5), 923–933.
62. Soares, B. V., Morais, S. M., Fontenelle, R. O. S., Queiroz, V. A., Vila-Nova, N. S., Pereira, C. M. C., Brito, E. S., Neto, M. A. S., Brito, E. H. S., Cavalcante, C. S. P., Castelo-Branco, D. S. C. M., & Rocha, M. F. G. (2012). Antifungal activity, toxicity and chemical composition of the essential oil of *Coriandrum sativum* L. fruits. *Molecules, 17*, 8439–8448.
63. Soll, D. R. (2004). Mating-type locus homozygosis, phenotypic switching and mating: A unique sequence of dependencies in *Candida albicans*. *BioEssays, 26*, 10–20.
64. Souza-Moreira, T. M., Severi, J. A., Lee, K., Preechasuth, K., Santos, E., Gow, N. A. R., Munro, C. A., Vilegas, W., & Pietro, R. C. L. R. (2013). Anti-*Candida* targets and cytotoxicity of casuarinin isolated from *Plinia caulifora* leaves in a bioactivity-guided study. *Molecules, 18*, 8095–8108.
65. Sudbery, P., Gow, N., & Berman, J. (2004). The distinct morphogenic states of *Candida albicans*. *Trends in Microbiology, 12*, 317–324.
66. Tsang, C. S., Chu, F. C., Leung, W. K., Jin, L. J., Samaranayake, L. P., & Siu, S. C. (2007). Phospholipase, proteinase and haemolytic activities of *Candida albicans* isolated from oral cavities of patients with type 2 diabetes mellitus. *Journal of Medical Microbiology, 56*, 1393–1398.
67. WHO. (2013). *Regulatory situation of herbal medicines: A worldwide review*. Geneva: World Health Organization.

68. Yan, Z., Hua, H., Xu, Y., & Samaranayake, L. P. (2012). Potent antifungal activity of pure compounds from traditional Chinese medicine extracts against six oral *Candida* species and the synergy with fluconazole against azole-resistant *Candida albicans*. *Evidence-Based Complementary and Alternative Medicine, 106583*, 1–6.
69. Zida, A., Bamba, S., Yacouba, A., Ouedraogo-Traore, R., & Guiguemdé, R. T. (2017). Anti-*Candida albicans* natural products, sources of new antifungal drugs: A review. *Journal of Mycology Medical, 27*(1), 1–19.
70. Zhimin, Y., Hua, H., Xu, Y., & Samaranayake, L. P. (2012). Potent antifungal activity of pure compounds from traditional Chinese medicine extracts against six oral *Candida* species and the synergy with fluconazole against azole-resistant *Candida albicans*. *Evidence-based Complementary and Alternative Medicine, 2012*(106583), 1–6.

Role of Mitochondria in Fungal Drug Resistance

Ritu Pasrija and Deepika Kundu

Abstract

Incidences of opportunistic fungal infections and related mortality are growing threats to immunosuppressed patients. Limited arsenal in form of antifungals, along with emerging resistance, their toxicity and efficacy are the bottlenecks in successful treatment. Thus there is an urgent need to develop new antifungal agents with no adverse effects and resistance emergence. This is possible with a better understanding of the physiology of fungi as well as mechanisms resulting in the development of multidrug resistance. Among the different mechanisms contributing to drug resistance, recently mitochondrion has been found to be crucial in determining the resistance in fungal cells. This includes mitochondrial ATP status, respiration, fission, fusion mutants, etc. The literature related to mitochondrial functioning in fungi suggests that this organelle can be targeted for future antifungal discovery.

Keywords

Drug resistance · Fungi · Mitochondria · Dysfunction · *Candida* · ERMES · Ca^{2+} · Phospholipid exchange

Introduction

The rise in incidences of fungal infections and multidrug resistance (MDR) has increased considerably and necessitates the development of novel antifungal agents and a relook on the other potential drug targets that could be further developed in the future. The vast majority of fungal infections reported are primarily caused by

R. Pasrija (✉) · D. Kundu
Department of Biochemistry, Maharshi Dayanand University, Rohtak, Haryana, India
e-mail: ritupasrija@yahoo.com

S. Hameed, Z. Fatima (eds.), *Pathogenicity and Drug Resistance of Human Pathogens*,
https://doi.org/10.1007/978-981-32-9449-3_8

Candida sp., *Aspergillus* sp., *Cryptococcus* sp., etc. Multiple pathways have been reported that contribute to the development of MDR in fungal pathogens, including over-expression of plasma membrane (PM) drug pumps, mutation or alteration in target proteins, alteration in membrane lipids, etc. [1, 2]. Adding to this list, the last few decades have also seen the mitochondrial morphology, its genome, and intracellular signaling dictating the drug resistance/sensitivity profile of different fungi [3–7]. However, a lot more needs to be understood about mitochondria, before it can be developed as an antifungal target in treating fungal infections.

Mitochondria are suggested to have endo-symbiotic origins, which are now integral to eukaryotic cells and perform many essential functions. Structurally, it is a double membranous organelle, where the outer mitochondrial membrane (OMM) forms a uniform, relatively smooth shell, and the inner mitochondrial membrane (IMM) is rather organized into many invaginations (cristae) in the matrix. The OMM and IMM are separated via inter-membrane space. Text-books generally portray it as a static organelle, but in reality, it is mobile and dynamic as it can both join by fusion during stress and divide by fission for production of new mitochondria [6, 8]. The energy demand of the cell maintains mitochondrial population and effective transfer during cell division. Mitochondrial multiplication is also known to be facilitated by different dynamin-related membrane (fission) proteins, with cytoskeleton-dependent transport mechanisms and ensuring their appropriate intracellular distribution [9]. Functionally, mitochondria are the main producers of ATP and site for many catabolic and anabolic reactions, including intracellular signaling, proliferation, aging, thermogenesis, iron-sulfur [Fe-S] cluster biosynthesis, formation of reactive oxygen species (ROS), etc. [10–12]. But mitochondria might dysfunction due to multiple factors including elevated ROS (as it is a mitochondrial waste product), mitochondrial DNA (mtDNA) damage, increase in mitochondria membrane potential, inhibition of electron transport chain (ETC), depletion of ATP in damaged mitochondria, and release of cytochrome *c* molecules, leading to activation of caspases and finally resulting in apoptosis, etc. [13, 14]. Its well-being requirement could also be estimated from the fact that mitochondrial dysfunction is one of the cause in many diseases and eventually can result in cell death [15]. Apart from energy status, the multiplication of mitochondria depends on replication as well as expression of the mitochondrial genome, which also involves the uptake of nuclear-encoded proteins. So, the cell requires constant feedback from mitochondria and regulates mitochondrial fusion, bio-genesis, and mitophagy during the cell cycle [9, 16]. Thus, it can be concluded that both nuclei to mitochondria and mitochondria to nucleus (retrograde) communication occurs in the cell.

In fungi, mitochondria are also found to be important in hyphal differentiation, biofilm formation, stress adaptation, cell wall biosynthesis, innate immune interactions, etc. [17, 18]. Thus, it would not be an exaggeration to implicate mitochondria in virulence of various human fungal pathogens. This chapter is an attempt to summarize the current knowledge about the mitochondrial genome and its importance in attributing drug resistance/sensitivity in different fungi.

Mitochondrial Genome in Different Fungi

Although mitochondrial function is essentially conserved in different fungi, but genome sizes and gene synteny are found to be highly inconsistent. The mitochondrial genome size varies among different species and can range from 20 kb (approx) for *Candida glabrata* to ~85 kb in non-pathogenic *Saccharomyces cerevisiae*. Table 1 shows the comparison of the mitochondrial genomes of *S. cerevisiae* and different pathogenic fungi. It is clearly evident that although gene content difference is not wide in fungi, variations in mitochondrial genome sizes are rather due to the disparity in the number and length of intronic region as well as organization of intergenic sequences. The mtDNA of baker's yeast, *S. cerevisiae*, is extensive, least GC % (17–18% only), with immoderate intergenic regions, constituting up to 62.2% of total [19]. Fungal mitochondrial genomes are generally circular dsDNA molecule, but still, controversy exists in most cases. Mitochondrial genome encodes for limited gene products, required for ETC and ATP synthesis, small (*runs*) and large (*rnl*) subunit mitochondrial rRNAs, and tRNA as shown in Table 1. Interestingly, genes in most ascomycetes are unequivocally encoded by sense mtDNA strand, while in basidiomycetes encode by either one of mtDNA strand [20]. The genetic code used in yeast mitochondrial translation also differs from the universal code, and AUA is translated as methionine, CUN as threonine, and UGA as tryptophan.

Some yeast can survive the loss of mtDNA, and so fungi serve as an outstanding prototype to genetically scrutinize the cellular and biochemical pathways required for maintenance of respiratory activity because it is also proficient of fulfilling its energy requirements with ATP generated by fermentation [9]. Thus, ATP synthesis via oxidative phosphorylation and the presence of the mitochondrial genome is redundant as long as fermentable carbon sources are present for growth. Even if oxygen is obtainable, yeast cells can generate ATP by glycolysis with ethanol as the end product. On the other hand, when yeast is cultured with non-fermentable carbon sources, like glycerol/lactose, respiration and the presence of a complete mitochondrial genome become vital [21]. The capacity to bear mtDNA loss also varies between species and is discussed in detail later.

Role of Mitochondria in Drug Resistance

Mitochondria perform varied functions, and so there are different mechanisms through which mitochondria can alter drug resistance in fungi. In the next sections, we have discussed different mitochondrial functions and how any alteration can be crucial in fungal virulence and pathogenesis.

Table 1 Comparison of the mitochondrial genome of different fungi

S. no	Genome features	*Saccharomyces cerevisiae*	*Candida glabrata*	*Candida albicans*	*Candida parapsilosis*	*Aspergillus fumigatus*	*Aspergillus niger*	*Aspergillus tubingensis*	*Talaromyces marneffei*	*Fusarium oxysporum*
1.	Strain name	S288C	CBS 138	SC5314	CBS 7157 (SR 23)	A1163	N909	932	MP1	F11
2.	Accession no.	NC_001224	NC_004691	NC_002653	NC_005253	NC_017016	NC_007445	NC_007597	NC_005256	AY945289
3.	Genome size (kb)	85.7	20.06	40.42	32.74	30.7	31.1	33.7	35.4	34.5
4.	Map	Circular	Circular	Circular	Linear	Circular	Circular	Circular	Circular	Circular
5.	G+C content (%)	17.11	17.64	32.2	23.8	25.5	26.9	26.8	24.63	31.0
6.	Protein-coding gene	–	8	14	20	19	16	16	15	16
7.	tRNA	24	23	30	24	31	25	25	28	25
8.	rRNA	2	2	2	2	2	2	2	2	2
9.	No. of introns	12	3	5	7	5	3	5	11	2
10.	Total length of introns (bp)	17,253	3995	6221	8965	4080	3308	5494	11,484	2742
11.	Genome length occupied by introns (%)	20.11	19.91	15.39	27.38	13.29	10.64	16.32	32.41	7.95
12.	Intergenic region (%)	62.2	14.9	36.1	7.7	–	21.5	21	8.8	32.12
13.	References	[72–74]	[74, 75]	[74, 76]	[74, 77, 78]	[79]	[78–80]	[78–80]	[7981	[78, 82]

Importance of mtDNA and Mitochondria Dysfunction in Drug Resistance

mtDNA-mediated dysfunction and its effect on drug resistance are well-studied in different fungal species. Interestingly the effect of mtDNA loss/mutation in *Candida* sp., *S. cerevisiae*, *Cryptococcus*, etc. is variable for separate classes of drugs [16]. Depending upon the changes in mitochondrial function, it can turn strains either hypo- or hypervirulent.

S. cerevisiae is not pathogenic but is the most well-studied fungi and can bear the partial or full loss of mtDNA, turning them to have small colony size, so-called as *petite* [16]. In some strains, there could be an extensive deletion of mtDNA (called rho^-) or no mtDNA at all (rho^0/ρ^0) [22]. Petite mutants cannot propagate on non-fermentable carbon sources (but do not die) as they are inadequate in the enzymes required for respiration. Respiratory deficiency can also arise from mtDNA mutations (cytoplasmic changes) or from the remodeling of nuclear genes and depend on fermentable carbon sources for survival [22]. *C. glabrata* also behaves similarly like *S. cerevisiae* and can bear the loss of mtDNA. These respiratory-deficient mutants of *S. cerevisiae* and *C. glabrata* turn resistant upon loss of mtDNA [16]. Both yeasts show increased expression of *ScPDR5* and *CgCDR1*, which are ATP-binding cassette (ABC) family of membrane transporter proteins. *C. glabrata* also show increased ergosterol content, turning them more susceptible to polyenes [23]. Another study show that petite mutants of *C. glabrata* are less virulent than parent strains in an animal model [24]. However, *Candida albicans* and *Cryptococcus neoformans* are more rigid and unable to survive mtDNA loss (partial or complete), and respiratory deficiency is caused by either loss of mitochondrial membrane potential or uncoupling of oxidative phosphorylation, etc. [16, 25, 26]. *C. albicans* petite strain (with uncoupled oxidative phosphorylation) shows over-expression of *MDR1* (MFS transporter) and reduced phagocytosis by macrophages and neutrophils [25]. This mutant strain is resistant to fluconazole and voriconazole, but sensitivity to itraconazole, ketoconazole, and amphotericin B remained largely unchanged [25]. In a separate study involving screening imidazole compounds against mitochondrial cytochrome bc_1, mutants of *C. albicans* show that its impairment causes a reversal of fluconazole resistance (by converting azoles to fungicidal) and reduces virulence [27]. *C. neoformans* is encapsulated yeast, and capsule is the prime virulence factor of this pathogen. Inhibitors of the respiration impair capsule expansion and thus reduce virulence [28]. However, the infamous Vancouver Island Outbreak (VIO) of cryptococcal disease (caused by *C. neoformans*, *Cryptococcus gatti)* study also shows that *C. gatti* lineage is hypervirulent and has mitochondrial morphological defects which appear tubular during intracellular growth [29].

The maintenance of mitochondrial function also depends on contribution of the nuclear genome, which includes proteins for oxidative phosphorylation and is jointly encoded by the mitochondrial and nuclear genome. So, cross-talk between mitochondria and the nucleus is of prime importance in the biogenesis of mitochondria and cellular growth. Some of the nuclear indicators gets transmitted through nuclear-encoded proteins which are imported into mitochondria and

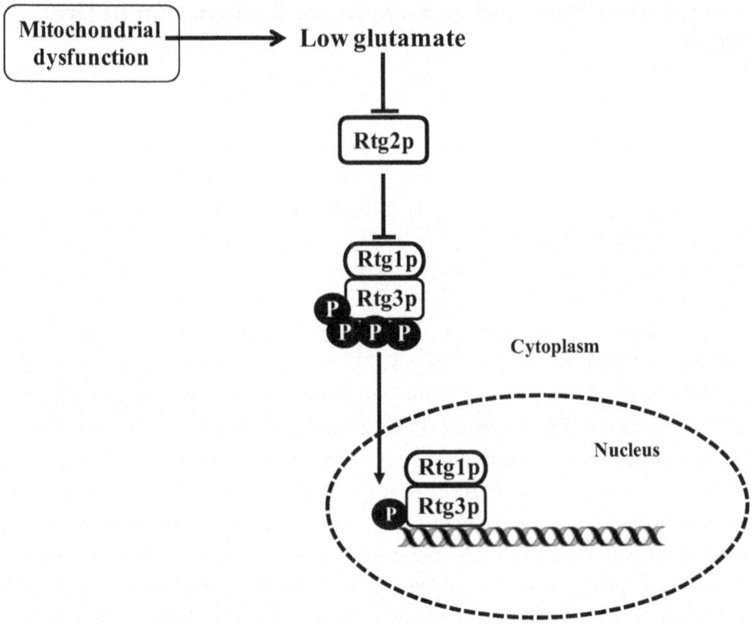

Fig. 1 **Retrograde signaling involving *RTG* genes**

influence the expression of mitochondrial genes. This flow of information from the nucleus to mitochondria is called anterograde regulation. However, mitochondria are also espoused in a retrograde mode of superintendence, in which cells try to counter the changes in organelle via modulation in nuclear gene expression [30]. Thus, retrograde signaling or *RTG* (reverse flow of direction from mitochondria to nucleus) in eukaryotes exists for normal feedback, as well as for stress signal communication to the nucleus. This, in turn, regulates the nuclear gene expression and changes in metabolism, which accommodates the defects in mitochondria [31, 32]. This could be the reason that mitochondria dysfunction can result in both sensitivity and resistance of fungal cells.

S. cerevisiae's retrograde signaling could be either *RTG*-dependent or alternate signaling pathway which is *RTG* independent [30]. *RTG*-dependent retrograde signaling depends on three different cytosolic proteins: Rtg1p, Rtg2p, and Rtg3p (Fig. 1). Here, Rtg1 and Rtg3 are transcription factors (basic helix-loop-helix (bHLH)/leucine zipper (LeuZip)) and bind to the DNA (at binding site sequence GTCAC) as heterodimers. When activated, the Rtg1/3p complex translocates from cytoplasm to nucleus and controls the translation of genes that encode mitochondrial proteins (Fig. 1). Rtg1/3 is generally retained in cytoplasm, and translocation requires Rtg3p dephosphorylation at specific sites [32]. Deprivation of nuclear-mitochondrial signaling gene *RTG1* reduces the *PDR5* (pleiotropic drug resistance) expression and drug resistance in ρ^0 cells [5].

Along with *RTG* signaling, *RTG*-independent retrograde signaling also confers Pdr5p-mediated resistance to cells (Fig. 2) [5, 33]. Loss of Oxa1p (an IMM protein

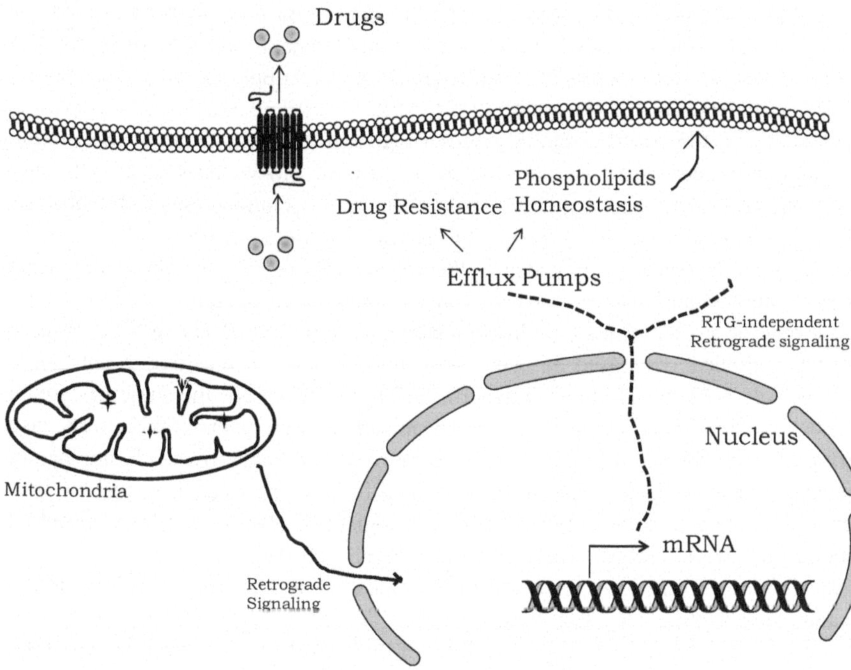

Fig. 2 **Pathway showing the role of mitochondria in drug resistance**

involved in export) results in an increase of Pdr3p activity, leading to the overexpression of *ScPDR5* [5]. *PDR5* of *S. cerevisiae* and *C. glabrata* is generally controlled by master regulator Pdr1 and Pdr3, known to upregulate ScPdr1, CgCdr1, and CgCdr2, resulting in drug resistance (Fig. 2) [16, 34–36]. ρ^0 cells also cause an up-regulation of Pdr3p and in-turn *PDR5* as described earlier.

In a separate study, *CaFZO1* (important for mitochondrial biogenesis) was studied in *C. albicans*. *fzo1Δ/Δ* mutant has fragmented mitochondria with mitochondrial genome loss and reduction in membrane potential and turns sensitive to azoles and amphotericin B due to reduced activity of *CDR* efflux pumps [37]. In addition, reduced mitochondrial ATP levels (due to Complex I mutants) also result in hypersensitivity to fluconazole [7].

In nutshell, it can be concluded that mitochondrial dysfunction can affect the fungal response to several drugs, which can be opposite for separate categories of drugs like azoles and polyenes.

Heme metabolism affecting Resistance/Sensitivity

Cytochrome P450 (*CYPs*/*P450s*) monooxygenases enzyme got their names after the fact that they absorb light at 450 nm under exposure to carbon monoxide. They are involved in oxidative metabolism in different kingdoms of life including fungi. P450s amino acid sequences could be variable between species, but their structures are highly conserved [38]. These structurally conserved proteins bind to cofactor heme, already found crucial for their function. Heme (iron-protoporphyrin IX) synthesis occurs in mitochondria and depends on iron supply just like other [Fe-S] proteins synthesized in mitochondria [39].

Azoles are the most commonly prescribed antifungals and target the unique fungal membrane sterol, i.e., ergosterol. Ergosterol biosynthesis is a multi-step pathway, and each step requires separate enzyme including one P450 protein, *CYP51* (also known as *ERG11* or sterol 14α-demethylase) [40]. *ERG11* de-methylate lanosterol, a key step in the ergosterol biosynthesis, and is the target of azole antifungals. Another protein Dap1 (damage resistance protein 1) is also a heme-binding protein (outside mitochondria) and activates *ERG11* in *S. cerevisiae*. *dap1*Δ cells have decreased mitochondrial function and are sensitive to antifungal itraconazole [41]. In addition, ergosterol and heme also have the same up-stream precursors, and their synthesis is synchronized as well [41]. Doxycycline also promotes the efficacy of azoles like fluconazole, probably by depleting heme-associated iron [42]. Thus there exists a possible link between iron metabolism, heme, and ergosterol biosynthesis. So, it is logical to observe that mitochondrial mutants also have altered cellular PM sterol levels [16, 43].

In *C. glabrata*, in vitro exposure to fluconazole can give rise to mitochondrial mutants at a significantly high frequency [23, 44, 45]. *C. neoformans* Atm1 (mitochondrial ABC transporter) is responsible for the export of iron-sulfur cluster (ISC) precursors to the cytoplasm, as well as heme metabolism. The *atm1* mutants are avirulent in cryptococcosis murine model [46]. The cytochrome b_5 CybE of *Aspergillus fumigatus* is also controlled by iron supply and is pivotal for azole resistance [38]. The findings suggest that mitochondrial defects involving iron/heme biosynthesis are associated with azole sensitivity/resistance and virulence in different fungi.

Endoplasmic Reticulum (ER)-Mitochondrial Calcium (Ca^{2+}) Crosstalk

Calcineurin (Cn) is a well-known regulator of Ca^{2+} homeostasis and also controls virulence and antifungal resistance in fungi [47–49]. Combinations of Ca^{2+} and Cn inhibitors show synergy in inhibiting the growth of resistant fungal strains and reduce their virulence (Fig. 3). Cn is actually a Ca^{2+}/calmodulin (CaM)-dependent Ser/Thr protein phosphatase and regulates intracellular Ca^{2+} concentration and ultimately signaling [48, 50]. In response to stimuli, Ca^{2+} channel pumps extracellular Ca^{2+} into cytoplasm, and ER also release the Ca^{2+} stores, both resulting in an increase in its concentration. Upon binding Ca^{2+} ions, the Ca^{2+}-binding protein CaM endures a conformational change and binds to Cn and increases its phosphatase activity (Fig. 3). In yeast, Cn dephosphorylates its downstream transcription factor Crz1 (Calcineurin-Responsive Zinc Finger 1), much like NFAT (Nuclear factor of activated T-cells) in other eukaryotes [51]. Interestingly, Crz1 has no mammalian counterpart. The immunosuppressive drugs, cyclosporin A (CsA) and FK506, complex with an "immunophilin" [cyclophilin (Cyp) and FK506-binding protein (FKBP) or FKBP12, respectively], and drug/immunophilin complex associates

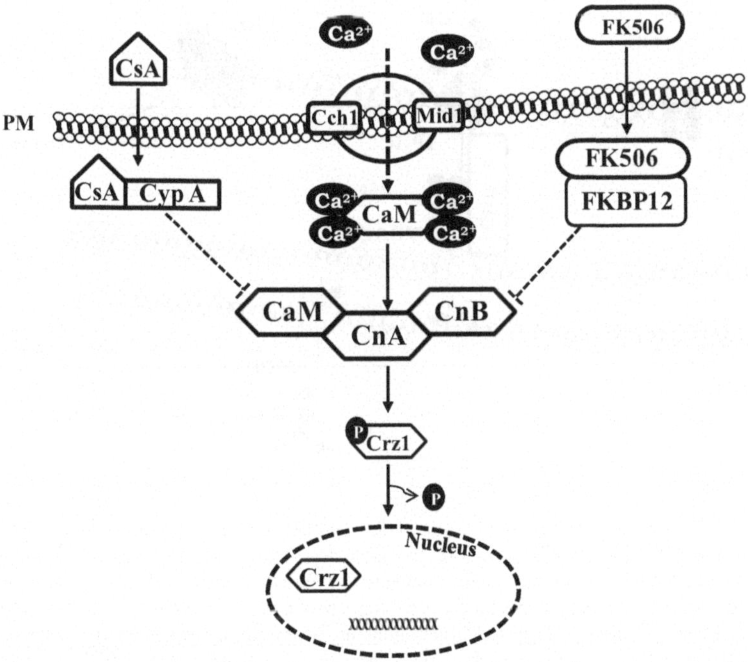

Fig. 3 **Ca^{2+}-calcineurin (Cn) signaling pathway:** In response to the external signal, the plasma membrane (PM) high-affinity Ca^{2+} channels (Cch1 and Mid1) are activated, resulting in a rapid influx of Ca^{2+} ions in the cell. The increased Ca^{2+} ions concentration is sensed by the Ca^{2+}-binding protein calmodulin (CaM) that undergoes a conformational change after binding to four Ca^{2+} ions. The activated Ca^{2+}-CaM complex binds to the inactive subunits of calcineurin heterodimer (CnA and CnB) and leads to its activation. The immunophilin-immunosuppressant complexes (FK506-FKBP12 and CsA-CypA) inhibit the activity of Cn. The activated Cn complex dephosphorylates Crz1 (Calcineurin-Responsive Zinc Finger 1), which translocates to the nucleus and further induces the expression of genes involved in the regulation of stress response, cell wall integrity, growth, and drug resistance

with Cn and inhibits Crz1 dephosphorylation (Fig. 3) [50]. That is the reason that cyclosporin A and FK506 show intrinsic antimicrobial effect. Also, it is well understood that in the absence of Ca^{2+} signaling, the fungistatic effect of fluconazole on *Candida* sp. rather becomes fungicidal [34]. However, Cn does become crucial for viability in *C. albicans* if the membrane is challenged with ergosterol biosynthesis inhibitors (azoles) and could be exploited in the development of potential drugs.

Here, mitochondrial role becomes significant, as intracellular Ca^{2+} is sectionalized in organelles including - ER, vacuole, and mitochondria [34]. G-protein-coupled receptors (GPCRs) lead to the generation of the second messenger inositol 1,4,5-trisphosphate (IP_3), which liberates Ca^{2+} from intracellular stores such as the ER having IP_3R (Inositol trisphosphate receptor) (Fig. 4) [52]. Ca^{2+} imaging techniques have shown that Ca^{2+} released from the ER is taken up by mitochondria, buffering excess of cytosolic Ca^{2+} and preventing its possible cytotoxic effects. It was reported that knockdown of *OPA1* (dynamin-related GTPase) located in the IMM enhances Ca^{2+} influx into mitochondria [53]. Immuno-histochemical studies

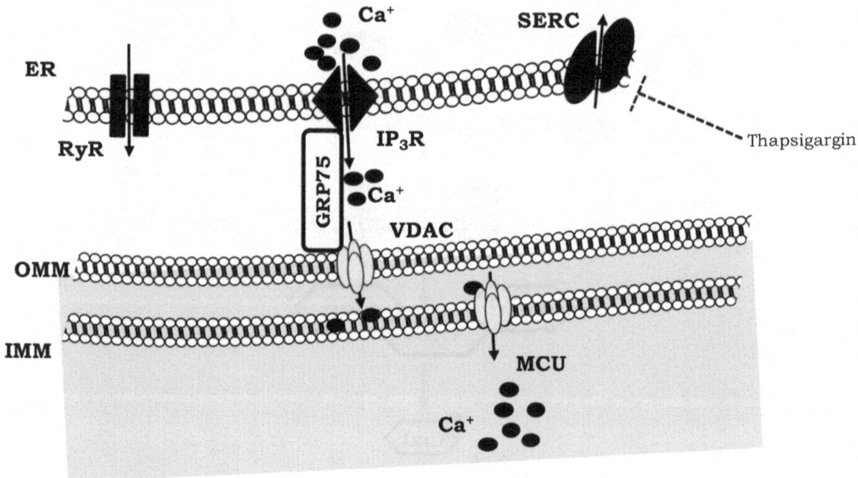

Fig. 4 Ca^{2+} crosstalk between endoplasmic reticulum (ER) and mitochondria. Close interactions between the ER and mitochondria are essential for rapid and sustained Ca^{2+} uptake by mitochondria. Voltage-dependent anion channels (VDACs), located at the outer mitochondrial membrane (OMM), are responsible for the rapid transfer of Ca^{2+} from ER-mitochondria apposition, and their function results in high Ca^{2+} microdomains in the mitochondria intermembrane space. Accumulation of Ca^{2+} into the mitochondrial matrix occurs via the mitochondrial Ca^{2+} uniporter (MCU), which rapidly accumulates Ca^{2+} across the steep electrochemical gradient. A number of chaperones and regulatory proteins control the formation of the ER-mitochondria junction, the clustering of signaling proteins and their modulation. A glucose-regulated protein (GRP75) facilitates Ca^{2+} uptake in mitochondria by stabilizing the interaction of VDAC with inositol 1,4,5-trisphosphate receptor (IP_3R)

show that IP_3R enriched sites of ER are apposed to mitochondria and these zones could be the "hotspots" of Ca^{2+} transfer from ER to mitochondria [54]. Along with this, IP_3R are physically linked to VDAC (voltage-dependent anion channel) of the OMM to mediate Ca^{2+} release through the molecular chaperone glucose-regulated protein 75 (GRP75) in rat liver and HeLa cells (Fig. 4) [54]. This ER-mitochondrial Ca^{2+} communication can be uncoupled by treatment with TGF-β in smooth muscle cells possibly by downregulation of IP_3R Ca^{2+} channels [55]. Further, mitochondrial Ca^{2+} can also activate Ca^{2+} sensitive matrix dehydrogenases of mitochondria and control efficiency of the respiratory chain [56]. Although whether these dehydrogenases affect drug resistance or not still needs to be verified. Apart from these, ERMES (Endoplasmic Reticulum (ER)-mitochondria encounter structure) complex connecting ER with mitochondria is also significant for Ca^{2+} exchange and implicated in drug resistance (discussed later).

Mitochondria association with ER

Different organelles were long presumed as separate units in eukaryotic cells, but electron microscope-derived images show electron-dense structures, bridging

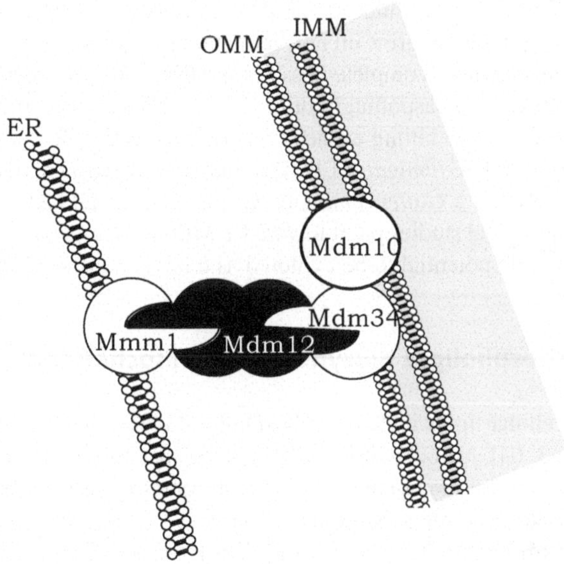

Fig. 5 ERMES (Endoplasmic Reticulum-mitochondria encounter structure) tethers the ER and mitochondria together in yeast and is a multi-subunit complex. It consists of four core subunits: Mmm1, Mdm10, Mdm12, and Mdm34. Mmm1 is an integral ER membrane protein. Mdm 10 and Mdm34 are present in OMM, and Mdm12 is a cytosolic protein between ER and mitochondria

organelles including ER and mitochondria in rat liver tissue [57]. ER membranes co-sediment with mitochondria as contaminants in cell fractionation experiments [58]. These inter-organellar interactions are referred to as mitochondria-associated membranes (MAMs) and exist as physical tethers/links for communication and exchange of metabolites. Since then, inter-organelle communication is thought to be an important component of mitochondria and includes a recently identified protein complex called ERMES in yeast [21, 59]. ERMES is located at the communion of the ER and mitochondria and serves to zip them together (Fig. 5). This complex is composed of four core proteins called as either *MDM* (mitochondria distribution and morphology) or *MMM* (mitochondria morphology maintenance) and got their names after their role in maintaining morphology of mitochondria and their distribution during cell division. The four core subunits are named Mmm1, Mdm10, Mdm12, and Mdm34. Interestingly, standard homology searches have identified orthologs of ERMES components in fungi only, and since then work on ERMES gained momentum. Mmm1 is an integral ER membrane protein. Mdm10 (β-barrel) and Mdm34 are proteins of the OMM, and Mdm12 is a cytosolic protein as shown in Fig. 5 [21]. These appear stable structures as confirmed by live cell microscopy for over 40 min [60]. ERMES is the hub in the management of mitochondrial membrane biosynthesis, genome replication, macrophage killing, hyphal elongation, Ca^{2+} signaling, protein import, and mitochondrial phospholipids homeostasis. So, any mutations in the components of ERMES can have varying mitochondrial phenotypes, including morphological defects, loss of mitochondrial nucleoids, the incompetence of cells to grow on non-fermentable media, etc. [21].

Researchers have already studied and knocked out the ERMES subunits in different fungi including *S. cerevisiae*, *C. albicans*, *A. fumigatus*, *Aspergillus nidulans*, *Neurospora crassa*, etc. [21, 59, 61]. Deletion of ERMES subunits in

S. cerevisiae results in mtDNA conservation and inheritance defects along with incapability to grow on non-fermentable carbon sources [21]. *MMM1*-deficient yeast cells exhibit complete loss of mtDNA [61]. *S. cerevisiae* ERMES mutants are sensitive to caspofungin drug [62]. *C. albicans* ERMES *mmm1* mutant has reduced macrophage killing capacity and delayed cell wall restructuring during morphogenesis [59]. *A. fumigatus* ERMES mutants have growth defects, and *Mmm1* conditional mutant in a *Galleria mellonella* infection model had reduced virulence [61]. These individual studies point toward ERMES to be crucial in affecting drug resistance and has the potential to be explored as a novel antifungal target.

Phospholipid Biosynthesis in Mitochondria

Cellular lipids including phospholipid biosynthesis take place in both ER and IMM [63, 64]. Mitochondria harbor a wide spectrum of glycerophospholipids common to biological membranes. It also contributes to phospholipid biosynthesis in cells and produces phosphatidylethanolamine (PE), phosphatidylglycerol (PG), and cardiolipin (CL) (Fig. 6) [65]. But for other phospholipids including phosphatidylcholine (PC), phosphatidylinositol (PI) and phosphatidylserine (PS) have to be procured from the ER (Fig. 6). PS obtained from the ER is utilized by PS decarboxylase (Psd1) to produce PE at the IMM. Thus, Psd1 substrate and product have to come from and return to the ER, respectively, and recommend ERMES role in phospholipid transport [21].

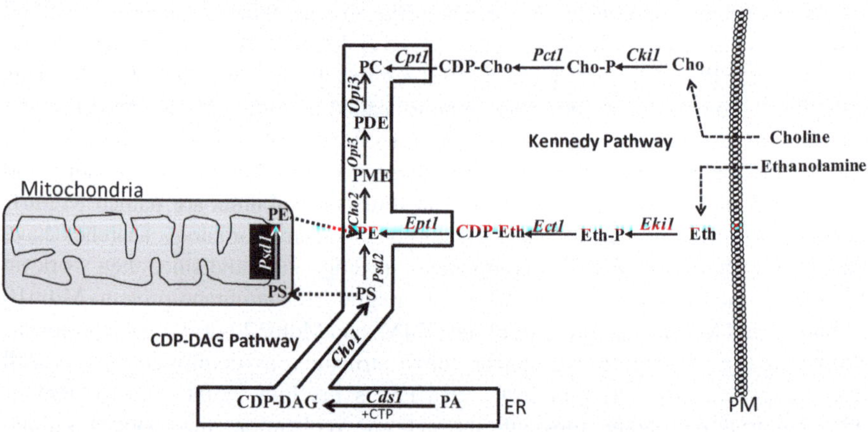

Fig. 6 Phospholipid trafficking between ER and mitochondria. In CDP-DAG pathway, the synthesis of PS is accomplished in the ER which is subsequently converted to PE in mitochondria (via Psd1p) and in ER (via Psd2p). The ER and mitochondrial interface fosters the trafficking of phospholipids between two organelles via ERMES/SAM complex, etc. The Kennedy pathway includes conversion of Etn and Cho from exogenous sources. **Abbreviations:** CDP-diacylglycerol (CDP-DAG); choline (Cho); ethanolamine (Etn); phosphatidate (PA); phosphatidylcholine (PC); phosphatidylethanolamine (PE); phosphatidylmonomethylethanolamine (PME); phosphatidylserine (PS)

Phospholipid biosynthesis impairment and mitochondria defects leading to alteration in drug resistance is studied in many fungi. *PGS1* is involved in the synthesis of mitochondrial anionic phospholipids PG and CL. Its loss in *S. cerevisiae* (*pgs1*Δ) results in defects in cell wall integrity and biogenesis due to reduced β-1,3-glucan, turning them sensitive to cell wall inhibitor [66, 67]. *C. glabrata CgPGS1* is also sensitive to azoles due to reduced mitochondrial function owing to decrease in level of cytochromes *b* and *a* [68]. Recently, *C. albicans* cell wall integrity and phospholipids imbalance were also found to be interconnected and crucial for caspofungin sensitivity [62]. The *C. albicans cho1*Δ/Δ mutant is also avirulent in a mouse model [69].

Psd1 overexpression activates *PDR5* transcription leading to drug resistance in a Pdr3-mediated manner, and disappearance of the *PSD1* gene from ρ^0 cells prevents the normal activation of *PDR5* expression [70]. It is already intended that mitochondria maintain a highly active exchange of phospholipids with ER [64]. The ER to mitochondria PS transfer slows down appreciably in cells missing both the ER-shaping reticulon proteins and the ERMES complex subunits [71]. Further, this defect could be restored by expression of an artificial protein tethering ER and the mitochondria [21].

Others

Besides these mechanisms, several other reports have also linked mitochondrial functioning with drug sensitivity. It includes various proteins in IMM and OMM, which affect mitochondrial performance. Loss of Oxa1p affecting resistance in fungi is already discussed earlier. Another protein complex SAM (sorting and assembly machinery) subunit Sam57 is required for multiple functions including the growth of *C. albicans* and virulence. *Sam57Δ/Δ* have got cell wall defects and show filamentation defects [7].

Concluding Comments

It can be concluded that as mitochondria are the prime site for biosynthesis of heme, ATP, and phospholipids and ROS generation. Therefore, working and efficiency of mitochondria affects many aspects of yeast biology and physiology and ultimately affects the drug resistance/sensitivity profile toward different drugs. Although a difference of response in different fungi could not be ruled out, still a common conserved aspect of mitochondrial physiology can be optimized for target validation. Even targeting yeast and mold separately would also be a good option to fill the gap in the development of long-awaited antifungals.

References

1. Robbins, N., Caplan, T., & Cowen, L. E. (2017). Molecular evolution of antifungal drug resistance. *Annual Review of Microbiology, 71*(1), 753–775.
2. Loeffler, J., & Stevens, D. A. (2003). Antifungal drug resistance. *Clinical Infectious Diseases, 36*(Supplement_1), S31–S41.
3. Ferrari, S., Sanguinetti, M., De Bernardis, F., Torelli, R., Posteraro, B., Vandeputte, P., et al. (2011). Loss of mitochondrial functions associated with azole resistance in *Candida glabrata* results in enhanced virulence in mice. *Antimicrobial Agents and Chemotherapy, 55*(5), 1852–1860.
4. Torelli, N. Q., Ferreira-Júnior, J. R., Kowaltowski, A. J., & da Cunha, F. M. (2015). *RTG1* – and *RTG2* -dependent retrograde signaling controls mitochondrial activity and stress resistance in *Saccharomyces cerevisiae. Free Radical Biology & Medicine, 81*, 30–37.
5. Hallstrom, T. C., & Moye-Rowley, W. S. (2000). Multiple signals from dysfunctional mitochondria activate the pleiotropic drug resistance pathway in *Saccharomyces cerevisiae. The Journal of Biological Chemistry, 275*(48), 37347–37356.
6. Verma, S., Shakya, V. P. S., & Idnurm, A. (2018). Exploring and exploiting the connection between mitochondria and the virulence of human pathogenic fungi. *Virulence, 9*(1), 426–446.
7. Li, D., & Calderone, R. (2017). Exploiting mitochondria as targets for the development of new antifungals. *Virulence, 8*(2), 159–168.
8. Archer, S. L. (2013). Mitochondrial dynamics – mitochondrial fission and fusion in human diseases. *The New England Journal of Medicine, 369*(23), 2236–2251.
9. Westermann, B. (2014). Mitochondrial inheritance in yeast. *Biochimica et Biophysica Acta BBA – Bioenergetics, 1837*(7), 1039–1046.
10. Starkov, A. A. (2008). The role of mitochondria in reactive oxygen species metabolism and signaling. *Annals of the New York Academy of Sciences, 1147*(1), 37–52.
11. Boniecki, M. T., Freibert, S. A., Mühlenhoff, U., Lill, R., & Cygler, M. (2017). Structure and functional dynamics of the mitochondrial Fe/S cluster synthesis complex. *Nature Communications, 8*(1), 1287.
12. Bratic, A., & Larsson, N.-G. (2013). The role of mitochondria in aging. *The Journal of Clinical Investigation, 123*(3), 951–957.
13. Li, D., Chen, H., Florentino, A., Alex, D., Sikorski, P., Fonzi, W. A., et al. (2011). Enzymatic dysfunction of mitochondrial complex I of the *Candida albicans* goa1 mutant is associated with increased reactive oxidants and cell death. *Eukaryotic Cell, 10*(5), 672–682.
14. Knorre, D., Sokolov, S., Zyrina, A., & Severin, F. (2016). How do yeast sense mitochondrial dysfunction? *Microbial Cell, 3*(11), 532–539.
15. Magalhães, J., Venditti, P., Adhihetty, P. J., Ramsey, J. J., & Ascensão, A. (2014). Mitochondria in health and disease. *Oxidative Medicine and Cellular Longevity, 2014*, 814042.
16. Shingu-Vazquez, M., & Traven, A. (2011). Mitochondria and fungal pathogenesis: Drug tolerance, virulence, and potential for antifungal therapy. *Eukaryotic Cell, 10*(11), 1376–1383.
17. She, X., Khamooshi, K., Gao, Y., Shen, Y., Lv, Y., Calderone, R., et al. (2015). Fungal-specific subunits of the *Candida albicans* mitochondrial complex I drive diverse cell functions including cell wall synthesis: Defective colonization in host tissues by mitochondrial mutants. *Cellular Microbiology, 17*(9), 1350–1364.
18. Calderone, R., Li, D., & Traven, A. (2015). System-level impact of mitochondria on fungal virulence: To metabolism and beyond. *FEMS Yeast Research, 15*(4), fov027.
19. de Zamaroczy, M., & Bernardi, G. (1985). Sequence organization of the mitochondrial genome of yeast – a review. *Gene, 37*(1–3), 1–17.
20. Li, Y., Hu, X.-D., Yang, R.-H., Hsiang, T., Wang, K., Liang, D.-Q., et al. (2015). Complete mitochondrial genome of the medicinal fungus *Ophiocordyceps sinensis. Scientific Reports, 5*, 13892.
21. Kornmann, B., Currie, E., Collins, S. R., Schuldiner, M., Nunnari, J., Weissman, J. S., et al. (2009). An ER-mitochondria tethering complex revealed by a synthetic biology screen. *Science, 325*, 477–481.

22. Chen, X. J., & Clark-Walker, G. D. (2018). Unveiling the mystery of mitochondrial DNA replication in yeasts. *Mitochondrion, 38*, 17–22.
23. Brun, S., Bergès, T., Poupard, P., Vauzelle-Moreau, C., Renier, G., Chabasse, D., et al. (2004). Mechanisms of azole resistance in petite mutants of *Candida glabrata*. *Antimicrobial Agents and Chemotherapy, 48*(5), 1788–1796.
24. Brun, S., Dalle, F., Saulnier, P., Renier, G., Bonnin, A., Chabasse, D., et al. (2005). Biological consequences of petite mutations in *Candida glabrata*. *The Journal of Antimicrobial Chemotherapy, 56*(2), 307–314.
25. Cheng, S., Clancy, C. J., Nguyen, K. T., Clapp, W., & Nguyen, M. H. (2007). A *Candida albicans* petite mutant strain with uncoupled oxidative phosphorylation overexpresses *MDR1* and has diminished susceptibility to fluconazole and voriconazole. *Antimicrobial Agents and Chemotherapy, 51*(5), 1855–1858.
26. Toffaletti, D. L., Nielsen, K., Dietrich, F., Heitman, J., & Perfect, J. R. (2004). *Cryptococcus neoformans* mitochondrial genomes from serotype A and D strains do not influence virulence. *Current Genetics, 46*(4), 193–204.
27. Vincent, B. M., Langlois, J.-B., Srinivas, R., Lancaster, A. K., Scherz-Shouval, R., Whitesell, L., et al. (2016). A fungal-selective cytochrome bc1 inhibitor impairs virulence and prevents the evolution of drug resistance. *Cell Chemical Biology, 23*(8), 978–991.
28. Trevijano-Contador, N., Rossi, S. A., Alves, E., Landín-Ferreiroa, S., & Zaragoza, O. (2017). Capsule enlargement in *Cryptococcus neoformans* is dependent on mitochondrial activity. *Frontiers in Microbiology, 8*, 1423.
29. Ma, H., & May, R. C. (2010). Mitochondria and the regulation of hypervirulence in the fatal fungal outbreak on Vancouver Island. *Virulence, 1*(3), 197–201.
30. Butow, R. A., & Avadhani, N. G. (2004). Mitochondrial signaling: The retrograde response. *Molecular Cell, 14*(1), 1–15.
31. Parikh, V. S., Morgan, M. M., Scott, R., Clements, L. S., & Butow, R. A. (1987). The mitochondrial genotype can influence nuclear gene expression in yeast. *Science, 235*(4788), 576–580.
32. Liu, Z., & Butow, R. A. (2006). Mitochondrial retrograde signaling. *Annual Review of Genetics, 40*(1), 159–185.
33. Traven, A., Wong, J. M., Xu, D., Sopta, M., & Ingles, C. J. (2001). Interorganellar communication. Altered nuclear gene expression profiles in a yeast mitochondrial DNA mutant. *The Journal of Biological Chemistry, 276*(6), 4020–4027.
34. Kaur, R., Castaño, I., & Cormack, B. P. (2004). Functional genomic analysis of fluconazole susceptibility in the pathogenic yeast *Candida glabrata*: Roles of calcium signaling and mitochondria. *Antimicrobial Agents and Chemotherapy, 48*(5), 1600–1613.
35. Balzi, E., Chen, W., Ulaszewski, S., Capieaux, E., & Goffeau, A. (1987). The multidrug resistance gene *PDR1* from *Saccharomyces cerevisiae*. *The Journal of Biological Chemistry, 262*(35), 16871–16879.
36. Delaveau, T., Delahodde, A., Carvajal, E., Subik, J., & Jacq, C. (1994). *PDR3*, a new yeast regulatory gene, is homologous to *PDR1* and controls the multidrug resistance phenomenon. *Molecular & General Genetics, 244*(5), 501–511.
37. Thomas, E., Roman, E., Claypool, S., Manzoor, N., Pla, J., & Panwar, S. L. (2013). Mitochondria influence *CDR1* efflux pump activity, Hog1-mediated oxidative stress pathway, iron homeostasis, and ergosterol levels in *Candida albicans*. *Antimicrobial Agents and Chemotherapy, 57*(11), 5580–5599.
38. Misslinger, M., Gsaller, F., Hortschansky, P., Müller, C., Bracher, F., Bromley, M. J., et al. (2017). The cytochrome b_5 CybE is regulated by iron availability and is crucial for azole resistance in *A. fumigatus*. *Metallomics, 9*(11), 1655–1665.
39. Lipiński, P., Starzyński, R. R., Styś, A., Gajowiak, A., & Staroń, R. (2014). Heme metabolism as an integral part of iron homeostasis. *Postepy Higieny Medycyny Doswiadczalnej Online, 68*, 557–570.
40. Cowen, L. E., Sanglard, D., Howard, S. J., Rogers, P. D., & Perlin, D. S. (2015). Mechanisms of antifungal drug resistance. *Cold Spring Harbor Perspectives in Medicine, 5*(7), a019752.

41. Craven, R. J., Mallory, J. C., & Hand, R. A. (2007). Regulation of iron homeostasis mediated by the heme-binding protein Dap1 (damage resistance protein 1) via the *P450* protein Erg11/Cyp51. *The Journal of Biological Chemistry, 282*(50), 36543–36551.
42. Fiori, A., & Van Dijck, P. (2012). Potent synergistic effect of doxycycline with fluconazole against *Candida albicans* is mediated by interference with iron homeostasis. *Antimicrobial Agents and Chemotherapy, 56*(7), 3785–3796.
43. Jawallapersand, P., Mashele, S. S., Kovačič, L., Stojan, J., Komel, R., Pakala, S. B., et al. (2014). Cytochrome P450 monooxygenase CYP53 family in fungi: Comparative structural and evolutionary analysis and its role as a common alternative anti-fungal drug target. Han K-H, editor. *PLoS One, 9*(9), e107209.
44. Sanglard, D., Ischer, F., & Bille, J. (2001). Role of ATP-binding-cassette transporter genes in high-frequency acquisition of resistance to azole antifungals in *Candida glabrata. Antimicrobial Agents and Chemotherapy, 45*(4), 1174–1183.
45. Brun, S., Aubry, C., Lima, O., Filmon, R., Berges, T., Chabasse, D., et al. (2003). Relationships between respiration and susceptibility to azole antifungals in *Candida glabrata. Antimicrobial Agents and Chemotherapy, 47*(3), 847–853.
46. Do, E., Park, S., Li, M.-H., Wang, J.-M., Ding, C., Kronstad, J. W., et al. (2018). The mitochondrial ABC transporter Atm1 plays a role in iron metabolism and virulence in the human fungal pathogen *Cryptococcus neoformans. Medical Mycology, 56*(4), 458–468.
47. Chen, Y.-L., Brand, A., Morrison, E. L., Silao, F. G. S., Bigol, U. G., Malbas, F. F., et al. (2011). Calcineurin controls drug tolerance, hyphal growth, and virulence in *Candida dubliniensis. Eukaryotic Cell, 10*(6), 803–819.
48. Juvvadi, P. R., Lee, S. C., Heitman, J., & Steinbach, W. J. (2017). Calcineurin in fungal virulence and drug resistance: Prospects for harnessing targeted inhibition of calcineurin for an antifungal therapeutic approach. *Virulence, 8*(2), 186–197.
49. Li, X., & Sun, S. (2016). Targeting the fungal calcium–calcineurin signaling network in overcoming drug resistance. *Future Medicinal Chemistry, 8*(12), 1379–1381.
50. Jin, L., & Harrison, S. C. (2002). Crystal structure of human calcineurin complexed with cyclosporin A and human cyclophilin. *Proceedings of the National Academy of Sciences, 99* (21), 13522–13526.
51. Aramburu, J., Heitman, J., & Crabtree, G. R. (2004). Calcineurin: A central controller of signalling in eukaryotes. *EMBO Reports, 5*(4), 343–348.
52. Bootman, M. D. (2012). Calcium signaling. *Cold Spring Harbor Perspectives in Biology, 4*(7), a011171.
53. Fülöp, L., Szanda, G., Enyedi, B., Várnai, P., & Spät, A. (2011). The effect of *OPA1* on mitochondrial Ca^{2+} signaling. *PLoS One, 6*(9), e25199.
54. Szabadkai, G., Bianchi, K., Várnai, P., De Stefani, D., Wieckowski, M. R., Cavagna, D., et al. (2006). Chaperone-mediated coupling of endoplasmic reticulum and mitochondrial Ca^{2+} channels. *The Journal of Cell Biology, 175*(6), 901–911.
55. Pacher, P., Sharma, K., Csordas, G., Zhu, Y., & Hajnoczky, G. (2008). Uncoupling of ER-mitochondrial calcium communication by transforming growth factor-β. *Renal Physiology, 295*, 11.
56. Denton, R. M. (2009). Regulation of mitochondrial dehydrogenases by calcium ions. *Biochimica et Biophysica Acta BBA – Bioenergetics, 1787*(11), 1309–1316.
57. Mannella, C. A., Buttle, K., Rath, B. K., & Marko, M. (1998). Electron microscopic tomography of rat-liver mitochondria and their interactions with the endoplasmic reticulum. *BioFactors, 8*(3–4), 225–228.
58. Giacomello, M., & Pellegrini, L. (2016). The coming of age of the mitochondria-ER contact: A matter of thickness. *Cell Death and Differentiation, 23*(9), 1417–1427.
59. Tucey, T. M., Verma-Gaur, J., Nguyen, J., Hewitt, V. L., Lo, T. L., Shingu-Vazquez, M., et al. (2016). The endoplasmic reticulum-mitochondrion tether ERMES orchestrates fungal immune evasion, illuminating inflammasome responses to hyphal signals. *mSphere, 1*(3), e00074–e00016.

60. Nguyen, T. T., Lewandowska, A., Choi, J.-Y., Markgraf, D. F., Junker, M., Bilgin, M., et al. (2012). Gem1 and ERMES do not directly affect phosphatidylserine transport from ER to mitochondria or mitochondrial inheritance. *Traffic, 13*(6), 880–890.
61. Geißel, B., Penka, M., Neubauer, M., & Wagener, J. (2017). The ER-mitochondria encounter structure contributes to hyphal growth, mitochondrial morphology and virulence of the pathogenic mold *Aspergillus fumigatus*. *International Journal of Medical Microbiology, 307*(1), 37–43.
62. Dagley, M. J., Gentle, I. E., Beilharz, T. H., Pettolino, F. A., Djordjevic, J. T., Lo, T. L., et al. (2010). Cell wall integrity is linked to mitochondria and phospholipid homeostasis in *Candida albicans* through the activity of the post-transcriptional regulator Ccr4-Pop2. *Molecular Microbiology, 79*(4), 968–989.
63. Kojima, R., Endo, T., & Tamura, Y. (2016). A phospholipid transfer function of ER-mitochondria encounter structure revealed in vitro. *Scientific Reports, 6*, 30777.
64. Cockcroft, S., & Raghu, P. (2018). Phospholipid transport protein function at organelle contact sites. *Current Opinion in Cell Biology, 53*, 52–60.
65. Daum, G., & Vance, J. E. (1997). Import of lipids into mitochondria. *Progress in Lipid Research, 36*(2–3), 103–130.
66. Zhong, Q., Gvozdenovic-Jeremic, J., Webster, P., Zhou, J., & Greenberg, M. L. (2005). Loss of function of KRE5 suppresses temperature sensitivity of mutants lacking mitochondrial anionic lipids. *Molecular Biology of the Cell, 16*(2), 665–675.
67. Zhong, Q., Li, G., Gvozdenovic-Jeremic, J., & Greenberg, M. L. (2007). Up-regulation of the cell integrity pathway in *Saccharomyces cerevisiae* suppresses temperature sensitivity of the *pgs1Δ* mutant. *The Journal of Biological Chemistry, 282*(22), 15946–15953.
68. Batova, M., Borecka-Melkusova, S., Simockova, M., Dzugasova, V., Goffa, E., & Subik, J. (2008). Functional characterization of the CgPGS1 gene reveals a link between mitochondrial phospholipid homeostasis and drug resistance in *Candida glabrata*. *Current Genetics, 53*(5), 313–322.
69. Chen, Y.-L., Montedonico, A. E., Kauffman, S., Dunlap, J. R., Menn, F.-M., & Reynolds, T. B. (2010). Phosphatidylserine synthase and phosphatidylserine decarboxylase are essential for cell wall integrity and virulence in *Candida albicans*. *Molecular Microbiology, 75*(5), 1112–1132.
70. Gulshan, K., Schmidt, J. A., Shahi, P., & Moye-Rowley, W. S. (2008). Evidence for the bifunctional nature of mitochondrial phosphatidylserine decarboxylase: Role in Pdr3-dependent retrograde regulation of PDR5 expression. *Molecular and Cellular Biology, 28*(19), 5851–5864.
71. Voss, C., Lahiri, S., Young, B. P., Loewen, C. J., & Prinz, W. A. (2012). ER-shaping proteins facilitate lipid exchange between the ER and mitochondria in *S. cerevisiae*. *Journal of Cell Science, 125*(20), 4791–4799.
72. Langkjær, R. B., Casaregola, S., Ussery, D. W., Gaillardin, C., & Piškur, J. (2003). Sequence analysis of three mitochondrial DNA molecules reveals interesting differences among *Saccharomyces* yeasts. *Nucleic Acids Research, 31*(12), 3081–3091.
73. Foury, F., Roganti, T., Lecrenier, N., & Purnelle, B. (1998). The complete sequence of the mitochondrial genome of *Saccharomyces cerevisiae*. *FEBS Letters, 440*(3), 325–331.
74. Pramateftaki, P. V., Kouvelis, V. N., Lanaridis, P., & Typas, M. A. (2006). The mitochondrial genome of the wine yeast *Hanseniaspora uvarum*: A unique genome organization among yeast/fungal counterparts. *FEMS Yeast Research, 6*(1), 77–90.
75. Koszul, R., Malpertuy, A., Frangeul, L., Bouchier, C., Wincker, P., Thierry, A., et al. (2003). The complete mitochondrial genome sequence of the pathogenic yeast *Candida (Torulopsis) glabrata*. *FEBS Letters, 534*(1–3), 39–48.
76. Anderson, J. B., Wickens, C., Khan, M., Cowen, L. E., Federspiel, N., Jones, T., et al. (2001). Infrequent genetic exchange and recombination in the mitochondrial genome of *Candida albicans*. *Journal of Bacteriology, 183*(3), 865–872.
77. Nosek, J., Novotna, M., Hlavatovicova, Z., Ussery, D. W., Fajkus, J., & Tomaska, L. (2004). Complete DNA sequence of the linear mitochondrial genome of the pathogenic yeast *Candida parapsilosis*. *Molecular Genetics and Genomics, 272*(2), 173–180.

78. Lin, R., Liu, C., Shen, B., Bai, M., Ling, J., Chen, G., et al. (2015). Analysis of the complete mitochondrial genome of *Pochonia chlamydosporia* suggests a close relationship to the invertebrate-pathogenic fungi in Hypocreales. *BMC Microbiology, 15*(1), 5.
79. Joardar, V., Abrams, N. F., Hostetler, J., Paukstelis, P. J., Pakala, S., Pakala, S. B., et al. (2012). Sequencing of mitochondrial genomes of nine *Aspergillus* and *Penicillium* species identifies mobile introns and accessory genes as main sources of genome size variability. *BMC Genomics, 13*(1), 698.
80. Juhasz, A., Pfeiffer, I., Keszthelyi, A., Kucsera, J., Vagvolgyi, C., & Hamari, Z. (2008). Comparative analysis of the complete mitochondrial genomes of *Aspergillus niger* mtDNA type 1a and *Aspergillus tubingensis* mtDNA type 2b: Mitochondrial genome organization in two *Aspergillus* species. *FEMS Microbiology Letters, 281*(1), 51–57.
81. Woo, P. C. Y., Zhen, H., Cai, J. J., Yu, J., Lau, S. K. P., Wang, J., et al. (2003). The mitochondrial genome of the thermal dimorphic fungus *Penicillium marneffei* is more closely related to those of molds than yeasts. *FEBS Letters, 555*(3), 469–477.
82. Pantou, M. P., Kouvelis, V. N., & Typas, M. A. (2008). The complete mitochondrial genome of *Fusarium oxysporum*: Insights into fungal mitochondrial evolution. *Gene, 419*(1–2), 7–15.

Preventive Potential and Action Mechanism of Essential Oils Against Dermatophytes

Narender Kumar and S. M. Paul Khurana

Abstract

Microsporum, *Trichophyton* and *Epidermophyton* are the most common potential dermatophytes responsible for skin infections. Their spread is fast, but available management is not up to the mark. This is mainly due to resistance developed by dermatophytes. The high manufacturing cost, non-ecofriendly nature, many side effects and even high toxicity of synthetic antidermatophytes necessitated the development of alternative natural drugs. These are in the form of herbal extracts and/or essential oils. Some are also from commercially available herbal formulations. This chapter compares the antidermatophytic action of common synthetic antifungals and different herbals useful for prevention of dominant dermatophytes, besides giving information on the preventive potential and mechanism of action of different essential oils vis-a-vis some synthetic antifungals.

Keywords

Dermatophytes · Antidermatophytic activity · Mechanism of action · Antifungal agents

Introduction

Dermatophytes are the fungi causing dermal infections. Their incidence increased in past decades especially in high-risk patients [1]. It has been observed that mycoses target more than 20% population of the world [5]. The major dermatophytic organisms responsible for most skin infections are, viz. *Microsporum, Trichophyton* and *Epidermophyton*. Sometimes these organisms cause lifetime problem.

N. Kumar (✉) · S. M. P. Khurana
Amity Institute of Biotechnology, Amity University Haryana, Manesar, Gurugram, India
e-mail: narendra.microbiology@rediffmail.com

S. Hameed, Z. Fatima (eds.), *Pathogenicity and Drug Resistance of Human Pathogens*,
https://doi.org/10.1007/978-981-32-9449-3_9

They come in keratinophilic fungi group which even infect keratinized tissues, such as hair, skin and even nails producing dermatophytoses [21]. Conventional antifungal drugs for their control are morpholine derivatives, allylamines and azoles.

However, there is an increase in development of resistance against the conventional compounds. This results in failure of the treatment. The life span of an effective classical antifungal agent is also inadequate. But research focused investigations on aromatic and medicinal plants mainly on leaf extracts and essential oils, has attracted researchers. Traditionally essential oils are used for antifungal activity [57]. The studies reveal their huge antifungal potential; hence they can be formulated as natural products [37, 42, 43]. They can be useful in cosmetics, foods and even pharmaceutical products. Essential oils fall in the category of the most effective group of natural products that can help develop cheaper, safer and broad-spectrum antifungal formulations.

The methods used for testing antidermatophytic activity in essential oils at in vitro level are broth micro-dilution, agar-based disk diffusion and vapour-phase tests. In vivo models have also been tried to confirm efficiency of the in vitro results. But their exact mode of action has yet to be worked out. The chapter summarizes the current knowledge on antidermatophytic mechanism of action of essential oils.

The Diseases Caused by Dermatophytes

The dermatophytes on the basis of habitat/host can be categorized as antropophilic, zoophilic and geophilic. The geophilic are abundant in soils and are associated with decomposition of keratinous forms, viz. horns, fur, feathers and hair. The zoophilic and antropophilic infectivity are more specific in action infecting animals and humans. The common ringworm infection-causing genera are, viz. *Microsporum, Trichophyton* and *Epidermophyton* [42, 43, 60, 62]. They colonize on keratinized human/animal tissues resulting in mild to very severe infections. The range of severity is dependent on the reaction of host due to production of metabolic products produced by fungi [13]. But sometimes it becomes tough to differentiate *dermatophytosis* from others with similar symptoms. For instance, *tinea corporis* can produce skin problems, viz. subacute cutaneous lupus erythematosus, nummular eczema, pustular psoriasis, dermatitis, subcorneal pustular dermatosis, varicella zoster and herpes simplex virus infections [36].

Dermatophytosis also occurs in domestic livestock, but it is exceptional among wild animals. The most frequent zoophilic species causing infections are *Trichophyton mentagrophytes* and *Trichophyton rubrum* [26] producing various abnormalities [35, 52]. The diseases due to dermatophytes are recorded in Table 1.

The superficial mycoses (e.g. *Tinea corporis T. pedis, T. cruris* and *T. manuum*) normally get relieved with topical antifungals [40]. *Trichophyton* is the most dominant fungus which results in dermatophytosis. For control the agents are various azoles (e.g. miconazole, clotrimazole, tioconazole, oxiconazole, econazole) and allylamines (e.g. naftifine and terbinafine). The derivatives of morpholine such as butenafine and amorolfine are used. Topical application at the site primarily results

Table 1 Diseases caused in human beings due to dermatophytes [21, 37]

Disease	Infection site	Involved dermatophyte
Tinea corporis (ringworm of the body)	Surface of exposed skin	*Microsporum canis, Trichophyton rubrum, T. verrucosum*
T. capitis (scalp ringworm)	Scalp, eyebrows, eyelashes	*Microsporum* spp., *Trichophyton* spp.
T. cruris (jock itch)	Inguinal region	*Epidermophyton floccosum, Trichophyton rubrum, T. mentagrophytes* var. *interdigitale*
T. manuum	Hands	*Trichophyton rubrum*
T. faciei	Face	*Trichophyton* species (zoophilic)
T. pedis (athlete's foot)	Feet	*Epidermophyton floccosum, Trichophyton mentagrophytes* var. *interdigitale, T. rubrum*
T. unguium (onychomycosis)	Toenails, fingernails	*Trichophyton rubrum, T. mentagrophytes* var. *interdigitale*
T. faciei	Face	Zoophilic *Trichophyton* species
T. barbae	Beard, moustache (adult man)	*T. verrucosum, T. mentagrophytes*

in mild skin reactions [38]. In patients having a wide range of infections, ketoconazole, griseofulvin, fluconazole and itraconazole are applied [40]. The antifungals inhibit the activity of enzymes, for example, itraconazole inhibits cytochrome (CYP), hence not given to patients who are administered quinidine, pimozide, triazolam, lovastatin, simvastatin and oral midazolam. The fluconazole inhibits CYP 3A4 and CYP 2C9. There should also be caution for patients taking oral sulfonylurea hypoglycemic agents, cyclosporine, warfarin and phenytoin. The terbinafine shows interaction with CYP 1A2. This is not given to patients when treated with nortriptyline, r-theophylline and warfarin [11, 12, 18]. The gastrointestinal interactions can occur with drugs which cause gastric acidity. The antacids and histamine-2 receptors act as inhibitors of proton pump [18]. So it can be clearly mentioned that treatment of dermatophytosis depends not only on knowledge of the disease but parameters, viz. strictness of the infection, causative agent and possible drug interactions need to be given due consideration. The medications for attendant as well as patient's needs should not be overlooked [11, 12].

Essential Oils in Control of Dermatophytic Problems

Essential oils are hydrophobic liquids. These are aromatic compounds and volatile. These can also be plant origin as aetherolea, volatile oils or ethereal oils. This is simply oil of plant extracted, for example, clove oil. The essential oil is called "essential" because it contains the "essence of" through using steam fragrance of the plant from which it is derived. These are obtained by steam distillation. Other processes include solvent extraction, expression, absolute oil extraction, wax embedding, cold pressing and resin tapping.

Essential oils are mixtures of non-terpenic and terpenic compounds. The most common major components are monoterpenes and sesquiterpenes and their oxygenated derivatives. Yet they may also have phenylpropanoids, fatty acids and their esters [19]. The secondary metabolites may be present in different plant parts such as roots, seeds, flowers, fruits, leaves and even stems. These are stored in various secretory structures [7]. It has been observed that they act as signals in defence of plants against microbes, insects and even herbivores [16, 19]. Essential oils obtained from aromatic plants were traditionally applied for their biological characteristic such as insecticidal, fungicidal, virucidal and bactericidal activity [57]. The in vitro screening programmes related to ethnobotanical knowledge proved highly effective when placed in record for traditional uses. This has provided new methods for search of active fractions [22]. The essential oils are highly valued in commercial market [19].

France, China, Germany, Japan, Italy, Spain, United States and United Kingdom are the markets at global level for aromatic and medicinal plants [2].

Several methods can be used to extract these oils from plants. But ISOE (International Standard Organization on Essential Oils) mentions that it must be obtained through distillation of plant material. The obtained oils are in the form of volatile liquids which present a strong odour. These are coloured and rarely insoluble in water and soluble in organic solvents. The cultivated and wild field-growing plants may also be used for extraction of essential oils and secondary metabolites. There should be a study on plant proliferation for production of a large quantity of plants for extraction of chemicals which in turn would prevent exploitation of wild populations [14]. This suggests large scale-up propagation under controlled situations at any time in a year [34]. It has been observed that in plants essential oil production is dependent on biochemical, metabolic, genetic and physiological regulation [9]. It results in diversity of chemical composition because of intrinsic (seasonal, sexual, genetic variations and ontogenetic) and extrinsic (environmental and ecological variations) factors [64].

Antidermatophytic Potential of Essential Oils

Many essential oils were tested for their antidermatophytic action (Table 2), and synergistic activity of essential oil and available antifungal drugs has also been assessed [15]. But only some studies have been done against dermatophytes. Shin and Lim [63] assessed the combination of *Pelargonium graveolens* essential oil and geraniol and citronellal. This was compared for ketoconazole against *Trichophyton* spp. There was an enhancement in ketoconazole antifungal activity when compared through natural compounds. There was reduction in minimal effective dose. Pyun and Shin [56] found showing significant synergism between *Allium sativum* oil and allicin.

Table 2 Antidermatophytic activity in essential oils of some higher plants

Oils	Organism tested	Fungitoxic efficacy	References
Aegle marmelos (L.)	*Aspergillus niger* and *Candida albicans*	Fungitoxicity was tested against *Aspergillus niger* (30 mm) and *Candida albicans* (30 mm) discs of culture	[23]
Ageratum houstonianum Mill	*Trichophyton mentagrophytes, Microsporum gypseum*	MIC of the essential oil 10 μg/ml	[41, 46]
Artemisia nilagirica	*Trichophyton violaceum, Epidermophyton floccosum*	MIC was 200 ppm, broadly fungistatic and widely fungitoxic	[30]
Caesulia axillaris	*Trichophyton rubrum* and *Microsporum gypseum*	Showed strong fungicidal activity	[31]
Callistemon lanceolatus DC	*Trichophyton tonsurans*	Minimum dose of essential oil for complete inhibition of mycelial growth was 8000 ppm	[4]
Cedrus deodara	*Trichophyton rubrum*	Strongly effective	[67]
Citrus bergamia	*Trichophyton, Microsporum* and *Epidermophyton* species	MIC ranged between 0.156% and 2.5% for natural essence, from 0.02% to 2.5% for distilled extract and 0.08–1.25% for the furocoumarin-free extract	[59]
Chenopodium ambrosioides	Against dermatophytes	Found highly effective	[55]
Cymbopogon martini	Against dermatophytes	Found highly effective	[55]
Cymbopogon flexuosus (steud)	*Fusarium oxysporum* and *Trichophyton mentagrophytes*	Completely inhibited the growth	[50]
Baccharis trimera Less (DC)	*Trichophyton rubrum* and *Microsporum canis*	MIC range was from 0.03 to 125 μg/ml	[8]
Eucalyptus citriodora and *Eucalyptus globulus*	*Candida lusitaniae, C. rugosa, C. tropicalis, C. utilis, C. krusei, C. guilliermondii, C. tropicalis, C. parapsilosis, C. albicans, C. parapsilosis, C. lusitaniae, C. glabrata*	Inhibitory up to 8 mg/ml were detected and inhibitory concentrations ranged from 0.5 to 0.125 mg/ml for *Eucalyptus citriodora* essential oil and 8–1 mg/ml for essential oil of *Eucalyptus globulus*	[24]
Hyptis ovalifolia Benth	20 *T. mentagrophytes,* 10 *Microsporum canis,* 10 *M. gypseum,* 20 *Trichophyton rubrum* and strains	Inhibited 100% of 60 dermatophytes tested	[65]

(continued)

Table 2 (continued)

Oils	Organism tested	Fungitoxic efficacy	References
Leonotis nepetaefolia	*T. mentagrophyte, M. gypseum, M. nanum, E. floccosum*	Found to be quite inhibitory to all the 4 dermatophytes tested	[17]
Lippia alba (Miller) N.E Brown	*Microsporum gypseum, Epidermophyton floccosum and T. rubrum*	Showed MICs of 39, 156 and 312 mg/ml, respectively	[10]
Melaleuca alternifolia essential oil (tea tree oil, TTO)	Against *Candida lipolytica, C. parapsilosis, C. glabrata, C. albicans, C. lusitaniae, C. guilliermondii, C. inconspicua, C. krusei, Blastoschizomyces capitatus*	Showed antifungal activity	[47]
Mentha arvensis L	*Fusarium oxysporum and Trichophyton mentagrophytes*	Completely inhibited the growth	[50]
Ocimum tenuiflorum	*T. mentagrophytes, M. gypseum, M. nanum*	Maximum inhibition zone of 37 mm against *T. mentagrophytes*, 31.67 mm for *M. gypseum* and 28.33 mm for *M. nanum*	[61]
Ocimum gratissimum L.	*T. rubrum, T. interdigitale, T. erinaceum, Microsporum canis* and *T. soudanense*	Recorded MICs (80 μl/1)	[32]
Three species of *Pogostemon*, viz. *P. vestitus, P. purpurascens* and *P. benghalensis*	*Candida albicans*	Significant microbicidal activity	[69]

The synergistic antifungal activity of *Ocimum sanctum* essential oil (OSEO) and established drugs, viz. azole antimycotics (ketoconazole and fluconazole), was studied. To carry forward, this antifungal activity was assessed against 74 fluconazole-sensitive and 16 fluconazole-resistant isolates of *Candida*. They observed selectively fungicidal activity. This provides a candidature of *O. sanctum* oil as an antifungal agent in the form of combinational therapy for candidosis [3].

Khan and Ahmad [29] studied combined effect of active fraction of oils with their main compounds with fluconazole against *T. rubrum*. There was maximum synergism between cinnamaldehyde and fluconazole. This fluconazole reduced MIC to eightfold. This also reduced own MIC by 32-fold. Use of oil obtained from *Syzygium aromaticum* registered resulted in high-level reduction in MIC up to 128-fold when used with fluconazole.

Clinical survey was done out in northeastern Uttar Pradesh, India, during years 2010–2014. Two hundred samples of nail infections were taken which showed presence of seven fungal species, viz. *Trichophyton rubrum*, *Candida albicans*, *Aspergillus flavus*, *A. fumigatus*, *A. niger*, *Fusarium oxysporum* and *Epidermophyton floccosum*. They were examined microscopically. There was dominance of *Trichophyton* and *Candida* based on per cent occurrence in infected nails in people of various age groups. These species were also present in sampling from both unsterilized and sterilized infected nails. Essential oils extracted from 32 higher plants separately were tested against *Candida* and *Trichophyton* species. The *Ageratum houstonianum* Mill (Asteraceae) essential oil was found to be the most effective one against the test fungi (inhibiting mycelial growth at 500 ppm). The minimum inhibitory concentration (MIC) of the oil was 400 ppm against *Candida albicans* and *Trichophyton rubrum*. However, it was fungicidal at 500 ppm against the test fungi. The oil was active on heavy doses of inoculums (ten discs 5 mm each) at 400 ppm. The minimum killing time was 30 s against *Candida albicans*. It was 40 s against *T. rubrum*. The oil was thermostable up to 100°C. It was safe for 180 days during storage. There was no loss of fungitoxicity of oil after autoclaving [33].

Oils of *Thymus vulgaris* and *Carum copticum* have been in use in ethnomedicine against various fungal infections. Khan et al. [28] evaluated their effects of oils on virulence and growth of drug-resistant strains of *Trichophyton rubrum* and *Aspergillus* spp. GC-MS investigation recorded thymol constituting 44.71% and 22.82% of *T. vulgaris* and *C. copticum*, respectively. It showed mycelial inhibition in order of thymol > *T. vulgaris* > *C. copticum*. Thymol showed high level of synergy when in combination with fluconazole for *Aspergillus fumigatus* MTCC2550 (FICI value 0.187) and *T. rubrum* IOA9 (0.156). *T. vulgaris* essential oils and thymol were equally effective against arthroconidia (MIC 72 μg/ml) and macroconidia. It showed reduction of activity of elastase above 80% for *A. fumigatus* by *C. copticum*, *T. vulgaris* oils and thymol. The power of oils against arthroconidia and synergistic interaction of *T. vulgaris* and thymol with fluconazole may be used to potentiate effects of fluconazole against drug-resistant strains of *Aspergillus* spp. and *T. rubrum* [28].

Essential Oils Mechanism of Action on Dermatophytes

It is very essential to find the mode and action mechanism of essential oils mainly due to the reason for use in various therapeutic applications. Although a lot of investigations have been carried out on its antifungal nature, still their mechanism of interaction is less explored. Mainly *Aspergillus* spp. and *Candida* spp. have been studied [58]. Scanty information is available on the mode and mechanism of action of essential oil on dermatophytes which are as follows:

Pinto et al. [54] studied ergosterol effect in *T. rubrum* reporting 0.08 μL/ml of *T. pulegioides* oil reduced the ergosterol content. There was an impairment in biosynthesis of ergosterol which also occurs in azole antifungal drugs [49]. Pereira et al. [53] through scanning electron microscopy observed that essential oil damaged cell wall and cell membrane in time and dose-dependent manner. Park et al. [51] studied action mechanism of eugenol which is a major compound of *Syzygium aromaticum* oil. There were alterations in *T. mentagrophytes* hyphal ultrastructure including cell walls destructions and inner mitochondrial membranes. This showed an expansion of endoplasmic reticulum near cell membranes. This suggests a mechanism going through alterations at membrane level in fungal cell structure. Bajpai et al. [6] conducted time-killing-dependent assay on *T. rubrum* IOA-9 for a comparison of potential active compounds with fluconazole and potent essential oil. It was observed through TEM that there were alterations, viz. leakage of cytoplasmic contents; lysis of cell wall and plasma membranes; excessive vacuolization; disintegration of cell walls, plasma membranes, nuclear and cytoplasmic contents, and mitochondria; endoplasmic reticulum expansion near cell membrane; and even abnormal distribution of polysaccharides (Fig. 1). It has been observed that the most active antifungal components in the *Carum copticum* essential oil are phenolic

Fig 1 Action mechanism of plant essential oils on the dermatophytes. Many synthetic antifungal drugs have been developed and mechanism of action and resistance of some drugs are presented (Table 3)

Table 3 Mode of action and mechanism of resistance of synthetic antifungal agents [39, 66]

SN	Antifungal drug	Mode of action	Mechanism of resistance	Reference
1.	Allylamines	Found fungicidal against species of *Candida* inhibit squalene epoxidase	Alteration in the gene ERG1	[45]
		(Checks ergosterol biosynthesis)	The enzyme squalene epoxidase	[27]
2.	Azoles	Broad-spectrum fungistatic, antifungal drug. It checks fungal cytochrome P450 14α-lanosterol demethylase	Drug efflux carried out due to decreased affinity in Erg11 protein by mutations upregulation of multidrug transporter genes, alteration in ergosterol biosynthetic pathway	[48, 45, 25]
		Inhibits ergosterol biosynthesis through 14-α demethylase and cytochrome P450 enzyme		[27]
		Alters integrity of fungal membranes morphology and inhibits growth		[27, 68]
3.	Polyenes (amphotericin B)	Broad-spectrum antifungal drug which is used in liposomal form having reduced toxicity. It binds to ergosterol, the major sterol of fungal membrane	Due to binding ergosterol content is decreased. Absence of ergosterol alters the specific steps in biosynthetic pathway	[48, 45, 25]
4.	Morpholines	Inhibition of sterol reductase and isomerase	Unknown	[25]
5.	5-Flurocytosine	Inhibition of nucleic acid synthesis by formation of fluorinated pyrimidine metabolites	Lack of enzyme essential in the metabolism of 5-Fc. Deregulation of the pyrimidine biosynthetic pathway. Defects in cytosine deaminase	[48, 45, 25]
6.	Echinocandins	Inhibition of cell wall synthase enzyme β-1, 3 glucan synthase	Alters affinity of echinocandins for β-1,3 glucan synthase	[48]
7.	Acylhydrazones	Targets the synthesis of fungal sphingolipids	Highly effective in vitro against 37 *Cryptococcus neoformans* and on mammalian cells had low toxicity	[39]

terpenes carvacrol and thymol. These compounds have the potential to attack membranes and cell walls. This affects permeability and releases intracellular constituents [70].

The herbal compounds have fungicidal activity against many fungal dermatophytes. It may be mentioned that potential of essential oils is not due to a single mechanism but complex effect of many components on several cell targets through different modes [20, 44].

Conclusions

The essential oils have a promising antidermatophytic potential which has been revealed through several in vitro assays. This is an alternative to conventional antifungal agents which are based on azole groups useful in management of dermatophytosis. The oils are obtained from ubiquitous plants that are environmentally safe and reliable as well as non-hazardous for human beings.

It is a general observation that fungal resistance may occur due to multiple mutations. Therefore it needs to study the mode of action of each component of essential oils. It needs a study before launching it in field/market to work out the safety parameters. Toxicity studies are also needed to find out mechanism of action and possible side-effects and as well as connections with marketed antibiotics. More formulations may also be developed for clinical applications.

Acknowledgements Authors are thankful to the Amity University, Haryana, authorities for the facilities, support and constant encouragement.

Conflict of Interest Statement We declare that we have no conflict of interest.

References

1. Achterman, R. R., & White, T. C. (2012). Dermatophyte virulence factors: Identifying and analyzing genes that may contribute to chronic or acute skin infections. *International Journal Microbiology, 2012*(1687–918X), 358305; 8.
2. Alviano, D. S., & Alviano, C. S. (2009). Plant extracts: Search for new alternatives to treat microbial diseases. *Current Pharmaceutical Biotechnology, 10*, 106–121.
3. Amber, K., Aijaz, A., Immaculata, X., Luqman, K. A., & Nikhat, M. (2010). Anticandidal effect of *Ocimum sanctum* essential oil and its synergy with fluconazole and ketoconazole. *Phytomedicine, 17*(12), 921–925.
4. Anita, K., & Misra, N. (2012). Efficacy of essential oils against dermatophytes associated with animals and human beings. *Proceeding of the National Academy of Sciences, India, Section B Biological Sciences, 82*(4), 517–523.
5. Arif, T., Mandal, T. K., & Dabur, R. (2011). Natural products: Antifungal agents derived from plants. In V. K. Tiwari (Ed.), *Opportunity, challenge and scope of natural products in medicinal chemistry* (pp. 283–311). India: Research Signpost.
6. Bajpai, V. K., Yoon, J. I., & Kang, S. C. (2009). Antifungal potential of essential oil and various organic extracts of *Nandina domestica* Thunb. Against skin infectious fungal pathogens. *Applied Microbiology and Biotechnology, 83*, 1127–1133.
7. Bakkali, F., Averbeck, S., Averbeck, D., & Idaomar, M. (2008). Biological effects of essential oils–a review. *Food and Chemical Toxicology, 46*, 446–475.

8. Caneschi, C. A., Martins, F. J., Larrudé, D. G., Romani, E. C., Brandão, M. A. F., & Raposo, N. A. R. B. (2015). In vitro antifungal activity of *Baccharis trimera* less (DC) essential oil against dermatophytes. *Tropical Journal of Pharmaceutical Research, 14*(11), 2083–2089.
9. Chaturvedi, H. C., Jain, M., & Kidwai, N. R. (2007). Cloning of medicinal plants through tissue culture-a review. *Indian Journal of Experimental Biology, 45*, 937–948.
10. Costa, M. C. M., Vermelho, A. B., Almeida, C. A., de Souza Dias, E. P., Cedrola, S. M. L., Blnk, M. F., Blank, A. F., Alviano, C. S., & Alviano, D. S. (2014). Inhibitory effect of linalool-rich essential oil from Lippia alba on the peptidase and keratinase activities of dermatophytes. *Journal of Enzyme Inhibition and Medicinal Chemistry, 29*(1), 12–17.
11. Del Rosso, J. Q., & Gupta, A. K. (2000). Drug interactions update. Focus on oral antifungal agents. *Today's Therapeutic Trends, 18*, 181–188.
12. Del Rosso, J. Q. (2000). Current management of onychomycosis and dermatomycoses. *Current Infectious Disease Reports, 2*, 438–445.
13. Doddamani, P. V., Harshan, K. H., Kanta, R. C., Gangane, R., & Sunil, K. B. (2013). Isolation identification and prevalence of dermatophytes. *People's Journal of Scientific Research, 6*(2), 10–13.
14. Elaissi, A., Salah, K. B. H., Mabrouk, S., Larbi, K. M., Chemli, R., & Harzallah, S. F. (2011). Antibacterial activity and chemical composition of 20 eucalyptus species' essential oils. *Food Chemistry, 129*, 1427–1434.
15. Ghannoum, M. (2015). Azole resistance in dermatophytes: Prevalence and mechanism of action. *Journal of the American Academy of Dermatology, 72*, AB133.
16. Ghasemi, P. A., Fatahi, V. M., Craker, L., & Shirmardi, H. (2014). Chemical composition and bioactivity of essential oils of *Hypericum helianthemoides, Hypericum perforatum* and *Hypericum scabrum. Pharmaceutical Biology, 52*(2), 175–181.
17. Gopal, R. H., Vasanth, S., & Vasudevan, S. V. (1994). Antimicrobial activity of essential oil of *Leonotis nepetaefolia. Ancient Science of Life, XIV*, 68–70.
18. Gupta, A. K., Katz, H. I., & Shear, N. H. (1999). Drug interactions with itraconazole, fluconazole and terbinafine and their management. *Journal of the American Academy of Dermatology, 41*, 237–249.
19. Gupta, A. K., Ryder, J. E., & Skinner, A. R. (2004). Treatment of onychomycosis: Pros and cons of antifungal agents. *Journal of Cutaneous Medicine and Surgery, 8*, 25–30.
20. Gupta, R., Ramteke, P. W., Pandey, H., & Pandey, A. C. (2013). Nano-structured herbal antimicrobials. *IJPSR, 4*(6), 2028–2034. https://doi.org/10.1088/2043-6262/4/3/035007.
21. Harvey, A., Stoppler, M. C.(2011). Fungal nails (Onychomycosis, *Tinea unguium*), www.medicinenet.com.
22. Hussain, A. I., Efstratiou, E., Moore, J. E., Singh-Nee, N., & Poonam. (2012). Antimicrobial activity of *Callendula officinalis* petal extracts against fungi, as well as gram negative and gram-positive clinical pathogens. *Complementary Therapies in Clinical Practice, 18*(3), 173–176.
23. Ibrahim, N., Sakhawy, F. E., Magdy, M. D. M., & Mohamed, F. (2015). Chemical composition, antimicrobial and antifungal activities of essential oils of the leaves of *Aegle marmelos* (L.) Correa growing in Egypt. *Journal of Applied Pharmaceutical Science, 5*(2), 001–005.
24. Janaina, P. B., de Thaís Rossini, O., de Daniele, G., Penteado Bragado, P., & Anderson Laerte, T. (2018). Anti candida activity of essential oils from *Eucalyptus* species. A preliminary study. *Advances in Dentistry and Oral Health*. 2018, *8*(2), 555740.
25. Johnson, M. D., & Perfect, J. R. (2010). Use of antifungal combination therapy: Agents, order, and timing. *Current Fungal Infection Reports, 4*(2), 87–95.
26. Kainthola, A., Gaur, P., Dobhal, A., & Sundriya, S. (2014). Prevalence of dermatophytoses in rural population of Garhwal Himalayan region, Uttarakhand. *International Research Journal of Medical Sciences, 2*(8), 9–12.
27. Kathiravan, M. K., Salake, A. B., Chothe, A. S., Dudhe, P. B., Watode, R. P., Mukta, M. S., et al. (2012). The biology and chemistry of antifungal agents: A review. *Bioorganic & Medicinal Chemistry, 20*, 5678–5698.

28. Khan, M. S. A., Iqbal, A., & Cameotra, S. S. (2014). *Carum copticum* and *Thymus vulgaris* oils inhibit virulence in *Trichophyton rubrum* and *Aspergillus* spp. *Brazilian Journal of Microbiology, 45*(2), 523–531.
29. Khan, M. S. A., & Ahmad, I. (2011). Antifungal activity of essential oils and their synergy with fluconazole against drug-resistant strains of *Aspergillus fumigatus* and *Trichophyton rubrum. Applied Microbial and Cell Physiology, 90*, 1083–1094.
30. Kishore, N., Dubey, N. K., & Chansouria, J. P. N. (2001). Antimycotic activity of the essential oil of *Artemisia nilagirica. Flavour and Fragrance Journal, 16*(1), 61–63.
31. Kishore, N., Mishra, A. K., & Chansouria, J. P. N. (1993). Fungitoxicity of essential oils against dermatophytes. *Mycoses, 36*, 211–215.
32. Koba, K., Wiyao, P. P., Christine, R., & Komla, S. (2009). Antifungal activity of the essential oils from *Ocimum gratissimum* L. grown in Togo. *Journal of Scientific Research, 1*(1), 164–171.
33. Kumar, N. (2014). Inhibition of nail infecting fungi of peoples of North Eastern U.P causing *Tinea unguium* through leaf essential oil of *Ageratum houstonianum* Mill. *IOSR Journal of Pharmacy, 4*(6), 36–42.
34. Kumar, P., Mishra, S., Malik, A., & Satya, S. (2012). Compositional analysis and insecticidal activity of *Eucalyptus globules* (family: Myrtaceae) essential oil against housefly (Muscadomestica). *Acta Tropica, 122*, 212–218.
35. Kumar, R., Shukla, S. K., Pandey, A., Srivastava, S. K., & Dikshit, A. (2015). Copper oxide nanoparticles: An antidermatophytic agent for *Trichophyton* spp. *Nanotechnology Reviews, 4* (5), 401–409.
36. Kumaran, G., & Jeya, M. (2014). Clinico-mycological profile of dermatophytic infections. *International Journal of Pharma and Bio Sciences, 5*(2), 1–5.
37. Laham, N. A., Abdelateef, N., & Naieem, M. (2011). Dermatophytosis among out patients in Gaza, particularly *Tinea capitis. Journal of Al Azhar University Gaza (Natural Sciences), 13*, 17–30.
38. Lakshmipathy, D. T., & Kannabiran, K. (2010). Review on dermatomycosis: Pathogenesis and treatment. *Natural Science, 2*(7), 726–731.
39. Lazzarinia, C., Haranahallic, K., Riegerd, R., Ananthulae, K. H., Desaie, P. B., Ashbaughf, A., Linkef, M. J., Cushionf, M. T., Ruzsicskac, B., Haleyc, H., Ojimac, I., & Del, M. P. (2018). Acylhydrazones as antifungal agents targeting the synthesis of fungal sphingolipids. *Antimicrobial Agents and Chemotherapy, 62*(5), e00156–e00118. https://doi.org/10.1128/AAC.00156-18.
40. Lecha, M., Effendy, I., Feuilhade, d., Chauvin, M., Di Chiac-chio, N., & Baran, R. (2005). Treatment options –development of consensus guidelines. *Journal of the European Academy of Dermatology and Venereology., 19*, 25–33.
41. Menut, C., Lamaty, G., Amvam, Z. P. H., Kuiate, J. R., & Bessiere, J. M. (1993). Chemical composition of the essential oils of *Ageratum houstonianum* Mill. and *Ageratum conyzoides* L. from Cameroon. *Flavour and Fragrance Journal, 8*, 1–60.
42. Mihali, C. V., Buruiana, A., Turcus, V., Covaci, A., & Ardelean, A. (2012). Comparative studies of morphology and ultra-structure in two common species of dermatophytes: *Microsporum canis* and *Microsporum gypseum. Annals of RSCB., 17*, 85–89.
43. Mikaili, A., Chalabi, M., Ghashghaie, A., & Mostafaie, A. (2012). Immunization against bovine dermatophytosis with live *Trichophyton verrucosum. African Journal of Microbiology Research, 6*(23), 4950–4953.
44. Mishra, S. B., Pandey, H., & Pandey, A. C. (2013). Nanosuspension of *Phyllanthus amarus* extract for improving oral bioavailability and prevention of paracetamol induced hepatotoxicity in Sprague-Dawley rats. IOP Publishing, Avt. *National Sciences: Nanosciences Nanotechnology, 4*(2013), 035007.
45. Moudgal, V., & Sobel, J. (2010). Antifungals to treat *Candida albicans. Expert Opinion on Pharmacotherapy, 11*(12), 2037–2048.

46. Njateng, G. S. S., Kuiate, J. R., Gatsing, D., Tamokou, J. D., Mouokeu, R. S., & Kuete, V. (2010). Antidermatophytic activity and dermal toxicity of essential oil from the leaves of *Ageratum houstonianum* (Asteraceae). *Journal of Biological Sciences, 10*, 448–454.
47. Oliva, B., Piccirilli, E., Ceddia, T., Pontieri, E., Aureli, P., & Ferrini, A. M. (2003). Antimycotic activity of *Melaleuca alternifolia* essential oil and its major components. *Letters in Applied Microbiology, 37*(2), 185–187.
48. Onyewu, C., & Heitman, J. (2007). Unique applications of novel antifungal drug combinations. *Anti-Infective Agents in Medicinal Chemistry., 6*(1), 3–15.
49. Palmeira, O. A., Salgueiro, L., Palmeira, R., Martinez, J., Pina, C., Queiroz, J. A., & Rodrigues, A. G. (2009). Anti-candida activity of essential oils. *Mini Reviews in Medicinal Chemistry, 9*, 1292–1305.
50. Pandey, A. K., Rai, M. K., & Acharya, D. (2003). Chemical composition and antimycotic activity of the essential oils of corn mint (*Mentha arvensis*) and lemon grass (*Cymbopogon flexuosus*) against human pathogenic fungi. *Pharmaceutical Biology, 41*(6), 421–425.
51. Park, M. J., Gwak, K. S., Yang, I., Choi, W. S., Jo, H. J., Chang, J. W., Jeung, E. B., & Choi, I. G. (2007). Antifungal activities of the essential oils in *Syzygium aromaticum* (L.) Merr. Et Perry and *Leptospermum petersonii* bailey and their constituents against various dermatophytes. *The Journal of Microbiology, 45*, 460–465.
52. Partha, P. M., Nandan, K., & Dey, S. (2014). Clinico-mycological profile of dermatophytosis in patients attending a tertiary care Hospital in Eastern Bihar. India||. *Journal of Evolution of Medical and Dental Sciences., 3*(29), 8263–8269. https://doi.org/10.14260/jemds/20.
53. Pereira, F. O., Wanderley, P. A., Viana, F. A. C., Lima, R. B., Sousa, F. B., & Lima, E. O. (2011). Growth inhibition and morphological alterations of *Trichophyton rubrum* induced by essential oil from *Cymbopogon winterianus* Jowitt ex Bor. *Brazilian Journal of Microbiology, 42*, 233–242.
54. Pinto, E., Vale-Silva, L., Cavaleiro, C., & Salgueiro, L. (2009). Antifungal activity of the clove essential oil from *Syzygium aromaticum* on Candida, *Aspergillus* and dermatophyte species. *Journal of Medical Microbiology, 58*, 1454–1462.
55. Prasad, C. S., Shukla, R., Kumar, A., & Dubey, N. K. (2009). In vitro and in vivo antifungal activity of essential oils of *Cymbopogon martini* and *Chenopodium ambrosioides* and their synergism against dermatophytes. *Mycoses, 53*(2), 123–129.
56. Pyun, M. S., & Shin, S. (2006). Antifungal effects of the volatile oils from *Allium* plants against *Trichophyton* species and synergism of the oils with ketoconazole. *Phytomedicine: International Journal of Phytotherapy and Phytopharmacology, 13*, 394–400.
57. Qidwai, A., Pandey, M., Shukla, S. K., Kumar, R., Pandey, A., & Dikshit, A. (2016). Antibacterial activity of *Mentha piperita* and *Citrus limetta* against *Propionibacterium acnes* (anaerobic bacteria). *IJPSR, 7*(7), 2917–2924.
58. Reichling, J., Schnitzler, P., Suschke, U., & Saller, R. (2009). Essential oils of aromatic plants with antibacterial, antifungal, antiviral and cytotoxic properties. *Forschende Komplementärmedizin, 16*, 79–90.
59. Sanguinetti, M., Posteraro, B., Romano, L., Battaglia, F., Lopizzo, T., De Carolis, E., & Fadda, G. (2007). In vitro activity of *Citrus ergamia* (bergamot) oil against clinical isolates of dermatophytes. *Journal of Antimicrobial Chemotherapy, 59*(2), 305–308.
60. Sharma, A., Sharma, V., Kumawat, T. K., & Seth, R. (2014). A review on antidermatophytic efficiency of plant essential oils. *International Journal Pure and Applied Biosciences, 2*(6), 265–278.
61. Sharma, V., Sharma, A., & Seth, R. (2016). A study on antidermatophytic potential of *Ocimum tenuiflorum* essential oil and chemical composition evaluation. *International Journal of PharmTech Research, 9*(11), 151–160.
62. Sharma, V., Kumawat, T. K., Sharma, A., Seth, R., & Chandra, S. (2015). Distribution and prevalence of dermatophytes in semi-arid region of India. *Advances in Microbiology, 5*, 93–106.

63. Shin, S., & Lim, S. (2004). Antifungal effects of herbal essential oils alone and in combination with ketoconazole against *Trichophyton* spp. *Journal of Applied Microbiology, 97*, 1289–1296.
64. Sigurgeirsson, B., & Baran, R. (2014). The prevalence of onychomycosis in the global population-a literature study. *Journal of the European Academy of Dermatology and Venereology, 28*, 1480–1491.
65. Souza, L. K. H., de Oliveira, C. M. A., Ferri, P. H., de Oliveira Júnior, J. G., de Souza Júnior, A. H., de Fátima Lisboa Fernandes, O., & Silva, M. d. R. R. (2003). Antimicrobial activity of *Hyptis ovalifolia* towards dermatophytes. *Memórias do Instituto Oswaldo Cruz, Rio De Janeiro, 98*(7), 963–965.
66. Supriya, P. M., & Sankunny, M. K. (2018). Review on combinatorial approach for inhibiting *Candida albicans* biofilm. *American Journal of Clinical Microbiology and Antimicrobials, 1* (5), 1022.
67. Takao, Y., Kuriyama, I., & Yamada, T. (2012). Antifungal properties of Japanese cedar essential oil from waste wood chips made from used sake barrels. *Molecular Medicine Reports.*, *5*(5), 1163–1168.
68. Tatsumi, Y., Nagashima, M., Shibanushi, T., Iwata, A., Kangawa, Y., Inui, F., et al. (2013). Mechanism of action of efinaconazole, a novel triazole antifungal agent. *Antimicrobial Agents and Chemotherapy, 57*, 2405–2409.
69. Thoppil, J. E., Tajo, A., Minija, J., Deena, M. J., Sreeranjini, K., Leeja, L., Sivadasan, M., & Alpharhan, A. H. (2014). Antimicrobial activity of the essential oils of three species of *Pogostemon*. *Journal of Environmental Biology, 35*, 795–798.
70. Zimmermam-Franco, D. C., Bolutari, E. B., Polonini, H. C., Carmo, A. M. R., Chaves, M. G. A. M., & Raposo, N. R. B. (2013). Antifungal activity of *Copaifera langsdorffii* Desf oleoresin against dermatophytes. *Molecules, 18*, 12561–12570.

Part III

Malaria Parasite Biology

Road Towards Development of New Antimalarial: Organelle Associated Metabolic Pathways in *Plasmodium* as Drug Targets and Discovery of Lead Drug Candidates

Mohd Asad, Azhar Muneer, Pradeep Kumar, Vandana Thakur, Sumit Rathore, Pawan Malhotra, and Asif Mohmmed

Abstract

Malaria remains a global threat with millions of deaths annually. Emergence of parasite strains resistant to widely used antimalarials, including the artemisinin combination therapy (ACT), and the absence of an effective vaccine makes treatment of malaria difficult than ever before. The need of the hour is to re-evaluate the chemotherapeutic approach and to identify new drug targets and develop new pharmacophores against the parasite. An important approach for antimalarial drug discovery is to understand critical metabolic pathways in the parasite which may help us to identify critical targets in the parasites and design specific inhibitors for these targets. Here, we have discussed proteins and pathways in different parasite organelles, i.e. apicoplast, mitochondrial and food vacuole, which have been suggested as potential drug targets; these unique parasite proteins can be targeted to develop new and novel antimalarials. In addition, we have also discussed several antimalarial projects currently under different stages of drug development pipeline. These promising antimalarial compounds have the potential to overcome multidrug resistance. Ongoing global efforts to develop new antimalarials and to identify drug targets suggest a promising future on malaria elimination and eradication.

Keywords

Plasmodium · Mitochondria · Apicoplast · Food vacuole · Antimalarial · Drug resistance · Drug targets · Proteases

---

M. Asad · A. Muneer · P. Kumar · V. Thakur · P. Malhotra · A. Mohmmed (✉)
International Centre for Genetic Engineering and Biotechnology, New Delhi, India
e-mail: amohd@icgeb.res.in

S. Rathore
Department of Biotechnology, All India Institute of Medical Sciences, New Delhi, India

S. Hameed, Z. Fatima (eds.), *Pathogenicity and Drug Resistance of Human Pathogens*,
https://doi.org/10.1007/978-981-32-9449-3_10

Introduction

Malaria is one of the oldest infectious diseases responsible for socio-economic burden on mankind. Malaria situation in India has undergone different phases from high prevalence to controllable condition with the use of insecticide and control measures, to resurgence in the era of insecticide and chloroquine resistance, to the use of artemisinin combination therapies from 1953 to 2008 [1–3]. One of the biggest challenges for malaria treatment is the rise of resistance against available antimalarials in the last decade [4]. Discovered in the early 1930s, chloroquine became the drug of choice for treatment against malaria [5]; however, the drug faced trouble of resistance in the late 1950s when reports of its ineffectiveness started to come from Colombia to Southeast Asia [6–8]. Later, discovery of artemisinin has helped treatment of malaria very effectively, although recently in 2009, *P. falciparum* resistant to artemisinin-based combination therapy (ACT) was reported [9]. Therefore, there is a constant requirement for the development of new lead molecules in the drug development pipeline. In this chapter, we have discussed parasite's metabolic pathways which could be novel key targets for drug development, recent advances in drug development and promising lead molecules which are in pipeline towards development of new antimalarials.

Antimalarials Targeting Haemoglobin Uptake, Degradation and the Food Vacuole

During the life cycle of parasite in red blood cell, the parasite internalizes a huge amount of haemoglobin for degradation, i.e. ring, trophozoite and schizont, internalizes a huge amount of haemoglobin for degradation. Haemoglobin uptake and its degradation has remained one of the most essential pathways for the survival of parasite and thus has been targeted with small molecule inhibitors as antimalarials. The haemoglobin uptake itself is being defined by four different processes including "Big Gulp", which is internalization of huge chunk of RBC cytoplasm with haemoglobin by ring-stage parasite. Big Gulp involves a cuplike shape formation by ring-stage parasite which does not involve actin polymerization. Another pathway defined for the uptake is somewhat similar to the classical endocytosis having cytostome, small vesicle-like structures marked with Rab5a, the early endosomal regulatory protein [10, 11]. There are reports of thin long tube-like structures formed through actin polymerization, termed as cytostomal tubes, which is considered as third pathway for the uptake by the parasite. Fourth process is termed as phagotrophic as it resembles phagocytosis-like process, but it lacks actin polymerization which is key event in the case of phagocytosis [12]. All of the defined four pathways of haemoglobin uptake are good targets for antimalarial therapy as they have some unique aspects with *Plasmodium*.

However, there are little efforts on development of new antimalarials targeting the endocytosis pathways. Antimalarials, like chloroquine, mefloquine and artemisinin, have been shown to have various effects on the endocytosis and endocytic pathways of the different mammalian cells [13, 14]. Mefloquine and artemisinin have been shown to inhibit phagocytosis in immune cells [14, 15], whereas chloroquine is shown to play role as lysosomotropic agents disrupting trafficking in the endocytic pathways, along with secretory pathway in different mammalian cells [16–18]. - Similarly, the 8-aminoquinoline primaquine is shown to disturb endocytic pathway in hep-G2 cells [19]. Quinolones and artemisinin have been also shown to disrupt haemoglobin uptake by inhibiting endocytosis mechanism in the parasite. Chloroquine can inhibit endocytosis-based uptake on 12 h treatment to the ring-stage parasite and leads to accumulation of small vesicles containing haemoglobin [20]. After the ingestion of host cytoplasm with haemoglobin by parasite, cytosomal-like vesicles, which are double membrane-bound structures, carry the ingested material to the food vacuole. Subsequently, haemoglobin gets degraded in the food vacuole by the action of several proteases under acidic conditions. Proteolytic digestion in the food vacuole leads to accumulation of toxic amounts of ferriprotoporphyrin IX prosthetic group of the protein (heme). Parasite crystallizes this heme into insoluble hemozoin crystals, which are nontoxic to the parasite. A number of proteases are suggested to play a role in haemoglobin degradation in food vacuole of *Plasmodium*; these include aspartic proteases (plasmepsins), cysteine proteases (falcipains), histo-aspartic proteases and some of diamino peptidyl aminopeptidases (DPAP1). As the haemoglobin degradation is one of the most important pathways for survival of the parasite, these sets of proteases have remained a target for development of various small molecule inhibitors and drugs. A number of small molecule compounds based on different scaffolds such as fluoromethyl ketones, leupeptin, vinyl sulfones and chalcones have been developed against these proteases [21–26].

Plasmepsins, aspartic proteases, are good targets for development of antimalarials as they have been associated with important metabolic functions during the parasite life cycle. Plasmepsin V has been shown to play a role in export of parasite proteins to the host cytoplasm and beyond and thus is an important molecule which can be targeted [27, 28]. Plasmepsins I, II, III and IV have been shown to be present in food vacuole and playing roles in the degradation of haemoglobin and thus poised as drug targets [29, 30]. However, initial efforts to develop antimalarial against these proteases have faced roadblocks due to redundant action of these proteases [29]. 2-Aminoquinazolin-4(3H)-ones are a novel class of malaria digestive vacuole plasmepsin inhibitors which are identified using NMR-based fragment screening against PM II. These compounds show antiplasmodial activity targeting plasmepsins I, II and IV although they show almost tenfold higher activity towards cathepsin D [29]. Recently, non-peptidomimetic inhibitors are being designed based on these scaffolds which specifically target plasmepsin IV [31]. Another approach targeting plasmepsins includes development of compounds on the basis of "double-drug" approach: primaquine, which has been linked to statine-based inhibitors of plasmepsins (PMs) [32]. Retroviral aspartic proteases RS370 and RS367 demonstrated inhibition against plasmepsin II at sub-micromolar concentrations. The 7-azabicyclo [2.2.1]heptane-based inhibitors have been reported to show high potency against PM I and IV. Recently, Dominique et al. showed multistage antimalarial activity of aspartic protease inhibitor hydroxyl-ethyl-amine-based scaffold compound, 49c [33].

Falcipains, which is a family of four papain-family proteases, are another set of cysteine proteases which are important for haemoglobin degradation [34–38].

Falcipain-2 and 3 work in combination with other proteases to hydrolyse haemoglobin in food vacuole. Falcipain-2 disruption has led to decrease in the rate of haemoglobin degradation, whereas falcipain-3 deletion was not successful pointing towards essential role of this protease for the parasite survival [38]. Due to their important role and no known homologue in humans, falcipains are valid drug targets, and constant efforts are going on to develop inhibitors against them. Small inhibitors ranging from peptidyl fluoromethyl ketone [39, 40], vinyl sulfone [22, 24] and aldehydic inhibitors are developed against falcipain and shown to have potent inhibitory activity against the parasite [25]. Drug development against falcipains has now been facilitated much with available structures of falcipains complexed with small molecule inhibitors, which leads to more potent compounds being developed [41]. Dihydroartemisinin derivative against Falcipain-2 has been developed and shown to have potent activity against parasite [42]. Similarly, some non-peptidomimetic inhibitors are also designed and used against falcipain-2 which showed a potential activity against the parasite [43]. E-64 which is a known cysteine protease inhibitor and has an important role in elucidating the role of falcipains in haemoglobin degradation has also been derivatized, and epoxy succinate E-64 compounds also have been developed [44].

Recently, the *Plasmodium* signal peptide peptidase (PfSPP), an aspartic protease, which is a member of family of intramembrane cleaving proteases, has been shown to be essential for parasite survival in both blood and liver stages [45]. In addition, some of the HIV protease inhibitors, e.g. saquinavir and ritonavir, are also being employed with chloroquine and mefloquine for the treatment [46].

Degradation of haemoglobin in the parasite food vacuole leads to generation of a large amount of toxic by product, i.e. heme, which can damage membranes due to its peroxidative properties. Parasite has devised a strategy of combating this toxic effect by converting heme to hemozoin crystals. This process involves oxidation of heme to hematin and then conversion to hemozoin. It has been shown recently that the process is carried out in coordination with the haemoglobin degradation through a protein complex, "degradosome" [47]. Chloroquine, one of the well-known antimalarials, is known to disrupt the process of hemozoin development. The heme-detoxification protein, HDP, and Histidine-rich protein (HRP-2) are the key proteins involved in the process, and these are suggested to be potential drug targets in the food vacuole linked pathways. In addition, interphase regions of protein-protein interaction in the haemoglobin degradation and hemozoin development complex can also be targeted to design new antimalarials.

Mitochondrial Metabolic Pathways as Targets for Anti-malarial Development

Mitochondria, vividly called as "powerhouse of the cell", are present in all eukaryotic organisms from protists to mammals. The origin of these conserved organelles is traced back to a single endosymbiotic event involving alpha proteobacteria, and all the mitochondria of eukaryotes have evolved as divergence from these early mitochondria [48]. Major role of mitochondria in most eukaryotic cells from fungi to mammals is the generation of ATPs by oxidative phosphorylation by complete oxidation of pyruvate generated from glycolysis pathway. During the process of glycolysis in cytosol, glucose is converted in to pyruvate, and two ATP equivalents are generated. The mitochondria harbours metabolic pathways for energy metabolisms including: TCA cycle for utilizing pyruvate to release CO_2 and high-energy electron carriers NADH or FADH2; and electron transport chain for harnessing energy of NADH/FADH2 to create a electrochemical potential across inner mitochondrial membrane and mitochondrial matrix called as mitochondrial membrane potential; membrane potential is then used to synthesize ATP by ATP synthase complex present in inner mitochondrial membrane. Along with energy metabolism, mitochondria also perform various other functions including, synthesis of mitochondrial proteins, replication of mitochondrial genome, provision of precursors and metabolites for various cellular processes, maintaining redox balance of the cell and making life or death decision by regulating PCD pathway and synthesis of Fe-S cluster complexes [49].

Similar to all the apicomplexan organisms, *Plasmodium* species also harbour a single tubular mitochondrion [50]. Although cristae are absent in plasmodial mitochondria, sometimes a few tubular membrane whorls are reported [51]. The mitochondrion of *Plasmodium* harbours multiple copies of ~6 Kb genome [52], encoding only three proteins, cytochrome b, cytochrome c oxidase subunit II and cytochrome c oxidase subunit III, all of which are components of electron transport chain residing in inner mitochondrial membrane. It also contains genes for rRNAs as fragments of 20–200 nucleotides distributed all over the mitochondrial genome [52, 53]. However, it doesn't encode for any tRNAs which are required for translation in mitochondrion. All the other mitochondrial proteins are encoded by nuclear genome of the parasite. During asexual erythrocytic cycle, the plasmodial mitochondria divides in synchrony with nuclear division, thus transferring a single mitochondrion with each nucleus to daughter cells (merozoites). Similar to all other eukaryotes, in *Plasmodium* also, only the female gametocyte provides mitochondrion to the zygote during sexual stages of parasite in mosquito vector [54]. Biochemical studies have revealed that during asexual intraerythrocytic stages of *Plasmodium*, CO_2 is not produced from oxidation of glucose [55]. Further studies revealed that indeed asexual stages of parasite derive its energy majorly from glycolysis which occurs in cytosol. This is in contrast to the mitochondria of other eukaryotes where mitochondrial oxidative phosphorylation is the main energy harnessing pathway. This reduces the role of *Plasmodium* mitochondrion to functions other than energy generation. *Plasmodium* mitochondria host whole repertoire of proteins and pathway components such as TCA cycle, ETC, ATP synthase complex, pyrimidine biosynthesis, Fe-S cluster biosynthesis, heme biosynthesis and protein turnover mechanisms. Some of these pathways/ mechanisms and their potential as drug targets for development of new antimalarials are being discussed below.

Electron Transport Chain (ETC)

Plasmodium mitochondria have functional ETC machinery in the inner mitochondrial membrane. Components of its ETC are complex II, complex III, complex IV and a type II NADH:ubiquinone oxidoreductase as complex I instead of NADH dehydrogenase [55]. ETC is proved to be indispensable for the parasite survival. Hence, a number of compounds are targeted to the ETC of *Plasmodium* mitochondria to contain malaria [56]. As NADH:ubiquinone oxidoreductase is more bacteria like and different from host's complex I, a number of compounds specifically target this complex [55]):

(a) Atovaquone: A hydroxynaphthoquinone derivative targets cytochrome bc1 of complex III in ETC and was the first mitochondria targeting drug for malaria [57]. Atovaquone binds to Qo site of cytochrome b active site where it acts as Ubiquinone analogue and ablates the binding of ubiquinone thus impairing the ETC [58]. It proved to be a potent antimalarial drug. Later on, in clinical settings resistant parasites appeared which limit the use of atovaquone alone as an antimalarial drug. Lack of good DNA editing mechanism in mitochondria together with drug pressure escalates the chances of resistance conferring mutations in cytochrome b gene which is encoded by mitochondrial genome itself and is the target of atovaquone [59, 60]. Although mutant cytochrome b is found to be less active, these parasites are still able to thrive through asexual stages, as it relies mostly on glycolysis for energy metabolism. However, in sexual stages of Plasmodium, where the availability of nutrients is limited, mitochondrial ETC and oxidative phosphorylation are of immense importance, and thus such mutant parasites are unable to pass through these stages, thereby limiting their transfer to another human host [61]. This restricts the resistance and does not allow it to spread any further. This kind of containment is unique to mitochondrial genes because it is only the female gametocyte, which transmits mitochondrion to the zygote and thus ablating any chance of crossover with wild-type gene of another gamete, which is common in nuclear-encoded genes [59]. Thus atovaquone also helps uniquely in containment of malaria within a patient [62].

(b) Malarone: To overcome the problem of resistance and bring down the dose of atovaquone, it is used in combination with a biguanide drug called proguanil under the trade name of "Malarone". Proguanil is a prodrug which in the liver is converted to active DHFR inhibitor, cycloguanil. Proguanil in monotherapy is not effective at all; however, in combination with atovaquone it significantly reduces the effective dose of atovaquone. The exact mechanism of action of proguanil is not yet known, but some studies point towards inhibition of reverse ATP synthase activity where ATP synthase pump is operated in opposite direction utilizing cytosolic ATPs to generate electrochemical potential, thereby maintaining the MOMP. As, in absence of atovaquone, the ETC is quite effective in maintaining the MOMP thus rendering the proguanil ineffective against the *Plasmodium* [62].

(c) ELQ-300 (in combination with atovaquone), pyridones, quinolones, acridones and acridinediones are other classes of compounds which target complex III of mitochondrial ETC in *Plasmodium* [63].

Components of TCA Cycle

Plasmodium mitochondria have most of the components of TCA cycle, but pyruvate dehydrogenase is exclusively localized in apicoplast, thereby limiting its ability to utilize pyruvate as precursor for acetyl-CoA [64]. Aconitase enzyme, along with its function in TCA cycle, also acts as cytosolic iron-response element-binding protein regulating mRNAs of iron homeostasis [65]. Isocitrate dehydrogenase is found to be upregulated under oxidative stress condition in parasite indicating its involvement in redox balance of the parasite [66]. However, some isolates of parasite from patients are found to have higher levels of TCA cycle components, which point towards nutrient limitation-induced expression of these proteins in parasite [67]. However, no potent compound has so far been found to be targeting TCA cycle that could be used as antimalarial.

Pyrimidine Biosynthesis Pathway

Plasmodium lacks salvage pathway for meeting its need for pyrimidine; hence, it solely depends on in vivo pyrimidine biosynthesis pathway making it indispensable for the parasite [68]. Mitochondria resident type II DHODH is needed to convert dihydroorotate to orotate which is a precursor for pyrimidine biosynthesis. *Plasmodium* DHODH transfers its electron to ubiquinone of mitochondrial ETC. Studies have pointed out that to act as sink for these electrons from DHODH is the sole essential purpose of *Plasmodium* mitochondrial ETC. This is further supported by reducing the effectiveness of ETC inhibitors such as atovaquone on parasites complemented with cytosolic type I DHODH from yeast, which uses fumarate instead of ubiquinone and thus does not require mitochondrial ETC [69]. DSM1 was the first drug found to be active against *Plasmodium* DHODH and proven antimalarial; however, resistance appeared quickly in DSM1 treated parasites as in case of atovaquone. A derivative of it, DSM 268 is a more potent compound which targets multiple stages and confers single-dose therapy [70, 71].

Protein Synthesis and Degradation Pathways in Mitochondria

The mitochondrial genome of *Plasmodium* encodes for three essential proteins of ETC, which needs to be translated inside mitochondrion itself [53]. Mitochondrial genome encodes for rRNAs which along with nuclear-encoded rRNAs form a mitoribosome for protein translation in mitochondrion. Recent reports have emphasized the role of ribosomal proteins in normal functioning of mitochondrion in *Plasmodium*. One of the well-characterized proteins of *Plasmodium* mitochondria is mitochondrial ribosomal protein L13 (PfMRPL13). PfMRPL13 is shown to be essential for parasite survival [72]. Its knockdown caused significant reduction in parasite survival beyond three cycles in asexual erythrocytic stages. Although under knockdown of PfMRPL13, the parasite shows increased sensitivity to proguanil similar to atovaquone treatment, but the parasite growth under PfMRPL13 knockdown could not be rescued by providing either decylubiquinone or by ectopically expressing type I yDHODH. The study suggested a domino effect of PfMRPL13 ablation in *Plasmodium*: (1) reduced translation efficiency in mitochondrial ribosomes; (2) shortage of cyt b, COXI and COXIII proteins; (3) failure to assemble functional mtETC complexes; (4) significant reduction of mitochondrial membrane potential; (5) inability to complete pyrimidine biosynthesis; and (6), ultimately, parasite death. This PfMRPL13 or mitochondrial ribosomes of the *Plasmodium* are potential drug targets for development of new, more potent antimalarials [72].

Mitochondria also maintain their proteostasis by utilizing a number of mitochondria-resident proteases. Most of the proteases in mitochondria are of prokaryotic-like including Lon protease FtsH and ClpQY protease systems. In *Plasmodium*, FtsH was shown to be present in mitochondria and suggested to be involved in mitochondrial division [73]. However, some recent studies suggested its role in apicoplast maintenance too [74]. The Clp (caseinolytic protease) proteases are prokaryotic counterparts of eukaryotic 26S proteasome. In *Plasmodium* ClpQY machinery is localized in mitochondrial matrix. PfClpQY is multimeric unit having six units of ClpQ and six units of ClpY. ClpY acts as chaperone to unfold the substrate protein by utilizing ATPs. This unfolded protein is then passed through the proteolytic barrel formed by ClpQ hexamer where ClpQ cleaves the protein by its protease activity. In *Plasmodium falciparum*, disruption of ClpQY activity is lethal for the parasite, while it is refractory to ClpQ knockout. This suggests the essential nature of ClpQY machinery in the parasite [75]. Further, a 12-amino acid synthetic peptide of PfClpY C-terminus specifically inhibits the interaction of PfClpQ with PfClpY and thus affects the overall activity of PfClpQY system. Same peptide when added to in vitro culture of *Plasmodium falciparum* asexual erythrocytic stages significantly abolishes the growth of parasite with signs of mitochondrial dysregulation such as loss of mitochondrial membrane potential and fragmentation of mitochondrion as a whole. PfClpQY system is thus essential for the parasite survival in asexual erythrocytic cycle. PfClpQY system poses as a good drug target for antimalarial development. One possible approach is designing peptide mimetic compounds which could ablate association of ClpQ with ClpY making it non-functional.

Fe-S Cluster Biosynthesis

The iron-sulphur cluster biosynthesis occurs in mitochondria. These are needed as cofactor of ABC cassette RNase L inhibitor for the maturation and incorporation of rRNAs into the ribosome. There are 18 putative proteins of Fe-S cluster synthesis pathway identified in *Plasmodium*, which could be potential drug targets owing to the essential nature of this pathway [50].

Metabolic Pathways in Apicoplast for Antimalarial Development

Apicoplast remains one of the most intriguing organelle of the parasite ever since it was reported [76]. The four-membrane-bound organelle is a site of various important pathways, which are essential for parasite survival. The prokaryotic nature of its genome and high homology of various apicoplast proteins to cyanobacteria has made it excellent target site for various antimalarials. Further, high sequence similarity between the plastid genomes of apicomplexan parasites *P. falciparum* and *Toxoplasma gondii* strongly suggests that the apicoplast biology is strongly conserved across all species of intracellular parasites and is inherited from a common ancestor [77]. Reverse genetics approaches helped to ascertain the essential role of this organelle in the parasite life cycle. It has been shown that blocking the apicoplast development does not inhibit parasite division and production of viable daughter merozoitese, which are able to invade fresh host cell; however, these parasites are no longer able to divide; since the effect is transferred to next cell cycle, the phenomenon has been termed as "delayed death" phenotype.

There are three key metabolic pathways which are functional in the apicoplast, viz. synthesis of fatty acids, de novo heme biosynthesis and isoprenoids biosynthesis. These are the three pathways which are majorly been targeted for drug development against the parasite.

Apicoplast-Associated Fatty Acid Biosynthesis

Fatty acids are required for the various essential features of the parasite life cycle including synthesis of lipids. Earlier it was believed that *Plasmodium* relies totally on the uptake of fatty acids from the host [78, 79]. Identification of three important nuclear-encoded apicoplast-targeted proteins, ACP, KAS III and β-hydroxyacyl-ACP dehydratase, provided the first-time evidence for the presence of FAS pathway in the apicoplast [77]. Fatty acid synthesis in *Plasmodium* apicoplast was found to be type II fatty acid synthesis which is altogether different from the type III fatty acid synthesis in eukaryotes or humans and thus has remained a good target for development of antimalarials for long time. Fatty acid synthesis starts by the carboxylation of acetyl-CoA to malonyl-CoA using bicarbonates as source of carboxyl group. Acetyl coenzyme carboxylase (ACC) carries out this reaction and initiates the fatty acid synthesis pathway. For a long time, *Plasmodium* ACC has remained as promising drug target as it is a discrete multidomain enzyme as compared to human ACC which is a part of multifunctional enzyme complex and thus became a suitable target for the action of aryloxyphenoxypropionate-based herbicides, e.g. fops and dims. Aryloxyphenoxypropionate-based herbicides are potent inhibitors of *Toxoplasma* ACC which harbours the same multidomain ACC as of *Plasmodium* [80]. As both fops and dims showed activity against blood-stage *Plasmodium*, the compounds were assumed to target the multifunctional apicoplast ACC [81]. However genetic manipulation studies highlighted the dispensability of FASII pathway in blood stage of *Plasmodium* [82–85], which suggested both herbicides have their effect on some other targets. Following the discovery that FASII was required for liver stage development in the rodent models, interest in the pathway then shifted towards its disruption for malaria prophylaxis. Several compounds thought to target FASII were tested against *Plasmodium* liver stages and showed promising activities in vitro [86–88]. FabD (malonyl-CoA:ACP transacylase) has also been confirmed to be localized to apicoplast and proposed to be a good target [89]. FabH, which carries out the ACP condensation with acetyl-CoA, was considered as target of thiolactomycin and its analogues [90, 91]. FabZ (β-hydroxyacyl-ACP dehydratase) which carries out formation of enoyl-ACP from β-hydroxyacyl-ACP are being targeted by synthesis NAS compounds [92]. Enoyl-ACP reductase, the enzyme carrying out reduction of enoyl-ACP to acyl-reductase, is also being targeted using compounds such as pyrazoles [93], rhodamines [94] and flavonoids [95].

Isoprenoids Biosynthesis as Target and Use of Antibiotics

As apicoplast have evolved from blue-green algae, it is seen as analogous to bacteria, and thus various important pathways of apicoplast are being targeted using antibiotics. *Plasmodium* differs from humans in synthesizing isoprenoids via 1-deoxy-D-xylulose 5-phosphate (DOXP). This non-mevalonate pathway resembles one form of bacteria and plants and thus has become an attractive target for designing inhibitors against the apicoplast development. Fosmidomycin has been shown to target DOXP reductoisomerase in isoprenoid biosynthesis pathway of apicoplast [96]. FR900098, a synthetic derivative of fosmidomycin, is found to be twice as active and is under clinical trials where it shows effective parasite clearance in human subjects in Gabon and Thailand [97, 98].

Inhibitors of Transcription and Translation Machinery in the Apicoplast

Families of antibiotics with antimalarial activity include tetracyclines, lincosamides, macrolides and ketolides. Similarly, ciprofloxacin, a DNA gyrase inhibitor, can specifically block plastid DNA synthesis of apicoplast by inhibiting the linearization of 35-Kb circular DNA [99]. However, in the clinics norfloxacin and ciprofloxacin have not shown any substantial effects during the trials [100, 101]. Some of the fluoroquinolones, e.g. grepafloxacin and norfloxacin, are effective against the parasite in vitro, and further development of the compounds based on these can be very effective [102]. Similarly, antibiotics inhibiting the transcription, e.g. rifampicin, a potent inhibitor of multisubunit RNA polymerase, show potential when used in combination therapy [103]. However, rifampicin was found not to be very effective in case of *Plasmodium vivax* infected patients [104]. Tetracycline-based antibiotic minocycline has been shown to reduce the transcript levels of rpoB and rpoC from plastid origin, while it imparts no effect on the nuclear-encoded rpoB/rpoC [105]. Clindamycin is also been used successfully in combination with other known antimalarials, e.g. quinine and atovaquone, to treat uncomplicated malaria [106, 107]. The site of action for clindamycin is shown to be domain V of 23S rRNA, in toxoplasma [108]. Another possible explanation for the antimalarial activity of clindamycin is that interruption of protein synthesis in the plastid blocks production of the Clp protein encoded on the plastid genome, which is required for the development of apicoplast. Protein translation inhibitors are also been shown as effective antimalarials when used in combinations. Treatment with prokaryotic translation inhibitor, doxycycline, targeted apicoplast [109, 110]. Doxycycline is now been used worldwide in prophylactic and combination chemotherapy against malaria [111]. Inhibitors targeting aminoacyl-tRNA synthetases, e.g. indolmycin and mupirocin, are recently being shown to inhibit parasite growth [112].

Heme Biosynthesis

The heme biosynthesis is one of the important pathways for the parasite development. Heme synthesis doesn't seem to be essential in the blood stages, but it seems to be important for the liver stages [113, 114]. Inhibitors targeting heme pathway includes succinylacetone which inhibits aminolevulinic acid dehydratase (ALAD). High concentration usage of succinylacetone (1–2 mM) points towards off-target effect of this compound [114].

Iron-Sulphur (Fe-S) Biosynthesis

Iron-sulphur (Fe-S) complex is known to play a role in electron transport and acts as a cofactor to enzymes in a variety of pathways, including fatty acid and isoprenoid synthesis. Disruption of Fe-S synthetic pathways (iron-sulphur cluster) leads to death of the parasite, which can be rescued by supplying isoprenoid by-product [115]. Fe-S cluster synthesis is also very essential in the sexual stage development of *Plasmodium.* Inhibition of SufS by D-cycloserine is the only known inhibitor of Suf pathway against the parasite till now [116].

As majority of proteins in apicoplast are nuclear encoded, transport of these proteins towards apicoplast can be a suitable target for antimalarials. A number of autophagy-related proteins, including ATG3 and ATG8 complex, are also suggested to be involved in protein trafficking and apicoplast maintenance. Thus protein import machineries can also be utilized to develop apicoplast targeting drugs. Till date only known inhibitor targeting the apicoplast protein transport machinery is deoxyspergualin (DSG), which has been shown to inhibit the interaction between transit peptide and HSP-70 [117, 118].

Apicoplast Proteases

Parasite proteases have remained as the front runner for the development of inhibitors in various systems. *Plasmodium* proteases have also been targeted in various studies for the development of antimalarials. Some of the important proteases playing role in the development of the apicoplast are caseinolytic proteases (i.e. PfClpP, PfClpC and PfClpA), Ftsh1 (mitochondria), ERAD(ER) and OTU. The ATP-dependent protease systems of Clp family have been shown to have essential role in the development of the parasite apicoplast. Disruption of *Pf*ClpP using chemical knockdown approach with a lactone-based inhibitor highlights the essential role of the protease in the parasite life cycle [119]. Further, a novel pyrimidine series of compounds inhibiting *P. falciparum* ClpP protease activity have been shown to be effective compounds which can be further developed into lead antimalarials [120].

Another important protease shown to be essential for the development of apicoplast is PfOTU. PfOTU is a cysteine protease which belongs to the family of deubiquitinating enzyme (DUB) family. It is localized in vesicular-like structures, which are found to be close to apicoplast. Downregulation of PfOTU has been shown to inhibit apicoplast protein transport along with regulation of apicoplast-bound PfATG8 [121]. The important role of PfOTU in apicoplast-directed protein import highlights the suitability of this protein as drug target.

As malaria parasite is highly evolved eukaryote, a number of proteins have different functions as is of their homologues in other systems. Endoplasmic reticulum (ER)-associated protein degradation (ERAD) has been "rewired" to provide a conduit for protein transport to the apicoplast [122]. LY-411575, an inhibitor targeting signal peptide peptidase component of this ERAD system, has been shown to be potent molecule for the parasite killing [123]. Further screening of libraries based on LY-411575 scaffold, two more compounds, e.g. NITD731 and NITD697, have been identified which are potent against *P. falciparum* at 17 nm and 65 nm, respectively. ERAD pathway in *Plasmodium* has also been targeted using HIV protease inhibitors. These inhibitors are two general aspartyl protease inhibitors and three AAA-p97 ATPase inhibitors which inhibit the zygote to ookinete transition of the *Plasmodium* parasite [124].

FtsH1 is an important metalloprotease in thylakoid membrane which plays a crucial role in the maintenance of thylakoid membranes [125]. In *Plasmodium* PfFtsH1 has been shown to be localized in mitochondria [73], but a recent report targeting TgFtsh1 and PfFtsH1 using actinonin drug points out towards its role in the development of apicoplast [74]. In addition, IPP-mediated "Chemical Rescue" of actinonin-treated *P. falciparum* resulted in "apicoplast-minus parasites" [126, 127].

Lipid Metabolism in *Plasmodium*: Potential Targets for Drug Development

During its development in the host cell, the parasite requires enormous amount of lipids to develop organelle (e.g. nucleus, mitochondria, food vacuole, apicoplast), tubulovesicular network (TVN) and also for subsequent replication. Additionally, emergence of new membranous structures (Maurer's cleft and transported vesicles) that involves trafficking of proteins and other factors to RBC surface require lipids [128]. Further, there is accumulation of lipid in form of lipid bodies in asexual stage and osmophilic bodies in sexual stage. There is huge dynamic change in the lipid content and composition in both parasite and in infected RBC. Phospholipids and fatty acid synthesis are the major pathways in lipid metabolism of the parasite. So the enzymes involved in these pathways are potential drug targets.

One of the key pathways of lipid homeostasis includes phospholipid metabolism in the cell. Earlier phospholipids were recognized as structural components only, but now phospholipids and their by-products are emerging as major signalling molecules [129] which control parasite development and differentiation in host cell. During the schizogony, lipid metabolism gets dramatically increased because parasite progeny requires huge amount of lipids for membrane synthesis and results in a sixfold increase in phospholipids in iRBCs. *P. falciparum* is dependent on phospholipids for multiplication, and there is some uniqueness in phospholipid metabolism which creates opportunities for identification of novel drug targets for the parasite. Parasite uses both exogenous source and also synthesizes phospholipids de novo to acquire the necessary lipids. About 300 different lipid species were identifies in asexual blood stages and gametocytes of *Plasmodium falciparum* which are essential for its growth, proliferation, transmission and sexual reproduction [130]. The major membrane lipid components are phosphatidylcholine (PC), phosphatidylethanolamine (PE) and phosphatidylserine (PS). In uninfected RBCs, PC, PE and PS constitute 30–40%, 25–35% and 10–20% of the total phospholipids, respectively [131], whereas in infected RBCs, these major phospholipids constitute 20–55%, 15–40% and 4–15% of the total phospholipids, respectively [132–134].

Plasmodium genome database and bioinformatic tools played an important role in identifying the components of PC, PE and PS biosynthetic machineries. Phosphatidylcholine (PC) is the major phospholipid making the parasite membranes and erythrocytes. Hydrolysis of PC by phospholipase leads to formation of signalling molecules; Diacylglycerols (DAG) and phosphatidic acid (PA) etc. One of the pathways for the PC synthesis is cytidine diphosphate- or CDP-choline branch of the Kennedy pathway. Metabolic studies have shown that 89% PC is synthesized from choline by using this pathway [135]. First, choline has to be transported to the parasite by a yet unidentified transporter, and then it is converted into phosphocholine after phosphorylation by choline kinase which is then used as a precursor for CDP-choline and finally phosphatidylcholine, by the 1,2-diacylglycerol cholinephosphotransferase [136]. Inhibition of choline kinase by hexadecyltrimethylammonium bromide (HDTAB) leads to decrease in PC synthesis and ultimately leads to parasite death [135]. Remaining 11% PC is synthesized via PMT pathway in which phosphoethanolamine (PE) is used as a substrate for the synthesis of PC. This phosphoethanolamine methyltransferase (PfPMT) does three sequential methylations of PE (via S-adenosyl methionine donors) to form PC [137]. PfPMT is also essential for the parasite; deletion of PfPMT affects gametogenesis in erythrocytes, their transmission to mosquitoes and oocysts development in mosquito gut. This essentiality of PfPMT for the parasite creates opportunities for identification of the novel drug targets against *Plasmodium falciparum*. Screening of more than 3000 molecules identified 28 lead compounds blocking the PfPMT activity at micromolar range, and 11 of them also inhibit asexual replication of the parasite. Out of these 11 compounds, NSC-158011 was found to be a competitive inhibitor of PfPMT [138]. Another study showed that 4-aminoquinoline, amodiaquine, was able to inhibit PfPMT activity [139].

Phosphatidylethanolamine (PE) and phosphatidylserine (PS) – PE is the second major membrane phospholipid essential for the cytokinesis and for membrane fusion and fission which increases the membrane curvature [140–142]. PE metabolism has been altered in various human diseases such as Parkinson's disease, Alzheimer's disease and non-alcoholic liver disease [143]. Mainly two pathways are responsible for PE biosynthesis. First is via Kennedy pathway in which parasite generates PE de novo from ethanolamine [144] and another is by decarboxylation of PS [131, 145, 146]. Recently, a detailed study shows the mode of action of two choline kinase inhibitors. Surprisingly these compounds inhibit the ethanolamine kinase activity of choline kinase leading to decrease in PE level in *P. falciparum* [147].

Recent Developments in Antimalarial Discovery

Artemisinin combination therapy has been the mainstay of malaria treatment in recent years; however, it has been threatened with the emergence and spread of resistance against the available drugs [148, 149]. Development and innovation of new, cheap, safe and effective drugs targeting various novel biochemical pathways and better mechanism of action are need of the hour. Over the past few years, vigorous effort in antimalarial drug discovery has generated several promising antimalarials, which are active across different stages of the parasite life cycle, offering hope for new treatments. Various organizations like Novartis, GSK and MMV are involved with malaria drug discovery. The Medicines for Malaria Venture (MMV) is one of the largest organizations which are involved in development of new antimalarial drugs. Today, several leading antimalarial candidates are compounds from Malaria Box and Pathogen Box of MMV [150, 151]. A number of antimalarial candidates are discussed below.

DSM265 Several antimalarial drugs like pyrimethamine and atovaquone target DNA synthesis by inhibiting the pyrimidine nucleotide biosynthesis. A vital step of this pathway is catalyzed by dihydroorotate dehydrogenase (DHODH). DSM 265 is a triazolopyrimidine-based inhibitor of DHODH [152]. It is the first DHODH inhibitor targeting *Plasmodium* DHODH to reach clinical development for treatment of malaria. DSM265 is highly selective towards *Plasmodium* DHODH and shows high efficacy against both blood and liver stage parasites [71, 153]. Pharmacokinetic studies of DSM 265 shows it to have a lengthy half-life in humans, thus making it a promising candidate for single-dose chemotherapy [154].

KAF 156 The KAF 156 or GNF156 is jointly developed by Novartis and STPHI with support from the Bill and Melinda Gates Foundation (MMV) and is currently undergoing phase IIb combination study with lumefantrine. It belongs to a novel class of antimalarial agent, imidazolopiperazines. It has been found to be effective against both asexual and liver-stage parasites and artemisinin-resistant parasites. Some studies have showed it to be effective against the mature gametocytes too [155–157]. The mode of action of the KAF156 is currently unknown because in vitro resistance to KAF156 is associated with mutations in three *P. falciparum* genes, acetyl-CoA transporters, UDP-galactose and *CARL* (cyclic amine resistance locus) which are not thought to be the target of KAF156 [155, 158, 159]. First in human pharmacokinetic data found it to be safe and well tolerated at even 1200 mg. Clinical trials showed KAF 156 to have a cure rate of around 67%. Clearance of parasitemia was reported from patients with *P. falciparum* as well as those with *P. vivax*. Even patients suffering with infections from artemisinin-resistant parasites were cured. However, many patients were reported to suffer from various adverse events like asymptomatic sinus bradycardia, thrombocytopenia, anaemia and hyperbilirubinemia [160].

Tafenoquine The tafenoquine or WR238605 is a new antimalarial drug jointly developed by the GlaxoSmithKline Pharmaceuticals (GSK) and Walter Reed Army Institute of Research (WRAIR) in association with MMV [161]. Tafenoquine, an 8-aminoquinoline, was developed as a possible alternative for primaquine. In vivo studies have shown it to be ten times more efficient and less toxic than primaquine when used as a prophylactic. It is active against the liver and blood and also blocks the sporozoite development in the mosquitoes [162–164]. It has completed phase III clinical trials and has recently been approved by the US FDA for prevention of relapse in the case of *P. vivax* malaria under the commercial name of Krintafel. PK studies for tafenoquine show that it has rapid absorption and a prolonged half-life of about 16 days, thus ideal for a single-dose regimen. The dose of tafenoquine has not been established. In general, a "fire and forget" strategy suggests three doses of 200 mg are sufficient to provide protection for up to 11 weeks idea for short-term travellers [165]. A single dose of 300 mg has been used for treatment of *P. vivax* under global *P. vivax* radical cure programme [166]. However, like primaquine, tafenoquine is reported to cause haemolysis in people with G6PD deficiency thus suggesting that care should be taken while administering drug to such patients [167]. Adverse effects include mild GI upset, headaches, nausea and myalgia.

Artefenomel OZ439 or artefenomel is a novel trioxolane being developed as a partner drug with ferroquine jointly by MMV and Sanofi [71]. It is one of the MMV's front-runner compounds with single-dose cure potential and is currently in phase IIb of clinical trials. Artefenomel is a synthetic ozonide which is based upon the 1, 2, 4-trioxolane pharmacophore which are similar to artemisinin. The first trioxolane, OZ277/arterolane, was developed by MMV and Ranbaxy India and was marketed in combination with piperaquine under the brand name Synriam [168]. Lately, it has now been licensed for clinical use in several African countries. However, studies have shown loss of potency against artemisinin-resistant parasites.

Artefenomel is the second trioxolane after arterolane to advance the clinical trials. Studies by Phyo et al. [9] showed artefenomel has high parasite clearance rate against both *P. falciparum* and *P. vivax* and was active against artemisinin-resistant parasites too. It offers a number of advantages like ease of synthesis, better ADME properties and a prolonged half-life of around 46–62 h making it ideal for single-dose therapy in combination with other drugs [169–172]. Combination studies with ferroquine are in the patient exploratory [173].

MMV0048 The MMV0048 or MMV 390048 is another promising drug candidate being developed by MMV in collaboration with H3D Cape Town. In 2014 it became the first drug to enter phase I in Africa; by 2017 FPFV phase IIa was completed in

Ethiopia. MMV0048 belongs to a novel class of inhibitors called 2-aminopyridine which was developed on the basis of a series of hits identified from a high-throughput screening of a commercially available BioFocus library [174].

Chibale et al., through various in vitro and in vivo studies, showed this compound to be highly efficacious against multiple stages of the parasite in host and mosquitoes (except liver hypnozoites) which not only provides protection against infection but also has the potential for transmission-blocking activity [175]. In order to understand the mechanism of action of MMV0048, Chibale et al. used various genomic and chemo-proteomic approaches; PI4k (a membrane-associated kinase, involved in cell signalling and trafficking, essential for parasite) was identified as the possible target of MMV0048. PKPD studies showed it to have a long half-life and good absorption which makes it a promising antimalarial which can further develop for single-dose combination therapy.

UCT943 The UCT943 is the second preclinical candidate developed jointly by H3D and Medicines for Malaria Venture (MMV). It was developed as a next-generation PI4K inhibitor, to address the issues of low aqueous solubility and anti-*Plasmodium* potency associated with MMV048 [176]. Modification of 2-aminopyridine to 2-aminopyrazine core along with incorporation of a piperazinylamide group not only improved the physiochemical properties of the compound but also significantly improved the efficacy across the parasite stages. Pharmacokinetic studies found it to be slow-acting compound with long half-life similar to mefloquine. Based on the studies by Chibale et al., UCT943 has potent activity against all stages of the malaria parasite and thus has the potential to form part of a single-exposure radical cure and prophylaxis treatment [177].

Spiroindolone/KAE609 (Cipargamin) The spiroindolone is a novel class of antimalarials discovered in a whole-cell screen by Novartis and STPHI [178]. KAE609 is a synthetic analogue of spiroindolone, which has shown to possess better antimalarial activity against multiple stages of malaria parasite than artemisinin [179]. The phase IIa trail of KAE609 has been completed, with a target set for next milestone – completion of phase II in 2019/2020. The likely molecular target of KAE609 has been identified as P-type Na^{+}ATpase, *PfATP4p* due to emergence of resistance mutations in the gene coding for the ATPase [180]. In vitro studies in *Plasmodium falciparum* show it to be active in nanomolar ranges including gametocytes stages [181]. Clinical trials data disclosed no safety concerns regarding KAE609 as well as had favourable pharmacokinetic properties needed for single-dose cure. Further studies showed it to have a fast parasite clearance time in patients with uncomplicated *Plasmodium falciparum* and *Plasmodium vivax* malaria [182]. It was also found to be highly effective in treatment of resistant malaria.

SJ733 The SJ733 is another clinical candidate identified through high-throughput screening that acts upon *pfATP4* [180]. It is a dihydroisoquinoline (DHIQ) and is being developed as single-exposure radical cure and prophylaxis (SERCaP) drug by

MMV in collaboration with St. Jude Children's Research Hospital and Rutgers University. In vitro and in vivo studies have shown it to be potent against both asexual and sexual blood stages in nanomolar ranges, with a rapid clearance rate [183]. Further studies showed it to be safe and highly orally bioavailable. Currently it is in phase I trial of clinical development.

Fosmidomycin Fosmidomycin is an antibacterial compound which was developed to treat urinary tract infections; however, it was later discontinued due to its lack of effectiveness against recurrent infections [184, 185]. It is a potent inhibitor of 1-deoxy-D-xylulose 5-phosphate reductoisomerase, an essential enzyme of the non-mevalonate pathway of isoprenoid pathway. The MEP pathway is shown to be essential for malarial parasites [186]. As this pathway is absent in humans, MEP pathway enzymes represent an attractive target for drug development [96, 187]. Clinical studies have showed that the drug is well tolerated and is effective in the treatment of clinical malaria. However, because of its highly charged nature, it has a poor pharmacokinetics. It also has a short plasma half-life of 3.5 h which is likely related to high rate of recrudescent malaria in children treated with the FSM combinations [188]. Few episodes of mild gastrointestinal side effects have also been reported. Currently the drug in phase IIb studies in combination with piperaquine [189].

AQ-13 Chloroquine was one of the most widely used antimalarials, but the emergence and spread of resistance against the drug has led to its abandonment. In recent years, various chemical approaches have been followed to develop new 4-aminoquinolines that are having similar chloroquine antimalarial potency as well as are active against CQ-resistant *Plasmodium* parasites. AQ-13 alternatively named Ro47–0543 is a CQ analogue with a shortened diaminoalkane side chain and is in development for treatment of resistant *P. falciparum* malaria. The molecule was discovered simultaneous by two research groups one at Roche, Switzerland, and other at Tulane University, USA [190, 191]. Because of the structural analogy, antiplasmodial activity of AQ-13 is probably similar to CQ, i.e. interferes with heme detoxification pathway [191]. Pk studies of AQ-13 showed it to have fast absorption and a half-life of about 13 days. Maximal concentration (Cmax) was achieved within 3–4 h. Phase 2 of clinical trial showed to it to be well tolerated and a safe drug [192, 193]. Thus, it possess all the excellent properties of being an ideal drug, namely, the prolonged coverage of protection provided by the long half-life leading to sustained cure and a post-treatment prophylactic effect. It is currently being considered for combination therapy.

Sevuparin Mortality due to severe malaria remains high. Death of severe malaria is characterized by hyperparasitaemia and microcirculatory flow obstruction in vital organs because of cytoadherence of infected RBCs and rosette formation [194, 195]. Sevuparin is being developed as an alternate treatment in severe *P. falciparum* malaria. It acts as a decoy receptor of heparan sulphate, during malaria infection, affecting both merozoite invasion and sequestration of infected erythrocytes [196, 197]. It is an acidic, negatively charged, anti-adhesive polysaccharide derived from heparin with eliminated antithrombin (AT) binding domain [198, 199]. A phase I/II clinical study showed sevuparin to be safe and well tolerated in the malaria patients. It is currently being tested for sickle cell disease.

SC83288 The SC83288 is an amicarbalide derivative which is being developed for the treatment of severe malaria. Amicarbalide was used in veterinary medicine as an antiprotozoal drug but was discontinued due to slow parasite clearance rates, high relapse frequencies, poor oral bioavailability, undesirable mutagenic and toxic side effects and manufacturing safety issues [200, 201]. Optimization of pharmacological and antiparasitic properties of amicarbalide yielded compound SC83288, which is fast-acting and is able to clear *P. falciparum* parasites at low nanomolar concentrations in vitro (IC50 valueso10nM). Studies in humanized NOD/SCID mouse model system showed it cured *P. falciparum* infection within 48 h following a dose of 2.5 mg/kg once per day over a period of 3 days. Preclinical pharmacokinetic and toxicological studies support the clinical development of SC83288. The compound is primarily active against trophozoites blood stage but also target early stage (I–III) gametocytes [202]. Considering its fast antiplasmodial activity and unique chemotype that does not reveal cross resistance to currently used antimalarials, SC83288 could be used in combination therapy against severe malaria.

Methylene Blue (Proveblue) The methylene blue is an old antimalarial introduced in the late nineteenth century which was later replaced by mepacrine and chloroquine because of the presence of heavy metals in its preparation. Recently, Provence Technologies have synthesized Proveblue which is a new methylene blue formulation that complies with the European Pharmacopoeia and contains limited organic impurities and heavy metals of recognized toxicity.

Methylene blue has an unusually high antimalarial potency (IC50 = 4 nM) and selectivity (with a cytotoxicity index of 450) [203]. Methylene blue is pleiotropic, interferes with heme metabolism and also inhibits the activity of *Pf* glutathione reductase leading into cytosolic depletion of glutathione which favours the activity of chloroquine [204]. Thus, methylene blue might be able to sensitize the parasite for chloroquine and even revert chloroquine resistance. Synergistic effects were observed when Proveblue was used with mefloquine, quinine and dihydroartemisinin [205]. Proveblue when combined with atorvastatin, showed high efficacy in preventing cerebral malaria [206]. However, further studies are needed to define its role in malaria treatment.

Summary

Malaria is an important disease in the aspect of mortality and socio-economic burden. Treatment strategies against malaria went through phases of success and failure for long time, and thus a constant effort for antimalarial development has remained a priority. Lack of effective vaccine for protection and the artemisinin and its derivatives as the only option has made the situation rather dangerous. Increase in drug resistance in recent times once again highlighted the utmost importance of newer small molecule inhibitor identification and development. Although lower success rates of lead molecules in the trials highlight the need for identification of more drug targets and rational drug design. Different organelles in parasite harbour some of the important pathways which are required for the survival of the parasite and thus pose as suitable drug targets. With advent of new technologies in genomics, proteomics, structure biology and medicinal chemistry, identification of such targets gained pace and the field of antimalarials development has progressed significantly in recent past. In recent past, a number of novel compounds targeting pathways related to haemoglobin degradation, lipid synthesis, fatty acid synthesis and pyrimidine synthesis have been developed as potential antimalarial. Development of these new candidates and designing of combination therapy will help to combat the drug resistance in malaria (Fig. 1 and Table 1).

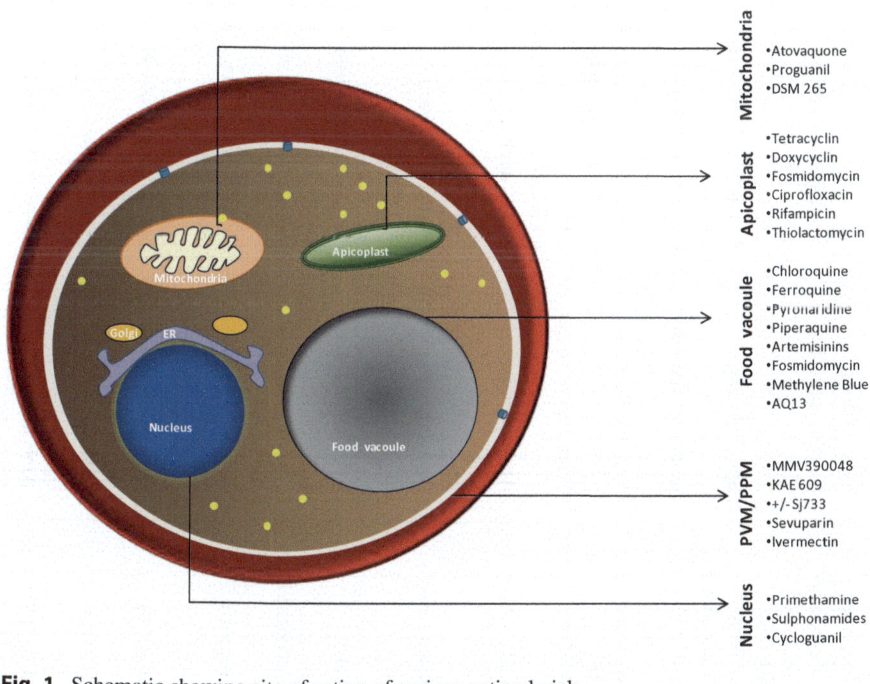

Fig. 1 Schematic showing site of action of various antimalarials

Table 1 New antimalarials under development

Drug name	Class of drug	Mechanism of action	Effected stage	Dosage	Current status
Tafenoquine/ WR238605/ Krintafel	8-Aminoquinoline	Heme polymerization	Pv liver, blood and transmission blocking	Single dose	Completed phase III clinical trials
KAE609 (Cipargamin)	Spiroindolone	Inhibit PfATP NA+-H pump	Blood stage Pf and Pv	Single dose	Phase IIb
KAF 156	Imidazolopiperazine	Currently unknown	Liver and blood stage, transmission blocking	Single dose with lumefantrine	Phase IIb
Artefenomel/ OZ439	1, 2, 4-Trioxolane	Oxidative degradation of membrane phospholipids	Blood stage	Single dose with ferroquine	Phase IIb
DSM265	Triazolopyrimidine	Inhibits pyrimidine nucleotide biosynthesis	Uncomplicated Pf blood and liver stage	Partner drug in combination therapy	Phase II
MMV390048	Aminopyridine	PI(4)K inhibitor	Blood stage, gametocytogenesis and oocyst formation	Single dose, chemoprevention	Phase II
Fosmidomycin	Propylphosphonic acid	Disrupts food vacuole integrity	Uncomplicated Pf blood stage	Combination therapy with piperaquine	Phase II
AQ-13	4-Aminoquinoline	Hemozoin metabolism	Blood stage Pf and Pv	Partner drug in combination therapy	Phase II
Methylene blue	Phenothiazine derivative	Hemozoin metabolism	Pf blood stage and gametocytogenesis	Combination therapy	Phase II
Sevuparin	Heparinoids (anti-adhesive polysaccharide)	Disrupts cytoadherence and rosette formation	Merozoite invasion and sequestration	Severe malaria	Phase II
+/–SJ733	Dihydroisoquinoline	Inhibit PfATP NA+-H pump	Pf and Pv blood stage and gametocytogenesis	Single dose	Phase I
UCT943	2-Aminopyrazine	PI(4)K inhibitor	Multistage Pf and Pv	Single dose, chemoprevention	Preclinical
SC83288	Amicarbalide derivative	Currently unknown	Blood stage	Severe malaria	Preclinical

References

1. Sharma, V. P. (1996). Re-emergence of malaria in India. *The Indian Journal of Medical Research, 103*, 26–45.
2. Sharma, V. P. (1999). Current scenario of malaria in India. *Parassitologia, 41*, 349–353.
3. Dash, A. P., Valecha, N., Anvikar, A. R., & Kumar, A. (2008). Malaria in India: Challenges and opportunities. *Journal of Biosciences, 33*, 583–592.
4. Cui, L., Mharakurwa, S., Ndiaye, D., Rathod, P. K., & Rosenthal, P. J. (2015). Antimalarial drug resistance: Literature review and activities and findings of the ICEMR network. *The American Journal of Tropical Medicine and Hygiene, 93*, 57–68.
5. Greenwood, D. (1992). The quinine connection. *The Journal of Antimicrobial Chemotherapy, 30*, 417–427.
6. Sweeney, A. W. (1996). The possibility of an "X" factor. The first documented drug resistance of human malaria. *International Journal for Parasitology, 26*, 1035–1061.
7. Baird, J. K. (2009). Resistance to therapies for infection by *Plasmodium vivax. Clinical Microbiology Reviews, 22*, 508–534.
8. Baird, J., & Chloroquine, K. (2004). Resistance in *Plasmodium vivax. Antimicrobial Agents and Chemotherapy, 48*, 4075–4083.
9. Phyo, A. P., et al. (2016). Antimalarial activity of artefenomel (OZ439), a novel synthetic antimalarial endoperoxide, in patients with Plasmodium falciparum and *Plasmodium vivax* malaria: An open-label phase 2 trial. *The Lancet Infectious Diseases, 16*, 61–69.
10. Milani, K. J., Schneider, T. G., & Taraschi, T. F. (2015). Defining the morphology and mechanism of the hemoglobin transport pathway in *Plasmodium falciparum* infected erythrocytes. *Eukaryotic Cell, **14***, 415–426.
11. Bakar, N. A., Klonis, N., Hanssen, E., Chan, C., & Tilley, L. (2010). Digestive-vacuole genesis and endocytic processes in the early intraerythrocytic stages of *Plasmodium falciparum. Journal of Cell Science, 123*, 441–450.
12. Elliott, D. A., et al. (2008). Four distinct pathways of hemoglobin uptake in the malaria parasite *Plasmodium falciparum. Proceedings of the National Academy of Sciences, 105*, 2463–2468.
13. Labro, M. T., & Babin-Chevaye, C. (1988). Effects of amodiaquine, chloroquine, and mefloquine on human polymorphonuclear neutrophil function in vitro. *Antimicrobial Agents and Chemotherapy, 32*, 1124–1130.
14. Tietze, C., Schlesinger, P., & Stahl, P. (1980). Chloroquine and ammonium ion inhibit receptor-mediated endocytosis of mannose-glycoconjugates by macrophages: Apparent inhibition of receptor recycling. *Biochemical and Biophysical Research Communications, 93*, 1–8.
15. Wenisch, C., Parschalk, B., Zedwitz-Liebenstein, K., Wernsdorfer, W., & Graninger, W. (1997). The effect of artemisinin on granulocyte function assessed by flow cytometry. *The Journal of Antimicrobial Chemotherapy, 39*, 99–101.
16. Schwartz, A. L., Bolognesi, A., & Fridovich, S. E. (1984). Recycling of the asialoglycoprotein receptor and the effect of lysosomotropic amines in hepatoma cells. *The Journal of Cell Biology, 98*, 732–738.
17. Moore, H.-P., & Kelly, R.B. Secretory protein targeting in a pituitary cell differential transport of foreign secretory proteins to distinct secretory pathways line. https://doi.org/10.1083/jcb.101.5.1773.
18. Smith, R. M., & Jarett, L. (1982). Ultrastructural basis for chloroquine-induced increase in intracellular insulin in adipocytes: Alteration of lysosomal function. *Proceedings of the National Academy of Sciences of the United States of America, 79*, 7302–7306.
19. Vanweert, A., Geuze, H. J., Groothuis, B., & Stoorvogel, W. (2000). Primaquine interferes with membrane recycling from endosomes to the plasma membrane through a direct interaction with endosomes which does not involve neutralisation of endosomal pH nor osmotic swelling of endosomes. *European Journal of Cell Biology, 79*, 394–399.

20. Rosenthal, P. J., Wollish, W. S., Palmer, J. T., & Rasnick, D. (1991). Antimalarial effects of peptide inhibitors of a *Plasmodium falciparum* cysteine proteinase. *The Journal of Clinical Investigation, 88*, 1467–1472.
21. Rosenthal, P. J., McKerrow, J. H., Rasnick, D., & Leech, J. H. (1989). *Plasmodium falciparum*: Inhibitors of lysosomal cysteine proteinases inhibit a trophozoite proteinase and block parasite development. *Molecular and Biochemical Parasitology, 35*, 177–183.
22. Rosenthal, P. J., et al. (1996). Antimalarial effects of vinyl sulfone cysteine proteinase inhibitors. *Antimicrobial Agents and Chemotherapy, 40*, 1600–1603.
23. Rockett, K. A., et al. (1990). Inhibition of intraerythrocytic development of *Plasmodium falciparum* by proteinase inhibitors. *FEBS Letters, **259***, 257–259.
24. Shenai, B. R., et al. (2003). Structure-activity relationships for inhibition of cysteine protease activity and development of *Plasmodium falciparum* by peptidyl vinyl sulfones. *Antimicrobial Agents and Chemotherapy, 47*, 154–160.
25. Lee, B. J., et al. (2003). Antimalarial activities of novel synthetic cysteine protease inhibitors. *Antimicrobial Agents and Chemotherapy, 47*, 3810–3814.
26. Mane, U. R., et al. (2013). Falcipain inhibitors as potential therapeutics for resistant strains of malaria: A patent review. *Expert Opinion on Therapeutic Patents, 23*, 165–187.
27. Russo, I., et al. (2010). Plasmepsin V licenses Plasmodium proteins for export into the host erythrocyte. *Nature, 463*, 632–636.
28. Boonyalai, N., Collins, C. R., Hackett, F., Withers-Martinez, C., & Blackman, M. J. (2018). Essentiality of *Plasmodium falciparum* plasmepsin V. *PLoS One, 13*, e0207621.
29. Rasina, D., et al. (2016). Fragment-based discovery of 2-aminoquinazolin-4(3 *H*)-ones as novel class nonpeptidomimetic inhibitors of the plasmepsins I, II, and IV. *Journal of Medicinal Chemistry, 59*, 374–387.
30. Ersmark, K., Samuelsson, B., & Hallberg, A. (2006). Plasmepsins as potential targets for new antimalarial therapy. *Medicinal Research Reviews, 26*, 626–666.
31. Wyatt, D. M., & Berry, C. (2002, Feb 27). Activity and inhibition of plasmepsin IV, a new aspartic proteinase from the malaria parasite, *Plasmodium falciparum FEBS Letters, 513*(2–3), 159–162.
32. Dell'Agli, M., et al. (2006). High antiplasmodial activity of novel plasmepsins I and II inhibitors. *Journal of Medicinal Chemistry, 49*, 7440–7449.
33. Nasamu, A. S., et al. (2017). A multistage antimalarial targets the plasmepsins IX and X essential for invasion and egress. *Science (80-.), 528*, 522–528.
34. Shenai, B. R., Sijwali, P. S., Singh, A., & Rosenthal, P. J. (2000). Characterization of native and recombinant falcipain-2, a principal trophozoite cysteine protease and essential hemoglobinase of *Plasmodium falciparum. The Journal of Biological Chemistry, 275*, 29000–29010.
35. Sijwali, P. S., Shenai, R. B., Gut, J., Singh, A., & Rosenthal, P. J. (2001). Expression and characterization of the *Plasmodium falciparum* haemoglobinase falcipain-3. *Biochemical Journal, 360*, 481.
36. Shenai, B. R., & Rosenthal, P. J. (2002). Reducing requirements for hemoglobin hydrolysis by *Plasmodium falciparum* cysteine proteases. *Molecular and Biochemical Parasitology, 122*, 99–104.
37. Sijwali, P. S., Koo, J., Singh, N., & Rosenthal, P. J. (2006). Gene disruptions demonstrate independent roles for the four falcipain cysteine proteases of *Plasmodium falciparum. Molecular and Biochemical Parasitology, 150*, 96–106.
38. Rosenthal, P. J. (2011). Falcipains and other cysteine proteases of malaria parasites. *Advances in Experimental Medicine and Biology, 712*, 30–48.
39. Olson, J. E., Lee, G. K., Semenov, A., & Rosenthal, P. J. (1999). Antimalarial effects in mice of orally administered peptidyl cysteine protease inhibitors. *Bioorganic & Medicinal Chemistry, 7*, 633–638.

40. Scheidt, K. A., et al. (1998). Structure-based design, synthesis and evaluation of conformationally constrained cysteine protease inhibitors. *Bioorganic & Medicinal Chemistry, 6*, 2477–2494.
41. Hernández González, J. E., Hernández Alvarez, L., Pascutti, P. G., & Valiente, P. A. (2017). Predicting binding modes of reversible peptide-based inhibitors of falcipain-2 consistent with structure-activity relationships. *Proteins: Structure, Function, and Bioinformatics, 85*, 1666–1683.
42. Liu, Y., et al. (2011). Synthesis and antimalarial activity of novel dihydro-artemisinin derivatives. *Molecules, 16*, 4527–4538.
43. Coterón, J. M., et al. (2010). Falcipain inhibitors: Optimization studies of the 2-pyrimidinecarbonitrile lead series †. *Journal of Medicinal Chemistry, 53*, 6129–6152.
44. Kerr, I. D., et al. (2009). Structures of falcipain-2 and falcipain-3 bound to small molecule inhibitors: Implications for substrate specificity ‡. *Journal of Medicinal Chemistry, 52*, 852–857.
45. Li, X., et al. (2009). *Plasmodium falciparum* signal peptide peptidase is a promising drug target against blood stage malaria. *Biochemical and Biophysical Research Communications, 380*, 454–459.
46. Mishra, L. C., Bhattacharya, A., Sharma, M., & Bhasin, V. K. (2010). HIV protease inhibitors, indinavir or nelfinavir, augment antimalarial action of artemisinin in vitro. *The American Journal of Tropical Medicine and Hygiene, 82*, 148–150.
47. Jani, D., et al. (2008). HDP—a novel heme detoxification protein from the malaria parasite. *PLoS Pathogens, 4*, e1000053.
48. Richards, T. A., & van der Giezen, M. (2006). Evolution of the Isd11–IscS complex reveals a single α-proteobacterial endosymbiosis for all eukaryotes. *Molecular Biology and Evolution, 23*, 1341–1344.
49. Kaul, S., & Wadhwa, R. (2012). *Mortalin biology: Life, stress and death*. Springer: Dordrecht.
50. Mather, M. W., & Vaidya, A. B. (2008). Mitochondria in malaria and related parasites: Ancient, diverse and streamlined. *Journal of Bioenergetics and Biomembranes, 40*, 425–433.
51. Krungkrai, J., Prapunwattana, P., & Krungkrai, S. R. (2000). Ultrastructure and function of mitochondria in gametocytic stage of *Plasmodium falciparum*. *Parasite, 7*, 19–26.
52. Vaidya, A. B., Akella, R., & Suplick, K. (1989). Sequences similar to genes for two mitochondrial proteins and portions of ribosomal RNA in tandemly arrayed 6-kilobase-pair DNA of a malarial parasite. *Molecular and Biochemical Parasitology, 35*, 97–107.
53. Feagin, J. E., Werner, E., Gardner, M. J., Williamson, D. H., & Wilson, R. J. M. (1992). Homologies between the contiguous and fragmented rRNAs of the two *Plasmodium falciparum extrachromosomal* DNAs are limited to core sequences. *Nucleic Acids Research, 20*, 879–887.
54. van Dooren, G. G., et al. (2005). Development of the endoplasmic reticulum, mitochondrion and apicoplast during the asexual life cycle of *Plasmodium falciparum*. *Molecular Microbiology, 57*, 405–419.
55. Biagini, G. A., Viriyavejakul, P., O'Neill, P. M., Bray, P. G., & Ward, S. A. (2006). Functional characterization and target validation of alternative complex I of *Plasmodium falciparum* mitochondria. *Antimicrobial Agents and Chemotherapy, 50*, 1841–1851.
56. Stocks, P. A., et al. (2014). Novel inhibitors of the *Plasmodium falciparum* electron transport chain. *Parasitology, 141*, 50–65.
57. Peters, J. M., et al. (2002). Mutations in cytochrome b resulting in atovaquone resistance are associated with loss of fitness in *Plasmodium falciparum*. *Antimicrobial Agents and Chemotherapy, 46*, 2435–2441.
58. Srivastava, I. K., Morrisey, J. M., Darrouzet, E., Daldal, F., & Vaidya, A. B. (1999). Resistance mutations reveal the atovaquone-binding domain of cytochrome b in malaria parasites. *Molecular Microbiology, 33*, 704–711.
59. Lynch, M. (2010). Evolution of the mutation rate. *Trends in Genetics, 26*, 345–352.

60. Wilson, R. J., Fry, M., Gardner, M. J., Feagin, J. E., & Williamson, D. H. (1992). Subcellular fractionation of the two organelle DNAs of malaria parasites. *Current Genetics, 21*, 405–408.
61. Goodman, C. D., et al. (2016). Parasites resistant to the antimalarial atovaquone fail to transmit by mosquitoes. *Science (80-.), 352*, 349–353.
62. Goodman, C. D., Buchanan, H. D., & McFadden, G. I. (2017). Is the mitochondrion a good malaria drug target? *Trends in Parasitology, 33*, 185–193.
63. Miley, G. P., et al. (2015). ELQ-300 prodrugs for enhanced delivery and single-dose cure of malaria. *Antimicrobial Agents and Chemotherapy, 59*, 5555–5560.
64. Foth, B. J., et al. (2004). The malaria parasite *Plasmodium falciparum* has only one pyruvate dehydrogenase complex, which is located in the apicoplast. *Molecular Microbiology, 55*, 39–53.
65. Hodges, M., et al. (2005). An iron regulatory-like protein expressed in *Plasmodium falciparum* displays aconitase activity. *Molecular and Biochemical Parasitology, 143*, 29–38.
66. Wrenger, C., & Müller, S. (2003). Isocitrate dehydrogenase of *Plasmodium falciparum. European Journal of Biochemistry, 270*, 1775–1783.
67. Daily, J. P., et al. (2007). Distinct physiological states of *Plasmodium falciparum* in malaria-infected patients. *Nature, 450*, 1091–1095.
68. Gutteridge, W. E., Dave, D., & Richards, W. H. (1979). Conversion of dihydroorotate to orotate in parasitic protozoa. *Biochimica et Biophysica Acta, 582*, 390–401.
69. Painter, H. J., Morrisey, J. M., Mather, M. W., & Vaidya, A. B. (2007). Specific role of mitochondrial electron transport in blood-stage *Plasmodium falciparum. Nature, 446*, 88–91.
70. Llanos-Cuentas, A., et al. (2018). Antimalarial activity of single-dose DSM265, a novel plasmodium dihydroorotate dehydrogenase inhibitor, in patients with uncomplicated *Plasmodium falciparum* or *Plasmodium vivax* malaria infection: A proof-of-concept, open-label, phase 2a study. *The Lancet Infectious Diseases, 18*, 874–883.
71. Phillips, M. A., et al. (2015). A long-duration dihydroorotate dehydrogenase inhibitor (DSM265) for prevention and treatment of malaria. *Science Translational Medicine, 7*, 296ra111–296ra111.
72. Ke, H., Dass, S., Morrisey, J. M., Mather, M. W., & Vaidya, A. B. (2018). The mitochondrial ribosomal protein L13 is critical for the structural and functional integrity of the mitochondrion in *Plasmodium falciparum. The Journal of Biological Chemistry, 293*, 8128–8137.
73. Tanveer, A., et al. (2013). An FtsH protease is recruited to the mitochondrion of *Plasmodium falciparum. PLoS One, 8*, e74408.
74. Amberg-Johnson, K., et al. (2017). Small molecule inhibition of apicomplexan FtsH1 disrupts plastid biogenesis in human pathogens. *eLife, 6*, e29865.
75. Jain, S. et al. (2013, Oct). The prokaryotic ClpQ protease plays a key role in growth and development of mitochondria in *Plasmodium falciparum. Cell Microbiology, 15*(10), 1660–1673.
76. McFadden, G. I., Reith, M. E., Munholland, J., & Lang-Unnasch, N. (1996). Plastid in human parasites. *Nature, 381*, 482–482.
77. Waller, R. F., et al. (1998). Nuclear-encoded proteins target to the plastid in Toxoplasma gondii and *Plasmodium falciparum. Proceedings of the National Academy of Sciences of the United States of America, 95*, 12352–12357.
78. Matesanz, F., Durán-Chica, I., & Alcina, A. (1999). The cloning and expression of Pfacs1, a *Plasmodium falciparum* fatty acyl coenzyme a synthetase-1 targeted to the host erythrocyte cytoplasm. *Journal of Molecular Biology, 291*, 59–70.
79. Maguire, P. A., & Sherman, I. W. (1990). Phospholipid composition, cholesterol content and cholesterol exchange in *Plasmodium falciparum*-infected red cells. *Molecular and Biochemical Parasitology, 38*, 105–112.
80. Zuther, E., Johnson, J. J., Haselkorn, R., McLeod, R., & Gornicki, P. (1999). Growth of Toxoplasma gondii is inhibited by aryloxyphenoxypropionate herbicides targeting acetyl-CoA carboxylase. *Proceedings of the National Academy of Sciences of the United States of America, 96*, 13387–13392.

81. Waller, R. F., et al. (2003). A type II pathway for fatty acid biosynthesis presents drug targets in *Plasmodium falciparum*. *Antimicrobial Agents and Chemotherapy, 47*, 297–301.
82. Yu, M., et al. (2008). The fatty acid biosynthesis enzyme FabI plays a key role in the development of liver-stage malarial parasites. *Cell Host & Microbe, 4*, 567–578.
83. Vaughan, A. M., et al. (2009). Type II fatty acid synthesis is essential only for malaria parasite late liver stage development. *Cellular Microbiology, 11*, 506–520.
84. Pei, Y., et al. (2010). Plasmodium pyruvate dehydrogenase activity is only essential for the parasite's progression from liver infection to blood infection. *Molecular Microbiology, 75*, 957–971.
85. López-Barragán, M. J., et al. (2011). Directional gene expression and antisense transcripts in sexual and asexual stages of *Plasmodium falciparum*. *BMC Genomics, 12*, 587.
86. Scala, F., et al. (2010). Bromopyrrole alkaloids as lead compounds against protozoan parasites. *Marine Drugs, 8*, 2162–2174.
87. Tarun, A. S., et al. (2008). A combined transcriptome and proteome survey of malaria parasite liver stages. *Proceedings of the National Academy of Sciences of the United States of America, 105*, 305–310.
88. Singh, A. P., Surolia, N., & Surolia, A. (2009). Triclosan inhibit the growth of the late liver-stage of *Plasmodium*. *IUBMB Life, 61*, 923–928.
89. Prigge, S. T., He, X., Gerena, L., Waters, N. C., & Reynolds, K. A. (2003). The initiating steps of a type II fatty acid synthase in *Plasmodium falciparum* are catalyzed by pfACP, pfMCAT, and pfKASIII. *Biochemistry, 42*, 1160–1169.
90. Jones, S. M., et al. (2004). Analogues of thiolactomycin as potential anti-malarial and anti-trypanosomal agents. *Bioorganic & Medicinal Chemistry, 12*, 683–692.
91. Jones, S. M., et al. (2005). Analogues of thiolactomycin as potential antimalarial agents. *Journal of Medicinal Chemistry, 48*, 5932–5941.
92. Sharma, S. K., et al. (2003). Identification, characterization, and inhibition of *Plasmodium falciparum* beta-hydroxyacyl-acyl carrier protein dehydratase (FabZ). *The Journal of Biological Chemistry, 278*, 45661–45671.
93. Kumar, S., Guha, M., Choubey, V., Maity, P., & Bandyopadhyay, U. (2007). Antimalarial drugs inhibiting hemozoin (beta-hematin) formation: A mechanistic update. *Life Sciences, 80*, 813–828.
94. Kumar, G., Banerjee, T., Kapoor, N., Surolia, N., & Surolia, A. (2010). SAR and pharmacophore models for the rhodanine inhibitors of *Plasmodium falciparum* enoyl-acyl carrier protein reductase. *IUBMB Life, 62*, 204–213.
95. Tasdemir, D., et al. (2006). Inhibition of *Plasmodium falciparum* fatty acid biosynthesis: Evaluation of FabG, FabZ, and FabI as drug targets for flavonoids. *Journal of Medicinal Chemistry, 49*, 3345–3353.
96. Jomaa, H., et al. (1999). Inhibitors of the nonmevalonate pathway of isoprenoid biosynthesis as antimalarial drugs. *Science, 285*, 1573–1576.
97. Missinou, M. A., et al. (2002). Fosmidomycin for malaria. *Lancet (London, England), 360*, 1941–1942.
98. Kremsner, P. G., et al. (2003). Fosmidomycin, a novel chemotherapeutic agent for malaria. *Antimicrobial Agents and Chemotherapy, 47*, 735–738.
99. Weissig, V., Vetro-Widenhouse, T. S., & Rowe, T. C. (1997). Topoisomerase II inhibitors induce cleavage of nuclear and 35-kb plastid DNAs in the malarial parasite *Plasmodium falciparum*. *DNA and Cell Biology, 16*, 1483–1492.
100. Watt, G., et al. (1991). Ciprofloxacin treatment of drug-resistant falciparum malaria. *The Journal of Infectious Diseases, 164*, 602–604.
101. McClean, K. L., Hitchman, D., & Shafran, S. D. (1992). Norfloxacin is inferior to chloroquine for falciparum malaria in northwestern Zambia: A comparative clinical trial. *The Journal of Infectious Diseases, 165*, 904–907.

102. Mahmoudi, N., et al. (2003). In vitro activities of 25 quinolones and fluoroquinolones against liver and blood stage Plasmodium spp. *Antimicrobial Agents and Chemotherapy, 47*, 2636–2639.
103. Goerg, H., Ochola, S. A., & Goerg, R. (1999). Treatment of malaria tropica with a fixed combination of rifampicin, co-trimoxazole and isoniazid: A clinical study. *Chemotherapy, 45*, 68–76.
104. Pukrittayakamee, S., et al. (1994). Antimalarial effects of rifampin in *Plasmodium vivax* malaria. *Antimicrobial Agents and Chemotherapy, 38*, 511–514.
105. Lin, Q., Katakura, K., & Suzuki, M. (2002). Inhibition of mitochondrial and plastid activity of *Plasmodium falciparum* by minocycline. *FEBS Letters, 515*, 71–74.
106. Adehossi, E., et al. (2003). Three-day quinine-clindamycin treatment of uncomplicated falciparum malaria imported from the tropics. *Antimicrobial Agents and Chemotherapy, 47*, 1173.
107. Lell, B., & Kremsner, P. G. (2002). Clindamycin as an antimalarial drug: Review of clinical trials. *Antimicrobial Agents and Chemotherapy, 46*, 2315–2320.
108. Fichera, M. E., & Roos, D. S. (1997). A plastid organelle as a drug target in apicomplexan parasites. *Nature, 390*, 407–409.
109. Dahl, E. L., et al. (2006). Tetracyclines specifically target the apicoplast of the malaria parasite *Plasmodium falciparum. Antimicrobial Agents and Chemotherapy, 50*, 3124–3131.
110. McConkey, G. A., Rogers, M. J., & McCutchan, T. F. (1997). Inhibition of *Plasmodium falciparum* protein synthesis. Targeting the plastid-like organelle with thiostrepton. *The Journal of Biological Chemistry, 272*, 2046–2049.
111. Gaillard, T., Madamet, M., & Pradines, B. (2015). Tetracyclines in malaria. *Malaria Journal, 14*, 445.
112. Pasaje, C. F. A., et al. (2016). Selective inhibition of apicoplast tryptophanyl-tRNA synthetase causes delayed death in *Plasmodium falciparum. Scientific Reports, 6*, 27531.
113. Nagaraj, V. A., et al. (2013). Malaria parasite-synthesized heme is essential in the mosquito and liver stages and complements host heme in the blood stages of infection. *PLoS Pathogens, 9*, e1003522.
114. Ke, H., et al. (2014). The heme biosynthesis pathway is essential for *Plasmodium falciparum* development in mosquito stage but not in blood stages. *The Journal of Biological Chemistry, 289*, 34827–34837.
115. Gisselberg, J. E., Dellibovi-Ragheb, T. A., Matthews, K. A., Bosch, G., & Prigge, S. T. (2013). The suf iron-sulfur cluster synthesis pathway is required for apicoplast maintenance in malaria parasites. *PLoS Pathogens, 9*, e1003655.
116. Charan, M., et al. (2017). [Fe-S] cluster assembly in the apicoplast and its indispensability in mosquito stages of the malaria parasite. *The FEBS Journal, 284*, 2629–2648.
117. Ramya, T. N. C., Mishra, S., Karmodiya, K., Surolia, N., & Surolia, A. (2007). Inhibitors of nonhousekeeping functions of the apicoplast defy delayed death in *Plasmodium falciparum. Antimicrobial Agents and Chemotherapy, 51*, 307–316.
118. Murty, U., Banerjee, A., & Arora, N. (2012). Analyzing a potential drug target N-myristoyltransferase of *Plasmodium falciparum* through in silico approaches. *Journal of Global Infectious Diseases, 4*, 43.
119. Rathore, S. et al. (2010). A cyanobacterial serine protease of *Plasmodium falciparum* is targeted to the apicoplast and plays an important role in its growth and development. *Molecular Microbiology, 77*, no–no.
120. Mundra, S., et al. (2017, Oct 15). A novel class of Plasmodial ClpP protease inhibitors as potential antimalarial agents. *Bioorganic & Medicinal Chemistry, 25*(20), 5662–5677.
121. Datta, G., Hossain, M. E., Asad, M., Rathore, S., & Mohmmed, A. (2017). *Plasmodium falciparum* OTU-like cysteine protease (PfOTU) is essential for apicoplast homeostasis and associates with noncanonical role of Atg8. *Cellular Microbiology, 19*, e12748.
122. Spork, S., et al. (2009). An unusual ERAD-like complex is targeted to the apicoplast of *Plasmodium falciparum. Eukaryotic Cell, 8*, 1134–1145.

123. Harbut, M. B., et al. (2012). Targeting the ERAD pathway via inhibition of signal peptide peptidase for antiparasitic therapeutic design. *Proceedings of the National Academy of Sciences of the United States of America, 109*, 21486–21491.
124. Goulielmaki, E., et al. (2018). Distinct effects of HIV protease inhibitors and ERAD inhibitors on zygote to ookinete transition of the malaria parasite. *Molecular and Biochemical Parasitology, 220*, 10–14.
125. Kato, Y., Kouso, T., & Sakamoto, W. (2012). Variegated tobacco leaves generated by chloroplast FtsH suppression: Implication of FtsH function in the maintenance of thylakoid membranes. *Plant & Cell Physiology, 53*, 391–404.
126. Yeh, E., & DeRisi, J. L. (2011). Chemical rescue of malaria parasites lacking an apicoplast defines organelle function in blood-stage *Plasmodium falciparum. PLoS Biology, 9*, e1001138.
127. He, C. Y., Striepen, B., Pletcher, C. H., Murray, J. M., & Roos, D. S. (2001). Targeting and processing of nuclear-encoded apicoplast proteins in plastid segregation mutants of toxoplasma gondii. *The Journal of Biological Chemistry, 276*, 28436–28442.
128. Ward, G. E., Miller, L. H., & Dvorak, J. A. (1993). The origin of parasitophorous vacuole membrane lipids in malaria-infected erythrocytes. *Journal of Cell Science, 106*(Pt 1), 237–248.
129. Bobenchik, A. M., et al. (2013). *Plasmodium falciparum* phosphoethanolamine methyltransferase is essential for malaria transmission. *Proceedings of the National Academy of Sciences of the United States of America, 110*, 18262–18267.
130. Gulati, S., et al. (2015). Profiling the essential nature of lipid metabolism in asexual blood and gametocyte stages of *Plasmodium falciparum. Cell Host & Microbe, 18*, 371–381.
131. Wein, S., et al. (2018). Contribution of the precursors and interplay of the pathways in the phospholipid metabolism of the malaria parasite. *Journal of Lipid Research, 59*, 1461–1471.
132. Flammersfeld, A., Lang, C., Flieger, A., & Pradel, G. (2017). Phospholipases during membrane dynamics in malaria parasites. *International Journal of Medical Microbiology, 308*, 129–141.
133. Flammersfeld, A., Lang, C., Flieger, A., & Pradel, G. (2018). Phospholipases during membrane dynamics in malaria parasites. *International Journal of Medical Microbiology, 308*, 129–141.
134. Vial, H. J., & Ancelin, M. L. (1992). Malarial lipids. An overview. *Sub-Cellular Biochemistry, 18*, 259–306.
135. Choubey, V., et al. (2007). Inhibition of *Plasmodium falciparum* choline kinase by hexadecyltrimethylammonium bromide: A possible antimalarial mechanism. *Antimicrobial Agents and Chemotherapy, 51*, 696–706.
136. Fagone, P., & Jackowski, S. (2013). Phosphatidylcholine and the CDP–choline cycle. *Biochimica et Biophysica Acta – Molecular and Cell Biology of Lipids, 1831*, 523–532.
137. Pessi, G., Choi, J.-Y., Reynolds, J. M., Voelker, D. R., & Mamoun, C. B. (2005). In vivo evidence for the specificity of *Plasmodium falciparum* phosphoethanolamine methyltransferase and its coupling to the Kennedy pathway. *The Journal of Biological Chemistry, 280*, 12461–12466.
138. Bobenchik, A. M., Augagneur, Y., Hao, B., Hoch, J. C., & Ben Mamoun, C. (2011). Phosphoethanolamine methyltransferases in phosphocholine biosynthesis: Functions and potential for antiparasite therapy. *FEMS Microbiology Reviews, 35*, 609–619.
139. Garg, A., et al. (2015). Structure, function and inhibition of the phosphoethanolamine methyltransferases of the human malaria parasites *Plasmodium vivax* and Plasmodium knowlesi. *Scientific Reports, 5*, 9064.
140. Emoto, K., et al. (1996). Redistribution of phosphatidylethanolamine at the cleavage furrow of dividing cells during cytokinesis. *Proceedings of the National Academy of Sciences, 93*, 12867–12872.
141. McMahon, H. T., & Boucrot, E. (2015). Membrane curvature at a glance. *Journal of Cell Science, 128*, 1065–1070.

142. Vance, J. E., & Tasseva, G. (2013). Formation and function of phosphatidylserine and phosphatidylethanolamine in mammalian cells. *Biochimica et Biophysica Acta – Molecular and Cell Biology of Lipids, 1831*, 543–554.
143. Calzada, E., Onguka, O., & Claypool, S. M. (2016). Phosphatidylethanolamine metabolism in health and disease. *International Review of Cell and Molecular Biology, 321*, 29–88.
144. Elabbadi, N., Ancelin, M. L., & Vial, H. J. (1992). Use of radioactive ethanolamine incorporation into phospholipids to assess in vitro antimalarial activity by the semiautomated microdilution technique. *Antimicrobial Agents and Chemotherapy, 36*, 50–55.
145. Choi, J.-Y., et al. (2016). Characterization of Plasmodium phosphatidylserine decarboxylase expressed in yeast and application for inhibitor screening. *Molecular Microbiology, 99*, 999–1014.
146. Elabbadi, N., Ancelin, M. L., & Vial, H. J. (1997). Phospholipid metabolism of serine in Plasmodium-infected erythrocytes involves phosphatidylserine and direct serine decarboxylation. *The Biochemical Journal, 324*, 435–445.
147. Serrán-Aguilera, L., et al. (2016). *Plasmodium falciparum* choline kinase inhibition leads to a major decrease in phosphatidylethanolamine causing parasite death. *Scientific Reports, 6*, 33189.
148. *For the treatment of malaria guidelines*. (2015).
149. Dondorp, A. M., et al. (2009). Artemisinin resistance in *Plasmodium falciparum* malaria. *The New England Journal of Medicine, 361*, 455–467.
150. Duffy, S., et al. (2017). Screening the medicines for malaria venture pathogen box across multiple pathogens reclassifies starting points for open-source drug discovery. *Antimicrobial Agents and Chemotherapy, 61*, e00379–e00317.
151. Antonova-Koch, Y., et al. (2018). Open-source discovery of chemical leads for next-generation chemoprotective antimalarials. *Science (80-.), 362*, eaat9446.
152. Phillips, M. A., & Rathod, P. K. (2010). Plasmodium dihydroorotate dehydrogenase: A promising target for novel anti-malarial chemotherapy. *Infectious Disorders Drug Targets, 10*, 226–239.
153. Coteron, J. M., et al. (2011). Structure-guided lead optimization of triazolopyrimidine-ring substituents identifies potent *Plasmodium falciparum* dihydroorotate dehydrogenase inhibitors with clinical candidate potential. *Journal of Medicinal Chemistry, 54*, 5540–5561.
154. Sulyok, M., et al. (2017). DSM265 for *Plasmodium falciparum* chemoprophylaxis: A randomised, double blinded, phase 1 trial with controlled human malaria infection. *The Lancet Infectious Diseases, 17*, 636–644.
155. Meister, S., et al. (2011). Imaging of Plasmodium liver stages to drive next-generation antimalarial drug discovery. *Science (80-.), 334*, 1372–1377.
156. Wu, T., et al. (2011). Imidazolopiperazines: Hit to lead optimization of new antimalarial agents. *Journal of Medicinal Chemistry, 54*, 5116–5130.
157. Kuhen, K. L., et al. (2014). KAF156 is an antimalarial clinical candidate with potential for use in prophylaxis, treatment, and prevention of disease transmission. *Antimicrobial Agents and Chemotherapy, 58*, 5060–5067.
158. LaMonte, G. et al. (2016). Mutations in the *Plasmodium falciparum* cyclic amine resistance locus (PfCARL) confer multidrug resistance. *MBio, 7*.
159. Lim, M. Y.-X., et al. (2016). UDP-galactose and Acetyl-CoA transporters as Plasmodium multidrug resistance genes. *Nature Microbiology, 1*, 16166.
160. Leong, F. J., et al. (2014, Nov). A first-in-human randomized, double-blind, placebo-controlled, single- and multiple-ascending oral dose study of novel imidazolopiperazine KAF156 to assess its safety, tolerability, and pharmacokinetics in healthy adult volunteers. *Antimicrobial Agents and Chemotherapy, 58*(11), 6437–6443.
161. Kitchener, S., Nasveld, P., & Edstein, M. D. (2007). Tafenoquine for the treatment of recurrent *Plasmodium vivax* malaria. *The American Journal of Tropical Medicine and Hygiene, 76*, 494–496.

162. Fisk, T. L., Millet, P., Collins, W. E., & Nguyen-Dinh, P. (1989). In vitro activity of antimalarial compounds on the exoerythrocytic stages of Plasmodium cynomolgi and *P. knowlesi*. *The American Journal of Tropical Medicine and Hygiene, 40*, 235–239.
163. Pradines, B., et al. (2006). In vitro activity of tafenoquine against the asexual blood stages of *Plasmodium falciparum* isolates from Gabon, Senegal, and Djibouti. *Antimicrobial Agents and Chemotherapy, 50*, 3225–3226.
164. Coleman, R. E. (1990). Sporontocidal activity of the antimalarial WR-238605 against Plasmodium berghei ANKA in Anopheles stephensi. *The American Journal of Tropical Medicine and Hygiene, 42*, 196–205.
165. Fda. *Highlights of prescribing information.*
166. Lell, B., et al. (2000). Malaria chemoprophylaxis with tafenoquine: A randomised study. *Lancet, 355*, 2041–2045.
167. Rueangweerayut, R., et al. (2017). Hemolytic potential of tafenoquine in female volunteers heterozygous for glucose-6-phosphate dehydrogenase (G6PD) deficiency (G6PD Mahidol variant) versus G6PD-normal volunteers. *The American Journal of Tropical Medicine and Hygiene, 97*, 702–711.
168. Patil, C., Katare, S., Baig, M., & Doifode, S. (2014). Fixed dose combination of arterolane and piperaquine: A newer prospect in antimalarial therapy. *Annals of Medical and Health Sciences Research, 4*, 466–471.
169. Dong, Y, † et al. (2005). Spiro and dispiro-1,2,4-trioxolanes as antimalarial peroxides: Charting a workable structure–activity relationship using simple prototypes. https://doi.org/10.1021/JM049040U
170. Wang, X., et al. (2013). Comparative antimalarial activities and ADME profiles of ozonides (1,2,4-trioxolanes) OZ277, OZ439, and their 1,2-dioxolane, 1,2,4-trioxane, and 1,2,4,5-tetraoxane isosteres. *Journal of Medicinal Chemistry, 56*, 2547–2555.
171. Moehrle, J. J., et al. (2013). First-in-man safety and pharmacokinetics of synthetic ozonide OZ439 demonstrates an improved exposure profile relative to other peroxide antimalarials. *British Journal of Clinical Pharmacology, 75*, 535–548.
172. McCarthy, J. S., et al. (2016). Efficacy of OZ439 (artefenomel) against early *Plasmodium falciparum* blood-stage malaria infection in healthy volunteers. *The Journal of Antimicrobial Chemotherapy, 71*, 2620–2627.
173. Held, J., et al. (2015). Ferroquine and artesunate in African adults and children with *Plasmodium falciparum* malaria: A phase 2, multicentre, randomised, double-blind, dose-ranging, non-inferiority study. *The Lancet Infectious Diseases, 15*, 1409–1419.
174. Younis, Y., et al. (2012). 3,5-diaryl-2-aminopyridines as a novel class of orally active antimalarials demonstrating single dose cure in mice and clinical candidate potential. *Journal of Medicinal Chemistry, 55*, 3479–3487.
175. Paquet, T., et al. (2017). Antimalarial efficacy of MMV390048, an inhibitor of Plasmodium phosphatidylinositol 4-kinase. *Science Translational Medicine, 9*, eaad9735.
176. Brunschwig, C., et al. (2018). UCT943, a next-generation *Plasmodium falciparum* PI4K inhibitor preclinical candidate for the treatment of malaria. *Antimicrobial Agents and Chemotherapy, 62*, 1–16.
177. Le Manach, C., et al. (2016). Identification of a potential antimalarial drug candidate from a series of 2-aminopyrazines by optimization of aqueous solubility and potency across the parasite life cycle. *Journal of Medicinal Chemistry, 59*, 9890–9905.
178. Yeung, B. K. S., et al. (2010). Spirotetrahydro β-carbolines (spiroindolones): A new class of potent and orally efficacious compounds for the treatment of malaria. *Journal of Medicinal Chemistry, 53*, 5155–5164.
179. van Pelt-Koops, J. C., et al. (2012). The spiroindolone drug candidate NITD609 potently inhibits gametocytogenesis and blocks *Plasmodium falciparum* transmission to Anopheles mosquito vector. *Antimicrobial Agents and Chemotherapy, 56*, 3544–3548.

180. Spillman, N. J., & Kirk, K. (2015). The malaria parasite cation ATPase PfATP4 and its role in the mechanism of action of a new arsenal of antimalarial drugs. *International Journal for Parasitology: Drugs and Drug Resistance, 5*, 149–162.
181. Rottmann, M., et al. (2010). Spiroindolones, a potent compound class for the treatment of malaria. *Science (80-.), 329*, 1175–1180.
182. White, N. J., et al. (2014). Spiroindolone KAE609 for falciparum and vivax malaria. *The New England Journal of Medicine, 371*, 403–410.
183. Jiménez-Díaz, M. B., et al. (2014). (+)-SJ733, a clinical candidate for malaria that acts through ATP4 to induce rapid host-mediated clearance of Plasmodium. *Proceedings of the National Academy of Sciences of the United States of America, 111*, E5455–E5462.
184. Kuemmerle, H. P., Murakawa, T., & De Santis, F. (1987). Pharmacokinetic evaluation of fosmidomycin, a new phosphonic acid antibiotic. *Chemioterapia, 6*, 113–119.
185. Kuemmerle, H. P., et al. (1985). Fosmidomycin, a new phosphonic acid antibiotic. Part II: 1. Human pharmacokinetics. 2. Preliminary early phase IIa clinical studies. *International Journal of Clinical Pharmacology, Therapy, and Toxicology, 23*, 521–528.
186. Odom, A. R., & Van Voorhis, W. C. (2010). Functional genetic analysis of the *Plasmodium falciparum* deoxyxylulose 5-phosphate reductoisomerase gene. *Molecular and Biochemical Parasitology, 170*, 108–111.
187. Hale, I., O'Neill, P. M., Berry, N. G., Odom, A., & Sharma, R. (2012). The MEP pathway and the development of inhibitors as potential anti-infective agents. *Medchemcomm, 3*, 418.
188. Na-Bangchang, K., Ruengweerayut, R., Karbwang, J., Chauemung, A., & Hutchinson, D. (2007). Pharmacokinetics and pharmacodynamics of fosmidomycin monotherapy and combination therapy with clindamycin in the treatment of multidrug resistant falciparum malaria. *Malaria Journal, 6*, 70.
189. Wells, T. N. C. (2011). New medicines to combat malaria: An overview of the global pipeline of therapeutics. In *Treatment and prevention of malaria* (pp. 227–247). Basel: Springer. https://doi.org/10.1007/978-3-0346-0480-2_12.
190. Ridley, R. G., et al. (1996). 4-aminoquinoline analogs of chloroquine with shortened side chains retain activity against chloroquine-resistant *Plasmodium falciparum*. *Antimicrobial Agents and Chemotherapy, 40*, 1846–1854.
191. De, D., Krogstad, F. M., Cogswell, F. B., & Krogstad, D. J. (1996). Aminoquinolines that circumvent resistance in *Plasmodium falciparum* in vitro. *The American Journal of Tropical Medicine and Hygiene, 55*, 579–583.
192. Mzayek, F., et al. (2007). Randomized dose-ranging controlled trial of AQ-13, a candidate antimalarial, and chloroquine in healthy volunteers. *PLoS Clinical Trials, 2*, e6.
193. Koita, O. A., et al. (2017). AQ-13, an investigational antimalarial, versus artemether plus lumefantrine for the treatment of uncomplicated *Plasmodium falciparum* malaria: A randomised, phase 2, non-inferiority clinical trial. *The Lancet Infectious Diseases, 17*, 1266–1275.
194. Udomsangpetch, R., et al. (1989). *Plasmodium falciparum*-infected erythrocytes form spontaneous erythrocyte rosettes. *The Journal of Experimental Medicine, 169*, 1835–1840.
195. Deen, J. L., von Seidlein, L., & Dondorp, A. (2008). Therapy of uncomplicated malaria in children: A review of treatment principles, essential drugs and current recommendations. *Tropical Medicine and International Health, 13*, 1111–1130.
196. Saiwaew, S., et al. (2017). Effects of sevuparin on rosette formation and cytoadherence of *Plasmodium falciparum* infected erythrocytes. *PLoS One, 12*, 1–15.
197. Leitgeb, A. M., et al. (2017). Inhibition of merozoite invasion and transient de-sequestration by sevuparin in humans with *Plasmodium falciparum* malaria. *PLoS One, 12*, e0188754.
198. Nde, P., et al. (2011). Low anticoagulant heparin disrupts *Plasmodium falciparum* rosettes in fresh clinical isolates. *The American Journal of Tropical Medicine and Hygiene, 84*, 390–396.
199. International Society on Thrombosis and Haemostasis. *Journal of Thrombosis and Haemostasis*. (Blackwell Pub., c2003-).

200. Ashley, J. N., Berg, S. S., & Lucas, J. M. (1960). 3:3′-diamidinocarbanilide: A new drug active against Babesial infections. *Nature, 185*, 461.
201. De Vos, A. J., Barrowman, P. R., Coetzer, J. A., & Kellerman, T. S. (1978). Amicarbalide: A therapeutic agent for anaplasmosis. *The Onderstepoort Journal of Veterinary Research, 45*, 203–208.
202. Pegoraro, S., et al. (2017). SC83288 is a clinical development candidate for the treatment of severe malaria. *Nature Communications, 8*, 3–4.
203. Pascual, A., et al. (2011). In vitro activity of Proveblue (methylene blue) on *Plasmodium falciparum* strains resistant to standard antimalarial drugs. *Antimicrobial Agents and Chemotherapy, 55*, 2472–2474.
204. Farber, P. M., Becker, K., Muller, S., Schirmer, R. H., & Franklin, R. M. (1996). Molecular cloning and characterization of a putative glutathione reductase gene, the PfGR2 gene, from *Plasmodium falciparum*. *European Journal of Biochemistry, 239*, 655–661.
205. Dormoi, J., et al. (2012). Proveblue (methylene blue) as an antimalarial agent: In vitro synergy with dihydroartemisinin and atorvastatin. *Antimicrobial Agents and Chemotherapy, 56*, 3467–3469.
206. Pradines, B., et al. (2007). Atorvastatin is 10-fold more active in vitro than other statins against *Plasmodium falciparum*. *Antimicrobial Agents and Chemotherapy, 51*, 2654–2655.

Cellular Homoeostasis and Cell Signalling in Malaria Parasite: Role of Autophagy

Azhar Muneer, Shweta Singh, Monika Narwal, Pawan Malhotra, Asif Mohmmed, and Sumit Rathore

Abstract

Autophagy has a direct and indirect role in health and disease. By definition, it is an exceedingly complex process which degrades and modifies damaged and surplus macromolecules in the cell. Autophagy has also been defined as a process which plays a role in dynamics of organelles using enzymes in lysosomes. Malaria parasite, i.e. *Plasmodium*, is a deadly parasite which encounters various conditions during its life cycle, ranging from temperature fluctuations to drug pressures. Role of autophagy in coping with these changes in *Plasmodium* is not well defined; however, there is growing evidence for role of this mechanism in life cycle of the parasite. This chapter highlights the link between different homologues of ATG protein present in *Plasmodium* and explores the mechanisms underlying these connections and their implications for cell physiology and survival of the parasite.

Keywords

Autophagy · Malaria · *Plasmodium* · ATG proteins · Programmed cell death · Stress

Introduction

Last decade has seen tremendous gains in knowledge regarding autophagy pathways. Autophagy plays a role in a variety of processes, which range from its cellular role of degrading unwanted material to more sophisticated phenomenon of

A. Muneer · S. Singh · M. Narwal · P. Malhotra · A. Mohmmed
International Centre for Genetic Engineering and Biotechnology, New Delhi, India

S. Rathore (✉)
Department of Biotechnology, All India Institute of Medical Sciences, New Delhi, India
e-mail: rathoresumit@aiims.edu

S. Hameed, Z. Fatima (eds.), *Pathogenicity and Drug Resistance of Human Pathogens*,
https://doi.org/10.1007/978-981-32-9449-3_11

organellar homoeostasis. Autophagy now represents an emerging role in various human diseases including infectious disease and cancer. Autophagy also plays crosstalk and overlapping function with a number of cellular processes such as apoptosis and senescence. A lot of information is available about autophagy events; however, a constant effort to understand process of autophagy is ongoing. Cellular homoeostasis in malaria pathogen, i.e. *Plasmodium*, is not a much explored area, but it gives a challenging opportunity for researchers to understand this complex phenomenon in this highly evolved parasite. In this chapter, we review three main aspects: autophagy in the *Plasmodium*; autophagy in cellular homoeostasis in *Plasmodium*; and autophagy of host affecting the parasite life cycle. *Plasmodium* autophagy is an upcoming area of research; thus, we have tried to compile all the information regarding homologues of autophagy pathways and proteins in *Plasmodium*. Finally, we describe the role of host autophagy in limiting the spread of *Plasmodium* infection and crosstalk between host and pathogen autophagy mechanisms.

Autophagy and Signalling in *Plasmodium*

Plasmodium falciparum is a deadly pathogen which causes malaria in millions of people globally, leading to death of thousands annually. It also causes a socio-economic impact on the well-being of the human society as it leads to poverty and neurological sequelae for an undetermined number [1]. There have been extensive studies defining cellular signalling during the life cycle of the parasite, which clearly defines its mechanism for survival and pathogenesis under normal conditions. However, little is been studied to understand the mechanism parasite follows to cope with the stress conditions it faces during the life cycle, which can include "Crisis form" morphology of the parasite. In crisis form morphology, parasite appears as punctuated and condensed parasite [2, 3]. This crisis form is attributed to the immunological response of human system against the parasite. Later the crisis form morphology was also observed in chloroquine-treated parasites, which subsequently died after the treatment [4]. Similarly, different cellular stress on the parasite led to induction of specific sequence of cellular events which ultimately caused apoptosis-like cell death [5–7]. This apoptosis-like death of the parasite opened a big debate about regulated cell death mechanism of the parasite. Regulated cell death is a balance between different mechanisms involving survival strategies, e.g. autophagy, and those involved in cell death process. Autophagy-like events have been observed both during the cell adaptation to stress and during the regulated cell death process, and it could be playing role in cell adaptation to stress [8]. Interestingly, the cellular stress-induced effects on parasite are reversible for a particular period till the apoptosis pathways are induced, suggesting an anti-apoptotic survival mechanism in the parasite during initial phase of stress [5–7]. It will be crucial to understand the metabolic pathways induced for survival mechanism which may guide towards key proteins involved in cellular homoeostasis in the parasite.

Another important aspect which needs investigation is about cellular homoeostasis in the parasite during various stages of its life cycle. *Plasmodium* enters a human when an *Anopheles* mosquito bites and releases sporozoite in the bloodstream from where it travels to the hepatocyte. After rounds of inter-hepatic cycle, the merozoites get released in the blood and invade erythrocytes. During this life cycle, the parasites transit between various stages, ring, trophozoites and schizonts, which also involve changes at the organellar level. Mechanism involved in this transition is not known, but autophagy can be speculated as one of the possible pathways which can govern this phenomenon. One such example comes from the observation that post-invasion of hepatocytes, *P. berghei* ATG8-decorated micronemes (an invasion-related organelle) are expelled from the parasite and degraded by enzymes present in the parasitophorous vacuole (PV) lumen [9]. Therefore, formation of the "crisis form" and degradation of specific organelles during parasite differentiation need exploration of a mechanism which can explain these phenomena. Autophagy is one of such mechanism which is suggested to play similar roles in other organisms [8, 10]. Autophagy is a catabolic pathway conserved among eukaryotes that allows cells to rapidly eliminate large unwanted structures such as aberrant protein aggregates, superfluous or damaged organelles and invading pathogens. Autophagy is considered to be a survival mechanism and an important pathway in the maintenance of the cell homoeostasis. Disruption in the normal autophagy machinery leads cells towards non-apoptotic pathways. Autophagy can be both specific and non-specific towards elimination of damaged organelles or aggregates of unfolded proteins.

Autophagy was described for the first time almost five decades back. Various studies based on lysosomal, biochemical and morphological changes provided the role of such a mechanism. Further research in the field of yeast biology allowed a better understanding of the phenomenon at molecular and genetic level. Identification, isolation and characterization of 27 genes related to the autophagy mechanism, which now have been termed as autophagy-related genes (ATGs), have paved the path for better understanding of the phenomenon [11]. Identification of autophagy started with acid phosphatase-stained electron microscopy images showing presence of lysosomes in rat liver [12]. Purification of these lysosomes leads to identification of other structures termed as dense bodies which have almost the same properties of lysosomes including presence of hydrolases [13]. Later these dense bodies were named as autophagosome vacuoles or AVs, which are distinct from lysosomes. The presence of whole organelles like mitochondria in these autophagosome vacuoles from newborn mouse kidney cells highlighted the organelle degradative role of these structures [14]. It is becoming highly evident that autophagy is highly selective quality control mechanism whose basal levels are important to maintain cellular homoeostasis [15]. A number of organelles have been found to be selectively turned over by autophagy, and cargo-specific names have been given to distinguish the various selective pathways including the ER (reticulophagy or ER-phagy), peroxisomes (pexophagy), mitochondria (mitophagy), lipid droplets (lipophagy), secretory granules (zymophagy) and even parts of the nucleus (nucleophagy) [16–18]. In addition, it has been proposed for a long time that, although generally a cytoprotective mechanism, autophagy can initiate or execute cell death under certain conditions such as cellular stress or starvation [19].

The basic feature of autophagy is generation of autophagosome, which is the site where degradation of macromolecules to their native constituents takes place. Different stages for the autophagosome generation have been listed as induction, nucleation, expansion, fusion and cargo recycling. All of these stages involve different autophagy proteins along with different signalling molecules. Out of ~30 autophagy-related genes (ATGs) which were identified in yeast, the *Plasmodium* genome contains 15 homologues of autophagy-related genes (ATGs). Here we have discussed role of different ATGs in various steps of the autophagy process, their homologue in *Plasmodium* and their possible role in cellular homoeostasis of parasite life cycle.

Induction and Nucleation of Autophagosome Vacuoles

The induction of autophagy vacuoles and the source of membranes for their formation has remained a phenomenon which has been investigated and has remained complex for long time. Endoplasmic reticulum and Golgi bodies were pointed out as the source for the generation of membranes for these autophagosome vacuoles, although no confirmatory source of AVs are suggested till now [20–22]. Once these AVs are formed, irrespective of their source of generations, morphologic studies showed that these structures acquire degradative enzymes through fusion with mature lysosomes [23]. Different steps and various ATG proteins are involved in the generation of mature autophagolysosome [24, 25].

Change in the environment, i.e. nutritional starvation or temperature changes, are sensed by the cells using different receptors on their surfaces, which is further relayed to the downstream effector molecules using different signalling mechanisms. Autophagy initiation occurs with development of single site called the pre-autophagosomal structure (PAS; also called phagophore assembly site). Under a fluorescence microscope, the PAS is detected as a single, dot-like accumulation of Atg proteins next to the vacuole. In Mammal's PAS is missing, but small sites where assembly starts are found to be associated initially with ER membrane [26, 27], Golgi membranes [28, 29] and nuclear membranes [30]. Nutrition starvation is one of the triggers which can throw a cell into the mechanism of autophagy. In deprivation of some amino acids or proteins, autophagy has been induced in the cell through target of rapamycin complex 1 (TORC1) which can sense changes in the extracellular environment of cell. TORC1 induction further leads to activation of ATG1 kinase complex. *Plasmodium* lacks homologue of TORC1 complex; PfTOR has very poor homology or even extensive variations that set it apart from other conserved sequences, which may suggest that PfTOR is regulated differently or its activity may be variable. Interestingly, phosphorylation of PfeIF2α by PfeIKI in response to starvation is described by Fennell et al. [31], which suggest that this system (downstream of TOR) regulates response to amino acid starvation. Similarly, Williams et al. [32] described an autophagy pathway independent of mTOR that is induced in a Ca^{2+}-dependent fashion in neuronal precursor cells. This precedent suggests that it is possible for autophagy to operate independently of TOR. In *Plasmodium*, generation of autophagosomes has remained a very intriguing mechanism. As in other systems, source of phagophore membranes is not very clear and has remained a question of investigation.

Atg1 Kinase Complex

In yeast, which is a metazoan, autophagy has been inhibited by the TOR kinase, which phosphorylates components of Atg1 kinase complex. The ATG1 complex is composed of ATG1 and ATG13, along with a scaffold made up of ATG17-ATG31 and ATG29. This whole complex acts a nucleation site to bring all the ATG proteins required for complex functioning. ATG1 kinase activity in complex with ATG13-Atg17 is required for the phagophore formation. Further the activity of ATG1 has been regulated by TOR kinase which prevents interaction of ATG13 with ATG1 by phosphorylating the ATG13 [33]. TOR kinase is energy-sensing kinase and thus acts as a regulator for the initiation of autophagy sensitive to nutrient and growth factor availability. Similar function is carried out by ULK1 which is a homologue of Atg1 in reticulocytes [34]. In mammals, the complex comprises of ULK1, ATG13, ATG101 and FIP200. Out of the described Atg1 kinase complex defined in mammals, Atg1 and its mammalian binding partner FIP200 and Atg101 are present in the *Plasmodium* genome. ULK1 is the mammalian orthologue of Atg1, while FLIP200 is the mammalian counterpart of Atg17. Atg13 is an important partner of Atg1 complex; however it is not present in the *Plasmodium* genome [35].

PfAtg1 also differs from Atg1 from other organisms; it lacks C-terminal domain, which is required for interaction with ATG13. Another interesting observation about PfAtg1 is presence of Atg8-interacting domains, which points towards presence of Atg13-independent pathway of autophagy in *Plasmodium*, as has been shown in yeast [36].

ATG9 and Its Cycling System

Apart from the start point of phagosome formation, another important question about autophagy is the source of lipid used for formation of autophagosomes and their cycling to the site of initiation. Atg9 is an integral membrane protein which acts as a carrier of membrane during the assembly process [37]. Atg9 is little different from most of other Atg proteins, as most of Atg proteins has single localization on PAS, whereas ATG9 localizes to multiple puncta structures [38]. *Plasmodium* genome shows absence of PfAtg9, which points towards presence of other important protein compensating for the role of this important protein in the generation of autophagosomes.

Cycling of ATG9 between PAS and structure other than PAS is thought to be potential source of lipid for the formation of PAS. Anterograde movement of Atg9 to

PAS involves several other Atg proteins in mammals including PfATG23 and PfAtg27. *Plasmodium* genome also lacks presence of PfAtg23 and pfAtg27 homologues. Retrieval of Atg9 from PAS to peripheral locations depends on Atg1-Atg13 kinase complex and Atg2, Atg18 and PtdIns3K complex I [38]. Atg2 and Atg8 is a set of important interacting peripheral membrane proteins which can interact with Atg9 [39]. Interaction of Atg18 requires both Atg2 and Atg8 [40, 41]. Atg18 has been shown to interact with two phosphoinositides, PtdIns(3)P and PtdIns[3, 5]P2; however Atg18 interacting with PtdIns(3)P is required for autophagy [42]. In yeast, Atg1-Atg13 complex promotes interaction of Atg9 with Atg2 and Atg18, and formation of the ternary complex leads to release of Atg9 [43].

PfAtg18 also shows interaction with PtdIns(3)P and gets localized to vesicles near apicoplast and food vacuole [44]. PfAtg18 disruption leads to increase in the number of PfATG18-positive vesicles with *P. falciparum* maturation, which corroborated well with increases in PfATG18 expression in mature trophozoites and schizonts. The PfATG18 vesicles were observed in close vicinity of the food vacuole and there was no apparent colocalization of PfATG18 with the apicoplast; some PfATG18-labeled vesicles were detected near the branching apicoplast and in proximity of vesicles containing downstream autophagy protein PfATG8. PfATG18 interact with 3′-PIPs via the FRRG motif, and mutation of this motif resulted in a significant loss of its vesicular localization. PfATG18 having FRRG domain and binding potential towards PI3PK have speculated role in binding of Atg2 and lead to attachment of PfAtg18-PfAtg2 to PAS.

In other systems ATG18 interaction with PtdIns(3)P leads to attachment of ATG8 to phagophore and its elongation [45]. In *Plasmodium*, no direct interaction has been shown between PfATG8 and PfATG18. However, downregulation of PfATG18 leads to decrease in the number of PfATG8-positive vesicles under normal conditions [44]. A putative homologue of PfAtg2 has also been identified in the genome.

Atg8 and Atg12 Ubiquitin Conjugation System

Another important pathway for autophagy mechanism in yeast is ATG5-ATG12 pathway, which involves ubiquitin-like enzymes for its implication. *Plasmodium* contains homologues for Atg8, Atg3 and Atg7 from this branch of autophagy [46–48]. PfAtg8 remains the choice of molecule that has been targeted in *Plasmodium* to understand the autophagy and its role in the life cycle of parasite. Atg8 orthologues from most organisms require proteolytic processing of one or several amino acids to expose a C terminal glycine. PfAtg8 differs from its mammalian homologue, in having a C-terminal glycine. This exposed C-terminal glycine is required for attaching Atg8 to its E1 ubiquitin enzyme Atg7. In mammals ATG4-based processing is required for making the C-terminal glycine for conjugation of ATG8 towards phosphatidylethanolamine (PE). Lipidated form of Atg8 has been suggested to be a marker of autophagy in mammalian system, as ATG8-PE remains conjugated with autophagosome during fusion to the lysosome [49]. In *Plasmodium* various studies tried to decipher the localization of PfAtg8. PfAtg8 has been targeted in Plasmodium to understand the autophagy and its role in the life cycle of parasite [43–48, 50, 51]. C-terminal glycine of PfAtg8 has an important role in the association of PfAtg8 with these punctate structures and apicoplast as removal of this C-terminal glycine leads to diffuse localization of this protein in the cytoplasm in liver stage [47]. Recently structure of PfAtg8-PfAtg3 has been solved, which suggest that *Plasmodium* Atg8 contains an insertion of nine residues only conserved within Apicomplexa that comprise β3 and the turn in the loop [35].

In yeast, lipidation of Atg8 requires E1-type ligase Atg7, E2-type ligase Atg3 and a cysteine protease Atg4. *Plasmodium* genome has all of these three ATG proteins, Atg3, Atg4 and Atg7, which are shown to be transcribed in blood stages of the parasite life cycle [52]. Processing of C-terminus of Atg8 by Atg4, followed by activation using ATP, allows it to interact with Atg7. Atg8 is further transferred to Atg3, an E2-like conjugating enzyme, leading to formation of second thioester bond followed by conjugation to nitrogen of PE. This process also requires noncovalent interaction between Atg8 and Atg3 through a well-characterized Atg8-interacting motif (AIM) in Atg3 and two hydrophobic pockets, termed the W- and L-site, in Atg8. In *Plasmodium* Atg3 also contains Atg8-interacting motif along with the WLLP residues which are responsible for the interaction of Atg3-Atg8 [35]. Atg3-Atg8 interaction has been targeted well in *Plasmodium* as possible site for the action of small molecule inhibitors [53].

Deconjugation of PfAtg8 from its binding sites for recycling has been carried out by other cysteine proteases, i.e. PfOTU [55]. Depletion of PfOTU levels also affects PfAtg8 conjugated forms and thus affects transport of apicoplast proteins. PfAtg8 localization to the punctate structure during normal conditions and relocalization during nutritional starvation conditions or during drug treatment point towards important role in autophagy mechanism in the *Plasmodium* [56]. PfAtg8-associated double-membrane vesicular structures are also shown to contain PfRab7 [57]. During nutritional starvation conditions, these double-membrane vesicular structures are shown to be near to the food vacuole, which points towards food vacuole as the final site of their localization [57].

Atg7 Protein Complex

The PfAtg7 shows low identity (14.7%) and similarity (32.2%) to yeast Atg7; however, it shows conservation of key residues, including the catalytic cysteine and ATP-binding domain [54]. Atg7 is an E1-type activating enzyme, which is a ubiquitin-related modifier. It follows the mechanism of protein ubiquitination in order to lipidate Atg8. During the process, a thioester intermediate is formed between the E1 (Atg7) and ubiquitin (Atg8). Ubiquitin (Atg8) is then transferred to the catalytic cysteine residue of the ubiquitin-conjugating enzyme or E2 (Atg3). The final step includes transfer of ubiquitin (Atg8) to its target protein (PE) forming a covalent bond through an isopeptide linkage. This can occur directly by the E2 or through a third ubiquitin-protein ligase or E3 (Atg5-Atg12). Putative homologue of ATG12 and Atg5 has also been identified in the host.

PtdIns3K Complex

Another signalling mechanism which plays a role in the initiation process of autophagy is class III PI-kinase-based pathway. Vesicular protein sorting 34 or vps34 is a class III PI-kinase which carries out initiation of autophagy in yeast in complex with Atg6 and Beclin1 [58]. Interaction of Beclin1 and Vps34 promotes activity of Vps34, which leads to higher production of phosphatidylinositol-3-phosphate (PtdIns3p) required for phagophore formation and elongation. Beclin1-Vps34 complex has further been joined by various different regulatory molecules which leads to either promotion of autophagy or its inhibition. UVRAG, ATG14L, BIF-1 and AMBRA are shown to promote autophagy when they interact with beclin1-vps34 complex [22, 59], whereas Rubicon and Bcl-2 lead to inhibition of autophagy [60, 61]. Interaction of Bcl-2 with Beclin leads to disruption of beclin1-vps34 complex and leads to inhibition of autophagy [61, 62]. Class III PtdIns3K complex is comprised of five distinct proteins: Vps34, the regulatory kinase Vps15, Vps30/Atg6, Atg14 and Atg38 [63, 64]. This complex is responsible for the production of PtdIns3p, which plays a very important role for the correct localization of other ATG proteins including Atg18 and Atg2 enabling the recruitment of Atg8, Atg9 and Atg12 of the pre-autophagosomal site [65].

Plasmodium genome harbours homologue of vps34, which is also termed as PfPI3K (ref). PfPI3K has been identified and found to be localized at various places ranging from digestive vacuole (DV), the parasite membrane and vesicles in host erythrocytes [66]. PfPI3K is mainly shown to be involved in haemoglobin uptake and degradation pathways, as inhibition of pfPI3K with wortmannin causes accumulation of unutilized haemoglobin [66]. Pfvps34 do not contain pleckstrin homology domains and Ras binding sites that are usual features of class I and class II PI3Ks. PfVps34 is an essential protein for the survival of the parasite [67]. The Pfvps34, along with process of haemoglobin uptake and degradation, had been hypothesized to be a key regulatory checkpoint in the autophagy-like cascade of *P. falciparum* [35, 51]. As cellular material is likely to be digested within the digestive in parasite, Pfvps34 appears to be key component of *Plasmodium falciparum* autophagy cascade in absence of TOR kinase activity [35].

Phagophore Expansion

Ubiquitin-like systems also play key role in the autophagy phagophore expansion process. Two major key ubiquitin pathways are Ag5-Atg12 complex and the LC3 processing pathway. ATG12 and ATG8 are the most important ubiquitin-like proteins in the process of autophagy. They have high similarity to the ubiquitin at the structural level, but they are not the real homologues. ATG5-ATG12 pathway starts with ATG7 binding to the carboxyl terminal of ATG12 activating the later one in an ATP-dependent manner. This activated ATG12 then has been transferred to ATG10. ATG10 acts like an E2-ubiquitin-like enzyme which carries out attachment of ATG12 to residue 130 lysine of ATG5 leading to formation of ATG12-Atg5 complex. This conjugated ATG5-ATG12 pairs with ATG16L dimers to form bigger multimers of ATG5-ATG12-ATG16L which leads to the extension of phagophore membrane. This growing multimer complex then further has been joined with LC3B-II, the product of second important ubiquitin system playing a role in autophagy in mammals. Microtubule-associated protein light chain 3 (LC3B) is full-length cytosolic protein, which has been acted upon by ATG4 which is a cysteine protease, leading to formation of LC3B-I, which contains a C-terminally exposed glycine residue. This exposed glycine residue has then been activated by ATG7 in an ATP-dependent manner. Activated LC3B-I is then transferred to Atg3; a different E2-like carrier protein before phosphatidylethanolamine (PE) is conjugated to the carboxyl glycine to generate processed LC3B-II.

PfAtg5 is a recently characterized ATG protein in *Plasmodium* [52]; it has been found to be associated with punctuate structures throughout the parasite cytosol, which may point towards presence of autophagosome-like structures near the mitochondria and ER as observed in mammalian systems [52]. The putative PfAtg5 is a much longer protein due to insertions as compared to its counterparts. The lysine amino acid residues present at position 479 are involved in conjugation with PfAtg12; in addition, a lysine residue is present at position 480.

Roles of Autophagy in *Plasmodium*

After being released in the blood after mosquito bites, sporozoite moves to the hepatic cells and tries to establish liver stage development of the parasite. During this process, sporozoite gets converted into metabolically active trophozoites. This whole metamorphosis involves lot of activity at the cellular level and involves changes in the shape of the parasite. Along with this shape changes, they start losing some of the organelles [68], and at the end of this metamorphosis, parasite is only left with organelles required for replication. The mechanism that underlies this process was speculated to be autophagy at the time, although much of ATGs identified in yeast have not been recognized in the parasite. *Plasmodium* cellular homoeostasis is a very important but complex mechanism in the life cycle of the parasite. *Plasmodium* sporozoites treated with 3-methyladenine (3-MA), which targets Vps34, needed for PI(3)P production and phagophore formation, delay conversion into trophozoites [69].

Autophagy in Protein Secretion and Trafficking

Protein secretion in *Plasmodium* is an important process which is required for its survival and pathogenesis. Parasite remodels the host and exports a number of proteins to the host cytoplasm and to the surface [70]. *Plasmodium* stays in the host encapsulated in the parasitophorous vacuole, from which parasite proteins have to travel across the PV and plasma membrane, which require novel pathways for secretion. Exophagy is one of the mechanisms through which these kinds of protein secretion are been observed in secretion of Acb1 protein in yeast, which utilizes autophagy mechanism involving GRASP65, t-SNARE and sso1 [71]. *Plasmodium* genome contains homologue of GRASP65 and t-SNARE [72], which points towards possibility of exophagy as part of transport mechanism in the parasite.

Another example regarding the role of autophagy in protein secretion comes from type III PI3K, e.g. vps34, a lipid kinase at the ER that produce PI(3)P on the outer leaflet, and recruit proteins to the phagophore and formation of autophagosome. *Plasmodium* homologue of the Vps34, PfVps34, from lysate has also been shown to have PI3K activity and can be inhibited by wortmannin class III PI3K. PfVps34 localizes to the food vacuole and different vesicular structures near the food vacuole and the plasma membrane in the blood stage [73]. Further, Vps34 inhibition in sporozoites delays its development to trophozoite [73]. The electron microscopy studies of liver stage parasite showed presence of double-membrane structures resembling autophagosomes having microneme point towards role of exophagy in protein secretion in the parasite.

PfAtg8 role in protein trafficking has also been highlighted in early-mid trophozoite, where it is suggested to be linked to haemoglobin uptake [57]. Chloroquine treatment to the early-mid trophozoite leads to increase in vesicles, which are assumed to be autophagosomic vesicles. Prolonged starvation or high cytocidal concentration of chloroquine (CQ) leads to relocalization of PfAtg8-containing vesicles from the parasite to host cytoplasm [51]. However, Cytostatic concentration of chloroquine (100 μm and 300 μM) and shorter treatment duration doesn't show any relocalization of PfAtg8 to host cytoplasm [74]. PfAtg8-containing vesicles are also shown to harbour Rab7 molecule on their surfaces, which also points towards the role of PfAtg8 in protein trafficking [57].

Autophagy in Programmed Cell Death

In the early 2000s, studies about effect of drugs on the development on the parasite pointed out towards probable role of programmed cell death in the *Plasmodium*. Treatment of the parasite with chloroquine, S-nitroso-N-acetylpenicillamine (SNAP) or staurosporine showed presence of apoptosis-like features during the death of the parasite [75]. In absence of canonical apoptosis pathway in the parasite, these features of death lead to investigate the role of other mechanisms in the parasite life cycle. Programmed cell death in unicellular organisms has not been supported earlier as the apoptotic pathway machinery is lacking as has been shown in yeast [76, 77]. Although presence of metacaspases in some of the unicellular eukaryotes including *Plasmodium* suggested a primordial form of apoptosis to exist, the cannonical pathway of apoptosis remains missing [78].

It is suggested that *Plasmodium berghei* invasion of hepatocytes involves death of a lot of parasites through autophagic mechanism [79]. This phenomenon was observed in mice model and thus seems to be a physiological mechanism of host defence. This phenomenon is seen to be associated with development of later stages of parasite in which double-membrane structures have been observed which are termed as "membranous whorls" [80]. PbAtg8 does not localize with these structures and remains associated with apicoplast with no changes in localization [9]. Another interesting observation is that treatment with rapamycin, a known TOR kinase inhibitor which has been used to induce autophagy in yeast, leads to decrease in parasite size and morphological changes in apicoplast in *Plasmodium*.

Host Autophagy Machinery During *Plasmodium* Life Cycle

During the life cycle, parasite transits between two hosts, mosquito and human. In human host the erythrocyte doesn't play any role in the control of the parasite infection in blood stage; however, liver cells employ its host cell autophagy machinery to stop the infection [87]. During its stay in the liver cell, *Plasmodium* parasite is protected from its host by parasite vacuole membrane (PVM). PVM has been generated from host cell membrane during invasion but has been modified by parasite using its own protein [81–83]. Some of these proteins, e.g. UIS3 and UIS4, have been shown to be important in the survival of the parasite in the liver cell. Disruption of UIS3 and UIS4 showed to cause growth arrest of the parasite [84]. For a long time, liver stage infection of the malaria parasite has been considered as "silent stage". It is becoming clear now that host cell can sense the parasite and take care of it very effectively using autophagy [79]. Induction of canonical nonselective autophagy with rapamycin or starvation enhances the development of parasite in hepatocytes and leads to increase in the number of liver stage parasite [85]. In addition, if host autophagy machinery is disrupted genetically, the parasite development is affected in liver stage [85]; however, there are contradictory reports as well, which may have arisen due to different cell types used in the experiment [79]. These different opinions on the role of host autophagy in *Plasmodium* led to intense discussion whether it is an advantage or disadvantage regarding the malaria infection [87]. Recent views on liver stage development could represent a non-canonical form of autophagy, termed as *Plasmodium* Associated Autophagic-like Response (PAAR) [85–87].

The process and the proteins involved in host autophagy during the hepatic stages of *Plasmodium* are well elucidated. One of the signature processes of parasite liver stage development is the rapid acquisition of host LC3 and other interacting proteins such as P62, NBR1, NDP52 and ubiquitin on the PVM [79]. The process of LC3 acquisition is so rapid that it suggests that either the parasite hijacks the host autophagy machinery or it is readily sensed by the host. Association of LC3 with the PVM itself has some striking aspects, such as LC3 decoration on the PVM does not involve the formation of new canonical double-membrane autophagosome, as it has been observed in other systems, rather LC3 associates with the existing PVM. This is also different from LC3-associated phagocytosis phenomenon as sporozoite invasion to the liver cells is an active process which is different from conventional phagocytosis. Another aspect which is different from canonical xerophagy pathway is the association of the LC3-binding proteins, including ubiquitin, to the PVM that is to a large extent directly mediated by LC3 [79]. Recruitment of LC3-binding proteins to the PVM appears to be in reverse order leading to the idea of an "inverted" recruitment of LC3-associated proteins on the PVM [76]. Further LC3 association with the PVM is temporary [85], and it gets dissociated in later stages,

Table 1 Homologues of autophagy proteins and their putative functions

Saccharomyces cerevisiae	*Plasmodium falciparum*	*Homo sapiens*	Probable functions
Atgl	PF3D7_1450000	Ulk1/2	Protein kinase
Atg2	PF3D7_1428300	Atg2A/B	Interacts with WIPI4
Atg3	PF3D7_0905700.2	Atg3	E2-like enzyme
Atg4	PF3D7_1417300	Atg4A/B/C/D	Cysteine protease processing of Atg8
Atg5	PF3D7_1430400	Atg5	E3-like enzyme for Atg5-Atg12 complex
Atg7	PF3D7_1126100	Atg7	E2-like enzyme
Atg8	PF3D7_1019900	LC3A/B/B2/C	Ubiquitin-like protein required for autophagosomal structure
Atg9	–	Atg9A/B	Transmembrane protein
Atgl0	–	Atg10	Ubiquitin-like ligase
Atgll	PF3D7_0216700	–	
Atg12	PF3D7_1470000	Atg12	Ubiquitin-like
Atg13	PF3D7_1201400	Atg13	Protein kinase binding
Atgl4	–	Atg14 (L)	Protein binding
Atg17	PF3D7_0203000	FIP200 Atg101	Protein kinase binding
Atg29	–		PAS formation
Atg3l	–		Autophagy induction in complex with Atg17-Atg31-Atg29
Atg18	PF3D7_1012900	WIPI 1/2/3/4	PIP2 binding
Atg19	PF3D7_0502000	–	Receptor-mediated selective autophagy
Atg22	PF3D7_0511300	–	Receptor-mediated selective autophagy
Vps15	PF3D7_0823000	Vps15	PI3-kinase regulator
Vps34	PF3D7_0515300	Vps34	Endosome trafficking
Vps30/Atg6	–	Beclin1	Endocytic retrograde transport

which is shown to be important for the proper development of the parasite [85]. LC3 recruitment to the PVM is governed by lipidation suggesting the role of conjugation machinery involving upstream ATGs such as ATG5. However, initiation complexes of autophagy such as FIP200 are not required [83].

The pathway and the factors that initiate the conjugation of LC3 to the PVM due to *Plasmodium* infection are not known in detail. UIS3 binds directly to and retains LC3 on the PVM as evident by uis3(−) parasite, which are arrested in development in wild-type hepatocytes. These uis3(−) parasite develop normally in ATG5−/- MEFs, pointing towards central role of UIS3 in interaction with host autophagy machinery (Table 1).

References

1. White, N. J. (2014). Malaria: A molecular marker of artemisinin resistance. *Lancet (London, England) [Internet]., 383*(9927), 1439–1440. Available from: http://www.ncbi.nlm.nih.gov/pubmed/24766952.
2. Ockenhouse, C. F., Schulman, S., Shear, H. L., et al. (1984). *Journal of Immunology [Internet], 133*(3), 1601–1608. Available from: http://www.ncbi.nlm.nih.gov/pubmed/6431003.
3. Jensen, J. B., Boland, M. T., & Akood, M. (1982). Induction of crisis forms in cultured *Plasmodium falciparum* with human immune serum from Sudan. *Science [Internet], 216* (4551), 1230–1233. Available from: http://www.ncbi.nlm.nih.gov/pubmed/7043736.
4. Picot, S., Burnod, J., Bracchi, V., Chumpitazi, B. F., & Ambroise-Thomas, P. Apoptosis related to chloroquine sensitivity of the human malaria parasite *Plasmodium falciparum. Transactions of The Royal Soceity Tropical Medicine Hygiene [Internet], 91*(5), 590–591. Available from: http://www.ncbi.nlm.nih.gov/pubmed/9463676.
5. Rathore, S., Datta, G., Kaur, I., Malhotra, P., Mohmmed, A., et al. (2015). *Cell Death and Disease [Internet], 6*, e1803. Available from: http://www.ncbi.nlm.nih.gov/pubmed/26136076.
6. Rathore, S., Jain, S., Sinha, D., Gupta, M., Asad, M., Srivastava, A., et al. (2011). Disruption of a mitochondrial protease machinery in *Plasmodium falciparum* is an intrinsic signal for parasite cell death. *Cell Death and Disease [Internet], 2*, e231. Available from: http://www.ncbi.nlm.nih.gov/pubmed/22113196.
7. Jain, S., Rathore, S., Asad, M., Hossain, M. E., Sinha, D., Datta, G., et al. (2013). The prokaryotic ClpQ protease plays a key role in growth and development of mitochondria in *Plasmodium falciparum. Cellular Microbiology [Internet], 15*(10), 1660–1673. Available from: http://www.ncbi.nlm.nih.gov/pubmed/23521916.
8. Denton, D., Aung-Htut, M. T., Lorensuhewa, N., Nicolson, S., Zhu, W., Mills, K., et al. (2013). UTX coordinates steroid hormone-mediated autophagy and cell death. *Nature Communications [Internet], 4*, 2916. Available from: http://www.ncbi.nlm.nih.gov/pubmed/24336022.
9. Voss, C., Ehrenman, K., Mlambo, G., Mishra, S., Kumar, K. A., Sacci, J. B., et al. (2016). Overexpression of Plasmodium berghei ATG8 by liver forms leads to cumulative defects in organelle dynamics and to generation of noninfectious Merozoites. *MBio [Internet], 7*(3). Available from http://www.ncbi.nlm.nih.gov/pubmed/27353755.
10. Betin, V. M. S., Singleton, B. K., Parsons, S. F., Anstee, D. J., & Lane, J. D. (2013). Autophagy facilitates organelle clearance during differentiation of human erythroblasts: Evidence for a role for ATG4 paralogs during autophagosome maturation. *Autophagy [Internet], 9*(6), 881–893. Available from: http://www.ncbi.nlm.nih.gov/pubmed/23508006.
11. Klionsky, D. J., Cregg, J. M., Dunn, W. A., Emr, S. D., Sakai, Y., Sandoval, I. V., et al. (2003). A unified nomenclature for yeast autophagy-related genes. *Developmental Cell [Internet], 5*(4), 539–545. Available from: http://www.ncbi.nlm.nih.gov/pubmed/14536056.
12. De Duve, C., Pressman, B. C., Gianetto, R., Wattiaux, R., & Appelmans, F. (1955). Tissue fractionation studies. 6. Intracellular distribution patterns of enzymes in rat-liver tissue. *The Biochemical Journal [Internet], 60*(4), 604–617. Available from: http://www.ncbi.nlm.nih.gov/pubmed/13249955.
13. Novikoff, A. B. (1955). Some aspects of hepatoma NK. *Journal of the National Cancer Institute [Internet], 15*(5, Suppl), 1533–1534. Available from: http://www.ncbi.nlm.nih.gov/pubmed/13243091.
14. Clark, S. L. (1957). Cellular differentiation in the kidneys of newborn mice studies with the electron microscope. *The Journal of Biophysical and Biochemical Cytology [Internet], 3*(3), 349–362. Available from: http://www.ncbi.nlm.nih.gov/pubmed/13438920.
15. He, C., & Klionsky, D. J. (2009). Regulation mechanisms and signaling pathways of autophagy. *Annual Review of Genetics [Internet], 43*, 67–93. Available from: http://www.ncbi.nlm.nih.gov/pubmed/19653858.

16. Beau, I., Esclatine, A., & Codogno, P. (2008). Lost to translation: When autophagy targets mature ribosomes. *Trends in Cell Biology [Internet], 18*(7), 311–314. Available from: http://www.ncbi.nlm.nih.gov/pubmed/18508269.
17. Kraft, C., Reggiori, F., & Peter, M. (2009). Selective types of autophagy in yeast. *Biochimica et Biophysica Acta [Internet], 1793*(9), 1404–1412. Available from: http://www.ncbi.nlm.nih.gov/pubmed/19264099.
18. van der Vaart, A., Mari, M., & Reggiori, F. (2008). A picky eater: Exploring the mechanisms of selective autophagy in human pathologies. *Traffic [Internet], 9*(3), 281–289. Available from: http://www.ncbi.nlm.nih.gov/pubmed/17988219.
19. Gordy, C., & He, Y.-W. (2012). The crosstalk between autophagy and apoptosis: Where does this lead? *Protein and Cell [Internet], 3*(1), 17–27. Available from: http://www.ncbi.nlm.nih.gov/pubmed/22314807.
20. De Duve, C., & Wattiaux, R. (1966). Functions of lysosomes. *Annual Review of Physiology [Internet], 28*, 435–492. Available from: http://www.ncbi.nlm.nih.gov/pubmed/5322983.
21. Novikoff, A. B., & Shin, W. Y. (1978). Endoplasmic reticulum and autophagy in rat hepatocytes. *Proceedings of the National Academy of Sciences of United States America [Internet], 75*(10), 5039–5042. Available from: http://www.ncbi.nlm.nih.gov/pubmed/283412.
22. Liang, C., Feng, P., Ku, B., Dotan, I., Canaani, D., Oh, B.-H., et al. (2006). Autophagic and tumour suppressor activity of a novel Beclin1-binding protein UVRAG. *Nature Cell Biology [Internet], 8*(7), 688–699. Available from: http://www.ncbi.nlm.nih.gov/pubmed/16799551.
23. Deter, R. L., Baudhuin, P., De Duve, C., et al. (1967). *The Journal of Cell Biology [Internet], 35* (2), C11–C16. Available from: http://www.ncbi.nlm.nih.gov/pubmed/6055998.
24. Dunn, W. A. (1994). Autophagy and related mechanisms of lysosome-mediated protein degradation. *Trends Cell Biol [Internet], 4*(4), 139–143. Available from: http://www.ncbi.nlm.nih.gov/pubmed/14731737.
25. Levine, B. (2005). Eating oneself and uninvited guests: Autophagy-related pathways in cellular defense. *Cell [Internet], 120*(2), 159–162. Available from: http://www.ncbi.nlm.nih.gov/pubmed/15680321.
26. Hayashi-Nishino, M., Fujita, N., Noda, T., Yamaguchi, A., Yoshimori, T., & Yamamoto, A. (2009). A subdomain of the endoplasmic reticulum forms a cradle for autophagosome formation. *Nature Cell Biology [Internet], 11*(12), 1433–1437. Available from: http://www.ncbi.nlm.nih.gov/pubmed/19898463.
27. Ylä-Anttila, P., Vihinen, H., Jokitalo, E., & Eskelinen, E.-L. (2009). 3D tomography reveals connections between the phagophore and endoplasmic reticulum. *Autophagy [Internet], 5*(8), 1180–1185. Available from: http://www.ncbi.nlm.nih.gov/pubmed/19855179.
28. Axe, E. L., Walker, S. A., Manifava, M., Chandra, P., Roderick, H. L., Habermann, A., et al. (2008). Autophagosome formation from membrane compartments enriched in phosphatidylinositol 3-phosphate and dynamically connected to the endoplasmic reticulum. *Journal of Cell Biology [Internet], 182*(4), 685–701. Available from: http://www.ncbi.nlm.nih.gov/pubmed/18725538.
29. Mizushima, N. (2007). The role of mammalian autophagy in protein metabolism. *Proceedings of the Japan Academy Series B Physical and Biology Sciences [Internet], 83*(2), 39–46. Available from: http://www.ncbi.nlm.nih.gov/pubmed/24019583.
30. English, L., Chemali, M., Duron, J., Rondeau, C., Laplante, A., Gingras, D., et al. (2009). Autophagy enhances the presentation of endogenous viral antigens on MHC class I molecules during HSV-1 infection. *Nature Immunology [Internet], 10*(5), 480–487. Available from: http://www.ncbi.nlm.nih.gov/pubmed/19305394.
31. Fennell, B. J., Al-shatr, Z. A., & Bell, A. (2008). Isotype expression, post-translational modification and stage-dependent production of tubulins in erythrocytic *Plasmodium falciparum*. *International Journal Parasitology [Internet], 38*(5), 527–539. Available from: http://www.ncbi.nlm.nih.gov/pubmed/17977543.
32. Williams, A., Sarkar, S., Cuddon, P., Ttofi, E. K., Saiki, S., Siddiqi, F. H., et al. (2008). Novel targets for Huntington's disease in an mTOR-independent autophagy pathway. *Nature*

Chemical Biology [Internet], 4(5), 295–305. Available from: http://www.ncbi.nlm.nih.gov/pubmed/18391949.

33. Díaz-Troya, S., Pérez-Pérez, M. E., Florencio, F. J., & Crespo, J. L. (2008). The role of TOR in autophagy regulation from yeast to plants and mammals. *Autophagy [Internet], 4*(7), 851–865. Available from: http://www.ncbi.nlm.nih.gov/pubmed/18670193.
34. Kundu, M., Lindsten, T., Yang, C.-Y., Wu, J., Zhao, F., Zhang, J., et al. (2008). Ulk1 plays a critical role in the autophagic clearance of mitochondria and ribosomes during reticulocyte maturation. *Blood [Internet], 112*(4), 1493–1502. Available from: http://www.ncbi.nlm.nih.gov/pubmed/18539900.
35. Hain, A. U. P., & Bosch, J. (2013). Autophagy in Plasmodium, a multifunctional pathway? *Computational and Structural Biotechnology Journal [Internet], 8*, e201308002. Available from: http://www.ncbi.nlm.nih.gov/pubmed/24688742.
36. Kraft, C., Kijanska, M., Kalie, E., Siergiejuk, E., Lee, S. S., Semplicio, G., et al. (2012). Binding of the Atg1/ULK1 kinase to the ubiquitin-like protein Atg8 regulates autophagy. *EMBO Journal [Internet], 31*(18), 3691–3703. Available from: http://www.ncbi.nlm.nih.gov/pubmed/22885598.
37. Noda, T., Kim, J., Huang, W. P., Baba, M., Tokunaga, C., Ohsumi, Y., et al. (2000). Apg9p/Cvt7p is an integral membrane protein required for transport vesicle formation in the Cvt and autophagy pathways. *Journal of Cell Biology [Internet], 148*(3), 465–480. Available from: http://www.ncbi.nlm.nih.gov/pubmed/10662773.
38. Reggiori, F., Shintani, T., Nair, U., & Klionsky, D. J. (2005). Atg9 cycles between mitochondria and the pre-autophagosomal structure in yeasts. *Autophagy [Internet], 1*(2), 101–109. Available from: http://www.ncbi.nlm.nih.gov/pubmed/16874040.
39. Suzuki, K., Kubota, Y., Sekito, T., & Ohsumi, Y. (2007). Hierarchy of Atg proteins in pre-autophagosomal structure organization. *Genes to Cells [Internet], 12*(2), 209–218. Available from: http://www.ncbi.nlm.nih.gov/pubmed/17295840.
40. Wang, C. W., Kim, J., Huang, W. P., Abeliovich, H., Stromhaug, P. E., Dunn, W. A., et al. (2001). Apg2 is a novel protein required for the cytoplasm to vacuole targeting, autophagy, and pexophagy pathways. *Journal of Biology Chemistry [Internet], 276*(32), 30442–30451. Available from: http://www.ncbi.nlm.nih.gov/pubmed/11382760.
41. Reggiori, F., Wang, C.-W., Nair, U., Shintani, T., Abeliovich, H., & Klionsky, D. J. (2004). Early stages of the secretory pathway, but not endosomes, are required for Cvt vesicle and autophagosome assembly in *Saccharomyces cerevisiae*. *Molecular Biology of the Cell [Internet], 15*(5), 2189–2204. Available from: http://www.ncbi.nlm.nih.gov/pubmed/15004240.
42. Strømhaug, P. E., Reggiori, F., Guan, J., Wang, C.-W., & Klionsky, D. J. (2004). Atg21 is a phosphoinositide binding protein required for efficient lipidation and localization of Atg8 during uptake of aminopeptidase I by selective autophagy. *Molecular Biology of the Cell [Internet], 15*(8), 3553–3566. Available from: http://www.ncbi.nlm.nih.gov/pubmed/15155809.
43. Reggiori, F., Tucker, K. A., Stromhaug, P. E., & Klionsky, D. J. (2004). The Atg1-Atg13 complex regulates Atg9 and Atg23 retrieval transport from the pre-autophagosomal structure. *Developmental Cell [Internet], 6*(1), 79–90. Available from: http://www.ncbi.nlm.nih.gov/pubmed/14723849.
44. Bansal, P., Tripathi, A., Thakur, V., Mohmmed, A., & Sharma, P. (2017). Autophagy-related protein ATG18 regulates Apicoplast biogenesis in apicomplexan parasites. *MBio [Internet], 8*(5). Available from: http://www.ncbi.nlm.nih.gov/pubmed/29089429.
45. Papinski, D., & Kraft, C. (2014). Atg1 kinase organizes autophagosome formation by phosphorylating Atg9. *Autophagy [Internet], 10*(7), 1338–1340. Available from: http://www.ncbi.nlm.nih.gov/pubmed/24905091.
46. Kitamura, K., Kishi-Itakura, C., Tsuboi, T., Sato, S., Kita, K., Ohta, N., et al. (2012). Autophagy-related Atg8 localizes to the apicoplast of the human malaria parasite *Plasmodium falciparum*. *PLoS One [Internet], 7*(8), e42977. Available from: http://www.ncbi.nlm.nih.gov/pubmed/22900071.

47. Eickel, N., Kaiser, G., Prado, M., Burda, P.-C., Roelli, M., Stanway, R. R., et al. (2013). Features of autophagic cell death in Plasmodium liver-stage parasites. *Autophagy [Internet], 9* (4), 568–580. Available from: http://www.ncbi.nlm.nih.gov/pubmed/23388496.
48. Jayabalasingham, B., Voss, C., Ehrenman, K., Romano, J. D., Smith, M. E., Fidock, D. A., et al. (2014). Characterization of the ATG8-conjugation system in 2 Plasmodium species with special focus on the liver stage: Possible linkage between the apicoplastic and autophagic systems? *Autophagy [Internet], 10*(2), 269–284. Available from: http://www.ncbi.nlm.nih.gov/pubmed/24342964.
49. Xie, Z., Nair, U., & Klionsky, D. J. (2008). Dissecting autophagosome formation: The missing pieces. *Autophagy [Internet], 4*(7), 920–922. Available from: http://www.ncbi.nlm.nih.gov/pubmed/18719358.
50. Cervantes, S., Bunnik, E. M., Saraf, A., Conner, C. M., Escalante, A., Sardiu, M. E., et al. (2014). The multifunctional autophagy pathway in the human malaria parasite, *Plasmodium falciparum. Autophagy [Internet], 10*(1), 80–92. Available from: http://www.ncbi.nlm.nih.gov/pubmed/24275162.
51. Gaviria, D., Paguio, M. F., Turnbull, L. B., Tan, A., Siriwardana, A., Ghosh, D., et al. (2013). A process similar to autophagy is associated with cytocidal chloroquine resistance in *Plasmodium falciparum. PLoS One [Internet], 8*(11), e79059. Available from: http://www.ncbi.nlm.nih.gov/pubmed/24278114.
52. Joy, S., Thirunavukkarasu, L., Agrawal, P., Singh, A., Sagar, B. K. C., Manjithaya, R., et al. (2018). Basal and starvation-induced autophagy mediates parasite survival during intraerythrocytic stages of *Plasmodium falciparum. Cell Death Discovery [Internet], 4*, 43. Available from: http://www.ncbi.nlm.nih.gov/pubmed/30302277.
53. Hain, A. U. P., Bartee, D., Sanders, N. G., Miller, A. S., Sullivan, D. J., Levitskaya, J., et al. (2014). Identification of an Atg8-Atg3 protein-protein interaction inhibitor from the medicines for malaria venture malaria box active in blood and liver stage *Plasmodium falciparum* parasites. *Journal of Medicinal Chemistry [Internet]., 57*(11), 4521–4531. Available from: http://www.ncbi.nlm.nih.gov/pubmed/24786226.
54. Walker, D. M., Mahfooz, N., Kemme, K. A., Patel, V. C., Spangler, M., & Drew, M. E. (2013). *Plasmodium falciparum* erythrocytic stage parasites require the putative autophagy protein PfAtg7 for normal growth. *PLoS One [Internet], 8*(6), e67047. Available from: http://www.ncbi.nlm.nih.gov/pubmed/23825614.
55. Datta, G., Hossain, M. E., Asad, M., Rathore, S., & Mohmmed, A. (2017). *Plasmodium falciparum* OTU-like cysteine protease (PfOTU) is essential for apicoplast homeostasis and associates with noncanonical role of Atg8. *Cell Microbiology [Internet], 19*(9), e12748. Available from: http://www.ncbi.nlm.nih.gov/pubmed/28423214.
56. Sinai, A. P., & Roepe, P. D. (2012). Autophagy in Apicomplexa: A life sustaining death mechanism? *Trends in Parasitology [Internet], 28*(9), 358–364. Available from: http://www.ncbi.nlm.nih.gov/pubmed/22819059.
57. Tomlins, A. M., Ben-Rached, F., Williams, R. A., Proto, W. R., Coppens, I., Ruch, U., et al. (2013). *Plasmodium falciparum* ATG8 implicated in both autophagy and apicoplast formation. *Autophagy [Internet], 9*(10), 1540–1552. Available from: http://www.ncbi.nlm.nih.gov/pubmed/24025672.
58. Backer, J. M. (2008). The regulation and function of class III PI3Ks: Novel roles for Vps34. *Biochemical Journal [Internet], 410*, 1):1–1)17. Available from: http://www.ncbi.nlm.nih.gov/pubmed/18215151.
59. Fimia, G. M., Stoykova, A., Romagnoli, A., Giunta, L., Di Bartolomeo, S., Nardacci, R., et al. (2007). Ambra1 regulates autophagy and development of the nervous system. *Nature [Internet], 447*(7148), 1121–1125. Available from: http://www.ncbi.nlm.nih.gov/pubmed/17589504.
60. Matsunaga, K., Saitoh, T., Tabata, K., Omori, H., Satoh, T., Kurotori, N., et al. (2009). Two Beclin 1-binding proteins, Atg14L and Rubicon, reciprocally regulate autophagy at different stages. *Nature Cell Biology [Internet], 11*(4), 385–396. Available from: http://www.ncbi.nlm.nih.gov/pubmed/19270696.

61. Pattingre, S., Tassa, A., Qu, X., Garuti, R., Liang, X. H., Mizushima, N., et al. (2005). Bcl-2 antiapoptotic proteins inhibit Beclin 1-dependent autophagy. *Cell [Internet], 122*(6), 927–939. Available from: http://www.ncbi.nlm.nih.gov/pubmed/16179260.
62. Maiuri, M. C., Zalckvar, E., Kimchi, A., & Kroemer, G. (2007). Self-eating and self-killing: Crosstalk between autophagy and apoptosis. *Nature Reviews Molecular Cell Biology [Internet], 8*(9), 741–752. Available from: http://www.ncbi.nlm.nih.gov/pubmed/17717517.
63. Schu, P. V., Takegawa, K., Fry, M. J., Stack, J. H., Waterfield, M. D., & Emr, S. D. (1993). Phosphatidylinositol 3-kinase encoded by yeast VPS34 gene essential for protein sorting. *Science [Internet], 260*(5104), 88–91. Available from: http://www.ncbi.nlm.nih.gov/pubmed/8385367.
64. Kihara, A., Noda, T., Ishihara, N., & Ohsumi, Y. (2001). Two distinct Vps34 phosphatidylinositol 3-kinase complexes function in autophagy and carboxypeptidase Y sorting in *Saccharomyces cerevisiae. Journal of Cell Biology [Internet], 152*(3), 519–530. Available from: http://www.ncbi.nlm.nih.gov/pubmed/11157979.
65. Obara, K., & Ohsumi, Y. (2008). Dynamics and function of PtdIns(3)P in autophagy. *Autophagy [Internet], 4*(7), 952–954. Available from: http://www.ncbi.nlm.nih.gov/pubmed/18769109.
66. Vaid, A., Thomas, D. C., & Sharma, P. (2008). Role of Ca2+/calmodulin-PfPKB signaling pathway in erythrocyte invasion by *Plasmodium falciparum. Journal of Biological Chemistry [Internet], 283*(9), 5589–5597. Available from: http://www.ncbi.nlm.nih.gov/pubmed/18165240.
67. Tawk, L., Chicanne, G., Dubremetz, J.-F., Richard, V., Payrastre, B., Vial, H. J., et al. (2010). Phosphatidylinositol 3-phosphate, an essential lipid in Plasmodium, localizes to the food vacuole membrane and the apicoplast. *Eukaryotic Cell [Internet], 9*(10), 1519–1530. Available from: http://www.ncbi.nlm.nih.gov/pubmed/20709789.
68. Jayabalasingham, B., Bano, N., & Coppens, I. (2010). Metamorphosis of the malaria parasite in the liver is associated with organelle clearance. *Cell Research [Internet], 20*(9), 1043–1059. Available from: http://www.ncbi.nlm.nih.gov/pubmed/20567259.
69. Coppens, I. (2011). Metamorphoses of malaria: The role of autophagy in parasite differentiation. *Essays in Biochemistry [Internet], 51*, 127–136. Available from: http://www.ncbi.nlm.nih.gov/pubmed/22023446.
70. Rug, M., Cyrklaff, M., Mikkonen, A., Lemgruber, L., Kuelzer, S., Sanchez, C. P., et al. (2014). Export of virulence proteins by malaria-infected erythrocytes involves remodeling of host actin cytoskeleton. *Blood [Internet], 124*(23), 3459–3468. Available from: http://www.ncbi.nlm.nih.gov/pubmed/25139348.
71. Manjithaya, R., Anjard, C., Loomis, W. F., & Subramani, S. (2010). Unconventional secretion of Pichia pastoris Acb1 is dependent on GRASP protein, peroxisomal functions, and autophagosome formation. *Journal of Cell Biology [Internet], 188*(4), 537–546. Available from: http://www.ncbi.nlm.nih.gov/pubmed/20156962.
72. Ayong, L., Pagnotti, G., Tobon, A. B., & Chakrabarti, D. (2007). Identification of *Plasmodium falciparum* family of SNAREs. *Molecular and Biochemical Parasitology [Internet], 152*(2), 113–122. Available from: http://www.ncbi.nlm.nih.gov/pubmed/17240462.
73. Hassett, M. R., & Roepe, P. D. (2018). PIK-ing new malaria chemotherapy. *Trends in Parasitology [Internet], 34*(11), 925–927. Available from: http://www.ncbi.nlm.nih.gov/pubmed/29934102.
74. Navale, R., Atul, A. A. D., & Sijwali, P. S. (2014). Characterization of the autophagy marker protein Atg8 reveals atypical features of autophagy in *Plasmodium falciparum. PLoS One [Internet], 9*(11), e113220. Available from: http://www.ncbi.nlm.nih.gov/pubmed/25426852.
75. Totino, P. R. R., Daniel-Ribeiro, C. T., Corte-Real, S., de Fátima Ferreira-da-Cruz, M., et al. (2008). *Experimental Parasitology [Internet], 118*(4), 478–486. Available from: http://www.ncbi.nlm.nih.gov/pubmed/18226811.
76. Carmona-Gutierrez, D., Ruckenstuhl, C., Bauer, M. A., Eisenberg, T., Büttner, S., & Madeo, F. (2010). Cell death in yeast: Growing applications of a dying buddy. *Cell Death and*

Differentiation [Internet], 17(5), 733–734. Available from: http://www.ncbi.nlm.nih.gov/pubmed/20383156.
77. Madeo, F., Eisenberg, T., & Kroemer, G. (2009). Autophagy for the avoidance of neurodegeneration. *Genes and Development [Internet], 23*(19), 2253–2259. Available from: http://www.ncbi.nlm.nih.gov/pubmed/19797764.
78. Meslin, B., Beavogui, A. H., Fasel, N., & Picot, S. (2011). *Plasmodium falciparum* metacaspase PfMCA-1 triggers a z-VAD-fmk inhibitable protease to promote cell death. *PLoS One [Internet], 6*(8), e23867. Available from: http://www.ncbi.nlm.nih.gov/pubmed/21858231.
79. Schmuckli-Maurer, J., Reber, V., Wacker, R., Bindschedler, A., Zakher, A., & Heussler, V. T. (2017). Inverted recruitment of autophagy proteins to the Plasmodium berghei parasitophorous vacuole membrane. *PLoS One [Internet], 12*(8), e0183797. Available from: http://www.ncbi.nlm.nih.gov/pubmed/28841718.
80. Stewart, M. J., Schulman, S., & Vanderberg, J. P. (1985). Rhoptry secretion of membranous whorls by Plasmodium berghei sporozoites. *Journal of Protozoology [Internet], 32*(2), 280–283. Available from: http://www.ncbi.nlm.nih.gov/pubmed/3925131.
81. Meis, J. F., Verhave, J. P., Jap, P. H., Sinden, R. E., Meuwissen, J. H., et al. (1983). *Journal of Protozoology [Internet], 30*(2), 361–366. Available from: http://www.ncbi.nlm.nih.gov/pubmed/6355454.
82. Lingelbach, K., & Joiner, K. A. (1998). The parasitophorous vacuole membrane surrounding Plasmodium and toxoplasma: An unusual compartment in infected cells. *Journal of Cell Science [Internet], 111*(Pt 1), 1467–1475. Available from: http://www.ncbi.nlm.nih.gov/pubmed/9580555.
83. Nyboer, B., Heiss, K., Mueller, A.-K., & Ingmundson, A. (2018). The Plasmodium liver-stage parasitophorous vacuole: A front-line of communication between parasite and host. *International Journal of Medical Microbiology [Internet], 308*(1), 107–117. Available from: http://www.ncbi.nlm.nih.gov/pubmed/28964681.
84. Real, E., Rodrigues, L., Cabal, G. G., Enguita, F. J., Mancio-Silva, L., Mello-Vieira, J., et al. (2018). Plasmodium UIS3 sequesters host LC3 to avoid elimination by autophagy in hepatocytes. *Nature Microbiology [Internet], 3*(1), 17–25. Available from: http://www.ncbi.nlm.nih.gov/pubmed/29109477.
85. Prado, M., Eickel, N., De Niz, M., Heitmann, A., Agop-Nersesian, C., Wacker, R., et al. (2015). Long-term live imaging reveals cytosolic immune responses of host hepatocytes against Plasmodium infection and parasite escape mechanisms. *Autophagy [Internet], 11*(9), 1561–1579. Available from: http://www.ncbi.nlm.nih.gov/pubmed/26208778.
86. Wacker, R., Eickel, N., Schmuckli-Maurer, J., Annoura, T., Niklaus, L., Khan, S. M., et al. (2017). LC3-association with the parasitophorous vacuole membrane of Plasmodium berghei liver stages follows a noncanonical autophagy pathway. *Cellular Microbiology [Internet], 19* (10). Available from: http://www.ncbi.nlm.nih.gov/pubmed/28573684.
87. Coppens, I. (2017). How toxoplasma and malaria parasites defy first, then exploit host autophagic and endocytic pathways for growth. *Current Opinionin Microbiology [Internet], 40*, 32–39. Available from: http://www.ncbi.nlm.nih.gov/pubmed/29102900.

Part IV

Emerging Viral Diseases

Hepatitis C Virus and Innate Interferon Response: Pathogen Biology, Drug Resistance, Novel Drug Targets, and Therapeutic Strategies

Khursheed ul Islam and Jawed Iqbal

Abstract

Innate immune response is a front line of defense against viral infection that depends on the production of interferon and other cytokines, which act by binding to their respective receptors and bring an antiviral effect in the cells. Hepatitis C virus (HCV) is a single-stranded (ss), positive (+ve)-sense RNA virus that infects hepatocytes and can lead chronic infection to hepatocellular carcinoma if left undiagnosed at earlier stage. HCV genome is recognized as nonself by different pathogen recognition receptors like RLRs and TLRs. After sensing of viral genome, innate immune response starts clearing the virus by producing interferons that in turn stimulate interferon stimulatory genes (ISGs) which actually block the virus at various stages of life cycle. HCV has evolved in such a way that it can block interferon response by using different strategies. HCV NS3/4A is a well-known protease that cleaves a mitochondrial antiviral signaling protein (MAVS) and abrogates innate IFN-β signaling. Similarly, some other viral proteins are also involved in inhibiting antiviral signaling pathways that have been discussed in this chapter. This chapter will provide an update on HCV pathogenesis, crosstalk between different sensing pathways, interferon response, current therapeutic strategies, and resistant mutations. It also focuses on the different strategies of the virus to evade innate interferon response.

Keywords

Hepatocellular carcinoma · Innate immune response · Resistant mutations · Sensing pathways

K. u. Islam · J. Iqbal (✉)
Multidisciplinary Centre for Advanced Research and Studies, Jamia Millia Islamia, New Delhi, India
e-mail: jiqbal1@jmi.ac.in

S. Hameed, Z. Fatima (eds.), *Pathogenicity and Drug Resistance of Human Pathogens*,
https://doi.org/10.1007/978-981-32-9449-3_12

Introduction

Interferons are the kind of natural defense against the invasion of pathogenic organisms like viruses, microbes, and tumor cells. Interferons belong to a large class of proteins called cytokines. On the basis of different types of receptors, human interferons are grouped in I, II, and III classes. Type I consists of interferon-α and interferon-β; however interferon-ϒ belongs to type II interferon class, whereas interferon-λ belongs to type III class of interferons [1]. Rapid induction of interferons and cytokines is a hallmark of innate immune response against viral pathogens, in mammalian hosts. Interferons neutralize the viral infection by inhibiting the viral replication and also develop an antiviral state in the neighboring cells by activating interferon stimulatory genes. Various immune effector cells are also activated by interferons which connect the adaptive arm of immune system to innate immune system [1].

Interferons play a central role by activating an antiviral effect in the cells and stimulate cellular components of the innate immune response [2]. There are mainly two pathways such as Toll-like receptor (TLR) pathway and cytosolic pathway by which viral infections are sensed and triggered due to the binding of viral RNA to retinoic inducible gene-1 (RIG-I), an RNA helicase, and to melanoma differentiation antigen-5 (MDA5) [3]. These two pathways congregate on activation of important TF (transcription factors), NF-κB, and interferon regulatory factors (IRF) 3 and 7. Activated transcription factors in turn activate interferon type I and III genes by binding to their corresponding promoter sites. Moreover, double-stranded DNA viruses are sensed by TLR9 in endosomes, where there is no requirement of either sequence-specific motif or viral replication to sense the dsDNA by TLR9. TLR9 also senses genome of single-stranded DNA (ssDNA) viruses like adeno-associated viruses [4]. TLR7 recognizes ssRNA viral genome in endosomes. In humans, myeloid dendritic cells expressed TLR8 which has been shown to be responsible for recognition of ssRNA viral genome in endosomes [5]. TLR3 senses the dsRNA that is either the genome of certain viruses or a replication intermediate of ssRNA viruses [6].

During HCV infection, the host cell senses the viral RNA by pattern recognition receptors (PRRs). PRRs mediate the recruitment of adaptor proteins to activate the downstream signaling that in turn the production of interferons [7]. Interferons bind to their interferon-α/β receptor (IFNAR), and this binding induces JAK (Janus kinase)-STAT (signal transducer and activator of transcription) pathway to impel the synthesis of >300 ISGs (interferon-stimulated genes) that help the host cell to block the replication of the virus at various stages of its life cycle [8, 9]. Antiviral resistance is an important topic of discussion in the modern healthcare system. HCV antiviral drug resistance gained its clinical relevance from 2011 onward, when direct-acting antivirals (DAAs) got first approval. High rate of HCV replication and high production of virus in chronic hepatitis correspond to the error-prone RNA polymerase of HCV that provides an encouraging setting for emergence of resistance-associated substitutions (RAS). Current approval of pangenotypic drugs that show an efficient genetic barrier may change the topic of antiviral resistance against HCV drugs in the future [10].

HCV Biology and Pathogenesis

HCV is an enveloped, +ssRNA virus which belongs to the Flaviviridae family and infects only hepatic cells. It has a genome of 9.6 kb that is translated into a single polyprotein which is around 3000 amino acids long [11]. The polyprotein is further processed either posttranslationally or co-translationally by the assistance of viral and host proteases to form ten different mature proteins [12], in which core (C), envelope glycoproteins (E1 & E2) are the structural proteins (SP), P7 is a viroporin that behaves as an ion channel protein [13], whereas NS2 to NS5B are the nonstructural proteins (NS) [11]. Flanking regions of two polyprotein termini are the 5′ and 3′ NTRs (nontranslated regions). 5′ NTR contains IRES (internal ribosome entry site) that is important for translation of HCV genome through cap-independent mechanism [14]. HCV core protein, E1, and E2 are involved directly in the virus maturation. Core protein is important for the development of viral capsid, and the envelope glycoproteins E1 and E2 are involved in the entry of virus into the hepatocytes (Fig. 1) [15]. Nonstructural proteins are important for other lifecycle aspects of HCV including replication and assembly of new virion particles (Fig. 2) [15]. P7 is an ion channel protein that mediates the release of virion particles. NS2 protein is believed to be a cysteine autoprotease that cleaves in between NS2 and NS3 in a polyprotein. Proteins like NS3, NS4A, NS4B, NS5A, and NS5B are the major components of replication complex where they help in replication of the viral genome at membranous web, a specialized site of ER in the cytoplasm of host cell [15].

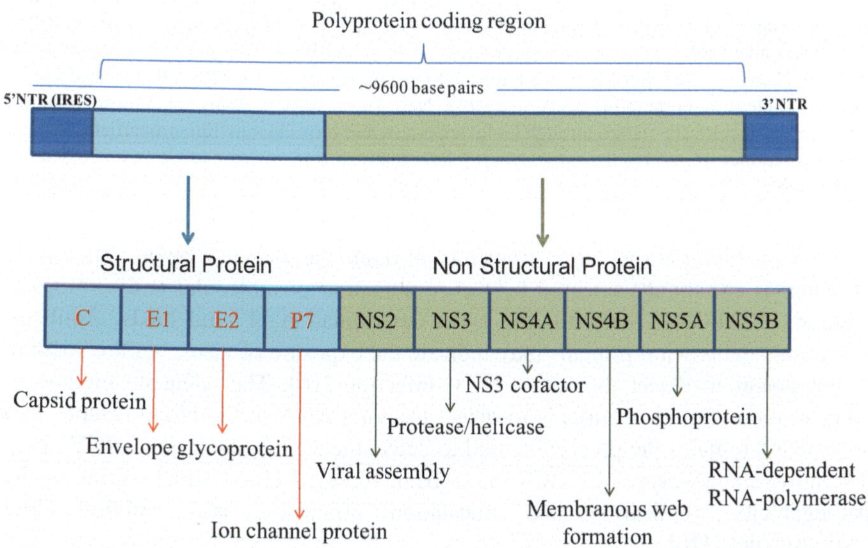

Fig. 1 Genome organization and translation: HCV genome codes for a polyprotein of 3000 amino acid (aa) length with a single ORF (open reading frame). After translation, polyprotein is processed with the help of viral and host proteases into ten different structural and nonstructural proteins. Different proteins with their respective function have been described

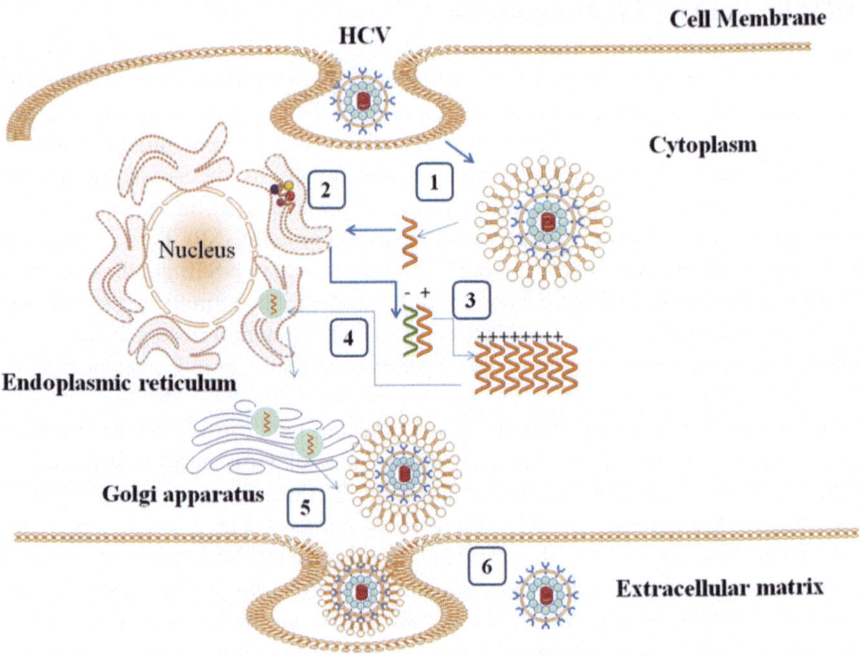

Fig. 2 Schematic representation of HCV life cycle [1]. HCV E1 and E2 bind to receptors on hepatocyte membrane and assist in virion entry via endosomes [2]. Polyprotein translation: After releasing the genome from endosome to cytoplasm, HCV RNA moves to rough ER (RER) and starts translation of a single polyprotein of around 3000 aa [3]. Viral replication: NS5B, an RdRp (RNA-dependent RNA polymerase), synthesizes a -ve strand RNA from the +ve strand of HCV genome. Negative (-ve) sRNA acts as a template to synthesize lots of +ve sRNA that is used for the virion particle development [4]. HCV assembly: New virion particle forms [5]. Virion combines with lipid molecules and passes through Golgi apparatus and forms mature lipoviroparticles (LVPs) which are further released in the cytoplasm [6]. Eventually, mature virion particles are released out of the cells

HCV, a blood-borne virus, transmits through the infected blood. This virus infection is commonly assessed by the combination of molecular and serological methods. RT-PCR is used for the exact determination of viral loads. Antibody screening against viral proteins may indicate the exposure of virus, but this method is not potent to detect the active HCV infection [16]. The adequate method to confirm the presence of virus is to detect the viral RNA in the blood sample, and sequencing remains the precise method to detect the actual genotype of HCV virus. Till now, six genotypes of HCV have been reported (1–6) which differ in the pathogenicity, replication and translation efficiency, and antiviral drug responsiveness [16].

Among the six reported genotypes, genotypes 1, 2, and 3 have been documented to be the major genotypes observed in Japan, North America, and Western Europe. Central Africa, Northern Africa, and the Middle East witness mostly genotype

4. Genotype 5 and genotype 6 have been observed in Southeast Asia and South Africa [16]. India accounts for 1–1.5% total HCV-infected population. The predominant HCV genotype unlike other countries is genotype 3 which is followed by genotype 1 [17]. HCV infection leads to the stepwise deterioration of liver cells that ultimately result in hepatocellular carcinoma (HCC) if not diagnosed at early stage of disease progression. The different phases after infection are characterized as acute phase, steatosis, fibrosis, cirrhosis, and hepatocellular carcinoma [18]. After HCV infection, acute infection period lasts for about 6 months. Most of the acute infected patients show no symptoms, and 20–30% of them may show clinical symptoms within 3–12 weeks. Symptoms at this stage may consist of jaundice, weakness, as well as anorexia. Serum ALT (alanine aminotransferase) levels start rising in 2–8 weeks after the HCV exposure [19].

After crossing the acute period, HCV leads to chronic infection which is noticeable by the persistence of viral load in the blood for a longer time. Chronic infection usually develops in 75–85% of the patients which encounter the virus infection and are unable to clear it within the 6-month period. This stage is characterized by activation of inflammatory cells; these cells bring a persistent inflammatory state in the liver tissue leading to liver fibrosis [20]. Liver fibrosis results from the unbalanced production of extracellular matrix and its degradation [21]. There are much possibilities of progression from fibrosis to cirrhosis. Approximately 10–15% of individuals having chronic HCV infection will progress to cirrhosis. Cirrhosis is the end stage of liver fibrosis where a liver develops lots of scar tissues that are not repaired by the host innate response. The most important consequence of untreated chronic infection causes the development of liver fibrosis followed by cirrhosis and ultimately HCC [18].

HCV Genome Sensing and Innate Immune System

HCV genome is recognized as PAMP (pathogen-associated molecular pattern) molecule by different PRRs including RLRs, TLRs, and MDA5 [22]. After the endocytosis of the viral particle, its genome can be sensed either in endosomal membrane or in the cytoplasm by different PRRs. Moreover, during HCV replication, double-stranded RNA intermediates are produced that also act as special PAMPs for its sensing [23]. After recognition of viral genome as nonself, innate response starts a cascade of reactions that mediate the interferons and cytokines, bringing an antiviral effect in the cells infected by virus. Both type I and III interferons are induced as a result of the downstream signaling of different pathways like MDA5, TLR, and RIG-I pathways [1].

HCV Genome Recognition by Retinoic Inducible Gene-I

RIG-I-like receptor for dsRNA is an RNA helicase enzyme that is encoded by DDX58 gene in humans. It is one of the family members of RIG-I-like receptors (RLRs) that also include LGP2 and MDA5 and function as PRRs during virion replication process [24]. During replication of HCV, double-stranded RNA is recognized within few hours of infection. It induces downstream signaling before synthesis of viral proteins in large extents. Initiation of RIG-I signaling occurs with its binding with HCV PAMP RNA that has a 5′triphosphate and polyU/UC tract at 3′end, which provides a nonself signature to HCV RNA [24]. This binding brings a conformational change in the structure of RIG-I receptor which induces its oligomerization and transportation from cytoplasm to intracellular membranes [25]. To ease the association of RIG-I with MAVS (mitochondrial antiviral signaling protein), a translocon is made by the interaction of e3 ubiquitin ligase (TRIM2) with $14 - 3 - 3^{\varepsilon}$ a chaperon together with RIG-I. Association of MAVS with RIG-I promotes the formation of MAVS signalosome that mediates induction of various downstream effector molecules, including NF-κB and IRF-3 and variety of proinflammatory cytokines (Fig. 3) [26].

HCV Genome Recognition by MDA5

MDA5 is another class of RLRs that sense the cytoplasmic dsRNA. In humans, MDA5 is encoded by *IFIH1* gene. MDA5 binds to intracellular or cytoplasmic long double-stranded RNA (>1000 bp) with no end specificity [24, 27]; in contrast, RIG-I specifically binds to the short dsRNA which has 5′triphosphate in it. MDA5 and RIG-I have in common the presence of a tandem N-terminal caspase recruitment domain with death domain folds, a DExD/H box helicase, and a C-terminal domain (CTD). As discussed earlier, when RIG-I binds with dsRNA through helicase and CTD domain, it releases its CARDs which then recruit and bind with MAVS. In contrast, MDA5 does not withdraw its CARDs but cooperatively assembles in the ATP-sensitive filaments present on dsRNA [28]. Further, it has been hypothesized that MDA5 CTD is necessary for supportive filament binding and not for RNA binding [29]. CARDS of MDA5 have been projected to nucleate MAVS assembly into its active polymeric form in a process that is stimulated by K63 polyubiquitin chains. After MAVS is activated, all other downstream signaling happens in the same way as RIG-I signaling pathway.

HCV Genome Recognition by TLRs

TLRs, being innate immune recognition receptors, sense some specific PAMPs expressed by various pathogenic organisms [30]. TLRs are the TM (transmembrane) glycoprotein receptors having an extracellular N-terminal binding domain with PAMP and an intracellular C-terminal domain. C-terminal domain is called Toll/IL-1R homology domain because of its resemblance with intracellular domain of the inteleukin-1 receptor [31]. Upon activation of the receptor, Toll/Interleukin-I receptor (TIR) domain mediates the downstream signaling events [31]. Family of Toll-like receptors can sense diverse kind of pathogens where TLR 3, 7, 8, and 9 sense viral nucleotides. TLR9 senses viral cPG-DNA motif containing genome of DNA viruses

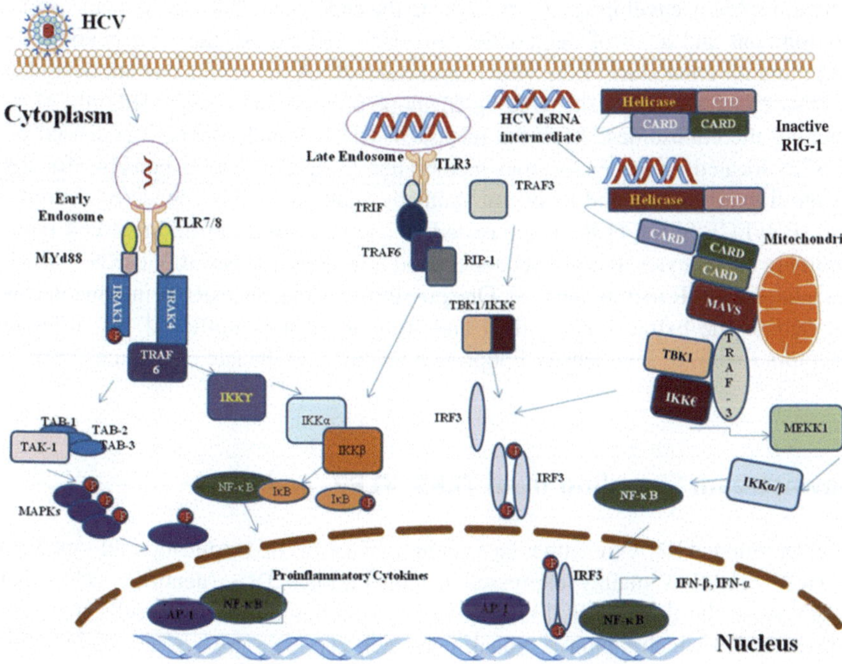

Fig. 3 Downstream signaling of different pathogen recognition receptors (PRRS). **(I)** During HCV entry into the hepatocytes, virion is endocytosed by an endocytic vesicle and recognized by TLR7/8 located in the endosomal membrane. HCV genome binding with TLR7/8 activates TLR7/8 and recruits an adaptor protein MyD88 that binds to TLR from its C-terminal end and to IRAK via its N-terminal domain. Stimulation of IRAK4 leads to phosphorylation of IRAK 1 which binds to TRAF6. Activated TRAF6 leads to polyubiquitination of TAK1 and IκB kinase-Υ. Subsequent association of IKKα/β/Υ leads to the phosphorylation of IκB that permits NF-κB to transport into the nucleus. TAK-1 in association with TAK-1 binding proteins trigger MAPK pathway that activates activator protein-1 (AP-1) which also translocate to nucleus. AP-1 and NF-κB both activate the proinflammatory cytokine genes. **(II)** TLR3 recruits an adaptor TRIF, which after activation binds to receptor- interacting protein-1 (RIP) that leads to the induction of NF-κB and IRF3 dimerization and phosphorylation. Activated NF-κB and IRF3 translocate to nucleus and activate IFN-α/β genes. **(III)** HCV synthesizes dsRNA intermediate during its replication process that may be responsible for the activation of STING followed by TBK1 which in turn IRF3 phosphorylation thereafter translocation into the nucleus to induce IFN-β. Similarly, dsRNA sensed by RIG-I brings a conformational change in CARD domains of RIG-I which binds to CARD-containing mitochondrial protein MAVS. This association leads to recruitment of TBK1 and IKKϵ which in turn activate IFN gene transcription through phosphorylation and nuclear transfer of IRF3. MEKK1, TRAF3, and TRAF6 are also recruited to MAVS for activation of NF-κB. Activated NF-κB translocates to the nucleus for IFN gene transcription

like MCMV (murine cytomegalovirus), HSV-1 (herpes simplex virus-1), and HSV-2. TLR 7 and TLR 8 are the dsRNA-sensing TLRs. They show high homology with each other at genomic level. Both of these TLR genes are located on X chromosomes [5]. Endosomal membranes of a cell express TLR7 and TLR8 that gives an insight that ssRNA access to the endomembrane is necessary for their activation. Many enveloped viruses traffic to the endosomes that constitute the acidic environment and a lot of degradation enzymes that expose the viral genome by degrading all other stuff of the virus. Unlike self-RNA molecules that are degraded by RNases, viral RNA is sheltered by envelope proteins and is degraded only when reaching the endosomes. After the release of RNA in endosomes, it is sensed by TLR7/8 located in the endosomal membranes [32]. TLR3 is responsible for the recognition of dsRNA and its potent synthetic analogue polyinosinedeoxycytidylic acid (PolyI:C). Since TLR3 is expressed in late endosomes in different cell types including hepatocytes, it is not yet understood how does it sense HCV dsRNA which lies adjacent to ER membrane [33]. One possibility is that the dsRNA intermediate is secreted to the extracellular milieu and from there it is trafficked via different receptors (oxidized low-density lipoprotein receptor) to the late endosomes [34].

Downstream Signaling from TLR3, TLR7, and TLR8

Double-stranded RNA can either be a genome of a virus or a replication intermediate of HCV. TLR3 is mainly expressed in conventional DCs (dendritic cells) that phagocytose the dying cells [35]. There are varieties of epithelial cells that also express TLR3 which include corneal, intestinal, cervical, airway, and vaginal epithelial cells. TLR3, unlike TLR7/8, recruit an adaptor protein called TIR-domain-containing adaptor-inducing interferon-β (TRIF) to the TIR domain of the receptor [36]. After activation, TRIF interacts with RIP-1, a receptor-interacting protein-1, and leads to the activation of NF-κB and further activation of IFN-β promoter inside the nucleus. When TLR7 and TLR8 get stimulated, they recruit an adaptor protein MyD88 to the receptor's TIR domain in the cytoplasm. The C-terminal domain of MyD88 binds to TLR, and N-terminal domain makes a complex with IL-1R-associated kinases (IRAKs) IRAK-4 and IRAK-1 [36, 37]. Stimulation of IRAK-4 phosphorylates IRAK-1that binds to the CTD of tumor necrosis factor receptor-associated factor 6 (TRAF6). Activated TRAF6 leads to the polyubiquitination of TGF-β-activated kinase 1 (TAK1) and IκB kinase-ϒ (IKKϒ); then there is subsequent association of IKKα and IKKβ with IKKϒ. TAK1 associated with its corresponding TAK1 binding proteins TAB1,TAB2, and TAB3 phosphorylate IKKβ which in turn result in IKK-mediated phosphorylation and degradation of IκB [38]. NFκB is bound to the unphosphorylated form of I-κB that is a key factor for its retention in the cytoplasm. Once IκB gets phosphorylated, it leaves NF-κB free to transport to the nucleus for induction of gene expression. Association of TAK-1 with TAB1, TAB2, and TAB3 triggers the MAPK signaling pathway which in turn leads to the formation of activator protein-1 (AP-1), whereas Ap-1 also translocate to nucleus and induce inflammatory gene expression along with NF-κB [38] (Fig. 3).

HCV Infection and Induction of Interferons

Regardless of interference created by HCV proteins, HCV-infected cells manage to produce interferons endogenously. According to the currently available literature, there are three types of interferons, among which type III interferons (IFN-λs) are the recently discovered interferons, and it has been elucidated that their mode of action is overlapping with that of type I interferons [39]. Once bound to their respective receptors, type I and III interferons start a signal through JAK-STAT pathway. This pathway further proceeds to the stimulation of cellular processes through the induction of interferon stimulatory genes (ISGs) and bring an immunomodulatory or antiviral effect in the cells (Fig. 4). Despite the interference created by HCV in interferon induction, there is a continuous ISG upregulation observed in HCV-infected liver of chimpanzee model [39] as well as HCV patient liver [40] that gives an insight that HCV infection stimulates the interferon production in infected hepatocytes. ISGs that are induced by interferons I and III are upregulated in HCV-infected liver [40]. Interferon stimulatory genes that are highly expressed in HCV infection include MX1, ISG15, IFI44, OAS-1, and IFI27 [40–42].

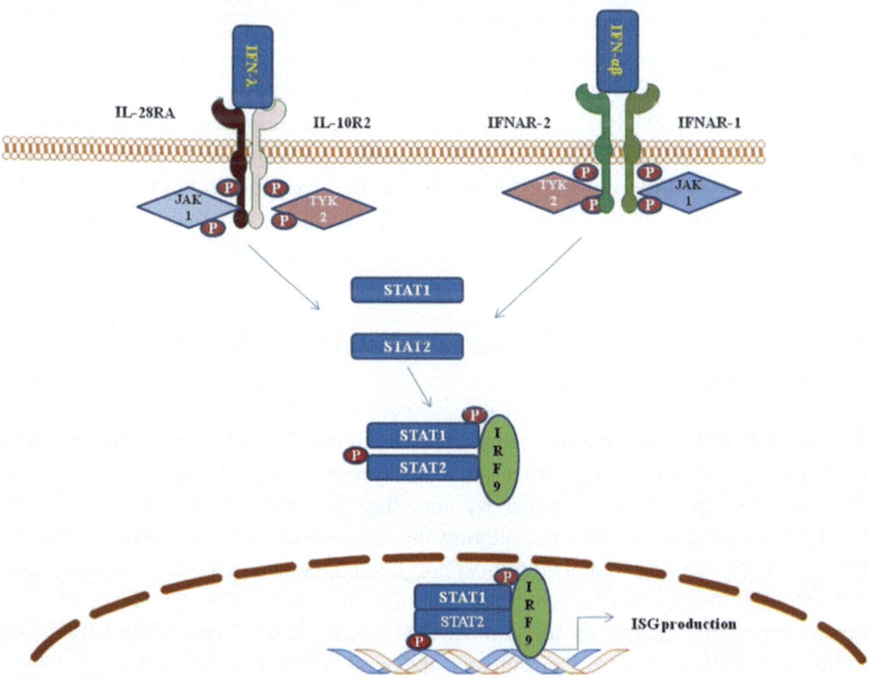

Fig. 4 Effect of interferon signaling during HCV infection. IFN produced by the HCV genome recognition by RIG-I in cytoplasm and TLRs in endosomes which further lead to product of interferon stimulatory genes (ISGs). IFN-α/β binds to interferon-α/β receptors (IFNAR1/2), while IFN-λ binds to IL-28RA and IL-10R2 which further direct JAK-STAT-mediated ISG transcription

Viral Proteins Known to Block Interferon Response

NS3/4A, a protease encoded by HCV genome, has been found to be potent inhibitor of interferon signaling in in vivo Study. NS34/A cleaves MAVS that is an adaptor protein in RIG-I signaling as discussed earlier. Except NS34/A, some other HCV proteins have also been documented to inhibit innate response. One of the structural proteins of HCV called core protein (C) can inhibit IFN response by interfering with STAT-1. This core protein is also involved in inducing expression of certain proteins called suppressor of cytokine signaling 1 (SOCS-1) and SOCS-3, which leads to the blockage of JAK-STAT signaling [43, 44]. Sayuri Nitta et al. (2013) observed that HCV protein, NS4B, blocks the IFN production by binding specifically to STING as the association of NS4B with STING abrogates the association between STING and MAVS that is critically important for robust activation of IFN-β [45]. Further interaction of NS4B with STING on ER membrane prevents its interaction with TBK, leading to the inhibition of interferon signaling [46]. Moreover, NS5A is another HCV nonstructural protein that helps in virus replication by modulating the host cell environment [47]. This protein binds to the important adaptor protein of TLR signaling called MyD88 resulting in the inhibition and recruitment of IRAK-1 to MyD88 thus abrogating the TLR signaling and diminishing the cytokine production [48]. Other structural protein of HCV envelop glycoprotein (E2) modulates the host innate immune system by a molecular mimicking strategy, in which viral genome encodes a protein similar to the host defense proteins to block interferon signaling. One of such proteins includes E2 glycoprotein that possesses a 12-amino-acid (aa) sequence which is similar to the eukaryotic initiation factor-2α (eIF2α) and protein kinase-R (PKR). PKR is a dsRNA-activated protein kinase, which abrogates synthesis of protein by phosphorylating eIF2-α. E2 glycoprotein binds with PKR to block its binding sites that usually bind to eIF2-α; thus virus opens a window to synthesize its proteins without any hurdle caused by PKR [49].

Role of Other Sensing Molecules in Interferon Response Against HCV

Except RLR and TLR signaling, other signaling molecules have also been identified to play important role in HCV infection. PKR (protein kinase-R) is a dsRNA sensor that has a role in antiviral response by impeding the host cell gene translation and synthesis of proteins by phosphorylating the eukaryotic initiation factor 2 (eIF2α) [50]. PKR gets dimerized and autophosphorylated in presence of low concentrations of dsRNA normally, but excess concentration of dsRNA doesn't allow PKR to dimerize properly that leads to its inactivation [51]. It has been documented that inhibition of PKR in vitro can enhance the viral replication by manifolds which highlights the importance of PKR in antiviral response [52].

Of course there are different sensing pathways for RNA and DNA viruses, as they depend on different receptors and adaptor proteins, but the crosstalk during the course of sensing and signaling pathway is also an undeniable fact. As highlighted by Brunette R.L. et al. (2012) in the reduction of IFN response to poly(dA:dT) was observed by >99% in MAVS and STING-deficient phagocytes and not in phagocytes deficient in STING alone [53]. Further, lots of different mechanisms that demonstrate the interplay between RNA and DNA sensing pathways have been found in which STING plays a pivotal role. It has also been demonstrated that the permissiveness to RNA viruses in primary cell cultures, cell lines, and mouse models can be enhanced by STING deficiency [54, 55]. For triggering of antiviral response, STING interacts with RIG-I and MAVS in a complex that is stabilized by RNA viral infection [56].

Novel Drugs and Therapeutic Strategies

Since the discovery of HCV, treatment strategies for chronic hepatitis C have evolved from using interferon to the use of direct-acting antivirals (DAAs). Interferon monotherapy was the first strategy used to cure chronic hepatitis C, but it was modified in the 1990s by adding ribavirin that is a nucleoside analogue. This modification leads to the increase in the sustained virological response (SVR) by 20% when compared with the interferon monotherapy [57]. After this pegylated interferon therapy was established as an effective drug than nonpegylated interferons, where SVR has increased from 10% to 15% [43, 58]. Until recently the standard of care for hepatitis C was the use of pegIFN-α with ribavirin for the period of 6–12 months. The SVR rates have been increased from 15% to 20% by the use of interferon monotherapy to over 90% by treatment of DAAs during this evolution of treatment strategies. DAAs are the medications that target different stages of virion life cycle and have shown promising SVR rates and low genetic barriers, as compared to other therapeutic approaches [59].

Boceprevir (Merck) and telaprevir (Vertex) were the first two DAAs approved by FDA in 2011 to be used against chronic hepatitis C; both of these drugs were NS3 protease inhibitors [60]. These two drugs are considered as the first-generation DAAs. Due to the development of resistant mutations, these drugs were given as a triple therapy with two other drugs including ribavirin and pegIFN α. This strategy led to the improved SVR rates despite some toxicity issues [61, 62]. Many other DAAs have got approval, and others are still in clinical trials, since their use as protease inhibitors to treat chronic HCV. Currently the major targets of DAAs are the three viral proteins NS5A, NS3, and NS5B. These inhibitors against NS5B, NS3/4A, and NS5A are the second-generation inhibitors which have the advantage of being all oral regimens and are interferon-free. They are thought to be the standard care for chronic hepatitis C replacing interferon monotherapy which has a deprived tolerability and severe side effects [63]. Dual- or triple-DAA therapy is the current regime of drugs against chronic hepatitis C that is thought to be a smart strategy to increase SVR rates and decrease the chances of resistant mutations. There is the addition of new DAAs in the clinical practice every year. Currently available DAAs are given in Table 1, and few of them have been voluntarily withdrawn from the market places, which include simeprevir that was withdrawn in May 2018 as well as Technivie and Viekira Pak, both of which were considered for discontinuation on 1 January 2019 [64].

Table 1 Currently available direct-acting antivirals (DAAs) [65, 66]

S/ No	Generic name	Tradename	Genotype
01	Daclatasvir	Daklinza	1, 3
02	Simeprevir	Olysio	1, 4
03	Sofosbuvir	Sovaldi	1, 2, 3, and 4
04	Paritaprevir/ombitasvir/ritonavir and dasabuvir	Viekira Pak	1
05	Ledipasvir/sofosbuvir	Harvoni	5, 6
06	Extended-release tablet (ombitasvir/dasabuvir/ritonavir/ paritaprevir)	Viekira XR	1
07	Velpatasvir/sofosbuvir	Epclusa	5, 6
08	Pibrentasvir/glecaprevir	Mavyret	5, 6
09	Paritaprevir/ombitasvir/ritonavir	Technive	4
10	Voxilaprevir/sofosbuvir/velpatasvir	Vosevi	5, 6
11	Grazoprevir/elbasvir	Zepatier	1 and 4

Antiviral Drug Resistance

Resistance to antiviral drugs has emerged as the important challenge in the field of HCV research. Modern therapeutic strategies like DAAs are at present the standard care for the treatment of chronic hepatitis C. Resistance to therapeutics against HCV got clinical relevance in 2011 when DAAs got first approval for clinical use; since then the management of RAVs and treatment optimization against HCV have been a matter of discussion [50]. Profiles of different resistant mutations against DAA treatment are already well documented. NS proteins (NS3, NS5A, and NS5B) are the major targets of DAAs that are currently in use.

DAAs targeting NS3 protease are the protease inhibitors, so their mechanism of action will be the inhibition of proteolytic cleavage of viral proteins from the newly made polyprotein. First DAAs against NS3 approved for clinical use were boceprevir and telaprevir, followed by asunaprevir, simeprevir, paritaprevir, and others. Besides the use of these inhibitors, several resistant mutations were observed in case of NS3 protein which are given in Table 2. NS5A, a nonstructural protein important for HCV replication and virion assembly, is also a frequent target of DAAs. Daclatasvir was the first NS5A-targeting DAA discovered [67]. It has an outstanding inhibitory potency, but virus has developed various resistant mutations against this protein as well.

HCV NS5B is an RdRp which is critical for virus replication process and also a vital therapeutic target. This protein lacks the proofreading capacity, which makes it vulnerable for genetic changes. Different NS5B inhibitors have been developed so far that differ in the mode of inhibition. They are classified as either nucleotide analogue (sofosbuvir) or nucleoside analogue (beclabuvir and dasabuvir). Among all of these inhibitors, sofosbuvir has shown an excellent profile of SVRs in various

Table 2 Commonly observed NS3-, NS5A-, and NS5B-resistant mutations [15]

S/No	Protein	Inhibitors	Genotype 1a	Genotype 1b	Other genotype (mutation)	References
01	NS3	Boceprevir	T54S, R155K, V36M	V55A, V170A, T54A/S, A156S		[15]
		Simeprevir	D168E/V, R155K	D168E/V, Q80R		[15]
		Paritaprevir	D168A/V/Y	D168V, Y56H		[15]
		Telaprevir	R155K, V36M	T54A, V36A, A156S		[15]
		Vaniprevir	D168T/V/Y, R155K	D168H/T/V		[15]
		Asunaprevir	D168E, R155K	D168E/Y/V		[15]
		Glecaprevir	V36M, Y56H/N, Q80K/R, R155T, A156G/T/V, and Q168A/K/L/R			[59]
		Grazoprevir	V36L/M,V158A, R155I/L/K/S, Q80K/L, Y56F/H, A156G/M/T/V, and D168A/C/E/G/K/N/V/Y			[68]
		Voxilaprevir	A156T/V	A156T/V		[50]
02	NS5A	Ombitasvir	Q30R, M28T	Y93H		[15]
		Daclatasvir	L31M, Y93H/N, H58D, M28T, Q30E/H/R	Y93H, L31M/V		[15]
		Ledipasvir	L31M,Q30E/R, Y93C/H/N	Y93H		[15]
		Elbasvir	M28/G/T/S,L31F/I/M/V, Q30H/K/R/Y,H58D, and Y93C/H/N/S	L31F/M/V and Y93H	GT-3 (S24F, A30G/K, P58T, L31F/I/M, M28G/K and Y93H)	[68, 69]
		Pibrentasvir	K24R, L31M, Q30KR, M28A/G, Y93H, and H58D		GT-2 (L31I/M/V,Y93H)	[70]
		Velpatasvir	Q30E/H/K/L/R, M28T/V, Y93H/N/R/S/W, and L31I/M/V	Y93C/H/N/S/T and L31M/V	GT-3 (A30K/V, E92K, L31 M/P/V	[50]

(continued)

Table 2 (continued)

S/ No	Protein	Inhibitors	Genotype 1a	Genotype 1b	Other genotype (mutation)	References
					and Y93H/ N/R)	
03	NS5B	Beclabuvir	P495L/S, A421V	None		[15]
		Dasabuvir	S556G, M414T	S556G		[15]
		Sofosbuvir	S282T	S282T	S282T (all GTs)	[71]

clinical trials. Further, it has also shown a high genetic barrier to the progress of resistant mutations. For the treatment of chronic hepatitis C, sofosbuvir has rendered it for the first choice of DAA due to its excellent profile. In spite of its outstanding antiviral profile, cost is still an important issue in many countries. Table 2 summarizes some commonly observed resistant mutations.

Conclusion

Interferon plays a key role in the innate response against different DNA and RNA viruses. In case of HCV infection, all the sensing pathways that include RLR (RIG-I and MDA5) and TLR signaling pathways ultimately lead to the production of interferons and cytokines. Interferons further stimulate the interferon stimulatory gene (MX1, ISG15, IF144, OAS-1, and IFI27) that brings an antiviral state in the cells and surrounding tissues. Despite the strong immunological processes that lead to virus clearance in the acute phase of HCV infection, virus still finds a way to enter into the chronic phase causing a persistent liver damage. There is a critical need for understanding these viral strategies due to which virus is able to hijack the innate response. Lots of literature is available in the context of strategies followed by HCV to evade the innate response of the host, which have led to the production of different antiviral drugs against this particular virus, but due to different reasons like high cost, ineffectiveness, and cytotoxicity, we are still trailing behind in developing a standard therapy against HCV that can eradicate this disease from its roots. Antiviral drug resistance is also an important issue to address because HCV being a positive-stranded RNA virus is very prone to mutations because of having an RNA-dependent RNA polymerase with less proofreading capacity. Modern therapeutic approaches like DAAs have been documented with various resistant mutations, but the current strategy of using dual or triple therapies has shown a good increase in SVR rate and low resistant mutations. There is the utter need to develop a pangenotypic inhibitor with high genetic barrier and low cost.

Acknowledgements The support of Ramalingaswami Fellowship grant (No. BT/RLF/Re-entry/09/2015), Government of India, Ministry of Science and Technology, Department of Biotechnology (DBT) is highly acknowledged.

References

1. Borden, E. C., et al. (2007). Interferons at age 50: Past, current and future impact on biomedicine. *Nature Reviews Drug Discovery, 6*, 975–990.
2. Stetson, D. B., & Medzhitov, R. (2006). Type I interferons in host defense. *Immunity, 25*, 373–381.
3. Yoneyama, M., & Fujita, T. (2007). Function of RIG-I-like receptors in antiviral innate immunity. *Journal of Biological Chemistry, 282*, 15315–15318.
4. Zhu, J., Huang, X., & Yang, Y. (2009). The TLR9-MyD88 pathway is critical for adaptive immune responses to adeno-associated virus gene therapy vectors in mice. *Journal of Clinical Investigation, 119*, 2388–2398.
5. Heil, F. (2004). Species-specific recognition of single-stranded RNA via toll-like receptor 7 and 8. *Science, 303*, 1526–1529.
6. Bell, J. K., Askins, J., Hall, P. R., Davies, D. R., & Segal, D. M. (2006). The dsRNA binding site of human Toll-like receptor 3. *Proceedings of the National Academy of Sciences, 103*, 8792–8797.
7. Kawai, T., & Akira, S. (2006). Innate immune recognition of viral infection. *Nature Immunology, 7*, 131–137.
8. Horner, S. M. (2014). Activation and evasion of antiviral innate immunity by hepatitis C virus. *Journal of Molecular Biology, 426*, 1198–1209.
9. Saito, T., & Gale, M. (2007). Regulation of innate immunity against hepatitis C virus infection. *Hepatology Research, 0*, 071117021731007. ???
10. Sorbo, M. C., et al. (2018). Hepatitis C virus drug resistance associated substitutions and their clinical relevance: Update 2018. *Drug Resistance Updates, 37*, 17–39.
11. Moradpour, D., & Penin, F. (2013). Hepatitis C virus proteins: From structure to function. *Hepatitis C Virus: From Molecular Virology to Antiviral Therapy, 369*, 113–142.
12. Catanese, M. T., et al. (2013). Ultrastructural analysis of hepatitis C virus particles. *Proceedings of the National Academy of Sciences, 110*, 9505–9510.
13. Pavlovic, D., et al. (2003). The hepatitis C virus p7 protein forms an ion channel that is inhibited by long-alkyl-chain iminosugar derivatives. *Proceedings of the National Academy of Sciences, 100*, 6104–6108.
14. Li, K., & Lemon, S. M. (2013). Innate immune responses in hepatitis C virus infection. *Seminars in Immunopathology, 35*, 53–72.
15. Kim, S., Han, K.-H., & Ahn, S. H. (2016). Hepatitis C virus and antiviral drug resistance. *Gut and Liver, 10*, 890–895.
16. Simmonds, P. (2004). Genetic diversity and evolution of hepatitis C virus – 15 years on. *Journal of General Virology, 85*, 3173–3188.
17. Acharya, S. K., Madan, K., Dattagupta, S., & Panda, S. K. (2006). Viral hepatitis in India. *National Medical Journal of India, 19*, 203–217.
18. de Andrade, L. O., et al. (2009). Association between hepatitis C and hepatocellular carcinoma. *Journal of Global Infectious Diseases, 1*, 33.
19. Thimme, R., et al. (2001). Determinants of viral clearance and persistence during acute hepatitis C virus infection. *The Journal of Experimental Medicine, 194*, 1395–1406.
20. Matsuzaki, K., et al. (2007). Chronic inflammation associated with hepatitis C virus infection perturbs hepatic transforming growth factor β signaling, promoting cirrhosis and hepatocellular carcinoma. *Hepatology, 46*, 48–57.
21. Ahmad, W., et al. (2011). A brief review on molecular, genetic and imaging techniques for HCV fibrosis evaluation. *Virology Journal, 8*, 53.
22. Kell, A., et al. (2015). Pathogen-associated molecular pattern recognition of hepatitis C virus transmitted/founder variants by RIG-I is dependent on U-core length. *Journal of Virology, 89*, 11056–11068.
23. Saito, T., & Gale, M. (2008). Differential recognition of double-stranded RNA by RIG-I–like receptors in antiviral immunity: Figure 1. *The Journal of Experimental Medicine, 205*, 1523–1527.

24. Hornung, V., et al. (2006). 5′-triphosphate RNA is the ligand for RIG-I. *Science, 314*, 994–997.
25. Liu, H. M., et al. (2012). The mitochondrial targeting chaperone 14-3-3ε regulates a RIG-I translocon that mediates membrane association and innate antiviral immunity. *Cell Host & Microbe, 11*, 528–537.
26. Loo, Y.-M., & Gale, M. (2011). Immune signaling by RIG-I-like receptors. *Immunity, 34*, 680–692.
27. Pichlmair, A., et al. (2006). RIG-I-mediated antiviral responses to single-stranded RNA bearing 5′-phosphates. *Science, 314*, 997–1001.
28. Peisley, A., et al. (2011). Cooperative assembly and dynamic disassembly of MDA5 filaments for viral dsRNA recognition. *Proceedings of the National Academy of Sciences, 108*, 21010–21015.
29. Takahasi, K., et al. (2009). Solution structures of cytosolic RNA sensor MDA5 and LGP2 C-terminal domains: IDENTIFICATION OF THE RNA RECOGNITION LOOP IN RIG-I-LIKE RECEPTORS. *Journal of Biological Chemistry, 284*, 17465–17474.
30. Szabo, G., Billiar, T. R., Machida, K., Crispe, I. N., & Seki, E. (2010). Toll-like receptor signaling in liver diseases. *Gastroenterology Research and Practice, 2010*, 1–2.
31. Bowie, A., & O'Neill, L. A. (2000). The interleukin-1 receptor/toll-like receptor superfamily: Signal generators for pro-inflammatory interleukins and microbial products. *Journal of Leukocyte Biology, 67*, 508–514.
32. Jensen, S., & Thomsen, A. R. (2012). Sensing of RNA viruses: A review of innate immune receptors involved in recognizing RNA virus invasion. *Journal of Virology, 86*, 2900–2910.
33. Yoneyama, M., et al. (2004). The RNA helicase RIG-I has an essential function in double-stranded RNA-induced innate antiviral responses. *Nature Immunology, 5*, 730–737.
34. Dansako, H., et al. (2013). Class A scavenger receptor 1 (MSR1) restricts hepatitis C virus replication by mediating toll-like receptor 3 recognition of viral RNAs produced in neighboring cells. *PLoS Pathogens, 9*, e1003345.
35. Akira, S., Uematsu, S., & Takeuchi, O. (2006). Pathogen recognition and innate immunity. *Cell, 124*, 783–801.
36. Yamamoto, M. (2003). Role of adaptor TRIF in the MyD88-independent toll-like receptor signaling pathway. *Science, 301*, 640–643.
37. Medzhitov, R., et al. (1998). MyD88 is an adaptor protein in the hToll/IL-1 receptor family signaling pathways. *Molecular Cell, 2*, 253–258.
38. Akira, S., & Takeda, K. (2004). Toll-like receptor signalling. *Nature Reviews Immunology, 4*, 499–511.
39. Wieland, S. F., & Chisari, F. V. (2005). Stealth and cunning: Hepatitis B and hepatitis C viruses. *Journal of Virology, 79*, 9369–9380.
40. Dill, M. T., et al. (2011). Interferon-induced gene expression is a stronger predictor of treatment response than IL28B genotype in patients with hepatitis C. *Gastroenterology, 140*, 1021–1031.e10.
41. Sarasin-Filipowicz, M., et al. (2008). Interferon signaling and treatment outcome in chronic hepatitis C. *Proceedings of the National Academy of Sciences, 105*, 7034–7039.
42. Sung, P. S., et al. (2015). Roles of unphosphorylated ISGF3 in HCV infection and interferon responsiveness. *Proceedings of the National Academy of Sciences, 112*, 10443–10448.
43. Suda, G., et al. (2010). IL-6-mediated intersubgenotypic variation of interferon sensitivity in hepatitis C virus genotype 2a/2b chimeric clones. *Virology, 407*, 80–90.
44. Funaoka, Y., et al. (2011). Analysis of interferon signaling by infectious hepatitis C virus clones with substitutions of core amino acids 70 and 91. *Journal of Virology, 85*, 5986–5991.
45. Nitta, S., et al. (2013). Hepatitis C virus NS4B protein targets STING and abrogates RIG-I-mediated type I interferon-dependent innate immunity. *Hepatology, 57*, 46–58.
46. Ding, Q., et al. (2013). Hepatitis C virus NS4B blocks the interaction of STING and TBK1 to evade host innate immunity. *Journal of Hepatology, 59*, 52–58.
47. Macdonald, A. (2004). Hepatitis C virus NS5A: Tales of a promiscuous protein. *Journal of General Virology, 85*, 2485–2502.

48. Abe, T., et al. (2007). Hepatitis C virus nonstructural protein 5A modulates the toll-like receptor-MyD88-dependent signaling pathway in macrophage cell lines. *Journal of Virology, 81*, 8953–8966.
49. Taylor, D. R., Shi, S. T., Romano, P. R., Barber, G. N., & Lai, M. M. (1999). Inhibition of the interferon-inducible protein kinase PKR by HCV E2 protein. *Science, 285*, 107–110.
50. Williams, B. R. G. (2001). Signal Integration via PKR. *Science Signaling, 2001*, re2–re2.
51. Vyas, J. (2003). Inhibition of the protein kinase PKR by the internal ribosome entry site of hepatitis C virus genomic RNA. *RNA, 9*, 858–870.
52. Oem, J.-K., et al. (2008). Hepatitis C virus non-structural protein-2 activates CXCL-8-transcription through NF-κB. *Archives of Virology, 153*, 293–301.
53. Brunette, R. L., et al. (2012). Extensive evolutionary and functional diversity among mammalian AIM2-like receptors. *The Journal of Experimental Medicine, 209*, 1969–1983.
54. Sun, W., et al. (2009). ERIS, an endoplasmic reticulum IFN stimulator, activates innate immune signaling through dimerization. *Proceedings of the National Academy of Sciences, 106*, 8653–8658.
55. Ishikawa, H., Ma, Z., & Barber, G. N. (2009). STING regulates intracellular DNA-mediated, type I interferon-dependent innate immunity. *Nature, 461*, 788–792.
56. Zevini, A., Olagnier, D., & Hiscott, J. (2017). Crosstalk between cytoplasmic RIG-I and STING sensing pathways. *Trends in Immunology, 38*, 194–205.
57. McHutchison, J. G., et al. (1998). Interferon alfa-2b alone or in combination with ribavirin as initial treatment for chronic hepatitis C. *New England Journal of Medicine, 339*, 1485–1492.
58. Fried, M. W., et al. (2002). Peginterferon alfa-2a plus ribavirin for chronic hepatitis C virus infection. *New England Journal of Medicine, 347*, 975–982.
59. Krishnan, P., et al. (2018). Pooled resistance analysis in patients with hepatitis C virus genotype 1 to 6 infection treated with glecaprevir-pibrentasvir in phase 2 and 3 clinical trials. *Antimicrobial Agents and Chemotherapy, 62*, AAC.01249-18.
60. Zhang, X. (2016). Direct anti-HCV agents. *Acta Pharmaceutica Sinica B, 6*, 26–31.
61. Poordad, F., et al. (2011). Boceprevir for untreated chronic HCV genotype 1 infection. *New England Journal of Medicine, 364*, 1195–1206.
62. Jacobson, I. M., et al. (2011). Telaprevir for previously untreated chronic hepatitis C virus infection. *New England Journal of Medicine, 364*, 2405–2416.
63. Lontok, E., et al. (2015). Hepatitis C virus drug resistance-associated substitutions: State of the art summary: VIRAL HEPATITIS. *Hepatology, 62*, 1623–1632.
64. Oberg, C. L., Hiensch, R. J., & Poor, H. D. (2017). Ombitasvir-paritaprevir-ritonavir-dasabuvir (Viekira Pak)–induced lactic acidosis. *Critical Care Medicine, 45*, e321–e325.
65. Ahmed, M. (2018). Era of direct acting anti-viral agents for the treatment of hepatitis C. *World Journal of Hepatology, 10*, 670–684.
66. AASLD-IDSA HCV Guidance Panel, et al. (2018). Hepatitis C guidance 2018 update: AASLD-IDSA recommendations for testing, managing, and treating hepatitis C virus infection. *Clinical Infectious Diseases, 67*, 1477–1492.
67. Gao, M., et al. (2010). Chemical genetics strategy identifies an HCV NS5A inhibitor with a potent clinical effect. *Nature, 465*, 96–100.
68. Komatsu, T. E., et al. (2017). Regulatory analysis of effects of hepatitis C virus NS5A polymorphisms on efficacy of elbasvir and grazoprevir. *Gastroenterology, 152*, 586–597.
69. Poordad, F., et al. (2016). High antiviral activity of NS5A inhibitor ABT-530 with paritaprevir/ritonavir and ribavirin against hepatitis C virus genotype 3 infection. *Liver International, 36*, 1125–1132.
70. Foster, G. R., et al. (2015). Sofosbuvir and velpatasvir for HCV genotype 2 and 3 infection. *New England Journal of Medicine, 373*, 2608–2617.
71. Xu, S., et al. (2017). In vitro selection of resistance to sofosbuvir in HCV replicons of genotype 1 to 6. *Antiviral Therapy, 22*, 587–597.

Epidemiological Trends and Current Challenges in Ebola: Pathogen Biology, Drug Targets, and Therapeutic Strategies

Sandeep Sharma and Jeena Gupta

Abstract

Ebola virus (EV) is a highly contagious infectious agent that causes hemorrhagic fever and bleeding in human and nonhuman primates. Ebola contains a single, non-segmented, single-stranded RNA and thus belongs to the order *Mononegavirales*. EVD is a highly transmittable and zoonotic disease, i.e., transmitted from animal to human and human to human through blood and body secretions. During replication of Ebola genome, it is highly prone to occurrence of mutations as RDRP (L protein) makes errors during replication and a proofreading mechanism is lacking in Ebola. At present, there is no FDA-approved drug or vaccine available agai5nst Ebola virus, although some promising drug candidates against Ebola virus are in clinical trial (favipiravir (T-705, AVI-7537)). The current book chapter will provide a general understanding to the researchers about the physiology, molecular genetics, its virulence, therapeutic options, and putative drug targets, repurposing the existing drugs for better prevention and understanding about the Ebola virus.

In addition to viral diseases like dengue, swine flu, chikungunya, and zika, Ebola virus disease (EVD) is emerging as a potential health threat to the public. Ebola virus is categorized as A category biothreat pathogen as its patients possess high fatality rates. It is a highly contagious virus on account of its easy human transmission via patient exudates and its continued presence in animal host at endemic areas and absence of therapeutics for the treatment and licensed vaccines for protection against EVD. At least 26 EVD outbreaks had been reported in several regions of Western and Central Africa since its first describable in 1976.

S. Sharma
Department of Medical Laboratory Sciences, Division of Research and Development, Lovely Professional University, Phagwara, India

J. Gupta (✉)
Department of Biochemistry, Division of Research and Development, Lovely Professional University, Phagwara, India
e-mail: jeena.20104@lpu.co.in

S. Hameed, Z. Fatima (eds.), *Pathogenicity and Drug Resistance of Human Pathogens*,
https://doi.org/10.1007/978-981-32-9449-3_13

An infected animal/person can spread the infection to crowded locations such as travelers and evacuates have carried the disease from Africa to Europe and North America during its 2014 outbreak in Congo and West Africa (WHO 2014). Intense research efforts on various aspects of EVD have been progressively growing against the fear of its increasing outbreak severity and emergence as a potent bioweapon in recent years.

Keywords
Ebola virus · Hemorrhagic disease · Favipiravir · VP35 · ZMapp · Ebola virus disease (EVD)

Ebola Virus Characteristics Taxonomy, Morphology, and Structure

Ebola contains a single, non-segmented, single-stranded RNA and thus belongs to the order *Mononegavirales* [1–3]. The order *Mononegavirales* contains four families: *Rhabdoviridae* (e.g., rabies virus), *Paramyxoviridae* (e.g., mumps and measles viruses), *Bornaviridae* (cause disease in mammalian animals), and *Filoviridae* (e.g., ebola virus) which share a common ancestor. The family *Filoviridae* also contains other genera like Cueva virus and Marburg virus of which Marburg virus causes hemorrhagic disease in infected humans which is different from EVD. The origin of these filoviruses coincides with the origin of great apes, around 16–23 million years ago [4]. Both Ebola and Marburg viral infections can cause infected human death in 6–16 days of infection and are thus considered as highly virulent pathogens [5–8]. However, the EVD patients unlike Marburg virus develop bleeding rarely and mostly in terminal phases [9]. Evidence suggests that these viruses are ever-changing, and new strains emerge during their passage in humans and animals [10].

The shape of Ebola virions is tubular with dimensions around 800 nM in length and 80 nM in diameter. They show the property of concatemerization which project the size of some Ebola virions to be as long as 14,000 nM. They show pleomorphic characteristic and adopt shapes like linear, branched, or V-shaped [11, 12]. The RNA genome (linear and negative sense) is enclosed in a nucleocapsid, further enveloped by another capsid followed by a lipid membrane on the outer side. The viral surface contains knob-shaped 10-nM-long projections which are placed 10 nM apart and appear as peplomes embedded in the outer lipid bilayer (Fig. 1).

The species of Ebola genus demonstrate sequence divergence in 3–41% of their genome and possess different geographical origins. These species are also identified by their geographical origin in their names [10, 13]. Based on their pathogenicity and virulence, they can be arranged in the following order: *Reston ebolavirus* (RESTV, nonpathogenic on humans) < *Sudan ebolavirus* (SUDV) < Tai Forest virus (TAFV) < Bundibugyo virus (BDBV, *Bundibugyo ebolavirus*) < *Zaire ebolavirus* (ZEBOV) [14, 15]. ZEBOV is regarded as type species and is generally referred as EBOV.

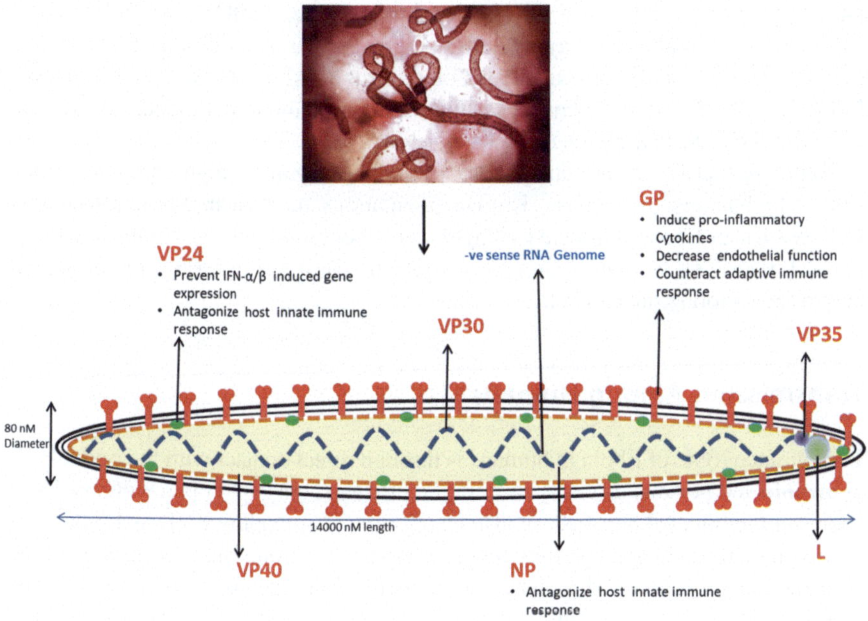

Fig. 1 Structure of Ebola virus

EVD is a highly transmittable and zoonotic disease. All the outbreaks of EVD have been started through transmission of Ebola virus from vertebrate animals to humans. Human to human transmission of Ebola virus only causes secondary infections. The direct contact with the blood or body fluids of infected animals or eating infected meat is responsible for transmitting the disease to humans. Factors like deforestation and population increase have increased the chances of contact between infected wild life and humans [16, 17].

Bats have been considered as the natural reservoir of Ebola virus in African countries, but they survive and lack overt disease [18–25]. Screening of African bats was found to possess Ebola-specific RNA and antibodies. Substantial amounts of Ebola antibodies were also detected in bat populations from China, the Philippines, and Bangladesh [26]. The coinhabiting vertebrates in the forest containing Ebola-infected bats were also thought to catch Ebola infection by coming in direct contact with Ebola diseased animals or eating food drooled or defecated with droppings of infected animals. Non-bat vertebrates like monkeys, foxes, cats, antelopes, hogs, rodents, and nonhuman primates were also found to possess Ebola RNA and antibodies. They can also be a source of spreading Ebola infection other than bats. The titer of Ebola virus is also high in lungs of pigs, transmitting the disease via air; however, this route of airborne transmission is less effective as lung virus titers are lower than in blood [27]. Out of five identified Ebola virus species, four species are known to infect and cause disease in humans, i.e., EBOV (ZEBOV, *Zaire ebolavirus*), Bundibugyo virus (BDBV, *Bundibugyo ebolavirus*), Sudan virus (SUDV, *Sudan ebolavirus*), and Tai Forest virus (TAFV, *Tai Forest ebolavirus*, formerly *Côte d'Ivoire ebolavirus*) while one is nonhuman pathogen, i.e., Reston virus (RESTV; *Reston ebolavirus*).

Being primarily an animal-specific virus, Ebola gained high virulence when shifted to humans (new host). Human populations in African Ebola-affected as well as African Ebola-unaffected villages were also found to carry Ebola antibodies (1.8–21.3%) [28]. Humans can also possibly transmit the disease through air but only over a short distance [29].

Transmission Among Humans

The entrance route of Ebola in humans is through direct contact with the body fluids of the infected persons or via eyes, nose, mouth, ears, cuts, and open wounds. Dead bodies of the infected persons can also be a source of spreading EVD as Ebola virus is present on the skin and touching them can cause infection. The transmission of the virus has been recorded 7 days postmortem by touching the diseased body [30]. The negative-sense RNA of the virus remains for months in cadavers, but native RNA is not infective. Saliva exposure during coughing or reusing contaminated needles/ medical equipment without sterilization can result in Ebola infection. Ebola virus can survive for several weeks on inanimate objects like bedding, clothing, eating utensils, furniture, door knobs, or electrical switches and also in water bodies where body fluids are drained or washed [31]. Ebola virus also remains for many months in the bodies of survivor human beings mostly in organs like testicles and eyes which are immunologically protected even after the subsidence of the disease. Active viruses were found in the aqueous humor of uveitis eye after the disappearance of Ebola viruses from the patient's blood [32]. The presence of Ebola in the semen of recovering men has been recorded even after 26 weeks [33–35]. The single introduction of Ebola virus was enough for causing an epidemic by human to human transmission [36].

Genetics Genome, Gene Functions, Transcription, and Replication

The genome of Ebolavirus contains a single-stranded negative-sense RNA molecule of size 19 kb. It encodes for seven genes which when lie in the order are 3′-Leader-NP-VP35- VP40-GP/sGP-VP30-VP24-L-Trailer-5′.These genes not only code for viral structural proteins but also regulate various processes required by virus to enter and multiply in the host cells like attachment of virion on host cells, entry of virion into cytoplasm of host cell, multiplication of virus, arresting antiviral responses of host cell, apoptosis of host cell, and spreading of infection by regulating pathophysiological changes in host cell. All these seven genes have their own initiation and termination signal flanking them. The coding region of each gene is flanked by some

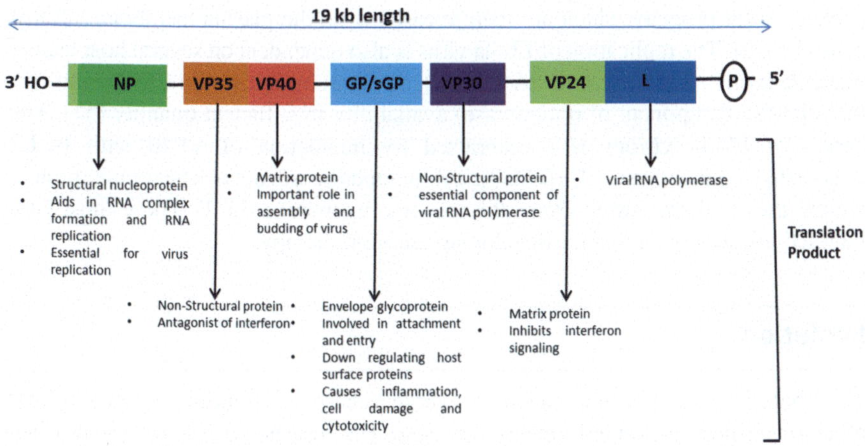

Fig. 2 Genetic makeup of Ebola virus with their translational roles

non-translated sequences of unknown function [37–39]. The cis-sequences present in Leader/Trailer regions are essential for regulating transcription, replication, and packaging of the viral genome into virus particles [40] (Fig. 2).

The transcription of Ebola genome transcription happens in the cytoplasm of host cell after partial uncoating of viral nucleocapsid. The RNA genome in nucleocapsid forms a ribonucleoprotein (RNP) complex with products of VP35, VP30, NP, and L genes. The RNA-dependent-RNA polymerase (RDRP) begins the transcription from L gene to a 5′-triphosphate Leader RNA and stops. The transcription then restarts by RDRP from NP gene start signal forming a capped NP-mRNA and terminating at the NP-gene transcription termination site. Before the release of NP-mRNA, the RDRP stutters at poly U stretch and add a polyadenylated tail on NP-mRNA. RDRP then moves on to start transcription of VP35-gene. All the seven genes on Ebola genome are transcribed sequentially according to their arrangement. Interestingly, the transcription of GP gene produces three different gene products, i.e., small GP (Pre-sGP), Pre-GP, and small secretary GP (Pre-ssGP), which gets translated into sGP, GP1/GP2, and ssGP. The primary product of GP gene is Pre-sGP, which is produced in bulk (>70%). Transcriptional editing by RDRP during which it causes a base addition by reading the template base more than once produces mRNAs for PreGP (<25%) and ssGP (~5%). The addition of an A residue at editing prone site generates mRNA for Pre-GP protein, whereas addition of two A residues generates ssGP-mRNA [41–47]. Genes which are located near the Leader sequence are more transcribed, whereas those near Trailer sequence are least transcribed.

The most expressed protein product is NP, and its concentration determines gene transcription to genome replication switching. During genome replication, RDRP binds to the 3′ leader region moving toward 5′ trailer end. This replication results in the formation of full-length positive strand called antigenomes. RDRP then again binds to the 5′ trailer of this antigenome accomplishing the production of negative-strand RNA genome copies. These genome copies are then encapsidated to produce packed virion progeny which are then released from the plasma membrane of host cells [48–50]. The replication of Ebola virus is also dependent on several host factors like eukaryotic initiation factor 5A (eIF5A), a polyamine spermidine, and control on the VP30 (a component of polymerase) availability in sufficient quantity [51]. The egress of Ebola virions is counteracted by interaction of VP40 with BCL2 Associated- Athanogene 3 (BAG3), an autophagy regulator-chaperone protein which throw them away from the plasma membrane [52]. These host-viral interactions are the targets for developing antiviral therapy.

Evolution

The Ebola genus has been assumed to diverge as early as 20 million years ago from other single negative-strand viruses' lineages. The insertion of filovirus genes into the genomes of old-world clades as well as new-world rodents, bats, and insectivorous mammals was evidenced around 25–18 million years ago [28]. Ebola genes like L, NP, and VP35 have been detected in several species of rodents, bats, and marsupials. The syntenous position of Ebola genes in fossils of the related species indicates the repeated interaction between Ebola and mammals [4, 53]. These Ebola genes may be present in their wild-type or mutant state in Ebola-sensitive species. The bat's Ebola-tolerant phenotype is thought to be because of the negative complementation between Ebola gene products from already present Ebola genes in animal genome and infecting Ebola virus specified proteins [28, 54].

During replication of Ebola genome, it is highly prone to occurrence of mutations as RDRP (L protein) makes errors during replication and proofreading mechanism is lacking in Ebola [55]. The frequency of mutation in the Ebola GP gene is estimated to be 3.6 × 10–5 nonsynonymous substitutions/site/year [56]. However, the mutation rates for the whole genome vary in different Ebola species (0.45×10^{-4}–1.25×10^{-3} nucleotide substitutions/site/year) [36]. Therefore, the Ebola genus possesses 3.5 × 104 times higher mutation rate than human genome [36, 57]. Thus the course of its passages between hosts, improves its contagiousness and extends its host range.. The induction of mutation by mutagens like ribavirin may cause bulk mutations of neutral or nonpositive type which may have a negative impact on virus multiplication and thus can be used as a therapeutic mechanism [58]. Three hundred and forty-one new mutations were revealed in Ebola genome in the 81 sequences originated from the Ebola 2014 outbreak in West Africa (Sierra Leone), and most of these mutations were localized in VP24, L, and NP genes [59].

The substitution mutations of nonsynonymous type in GP gene of Ebola genome had been shown to increase its tropism in human cells and decrease its tropism in bat cells with achievement of greater adaptability for human cells [60]. For example, GP gene mutation A82V significantly increases the infectivity of Ebola toward human cells [61]. Mutations in Ebola genome are reported not only during human-human or animal-human transmission but also during animal-animal transmission. The infection of Ebola virus is asymptomatic in guinea pigs where virus undergoes multiplication in infected animals without clinical symptoms; the clinical symptoms only occur after five to seven time passaging of Ebola virus in guinea pigs. The increase in Ebola virulence in guinea pigs was only observed by mutations in L, NP, GP, VP35, and VP24 genes but not in VP30 and VP40 genes. Mutation in VP24 (position 26) and increased GP editing was associated with significant increase in Ebola virulence [62, 63]. However, researchers are of the opinion that even if the treatments against Ebola may be available in future, the virus may survive by mutating and increasing its contagiousness.

Biology of Virus Entry and Multiplication in Host Cells

The Ebola virus entry into the human host mainly occurs through injured skin and mucosal surfaces like nostrils, ears, eyelids, lips, anus, and genitals of men and women from secretions of infected individuals and animals. The initial targets of Ebola virus are mononuclear phagocytic cells like macrophages, dendritic cells, and monocytes. These infected cells then migrate and make the infection systemic by releasing the virions into the bloodstream and lymphatic system. In the course other cell types like epithelial, endothelial, fibroblasts, adrenal cortical cells, and hepatocytes also get infected, and the necrosis of these infected cells makes the condition fatal [5, 49, 64–66].

To enter into the host cell, at first the virus envelope glycoprotein GP1 binds to the specific receptors on the susceptible host cell surface. This process is facilitated by the interaction of host cell membrane's phosphatidylserine with Ebola virus. Different host cells demonstrate different spectrum of Ebola receptors on their surface like folate receptor-á, G protein-coupled receptors, C-type lectin family proteins, integrin β, T-cell immunoglobulin mucin domain (TIM-1) protein, Tyro3/Axl/Mer (TAM) family proteins, etc [41, 67–75]. This single-molecule force spectroscopy demonstrated that TIM-EBOV interactions are mechanically strong (40–100 pN) and are comparable to adhesion molecule-ligand interactions [76]. After attachment, the formation of macropinosome by host cell's plasma membrane encloses the virus and internalizes it by macropinocytosis [72], which are then shuttled into endolysosomal pathway [77–81].

The endosomes possess acidic environment which activates certain proteases like cathespin-B and –L (cysteine proteases). These proteases cleave viral GP1 to expose its amino-terminal domain, which acts as a ligand for host intracellular receptor Niemann-Pick type C1 (NPC1 = cholesterol transporter). This interaction between NPC1 and GP1 is essential for the fusion of host and viral plasma membrane [78, 81–84]. GP2, a class 1 viral fusion protein, then undergoes a conformational change which unmasks a 45 amino acid loop near N terminus also called as fusogenic domain. GP2 was hypothesized to form a prehairpin conformation where N- and C-terminus heptad repeats of GP2 extended, comes closer to each other, forms a six-helix bundle, and completes the fusion process by forming a fusion pore [78, 85–87]. These events led to the release of viral ribonucleocapsid into host cell cytoplasm.

The virus genome is now free in the host cell cytoplasm where it serves as a template to get transcribed into a monocistronic mRNAs (sequentially transcribed seven viral genes). The mRNA is processed by capping and polyadenylation and then translated to produce different viral proteins. The production of enough NP viral starts genome replication. Viral VP35 protein acts as a chaperone for NP and helps it to coil around negative RNA strand to form a nucleocapsid shell [88, 89]. This makes RNA helicoidal and increases its interaction with L, VP30, and VP35 proteins by allowing self-assembly and encapsidation of viral particles [90–92]. VP24 controls the replication phase termination and virion assembly onset followed by their egress from host cell.

Viral proteins GP, VP40, and VP24 and host cell's endosomal-sorting-complex required for transport (ESCRT) are involved in the budding and egress of virus from host cell [38, 91, 93, 94]. The constitution of virions occurs with the arrival of nucleocapsids at multivacuolar bodies which contains VP40 and GP at the plasma membrane [39, 95, 96]. Phosphatidylserine aids in formation of viral particles at the plasma membrane [97, 98]. Virus particles acquire GP-studded plasma membrane lipid bilayer of host cell which links to ribonucleocapsid through matrix proteins VP24 and VP40 during egress [37, 49, 70, 71, 85, 96, 99–101].

The virus inoculum (both route and size) and genetic factors of host determine the asymptomatic viral incubation period (which varies from 2 to 21 days) for the development of Ebola disease and the success of mounting an innate/adaptive response in host against virus multiplication. The viral titer was found to be 100–1000 times lower in the survivors of Ebola virus disease as compared to non-survivors. Ten billion virus particles may be present in the body of EVD patients nearing death. The replicating viral RNAs were found in the upper and the lower respiratory tracts for many days in EVD patients under recovery even when viral RNA is not detectable in plasma [102]. Therefore, although all host tissues get affected, lung tissues play a major role in transmitting the disease through exhaled air.

Drug Targets and Therapeutics

Ebola virus infection in humans causes severe illness with a very high mortality rate. At present, there is no FDA-approved drug or vaccine available against Ebola virus. The severity of West African epidemic of 2014 marks our negligence and lack of understanding against pathogenesis and infection of Ebola virus. It also creates the urgency of drug discovery and development to combat this pathogen. The activities of scientific community then picked up to identify anti-Ebola therapeutics. The following section will cover the advancements in Ebola therapeutics identifying different drug targets (summarized in Table 1).

Table 1 Therapeutics under investigation against Ebola virus disease (EVD)

S. No.	Treatment name	Concentration used	Experimental model	Inference	References
1.	Convalescent whole blood	Transfusion: 150–400 ml blood	Human patients	Reduced mortality to 12.5%	Mupapa et al. [202]
2.	Polyclonal IgG	Intraperitoneal: 100 mg/Kg	Mice	Significant protection 24 h post challenge	Dye et al. [203]
3.	Viral entry inhibitor: MBX2254	10 μmol/l	A549 cells	Ebola entry inhibited at late stage	Basu et al. [204]
4.	Viral entry inhibitor: MBX2270	30 μmol/l	A549 cells	Ebola entry inhibited at late stage	Basu et al. [204]
5.	Tetrandrine	$IC_{50} = 55$ nM	HeLa cells	Inhibition of human macrophage infection	Sakurai et al. [105]
6.	MLS000078751 MLS000534476	0.39 up to 50 μM	HeLa cells	Inhibition of human macrophage infection	Anantpadma et al. [111]
7.	MLS000394177 MLS000730532 MLS000733230	0.39 up to 50 μM	HeLa cells	Inhibits viral uptake	Anantpadma et al. [111]
8.	MLS000555232	0.39 up to 50 μM	HeLa cells	Inhibits late endocytic trafficking	Anantpadma et al. [111]
9.	MLS000554255 MLS001101371	0.39 up to 50 μM	HeLa cells	Inhibits late trafficking of endocytes	Anantpadma et al. [111]
10.	3-hydroxyphthalic anhydride (HP)-modified human serum albumin	EC_{50}0.068–0.124, for Zaire and Sudan pseudoviruses respectively	Huh-7 cell	Inhibits cell surface attachment of pseudovirus	
11.	Benztropine mesylate	IC_{50}: 1.7 to 4.9 μM for different Ebola strains	A549 and vero cells	Screening 1200 FDA-approved drugs from Prestwick Chemical Library	Cheng et al. [205]
12.	*Prunella vulgaris* extract	2.5 μg/ml	HEK293T cells	Synergizes mAb 2G4 against EBOV-GP	Zhang et al. [113]

(continued)

Table 1 (continued)

S. No.	Treatment name	Concentration used	Experimental model	Inference	References
13.	Quercetin 3-β-*O*-D-glucoside (Q3G)	50 mg/kg	BALB/c or C57BL/6 mice	Inhibits glycoprotein-mediated entry of Ebola	Qiu et al. [112]
14.	Double-stranded RNA-binding protein 76 (DRBP76)		293 T cells	Inhibits Ebola polymerase activity	Shabman et al. [138]
15.	Silvestrol	IC_{50}: 96 nM	Huh-7 cells/primary human macrophages	Strongly reduce VP40 levels	Biedenkopf et al. [117]
16.	IFN-α	IC_{50}: 0.038 μM	HEK 293 T cells	Inhibits viral replication	McCarthy et al. [206]
17.	IFN-β	IC_{50}: 0.016 μM	HEK 293 T cells	Inhibits viral replication	McCarthy et al. [206]
18.	IFN β-1a	30 μg/day	Human patients	Decreases death rate by 1.5–1.9 folds in EVD patients	Konde et al. [207]
19.	ZMAb	0.1–100 μg/ml	VeroE6 cells	Targets GP1-GP2 interface and glycan cap	Audet et al. [208]
20.	ZMapp	5 mg/animal	Nonhuman primates and human patients	Significant improvement	[129]
21.	KL-2E5 and KL-2H7	10 mg/kg	$Stat2^{-/-}$ mice	Non-neutralizing but protective due to fc-FcR interactions	Duehr et al. [209]
22.	FVM04	Single IP injection: 10 mg/kg	Mice	100% protection against lethal challenge	Howell et al. [158]
23.	KZ52	50 mg/kg	Guinea pigs	Dose-dependent protection	Parren et al. [161]
24.	Q206, Q314, and Q411	100 μg of each mAb	BALB/c mice	mAbs cocktail administration at 1–2 days post infection, neutralized live Ebola virus	Zhang et al. [113]
25.	Cell-penetrable human VP40 binding scFvs (HuscFvs)	40 μg/well	Huh7 cells	Inhibited Ebola egress	Teimoori et al. [162]

26.	Cell-penetrable human scFvs to IFN-inhibitory domain of VP35	25 μg/well	HepG2 cells transduced with EBOV minigenome and VP35 expression cassette	Inhibit VP35 functions	Seesuay et al. [163]
27.	Bispecific antibody (FVM09~548 and FVM09~MR72 dual–variable domain immunoglobulin)	20 mg/kg	Female BALB/c mice	Neutralize recombinant VSV-Ebola GP in comparison with poor neutralizing parental mAbs FVM09, mAb-548, and MR72	Wec et al. [164]
28.	Okadaic acid	$IC_{50} = 130$ nM	BSR T7/5 cells	Blocks Ebola multiplication by inhibiting protein phosphatases PP1A and PP2A	Modrof et al. [210]
29.	Favipiravir	300 mg/kg/day	$IFNAR^{-/-}$ C57BL/6 mice	100% reduction in mortality when treatment initiated 6 days pre- or post-infection	Oestereich et al. [195]
		IC_{90} of 110 μM	Vero E6 cells	Suppress Ebola replication	
30.	Genistein and tyrphostin AG1478 cocktail	100 μM	HEK 293 cells	Ebola virus inhibition	Kolokoltsov et al. [211]

Inhibitors of Ebola Virus Entry

After cell entry of Ebola through macropinocytosis, it is processed by endosomal proteases and transport to endolysosomes containing NPC1 (Niemann–Pick C1: internal receptor for EBOV). The maturation of endosome requires phosphatidylinositol-3-phosphate 5-kinase which is a critical step for EBOV infection [103]. The phosphatidylinositol-3-phosphate 5-kinase antagonist, apilimod, was found to block the trafficking of viral particles to endolysosomes [104]. The screening of a library of small molecules identified a derivative of benzylpiperazine adamantane diamide which targets endosomal NPC1 and inhibits infection by vesicular stomatitis virus (VSV) Ebola pseudovirions [82].

After understanding that Ebola virus uses two-pore channels to enter into host cells, a calcium channel blocker, tetrandrine, a bis-benzylisoquinoline alkaloid, was evaluated to potentially inhibit the entry of Ebola virus into the cells [105]. Another ion channel blocker, amiodarone, was also found to inhibit Ebola entry into host cells [106, 107]. Selective estrogen receptor modulators (SERMs), like clomiphene and toremifene, were found to be the potent inhibitors of Ebola infection. These two estrogen receptors were found to interfere in late steps of viral entry, most likely to affect the triggering of viral fusion with host cells [108]. The compounds with unique symmetry and 3D globular structure like dendrimers and fullerene act as a biocompatible carbon platform with multivalent carbohydrate presentation. Fullerenes having 12–36 mannoses were found to possess antiviral activity against Ebola pseudotyped infection model but in low micromolar range. However, 12 sugar-containing fullerene units when attached to a central alkyne scaffold which is also known as tridecafullerenes (containing total 120 mannoses) were found to possess excellent antiviral activity in sub-nanomolar range and were found to effectively inhibit Ebola entry in vitro [109]. Antibiotics like clarithromycin and posaconazole (antifungal agent) exhibit anti-Ebola activity by inhibiting the calcium release from lysosomes, NPC1 protein functions, and acid sphingomyelinase activity which ultimately inhibit Ebola virus entry into host cell [110]. The drug macropinocytosis of Ebola was found to be blocked by a drug 5-(N-ethyl-N-isopropyl) amiloride, and synthetic compounds (MLS000394177 and MLS000733230) also inhibit the entry of Ebola into cells [111].

Derivatives of phytomolecules like Q3G (Quercetin 3-β-O-d-glucoside) had been demonstrated to target entry of Ebola virus into host cell and showed protective effects in Ebola-challenged mice [112]. A Chinese herb *Prunella vulgaris* was also found to inhibit entry of Ebola virus into cells [113]. Pseudovirions which contain Ebola virus-GP are emerging as an excellent model system to identify anti-Ebola drugs. In a study, these pseudovirions were used to screen a chemical library (Prestwick Chemical Library) to identify potential inhibitors of viral entry into cells. Out of screened 1200 FDA-approved drugs, 20 were found to 80% inhibit the entry of virus into cells of which 16 were identified as antagonists of G protein-coupled receptor (GPCR). The study highlights that GPCRs play an important role in Ebola virus entry into host cell and GPCR antagonists can be effectively used as anti-Ebola therapy. It was suggested by microscopic studies and time-of-addition experiment that GPCR antagonists block the viral entry after the initial attachment step but before viral/cell membrane fusion [74]. The 3-hydroxyphthalic anhydride-modified human serum albumin (HP-HSA) was found to effectively inhibit (nanomolar concentrations are 50% effective) the entry of Ebola virus (both Zaire pseudovirus and Sudan pseudovirus) and was found to be more potent than MIL77-2, EBOV-neutralizing antibody which is a ZMapp drug cocktail component antibody. In fact, the synergistic effect was seen between the combination of MIL77-2 and HP-HSA with no obvious toxic effects in vitro or in vivo. Furthermore, the inhibitory activity of HP-HSA remains intact for 8 weeks even when storing at 45 °C and thus has potential to be used worldwide including African tropical regions [114].

Replication of EBOV

During replication of Ebola virus, several proteases like cathepsin B led to the proteolytic cleavage of Ebola's surface GPs in the endosomes [115]. Therefore, proteases can prove a good target for inhibiting replication of Ebola virus. A synthetic inhibitor of serine proteases, nafamostat mesilate (NM), had been shown to reduce the release of CatB from rat pancreases. Furthermore, NM also possesses anticoagulant properties and thus can be effectively used to prevent disseminated intravascular coagulation caused during Ebola virus infections [114]. While screening of genes which are essential for Ebola infection, a gene GNPTAB encoding α and β subunits of N-acetylglucosamine-1-phosphate transferase was identified. Knockdown of GNPTAB expression was found to inhibit Ebola infection. Its activity is dependent on its proteolytic cleavage by protease SKI-1/S1P, and the inhibition of this protease by a small-molecule PF-429242 prevents entry and infection of Ebola virus [116]. Silvestrol, an organic-heterotricyclic compound enriched in *Aglaia foveolata*, had been demonstrated to treat acute infections of Ebola by inhibiting its replication [117]. A single siRNA molecule was found by employing a web-based identification approach for therapeutic agents to inhibit the transcription of three Ebola species mRNA. Ebola VCR, a web server which contains the details for identification of suitable therapeutic agents, has been developed [118].

Interferons

Ebola infection results in increased production/release of pro-inflammatory proteins, coagulation factors, and vasoactive molecules with dysregulation of antigen-presenting cells like dendritic cells and macrophages which plays an important role in sustained virus replication [119–122]. The release of pro-inflammatory molecules then attracts other target cells of the immune system which provides additional cells to the virus for spreading infection and also increases the circulation of these cells. The dysregulated inflammatory response thus led to uncontrolled amplification of viral infection with excessive cytokine accumulation, viral systemic spread, and even circulatory collapse in acute cases with fatal hemorrhagic fever [123]. These events are further contributed by sustained replication of Ebola virus in dendritic cells and macrophages which inhibit early innate immune responses and decreased T-cell numbers and also led to poor adaptive immune responses to viral infection [124, 125]. A recent study has demonstrated that macrophage activation but not T-cell activation contributes to EVD pathogenesis [126].

The disruption of Ebola infection in macrophages by eliciting an early immune response can help to control viral replication. The triggering of these responses prior to disruption of immune system by virus can potentiate the systemic control of Ebola replication. Studies have highlighted the potential of type I interferons (IFNs) in decreasing the mortality and morbidity of Ebola virus [127–129]. IFN treatment had been demonstrated to enhance host cell defenses, stimulate the expression of various interferon-stimulated genes (ISGs) with antiviral activity, and generate Ebola-resistant macrophages [130–133]. IFN-β was found to be more competent as compared to IFN-alpha2b against Ebola infection in cynomolgus macaques and prolonged the survival of rhesus macaques [127, 128]. Furthermore, type I IFN in combination with monoclonal antibodies against Ebola GP was found to be highly efficacious and provide robust protection against Ebola's lethal challenge [129].The administration of IFN-γ either before or after Ebola infection was found to reduce the mortality rate in mice and thus it as a promising therapeutic drug [134]. IFN-γ had already been approved by FDA for treating certain chronic medical conditions and thus can be readily adapted for controlling Ebola infections.

EBOV Gene Expression Inhibitors

Virus replication is critically dependent on the expression of viral genes by viral RNA polymerase using host cell machinery. A nucleoside analogue BCX4430 was found to inhibit viral RNA polymerase and protect mice against Ebola's lethal challenge [135–137]. VP35 which functions as a viral RNA polymerase cofactor is also responsible for inhibiting the production of IFN-α/β. Double-stranded RNA-binding protein 76 (DRBP76/NFAR-1/NF90), a cellular factor responsible for protein synthesis and host cell defense, can associate with the C-terminal interferon inhibitory domain of viral VP35. Reports have shown that overexpression of DRBP76 impairs the transcription/replication of Ebola genome in host cells without interfering in the inhibition of IFN production by VP35 and thus can be used as anti-Ebola agent [138]. A cationic porphyrin TmPyP4 had been reported to inhibit the expression of L gene [139]. The cellular polyamines were found critical for Ebola replication, and small molecular polyamine synthesis inhibitors were found to shut down gene expression by viral RNA polymerase. The knockdown of spermidine, a polyamine pathway enzyme by short hairpin RNA (shRNA), also reduced the replication of Ebola virus. Spermidine was found essential for the hypusination of eIF5A (eukaryotic initiation factor 5A), and hypusinated eIF5A is important for VP30 accumulation which is an essential component of viral RNA polymerase. The blocking of polyamine synthesis or hypusination of eIF5A can therefore be strategically used to inhibit Ebola gene expression by RNA polymerase [51].

Other Therapeutic Strategy: Transfusion of Convalescent Blood/ Serum

The World Health Organization (WHO) had approved the transfusion of convalescent serum/ whole blood during critical Ebola infections [140, 141]. Convalescent serum is taken from diseased survivors and is devoid of clotting factors and red blood cells but contains immunoglobulins like IgM and IgG against Ebola virus whose transfusion had been found to reduce the viral load [142, 143]. However, the plasma has to be screened to exclude the residual Ebola RNA and other pathogens (especially blood-borne) like HIV, hepatitis B, and C virus. The postexposure antibody therapy using either polyclonal or monoclonal antibodies was also found to confer the protection and is also approved by FDA (Food and Drug Administration) [144].

The passive immunization using neutralizing antibodies by sera transfusion (from Ebola recovering individuals) emerges as valuable emergency therapeutics during Ebola [145, 146]. It is although not 100% protective when transfusion occurs 3 days after exposure to Ebola virus, e.g., *Zaire ebolavirus* Makona [147]. Using equine serum which contains precise immunoglobulins against Ebola virus was found to be effective and safe in non-allergic patients [148]. The different mAb-cocktails had been developed recently for the treatment of Ebola viral disease like ZMapp, ZMAb, mAb114 and MB-003, and MIL-77E [149]. The administration of ZMAb which consist of three murine mAbs, i.e., 2G4, 4G7, and 1H3, (dose 25 mg/kg 3 times daily), was found to completely protect cynomolgus macaques from EVD, while ZMAb administration with IFN-α-vectored adenovirus resulted in increased survival of cynomolgus (75%) and rhesus macaques (100%) against EVD [150]. The mAbs which can bind to the GP base like 4G7 and 2G4 are neutralizing antibodies, while the ones which can bind the glycan cap like mAb114, 1H3, and 13C6 are non-neutralizing antibodies. Furthermore, 13C6 and 6D8 are chimeric antibodies possessing mice variable region and human Fc region which can neutralize Ebola virus during the presence of complement proteins [151]. mAb114, a single-monoclonal antibody, targets the Ebola GP receptor-binding domain. It prevents mortality in rhesus macaques from lethal challenge of *Zaire ebolavirus* and was found well tolerated in clinical trials [152].

The repeated immunization with filovirus glycoproteins in mice led to the generation of pan-Ebola-specific mAbs which were found effective against viruses like RESTV and SUDV [153]. ZMapp components include chimeric mAbs like c13C6 from MB-003 (already known antibody cocktail) and two mAbs (c2G4 and c4G7) from ZMab cocktail. ZMapp is approved by WHO for EVD treatment and was found to reverse the clinical signs of EVD in rhesus macaques (100%), even after late administration (5 days after viral exposure) [154]. Another mAb cocktail MB-003 was also found effective against Ebola virus in ZMapp resistant individuals [155]. Furthermore, there is a scope to exploring the synergistic effect of different mAbs (combination pairs of neutralizing and non-neutralizing mAbs) for designing anti-Ebola immunotherapeutics/vaccines [156].

These mAbs mechanistically bind the inter-protomer epitope present on GP fusion loop and thus inhibit the membrane fusion of virus and also neutralize its entry into host cell [157]. However, the potential limitation of using mAb-based therapies is that their high doses are required, viral epitope mutations reduce the effectiveness of these therapeutic mAbs, and these mAb mixtures are generally outbreak specific since the virus is constantly evolving becoming more furious with every outbreak. For example, ZMapp is only effective against *Zaire ebolavirus*; however, replacing a component of ZMapp with FVM04, a mAb with cross-neutralizing activity for SUDV (Sudan virus), increases the activity range of ZMapp against SUDV [158].

Ebola virus GP is cleaved by cathepsins which then fuses with host cell membrane and forms a fusion pore through which Ebola genome enters the host cell cytoplasm for replication. Human mAb KZ52 binds to a 23-residue non-glycosylated epitope at GP base of which 15 residues are in direct contact through Van der Waals interactions and 8 by hydrogen bonding [159]. KZ52 thus inhibit the GP cleavage by cathepsin and show 50% neutralization at dose 0.4 μg/ml (23). The FabKZ52 was originally derived from an Ebola disease survivor's bone marrow and was found to exhibit 50% neutralization at 8 nM concentrations [160]. It was found to protect guinea pigs from the lethal challenge of *Zaire ebolavirus* failed in rhesus macaques [161].

Human single-chain antibody variable fragments (HuscFvs) also known as transbodies or superantibodies (by Kohler and Paul (1921)) are cell-penetrable bodies which can cross cell membrane and get accumulated intracellularly only at the site of target antigen. HuscFvs, produced by phage display technology which can bind VP40 protein of Ebola virus, can potentially inhibit the egress of virus from hepatic cells [162]. These transbodies bind to C-terminal domain at several cationic patches of VP40 which a matrix protein is and plays a pivotal role in the assembly and budding of virus. They also show binding to WW binding motifs in L-domain peptide and thus can act directly as anti-Ebola agents [162]. HuscFvs against interferon-inhibitory domain (IID) of Ebola's VP35 was also found to inhibit both polymerase cofactor activity of VP35 and its antagonistic activity against host IFNs by forming contact with basic residues in its first/central patch, end cap, and the residues important for the formation of multimeric IID for binding dsRNA [163]. However, since these super antibodies accumulate intracellularly, their disappearance from blood circulation does not confirm their elimination from the body. Therefore, further evaluation in various animal models of EVD using authentic Ebola virus and clinical studies is required before considering them as a promising therapeutic alternative for the treatment of EVD. The bispecific Trojan horse antibodies which can neutralize other filoviruses were also found to be effective against multiple Ebola infections in mice [164].

Edible vaccines produced in plants are now emerging for easy access to vaccines. A recent study has highlighted the production of three mAbs targeting Ebola GP in tobacco plants which were tested and were found effective in humans [165]. The identification of cross-protective antibodies which are effective against multiple Ebola species can help in developing effective therapeutic strategies for treatment of EVD [166]. A mAbs isolated from survived patients of Uganda outbreak (2007) against BDBV-GP was found to neutralize multiple Ebola species [167]. Since the Ebola GPs are conserved across different species, four mABs – S3, S12, S17, and S33 as revealed by ELISA – were found to possess cross-reactivity against GPs of five different Ebola species [168]. Recently a pan-ebolavirus therapeutic MBP134AF, which comprise two broadly neutralizing human antibodies (bNAbs), had been developed whose single dose (25 mg/kg) was sufficient to protect ferrets against lethal challenges of EBOV, SUDV, and BDBV infection [169].

Another study has demonstrated the isolation of 349 EBOV GP mAbs from Zaire Ebola outbreak survivors of which 77% of mAbs were found effective in neutralizing Ebola virus [170]. Three mAbs isolates (Q206, Q314, and Q411) from West African Ebola outbreak against Ebola GP were found to recognize novel epitopes and of which Q206 and Q411 were found to protect mice from Ebola infection [171]. A mAbs-based therapeutic vaccine had been developed to inhibit replication of invasive EBOV, even when administered 4 days post-infection as a single dose [172]. Adeno-associated virus (AAV) expressing non-neutralizing mAb (5D2 or 7C9) was found to be 100% protective in mice by consistently releasing these anti-Ebola mAbs in the body. However, a neutralizing mAb 2G4 was found to be 83% protective alone, but antibody cocktail containing these two mAbs was 100% protective when provided 7 days pre-infection and provides sustained protection after challenging with Ebola 5 months post AAV-mAb immunization [173]. Prediction algorithm and molecular docking help to identify different peptides which contain overlapping T-cell epitopes with ability to interact with multiple HLA alleles. A peptide LANETTQALQLF (P5) was found to be 100% conserved in *Bundibugyo*, *Sudan*, *Zaire*, and *Tai Forest* species and is capable of inducing T- and B-cell response against EVD [174].

Small Molecules as Inhibitors of Ebola Virus Infection

Approaches of medicinal chemistry and biological screening using both in silico and in vitro experiments have identified small molecules to be used against infection of Ebola virus [175].Viral RNA inhibitor (combination of three siRNAs) packed in lipid nanoparticle called TKM-Ebola was formulated with Tekmira's lipid nanoparticle technology and targets L polymerase, VP24the membrane-associated protein, and VP35, polymerase-complex protein of ZEBOV. New drug discovery researchers are facing a global challenge of emergence of drug-resistant viral strains which hampered the development of novel therapeutic interventions adapted for emerging Ebola virus strains. So different strategies were employed of which the most popular is forming nanoconjugates which were found to provide 100% protection in rhesus monkeys against Ebola's lethal challenge when treatment starts postexposure at third day [176]. However, these approaches need to be further validated by in vivo studies for their efficacy and safety in human lethal EVD. Another experimental antiviral agent that was found effective against Ebola viral infection in animal models is AVI-7537 [177, 178]. AVI-7537 was found to act on gene silencing and also having some pharmacokinetic and safety data in humans [179]. However, these gene-silencing drugs are cumbersome to synthesize, and the scaling up of their industrial production to meet the public demands at the time of emergency is a challenging task.

Repurposed Drugs

The development of new therapeutic agents against this ever-changing infectious virus is a time-taking task. An important option till the time for development of new therapies is to depend upon "drug repurposing." Drug repurposing involves investigation of already approved drugs for new therapeutic indications. The lack of approved therapies for EVD and screening of efficacious approved drugs revealed that these drugs which have already proven safe can be potentially repurposed for combating EVD [180]. The drugs used for arrhythmias, tachycardia, and high blood pressure like dronedarone, amiodarone, and verapamil were found effective to inhibit filoviruses entry into cells in in vitro models [106]. Since the virus used the host cell for multiplication, drugs targeting host response like angiotensin receptor blockers, angiotensin-converting enzyme inhibitors, statins, phosphoinositide 3-kinases inhibitor (LY294002), and calcium/calmodulin kinase (CAMK2) inhibitor (KN-93) were reported to attenuate Ebola infection in in vitro conditions [181]. The virus-mediated increase in cytokine production was reduced by SB202190, a p38 mitogen-activated protein kinase (MAPK) inhibitor in human monocytes [182]. Clomiphene and toremifene, which are estrogen reuptake modulators, were found to inhibit virus entry but are loaded with side effects when administered at higher doses like clomiphene causes ocular adverse reaction and toremifene causes serious electrolytes derangements, therefore combination therapy is suggested for using them [180]. An analogue of cidofovir with lipid conjugation, brincidofovir, had been demonstrated to inhibit Ebola replication but needs to be evaluated using in vivo models [183]. Anti-protozoal agents emetine and cephaeline and its desmethyl analogue cephaeline are also potent inhibitors of Ebola replication with cephaeline that is more tolerable than emetine in patients [184].

Inflammatory inhibitors like atorvastatin, rosuvastatin, and pravastatin which reduce TNFα levels and C-reactive protein have been reported to hamper cholesterol-supported membrane biosynthesis in Ebola virus [185]. Ebola infection results in coagulation imbalance by overexpression of procoagulant tissue factor in monocytes/macrophages. An anticoagulant protein c2 from recombinant nematode was found to inhibit blood coagulation (tissue factor mediated) and improve macaque's survival from Ebola-hemorrhagic fever [186]. The lysomotropic agents which prevent acidification of endosomes/lysosomes were also reported to limit viral infections. These include antimalarial drugs like chloroquine-hydroxychloroquine, primaquine, pamaquine, and plasmoquine [187]. However, there are conflicting reports as chloroquine was found effective in in vitro models to inhibit virus replication but showed no protection in guinea pigs, hamsters, and mice against Ebola infection [188, 189]. Proton-pump inhibitors like esomeprazole and omeprazole inhibit entry of Ebola viral in vitro studies but required higher concentrations when used in vivo [190]. The mass administration of antimalaria drugs were found effective in reducing fever during Liberia outbreak of Ebola in 2014 [185].

Toyama Chemicals introduced a selective inhibitor of influenza virus replication, favipiravir (T-705), which shows minimum cell toxicity [191, 192]. It was found to induce level of lethal RNA mutations by interfering with viral RNA-dependent RNA polymerase activity [191, 193]. Recently, a study has highlighted the in vitro effectiveness of T-705 in inhibiting the replication of Ebola virus (IC-5067 μM) without showing any cytotoxicity under the set experimental conditions [194, 195]. Oral administration of T-705 twice daily, to type-I IFN-α/β-receptor knockout mice, was found to prevent mortality in all the animals. Favipiravir was found to prevent mortality in Ebola-infected mice even if the treatment starts post-infection of 1 week [194]. Favipiravir possesses various advantages for using during Ebola outbreak as it has already been under extensive use for its antiviral activity against influenza in Japan, easily available and effective via oral administration.

An antitrypanosomal agent, suramin, possesses anti-heparin potential and was demonstrated to interfere with the entry and replication of virus into host cells [196]. An antiflu medicine, the low concentrations (48–140 nM) of microtubule inhibitor drugs like vincristine, vinblastine, and vinorelbine/navelbine which are known for their anticancer activity were found effective by inhibiting the entry of Ebola virus into HeLa cells. Another microtubule modulator, colchicine, already in use for gout, also showed anti-Ebola activity [197, 198]. The RNA-directed RNA polymerase of Ebola virus had been reported to get inhibited by different antiviral drugs like cidofovir, maraviroc, abacavir, and telbivudine which target the MTase domain of Ebola virus [199].

An HIV protease inhibitor, indinavir, was found effective in reducing Ebola severity while screening of 1766 FDA-approved drugs and 259 experimental drugs [199]. Recently, a Computational Analysis of Novel Drug Opportunities platform had been developed for screening of drugs already approved by FDA with which the drugs like vancomycin, enfuvirtide, bleomycin, somatostatin, octreotide, lanreotide, and ubidecarenone (CoQ10) were found significantly effective against Ebola virus [200]. Through virtual screening of many thousands of Drug Bank molecules (repurposed drugs), ibuprofen was identified for its inhibitory potential on Ebola-induced infection. The antiviral activity of ibuprofen had been validated by cell culture studies, and thus it can be used as molecular template for developing anti-Ebola drugs [201].

Challenges and Prospects

Latest outbreak of Ebola virus which occurs between 2014 and 2016 arises an international public health emergency. Due to high mortality rate of Ebola virus disease (EVD), it becomes highly significant to develop vaccines and treatment strategies against this serious disease. Recent years have seen the emergence of several therapies to tackle lethal Ebola infections. The formulation of "ZMapp" a plant-derived humanized mAbs to treat EVD patients although shows promising results but its short supply warrants the need to develop and evaluate new mAbs. Repurposed drugs are emerging as an alternative approach as those drugs which have already proven safe can be potentially repurposed for combating EVD. Numerous active repurposed compounds had been evaluated against EVD, and their mechanism of action may be variable which include EV inhibitors, ion channel blockers, microtubule inhibitors, kinase inhibitors, estrogen receptors, reuptake modulators, and histamine antagonists. However, high throughput screening is required to prove the potency and efficacy of these repurposed drugs for EVD treatment. The need is to develop effective as well as economic antiviral agents against Ebola virus which should be easily approachable even for poor countries, and drug repurposing is the best solution for that. Reports have also analyzed the effectiveness of engineered nucleic acid molecules and nucleotide analogues like antisense phosphorodiamidate morpholino oligomers (PMOs) siRNA and miRNA against Ebola challenge. Computational approaches emerge as promising technology to screen large number of molecules for their inhibitory effects on Ebola virus and promoting human health. However, instead of these developments, more research efforts are required to develop potent anti-Ebola therapeutic/prophylactic agents and appropriate preventive measures to limit viral spread and prevent future outbreaks. Strict biosecurity principles should also be developed against the risk of using it as a bioweapon. The cross-reactive mAbs that can neutralize all five species of Ebola need to be developed and evaluated for their therapeutic effects and can be further engineered for developing human homogenous antibodies.

References

1. Kuhn, J. H., Becker, S., Ebihara, H., Geisbert, T. W., Johnson, K. M., Kawaoka, Y., Lipkin, W. I., Negredo, A. I., Netesov, S. V., Nichol, S. T., et al. (2010). Proposal for a revised taxonomy of the family Filoviridae: Classification, names of taxa and viruses, and virus abbreviations. *Archives of Virology, 155*, 2083–2103.
2. Li, Y. H., & Chen, S. P. (2014). Evolutionary history of Ebola virus. *Epidemiology and Infection, 142*, 1138–1145.
3. International Committee on Taxonomy of Viruses. (2013 Release). Virus Taxonomy. Ebolavirus. http://www.ictvonline.org/%20virusTaxonomy.asp
4. Taylor, D. J., Ballinger, M. J., Zhan, J. J., Hanzly, L. E., & Bruenn, J. A. (2014). Evidence that ebolaviruses and cuevaviruses have been diverging from marburgviruses since the Miocene. *Peer J, 2*, e556.
5. Mahanty, S., & Bray, M. (2004). Pathogenesis of filoviral haemorrhagic fevers. *The Lancet Infectious Diseases, 4*, 487–498.

6. Macleod, C. (2010). Towards a philosophical account of crimes against humanity. *European Journal of International Law, 21*, 281–302.
7. Leroy, E. M., Gonzalez, J. P., & Baize, S. (2011). Ebola and Marburg haemorrhagic fever viruses: Major scientific advances, but a relatively minor public health threat for Africa. *Clinical Microbiology and Infection, 17*, 964–976.
8. Funk, D. J., & Kumar, A. (2015). Ebola virus disease: An update for anesthesiologists and intensivists. *Canadian Journal of Anaesthesia, 62*, 80–91.
9. Fowler, R. A., Fletcher, T., Fischer, W. A., Lamontagne, F., Jacob, S., Brett-Major, D., et al. (2014). Caring for critically ill patients with ebola virus disease. Perspectives from West Africa. *American Journal of Respiratory and Critical Care Medicine, 190*, 733–737.
10. Kuhn, J. H., Andersen, K. G., Baize, S., et al. (2014). Nomenclature- and database-compatible names for the two Ebola virus variants that emerged in Guinea and the Democratic Republic of the Congo in 2014. *Viruses, 6*, 4760–4799.
11. Sanchez, A. (2001). Filoviridae: Marburg and Ebola viruses. In D. M. Knipe & P. M. Howley (Eds.), *Fields virology* (4th ed., pp. 1279–1304). Philadelphia: LippencottRavenpp.
12. Acha, P. N., & Szyfres, B. (2003). In Pan American Health Organization (Ed.), *Zoonoses and communicable diseases common to man and animals* (3rd ed., pp. 142–145). Washington DC: Pan American Health Organization.
13. Fauci, A. S. (2014). Ebola – underscoring the global disparities in health care resources. *The New England Journal of Medicine, 371*, 1084–1086.
14. Cheng, A. C., & Kelly, H. (2014). Are we prepared for Ebola and other viral haemorrhagic fevers? *Australian and New Zealand Journal of Public Health, 38*, 403–404.
15. Spickler, A. R. (2014). Ebolavirus and marburgvirus infections. The Centre for Food Security and Public Health. IOWA State University, Institute for International Cooperation in Animal Biologics.
16. Bausch, D. G., & Schwarz, L. (2014). Outbreak of ebola virus disease in Guinea: Where ecology meets economy. *PLoS Neglected Tropical Diseases, 8*, e3056.
17. Walsh, M. G., & Haseeb, M. A. (2015). The landscape configuration of zoonotic transmission of Ebola virus disease in West and Central Africa: Interaction between population density and vegetation cover. *Peer J, 3*, e735.
18. Swanepoel, R., Leman, P. A., Burt, F. J., Zachariades, N. A., Braack, L. E., Ksiazek, T. G., Rollin, P. E., Zaki, S. R., & Peters, C. J. (1996). Experimental inoculation of plants and animals with Ebola virus. *Emerging Infectious Diseases, 2*, 321–325.
19. Reiter, P., Turell, M., Coleman, R., Miller, B., Maupin, G., Liz, J., Kuehne, A., et al. (1999). Field investigations of an outbreak of Ebola hemorrhagic fever, Kikwit, Democratic Republic of the Congo, 1995: Arthropod studies. *The Journal of Infectious Diseases, 179*, S148–S154.
20. Leroy, E. M., Rouquet, P., Formenty, P., Souquiere, S., Kilbourne, A., Froment, J. M., Bermejo, M., et al. (2004). Multiple Ebola virus transmission events and rapid decline of central African wildlife. *Science, 303*, 387–390.
21. Pourrut, X., Delicat, A., Rollin, P. E., Ksiazek, T. G., Gonzalez, J. P., & Leroy, E. M. (2007). Spatial and temporal patterns of Zaire ebolavirus antibody prevalence in the possible reservoir bat species. *The Journal of Infectious Diseases, 196*, S176–S183.
22. Peterson, A. T., Papes, M., Carroll, D. S., Leirs, H., & Johnson, K. M. (2007). Mammal taxa constituting potential coevolved reservoirs of filoviruses. *Journal of Mammalogy, 88*, 1544–1554.
23. Olival, K. J., Islam, A., Yu, M., Anthony, S. J., Epstein, J. H., Khan, S. A., et al. (2013). Ebola virus antibodies in fruit bats. *Bangladesh Emerging Infectious Disease, 19*, 270–273.
24. Ng, M., Ndungo, E., Kaczmarek, M. E., Herbert, A. S., Binger, T., Kuehne, A. I., Jangra, R. K., et al. (2015). Filovirus receptor NPC1 contributes to species-specific patterns of ebolavirus susceptibility in bats. *eLife, 4*, e11785.
25. Baize, S., Pannetier, D., Oestereich, L., Rieger, T., Koivogui, L., Magassouba, N., Soropogui, B., et al. (2014). Emergence of Zaire Ebola virus disease in Guinea. *The New England Journal of Medicine, 371*, 1418–1425.

26. Olival, K. J., & Hayman, D. T. (2014). Filoviruses in bats: Current knowledge and future directions. *Viruses, 6*, 1759–1788.
27. Weingartl, H. M., Nfon, C., & Kobinger, G. (2013). Review of Ebola virus infections in domestic animals. *Developmental Biology, 135*, 211–218.
28. Leroy, E. M., Baize, S., Volchkov, V. E., Fisher-Hoch, S. P., Georges Courbot, M. C., Lansoud-Soukate, J., et al. (2000). Human asymptomatic Ebola infection and strong inflammatory response. *Lancet, 355*, 2210–2215.
29. Wong, G., Qiu, X., Richardson, J. S., Cutts, T., Collignon, B., Gren, J., Aviles, J., Embury-Hyatt, C., & Kobinger, G. P. (2015). Ebola virus transmission in Guinea pigs. *Journal of Virology, 89*, 1314–1323.
30. Prescott, J., Trenton Bushmaker, T., Fischer, R., Miazgowicz, K., Judson, S., & Munster, V. J. (2015). Postmortem stability of biology and control of Ebola virus disease 145 Ebola virus. *Emerging Infectious Diseases*. https://doi.org/10.3201/eid2105.150041.
31. Bibby, K., Casson, L. W., Stachler, E., Charles, N., & Haas, C. N. (2015). Ebola virus persistence in the environment: State of the knowledge and research needs. *Environmental Science & Technology Letters, 2*, 2–6.
32. Varkey, J. B., Shantha, J. G., Crozier, I., Kraft, C. S., Lyon, M., Mehta, A. K., Kumar, G., Smith, J. R., Kainulainen, M. H., Whitmer, S., et al. (2015). Persistence of Ebola virus in ocular fluid during convalescence. *New England Journal of Medicine*. https://doi.org/10.1056/NEJMoa1500306.
33. Rowe, A. K., Bertolli, J., Khan, A. S., Mukunu, R., Muyembe-Tamfum, J. J., Bressler, D., Williams, A. J., Peters, C. J., et al. (1999). Clinical, virologic, and immunologic follow-up of convalescent Ebola hemorrhagic fever patients and their household contacts, Kikwit, Democratic Republic of the Congo. *Commission de Lutte contre les Epidémies à Kikwit J Infect Dis, 179*, S28–S35.
34. Cardona-Maya, W. D., Hernandez, P. A. V., & Henao, D. E. (2014). Male Ebola survivors: Do not forget to use a condom. *Reproductive Sciences*. https://doi.org/10.1177/1933719114563733.
35. Mate, S. E., Kugelman, J. R., Nyenswah, T. G., Ladner, J. T., Wiley, M. R., Cordier-Lassalle, T., et al. (2015). Molecular evidence of sexual transmission of Ebola virus. *The New England Journal of Medicine*. https://doi.org/10.1056/NEJMoa1509773.
36. Park, D. J., Dudas, G., Wohl, S., Goba, A., Whitmer, S. L., Andersen, K. G., et al. (2015). Ebola virus epidemiology, transmission and evolution during seven months in Sierra Leone. *Cell, 161*, 1516–1526.
37. Klenk, H. D., & Feldmann, H. (2004). *Ebola and Marburg viruses: Molecular and cellular biology*. Wymondham: Horizon Bioscience.
38. Jasenosky, L. D., & Kawaoka, Y. (2004). Filovirus budding. *Virus Research, 106*, 181–188.
39. Hartlieb, B., Muziol, T., Weissenhorn, W., & Becker, S. (2007). Crystal structure of the C-terminal domain of Ebola virus VP30 reveals a role in transcription and nucleocapsid association. *Proceedings of the National Academy of Sciences of the United States of America, 104*, 624–629.
40. Crary, S. M., Towner, J. S., Honig, J. E., Shoemaker, T. R., & Nichol, S. T. (2003). Analysis of the role of predicted RNA secondary structures in Ebola virus replication. *Virology, 306*, 210–218.
41. Volchkov, V. E., Becker, S., Volchkova, V. A., Ternovoj, V. A., Kotov, A. N., Netesov, S. V., & Klenk, H. D. (1995). GP mRNA of Ebola virus is edited by the Ebola virus polymerase and by T7 and vaccinia virus polymerases. *Virology, 214*, 421–430.
42. Sanchez, A., Trappier, S. G., Mahy, B. W., Peters, C. J., & Nichol, S. T. (1996). The virion glycoproteins of Ebola viruses are encoded in two reading frames and are expressed through transcriptional editing. *Proceedings of the National Academy of Sciences of the United States of America, 93*, 3602–3607.
43. Mehedi, M., Falzarano, D., Seebach, J., Hu, X., Carpenter, M. S., Schnittler, H. J., & Feldmann, H. (2011). A new Ebola virus nonstructural glycoprotein expressed through RNA editing. *Journal of Virology, 85*, 5406–5414.

44. Mehedi, M., Hoenen, T., Robertson, S., Ricklefs, S., Dolan, M. A., Taylor, T., Falzarano, D., Ebihara, H., Porcella, S. F., & Feldmann, H. (2013). Ebola virus RNA editing depends on the primary editing site sequence and an upstream secondary structure. *PLoS Pathogens, 9*, e1003677.
45. Mohan, G. S., Li, W., Ye, L., Compans, R. W., & Yang, C. (2012). Antigenic subversion: A novel mechanism of host immune evasion by Ebola virus. *PLoS Pathogens, 8*, e1003065.
46. Shabman, R. S., Jabado, O. J., Mire, C. E., Stockwell, T. B., Edwards, M., Mahajan, M., Geisbert, T. W., & Basler, C. F. (2014). Deep sequencing identifies noncanonical editing of Ebola and Marburg virus RNAs in infected cells. *MBio, 5*, e02011.
47. Mohan, G. S., Ye, L., Li, W., Monteiro, A., Lin, X., Sapkota, B., Pollack, B. P., Compans, R. W., & Yang, C. (2015). Less is more: Ebola virus surface glycoprotein expression levels regulate virus production and infectivity. *Journal of Virology, 89*, 1205–1217.
48. Ascenzi, P., Bocedi, A., Heptonstall, J., Capobianchi, M. R., Di Caro, A., Mastrangelo, E., Bolognesi, M., & Ippolito, G. (2008). Ebolavirus and Marburgvirus: Insight the Filoviridae family. *Molecular Aspects of Medicine, 29*, 151–185.
49. Olejnik, J., Ryabchikova, E., Corley, R. B., & Muhlberger, E. (2011). Intracellular events and cell fate in filovirus infection. *Viruses, 3*, 1501–1531.
50. Choi, J. H., & Croyle, M. A. (2013). Emerging targets and novel approaches to Ebola virus prophylaxis and treatment. *Bio Drugs, 27*, 565–583.
51. Olsen, M. E., Filone, C. M., Rozelle, D., Mire, C. E., Agans, K. N., Hensely, L., & Connor, J. H. (2016). Polyamines and hypusination are required for Ebola virus gene expression and replication. *MBio, 7*, e00882–e00816.
52. Liang, J., Sagum, C. A., Bedford, M. T., Sidhu, S. S., Sudol, M., Han, Z., et al. (2017). Chaperone-mediated autophagy protein BAG3 negatively regulates Ebola and Marburg VP40-mediated egress. *PLoS Pathogens, 13*, e 1006132.
53. Taylor, D. J., Dittmar, K., Ballinger, M. J., & Bruenn, J. A. (2011). Evolutionary maintenance of filovirus-like genes in bat genomes. *BMC Evolutionary Biology, 11*, 336.
54. Leroy, E. M., Kumulungui, B., Pourrut, X., Rouquet, P., Hassanin, A., Yaba, P., et al. (2005). Fruit bats as reservoirs of Ebola virus. *Nature, 438*, 575–576.
55. Liu, G., Zhao, S., Bailey, J. A., Sahinalp, S. C., Alkan, C., Tuzun, E., Green, E. D., & Eichler, E. E. (2003). Analysis of primate genomic variation reveals a repeat-driven expansion of the human genome. *Genome Research, 13*, 358–368.
56. Suzuki, Y., & Gojobori, T. (1997). The origin and evolution of Ebola and Marburg viruses. *Molecular Biology and Evolution, 14*, 800–806.
57. Carroll, S. A., Towner, J. S., Sealy, T. K., McMullan, L. K., Khristova, M. L., Burt, F. J., Swanepoel, R., Rollin, P. E., & Nichol, S. T. (2013). Molecular evolution of viruses of the family Filoviridae based on 97 whole-genome sequences. *Journal of Virology, 87*, 2608–2616.
58. Alfson, K. J., Worwa, G., Carrion, R., Jr., & Griffiths, A. (2016). Determination of the spontaneous mutation frequency of Ebola virus and exploitation of this therapeutically. *Journal of Virology*, 02701–02715.
59. Gire, S. K., Goba, A., Andersen, K. G., Sealfon, R. S., Park, D. J., Kanneh, L., Jalloh, S., et al. (2014). Genomic surveillance elucidates Ebola virus origin and transmission during the 2014 outbreak. *Science, 345*, 1369–1372.
60. Urbanovicz, R. A., McClure, C. P., Sakuntabhai, A., Sall, A. A., Kobinger, G., Muller, M. A., et al. (2016). Human adaptation of Ebola virus during the west African outbreak. *Cell, 167*, 1079–1087.
61. Diehl, W. E., Lin, A. E., Gurubagh, N. D., Carvalho, L. M., Kim, K., Kyawe, P. P., et al. (2016). Ebola virus glycoprotein with increased infectivity dominated the 2013–2016 epidemic. *Cell, 167*, 1088–1098.
62. Subbotina, E., Dadaeva, A., Kachko, A., & Chepurnov, A. (2010). Genetic factors of Ebola virus virulence in guinea pigs. *Virus Research, 153*, 121–133.
63. Dowall, S. D., Matthews, D. A., Garcia-Dorival, I., Taylor, I., Kenny, J., Hertz-Fowler, C., Hall, N., Corbin-Lickfett, K., et al. (2014). Elucidating variations in the nucleotide sequence of Ebola virus associated with increasing pathogenicity. *Genome Biology, 15*, 540.

64. Feldmann, H., & Geisbert, T. W. (2011). Ebola haemorrhagic fever. *Lancet, 377*, 849–862.
65. Chiappelli, F., Bakhordarian, A., Thames, A. D., Du, A. M., Jan, A. L., Nahcivan, M., et al. (2015). Ebola: translational science considerations. *Journal of Translational Medicine, 13*, 11.
66. de La Vega, M. A., Wong, G., Kobinger, G. P., & Qiu, X. (2015). The multiple roles of sGP in Ebola pathogenesis. *Viral Immunology, 28*, 3–9.
67. Weissenhorn, W., Calder, L. J., Wharton, S. A., Skehel, J. J., & Wiley, D. C. (1998). The central structural feature of the membrane fusion protein subunit from the Ebola virus glycoprotein is a long triple-stranded coiled coil. *Proceedings of the National Academy of Sciences of the United States of America, 95*, 6032–6036.
68. Chan, S. Y., Empig, C. J., Welte, F. J., Speck, R. F., Schmaljohn, A., Kreisberg, J. F., & Goldsmith, M. A. (2001). Folate receptor-alpha is a cofactor for cellular entry by Marburg and Ebola viruses. *Cell, 106*, 117–126.
69. Takada, A., Feldmann, H., Ksiazek, T. G., & Kawaoka, Y. (2003). Antibody-dependent enhancement of Ebola virus infection. *Journal of Virology, 77*, 7539–7544.
70. Marzi, A., Akhavan, A., Simmons, G., Gramberg, T., Hofmann, H., Bates, P., Lingappa, V. R., & Pohlmann, S. (2006). The signal peptide of the ebolavirus glycoprotein influences interaction with the cellular lectins DC-SIGN and DCSIGNR. *Journal of Virology, 80*, 6305–6317.
71. Marzi, A., Wegele, A., & Pohlmann, S. (2006). Modulation of virion incorporation of ebolavirus glycoprotein: Effects on attachment, cellular entry and neutralization. *Virology, 352*, 345–356.
72. Hunt, C. L., Kolokoltsov, A. A., Davey, R. A., & Maury, W. (2011). The Tyro3 receptor kinase Axl enhances micropinocytosis of Zaire ebolavirus. *Journal of Virology, 85*, 334–347.
73. Lennemann, N. J., Rhein, B. A., Ndungo, E., Chandran, K., Qiu, X., & Maury, W. (2014). Comprehensive functional analysis of N-linked glycans on Ebola virus GP1. *MBio, 5*, e00862–e00813.
74. Cheng, H., Lear-Rooney, C. M., Johansen, L., Varhegyi, E., Chen, Z. W., Olinger, G. G., & Rong, L. (2015). Inhibition of Ebola and Marburg virus entry by G protein-coupled receptor antagonists. *Journal of Virology, 89*, 9932–9938.
75. Rhein, B. A., & Maury, W. J. (2015). Ebola virus entry into host cells: Identifying therapeutic strategies. *Virology, 2*, 115–124.
76. Dragovich, M. A., Fortoul, N., Jagota, A., Zhang, W., Schutt, K., Xu, Y., Sanabria, M., Moyer, D. M., Moller-Tank, S., Maury, W., & Zhang, X. F. (2019). Biomechanical characterization of TIM protein–mediated Ebola virus–host cell adhesion. *Scientific Reports, 9*(1), 267.
77. Bhattacharyya, S., Mulherkar, N., & Chandran, K. (2012). Endocytic pathways involved in filovirus entry: Advances, implications and future directions. *Viruses, 4*, 3647–3664.
78. Miller, E. H., & Chandran, K. (2012). Filovirus entry into cells – new insights. *Current Opinion in Virology, 2*, 206–214.
79. Nanbo, A., Imai, M., Watanabe, S., Noda, T., Takahashi, K., Neumann, G., Halfmann, P., & Kawaoka, Y. (2010). Ebolavirus is internalized into host cells via macropinocytosis in a viral glycoprotein-dependent manner. *PLoS Pathogens, 6*, e1001121.
80. Saeed, M. F., Kolokoltsov, A. A., Albrecht, T., & Davey, R. A. (2010). Cellular entry of Ebola virus involves uptake by a macropinocytosis-like mechanism and subsequent trafficking through early and late endosomes. *PLoS Pathogens, 6*, e1001110.
81. Carette, J. E., Raaben, M., Wong, A. C., Herbert, A. S., Obernosterer, G., Mulherkar, N., et al. (2011). Ebola virus entry requires the cholesterol transporter Niemann-Pick C1. *Nature, 477*, 340–343.
82. Cote, M., Misasi, J., Ren, T., Bruchez, A., Lee, K., Filone, C. M., Hensley, L., Li, Q., Ory, D., Chandran, K., & Cunningham, J. (2011). Small molecule inhibitors reveal Niemann-Pick C1 is essential for Ebola virus infection. *Nature, 477*, 344–348.
83. Herbert, A. S., Davidson, C., Kuehne, A. I., Bakken, R., Braigen, S. Z., Gunn, K. E., Sean, P., Whelan, S. P., et al. (2015). Niemann-Pick C1 is essential for ebolavirus replication and pathogenesis. *In Vivo, 6*, e00565–e00515. https://doi.org/10.1128/mBio.00565-1526.
84. Spence, J. S., Krause, T. B., Mittler, E., Jangra, R. K., & Chandran, K. (2016). Direct visualization of Ebola virus fusion triggering in the endocytic pathway. *MBio, 7*, e01857–e01815.

85. Chandran, K., Sullivan, N. J., Felbor, U., Whelan, S. P., & Cunningham, J. M. (2005). Endosomal proteolysis of the Ebola virus glycoprotein is necessary for infection. *Science, 308*, 1643–1645.
86. Hood, C. L., Abraham, J., Boyington, J. C., Leung, K., Kwong, P. D., & Nabel, G. J. (2010). Biochemical and structural characterization of cathepsin L-processed Ebola virus glycoprotein: Implications for viral entry and immunogenicity. *Journal of Virology, 84*, 2972–2982.
87. Gregory, S. M., Harada, E., Liang, B., Delos, S. E., White, J. M., & Tamm, L. K. (2011). Structure and function of the complete internal fusion loop from ebolavirus glycoprotein 2. *Proceedings of the National Academy of Sciences of the United States of America, 108*, 11211–11216.
88. Kirchdoerfer, R. N., Abelson, D. M., Li, S., Wood, M. R., & Saphire, E. O. (2015). Assembly of the Ebola virus nucleoprotein from a chaperoned VP35 complex. *Cell Reports, 12*, 140–149.
89. Leung, D. W., Borek, D., Basler, C. F., & Amarasinghe, G. K. (2015). An intrinsically disordered peptide from Ebola virus VP35 controls viral RNA synthesis by modulating nucleoprotein RNA interactions. *Cell Reports*. https://doi.org/10.1016/j.celrep.2015.03.034.
90. Watanabe, S., Noda, T., & Kawaoka, Y. (2006). Functional mapping of the nucleoprotein of Ebola virus. *Journal of Virology, 80*, 3743–3751.
91. Watanabe, S., Noda, T., Halfmann, P., Jasenosky, L., & Kawaoka, Y. (2007). Ebola virus (EBOV) VP24 inhibits transcription and replication of the EBOV genome. *The Journal of Infectious Diseases, 196*, S284–S290.
92. Bharat, T. A., Noda, T., Riches, J. D., Kraehling, V., Kolesnikova, L., Becker, S., Kawaoka, Y., & Briggs, J. A. (2012). Structural dissection of Ebola virus and its assembly determinants using cryo-electron tomography. *Proceedings of the National Academy of Sciences of the United States of America, 109*, 4275–4280.
93. Han, Z., Boshra, H., Sunyer, J. O., Zwiers, S. H., Paragas, J., & Harty Biology and Control of Ebola Virus Disease 139 R N. (2003). Biochemical and functional characterization of the Ebola virus VP24 protein: Implications for a role in virus assembly and budding. *Journal of Virology, 77*, 1793–1800.
94. Licata, J. M., Johnson, R. F., Han, Z., & Harty, R. N. (2004). Contribution of ebola virus glycoprotein, nucleoprotein, and VP24 to budding of VP40 virus-like particles. *Journal of Virology, 78*, 7344–7351.
95. Dolnik, O., Volchkova, V., Garten, W., Carbonnelle, C., Becker, S., Kahnt, J., Stroher, U., Klenk, H. D., & Volchkov, V. (2004). Ectodomain shedding of the glycoprotein GP of Ebola virus. *The EMBO Journal, 23*, 2175–2184.
96. Liu, Y., Cocka, L., Okumura, A., Zhang, Y. A., Sunyer, J. O., & Harty, R. N. (2010). Conserved motifs within Ebola and Marburg virus VP40 proteins are important for stability, localization, and subsequent budding of virus-like particles. *Journal of Virology, 84*, 2294–2303.
97. Soni, S. P., Adu-Gyamfi, E., Yong, S. S., Jee, C. S., & Stahelin, R. V. (2013). The Ebola virus matrix protein deeply penetrates the plasma membrane: An important step in viral egress. *Biophysical Journal, 104*, 1940–1949.
98. Soni, S. P., & Stahelin, R. V. (2014). The Ebola virus matrix protein biology and control of Ebola virus disease VP40 selectively induces vesiculation from phosphatidylserine-enriched membranes. *The Journal of Biological Chemistry, 289*, 33590–33597.
99. Manicassamy, B., Wang, J., Jiang, H., & Rong, L. (2005). Comprehensive analysis of ebola virus GP1 in viral entry. *Journal of Virology, 79*, 4793–4805.
100. Hoenen, T., Jung, S., Herwig, A., Groseth, A., & Becker, S. (2010). Both matrix proteins of Ebola virus contribute to the regulation of viral genome replication and transcription. *Virology, 403*, 56–66.
101. Adu-Gyamfi, E., Soni, S. P., Xue, Y., Digman, M. A., Gratton, E., & Stahelin, R. V. (2013). The Ebola virus matrix protein penetrates into the plasma membrane: A key step in viral protein 40 (VP40) oligomerization and viral egress. *The Journal of Biological Chemistry, 288*, 5779–5789.

102. Biava, M., Caglioti, C., Bordi, L., Castilleti, C., Colavite, F., Quartu, S., et al. (2017). Detection of viral RNA in tissues following plasma clearance from an Ebola virus infected. *PLOS Pathology, 13*, e1006065.
103. Qiu, S., Leung, A., Bo, Y., Kozak, R. A., Anand, S. P., Warkentin, C., et al. (2018). Ebola virus requires phosphatidylinositol (3,5) bisphosphate production for efficient viral entry. *Virology, 513*, 17–28. https://doi.org/10.1016/j.virol.2017.09.028.
104. Nelson, E. A., Dyall, J., Hoenen, T., Barnes, A. B., Zhou, H., Liang, J. Y., et al. (2017). The phosphatidylinositol-3-phosphate 5-kinase inhibitor apilimod blocks filoviral entry and infection. *PLoS Neglected Tropical Diseases, 11*(4), e0005540. https://doi.org/10.1371/journal.pntd.0005540.
105. Sakurai, Y., Kolokoltsov, A. A., Chen, C. C., Tidwell, M. W., Bauta, W. E., Klugbauer, N., et al. (2015). Two-pore channels control Ebola virus host cell entry and are drug targets for disease treatment. *Science, 347*, 995–998. https://doi.org/10.1126/science.1258758.
106. Gehring, G., Rohrmann, K., Atenchong, N., Mittler, E., Becker, S., Dahlmann, F., et al. (2014). The clinically approved drugs amiodarone, dronedarone and verapamil inhibit filovirus cell entry. *The Journal of Antimicrobial Chemotherapy, 69*, 2123–2131. https://doi.org/10.1093/jac/dku091.
107. Salata, C., Munegato, D., Martelli, F., Parolin, C., Calistri, A., Baritussio, A., et al. (2018). Amiodarone affects Ebola virus binding and entry into target cells. *The New Microbiologica, 41*(2), 162–164.
108. Johansen, L. M., Brannan, J. M., Delos, S. E., Shoemaker, C. J., Stossel, A., Lear, C., et al. (2013). FDA-approved selective estrogen receptor modulators inhibit Ebola virus infection. *Science Translational Medicine, 5*, 190ra79. https://doi.org/10.1126/scitranslmed.3005471.
109. Illescas, B. M., Rojo, J., Delgado, R., & Martín, N. (2017). Multivalent glycosylated nanostructures to inhibit Ebola virus infection. *Journal of the American Chemical Society, 139*(17), 6018–6025. https://doi.org/10.1021/jacs.7b01683.
110. Sun, W., He, S., Martínez-Romero, C., Kouznetsova, J., Tawa, G., Xu, M., et al. (2016). Synergistic drug combination effectively blocks Ebola virus infection. *Antiviral Research, 137*, 165–172. https://doi.org/10.1016/j.antiviral.2016.11.017.
111. Anantpadma, M., Kouznetsova, J., Wang, H., Huang, R., Kolokoltsov, A., Guha, R., et al. (2016). Large-scale screening and identification of novel Ebola virus and Marburg virus entry inhibitors. *Antimicrobial Agents and Chemotherapy, 60*(8), 4471–4481. https://doi.org/10.1128/AAC.00543-16.
112. Qiu, X., Kroeker, A., He, S., Kozak, R., Audet, J., Mbikay, M., et al. (2016). Prophylactic efficacy of quercetin 3-β-O-d-glucoside against Ebola virus infection. *Antimicrobial Agents and Chemotherapy, 60*(9), 5182–5188. https://doi.org/10.1128/AAC.00307-16.
113. Zhang, X., Ao, Z., Bello, A., Ran, X., Liu, S., Wigle, J., et al. (2016). Characterization of the inhibitory effect of an extract of *Prunella vulgaris* on Ebola virus glycoprotein (GP)-mediated virus entry and infection. *Antiviral Research, 127*, 20–31. https://doi.org/10.1016/j.antiviral.2016.01.001.
114. Li, H., Yu, F., Xia, S., Yu, Y., Wang, Q., Lv, M., et al. (2017). Chemically modified human serum albumin potently blocks entry of Ebola pseudoviruses and virus like particles. *Antimicrobial Agents and Chemotherapy, 61*(4), e2168–e2116. https://doi.org/10.1128/AAC.02168-16.
115. Bornholdt, Z. A., Ndungo, E., Fusco, M. L., Bale, S., Flyak, A. I., Crowe, J. E., Jr., et al. (2016). Host-primed Ebola virus GP exposes a hydrophobic NPC1 receptor-binding pocket, revealing a target for broadly neutralizing antibodies. *MBio, 7*(1), e2154–e2115. https://doi.org/10.1128/mBio.02154-15.
116. Flint, M., Chatterjee, P., Lin, D. L., McMullan, L. K., Shrivastava-Ranjan, P., Bergeron, É., Lo, M. K., Welch, S. R., Nichol, S. T., Tai, A. W., & Spiropoulou, C. F. (2019). A genome-wide CRISPR screen identifies N-acetylglucosamine-1-phosphate transferase as a potential antiviral target for Ebola virus. *Nature Communications, 10*(1), 285.

117. Biedenkopf, N., Lange-Grunweller, K., Schulte, F. W., Weißer, A., Müller, C., Becker, D., et al. (2017). The natural compound silvestrol is a potent inhibitor of Ebola virus replication. *Antiviral Research, 137*, 76–81. https://doi.org/10.1016/j.antiviral.2016.11.011.
118. Dhanda, S. K., Chaudhary, K., Gupta, S., Brahmachari, S. K., & Raghava, G. P. (2016). A web-based resource for designing therapeutics against Ebola virus. *Scientific Reports, 6*, 24782. https://doi.org/10.1038/srep24782.
119. Geisbert, T. W., Hensley, L. E., Larsen, T., Young, H. A., Reed, D. S., Geisbert, J. B., Scott, D. P., Kagan, E., Jahrling, P. B., & Davis, K. J. (2003). Pathogenesis of Ebola hemorrhagic fever in cynomolgus macaques: Evidence that dendritic cells are early and sustained targets of infection. *The American Journal of Pathology, 163*(6), 2347–2370.
120. Bray, M., & Geisbert, T. W. (2005). Ebola virus: The role of macrophages and dendritic cells in the pathogenesis of Ebola hemorrhagic fever. *The International Journal of Biochemistry & Cell Biology, 37*(8), 1560–1566.
121. Gupta, M., Mahanty, S., Bray, M., Ahmed, R., & Rollin, P. E. (2001). Passive transfer of antibodies protects immunocompetent and immunodeficient mice against lethal Ebola virus infection without complete inhibition of viral replication. *Journal of Virology, 75*(10), 4649–4654.
122. Hensley, L. E., Young, H. A., Jahrling, P. B., & Geisbert, T. W. (2002). Proinflammatory response during Ebola virus infection of primate models: Possible involvement of the tumor necrosis factor receptor superfamily. *Immunology Letters, 80*(3), 169–179.
123. Bray, M., & Mahanty, S. (2003). Ebola hemorrhagic fever and septic shock. *The Journal of Infectious Diseases, 188*(11), 1613–1617.
124. Ströher, U., West, E., Bugany, H., Klenk, H. D., Schnittler, H. J., & Feldmann, H. (2001). Infection and activation of monocytes by Marburg and Ebola viruses. *Journal of Virology, 75*(22), 11025–11033.
125. Sanchez, A., Lukwiya, M., Bausch, D., Mahanty, S., Sanchez, A. J., Wagoner, K. D., & Rollin, P. E. (2004). Analysis of human peripheral blood samples from fatal and nonfatal cases of Ebola (Sudan) hemorrhagic fever: Cellular responses, virus load, and nitric oxide levels. *Journal of Virology, 78*(19), 10370–10377.
126. McElroy, A. K., Shrivastava-Ranjan, P., Harmon, J. R., Martines, R. B., Silva-Flannery, L., Flietstra, T. D., Kraft, C. S., Mehta, A. K., Lyon, G. M., Varkey, J. B., & Ribner, B. S. (2019). Macrophage activation marker soluble CD163 associated with fatal and severe Ebola virus disease in humans1. *Emerging Infectious Diseases, 25*(2), 290–298.
127. Jahrling, P. B., Geisbert, T. W., Geisbert, J. B., Swearengen, J. R., Bray, M., Jaax, N. K., Huggins, J. W., LeDuc, J. W., & Peters, C. J. (1999). Evaluation of immune globulin and recombinant interferon-α2b for treatment of experimental Ebola virus infections. *The Journal of Infectious Diseases, 179*(Supplement_1), S224–S234.
128. Smith, L. M., Hensley, L. E., Geisbert, T. W., Johnson, J., Stossel, A., Honko, A., Yen, J. Y., Geisbert, J., Paragas, J., Fritz, E., & Olinger, G. (2013). Interferon-β therapy prolongs survival in rhesus macaque models of Ebola and Marburg hemorrhagic fever. *The Journal of Infectious Diseases, 208*(2), 310–318.
129. Qiu, X., Wong, G., Fernando, L., Audet, J., Bello, A., Strong, J., Alimonti, J. B., & Kobinger, G. P. (2013). mAbs and Ad-vectored IFN-α therapy rescue Ebola-infected nonhuman primates when administered after the detection of viremia and symptoms. *Science Translational Medicine, 5*(207), 207ra143.
130. Murray, B. E., Duchin, J. S., & Neill, M. A. (2014). IDSA Ebola summary—August 2014. InOpen forum infectious diseases 2014 Dec 1 (Vol. 1, No. 3). Oxford University Press.
131. Martinez, V. P., Bellomo, C. M., & Iglesias, A. A. (2014). Laboratory preparation for the diagnosis of Ebola virus disease in Argentina. *Medicina, 74*(6), 506.
132. Sica, A., & Mantovani, A. (2012). Macrophage plasticity and polarization: In vivo veritas. *The Journal of Clinical Investigation, 122*(3), 787–795.

133. Liu, Y., Shi, Z. X., Ma, Y. K., Wang, H. T., Wang, Z. Y., Shao, D. H., Wei, J. C., Wang, S. H., Li, B. B., Wang, S. M., & Liu, X. H. (2012). Development of SYBR Green I real-time RT-PCR for the detection of Ebola virus. *Bing du xue bao= Chinese Journal of Virology, 28* (5), 567–571.
134. Rhein, B. A., & Maury, W. J. (2015). Ebola virus entry into host cells: Identifying therapeutic strategies. *Current Clinical Microbiology Reports, 2*(3), 115–124.
135. Warren, T. K., Wells, J., Panchal, R. G., Stuthman, K. S., Garza, N. L., Van Tongeren, S. A., et al. (2014). Protection against filovirus diseases by a novel broad-spectrum nucleoside analogue BCX4430. *Nature, 508*, 402–405. https://doi.org/10.1038/nature13027.
136. Taylor, R., Kotian, P., Warren, T., Panchal, R., Bavari, S., Julander, J., et al. (2016). BCX4430—a broad-spectrum antiviral adenosine nucleoside analog under development for the treatment of Ebola virus disease. *Journal of Infection and Public Health, 9*, 220–226. https://doi.org/10.1016/j.jiph.2016.04.002204.
137. Cardile, A. P., Warren, T. K., Martins, K. A., Reisler, R. B., & Bavari, S. (2017). Will there be a cure for Ebola? *Annual Review of Pharmacology and Toxicology, 57*, 329–348. https://doi.org/10.1146/annurev-pharmtox-010716-105055.
138. Shabman, R. S., Leung, D. W., Johnson, J., Glennon, N., Gulcicek, E. E., Stone, K. L., Leung, L., Hensley, L., Amarasinghe, G. K., & Basler, C. F. (2011). DRBP76 associates with Ebola virus VP35 and suppresses viral polymerase function. *The Journal of Infectious Diseases, 204* (suppl_3), S911–S918.
139. Ruggiero, E., & Richter, S. N. (2018). G-quadruplexes and G-quadruplex ligands: Targets and tools in antiviral therapy. *Nucleic Acids Research, 46*(7), 3270–3283. https://doi.org/10.1093/nar/gky187.
140. Gulland, A. (2014). First Ebola treatment is approved by WHO. *British Medical Journal, 349*, g5539. https://doi.org/10.1136/bmj.g5539.
141. Geisen, C., Kann, G., Strecker, T., Wolf, T., Schüttfort, G., van Kraaij, M., et al. (2016). Pathogen-reduced Ebola virus convalescent plasma: First steps towards standardization of manufacturing and quality control including assessment of Ebola-specific neutralizing antibodies. *Vox Sanguinis, 110*(4), 329–335. https://doi.org/10.1111/vox.12376.
142. van Griensven, J., De Weiggheleire, A., Delamou, A., Smith, P. G., Edwards, T., Vandekerckhove, P., et al. (2015). The use of Ebola convalescent plasma to treat Ebola virus disease in resource constrained set-tings: A perspective from the field. *Clinical Infectious Diseases, 62*, 69–74. https://doi.org/10.1093/cid/civ680.
143. Garraud, O. (2017). Use of convalescent plasma in Ebola virus infection. *Transfusion and Apheresis Science, 56*(1), 31–34. https://doi.org/10.1016/j.transci.2016.12.014.
144. Dye, J. M., Herbert, A. S., Kuehne, A. I., Barth, J. F., Muhammad, M. A., Zak, S. E., et al. (2012). Postexposure antibody prophylaxis protects nonhuman primates from filovirus disease. *Proceedings of the National Academy of Sciences of the United States of America, 109*, 5034–5039. https://doi.org/10.1073/pnas.1200409109.
145. Saphire, E. O. (2013). An update on the use of antibodies against the filoviruses. *Immunotherapy, 5*, 1221–1233. https://doi.org/10.2217/imt.13.124175.
146. Corti, D., Misasi, J., Mulangu, S., Stanley, D. A., Kanekiyo, M., Wollen, S., et al. (2016). Protective monotherapy against lethal Ebola virus infection by a potently neutralizing antibody. *Science, 351*, 1339–1342. https://doi.org/10.1126/science.aad5224176.
147. Mire, C. E., Geisbert, J. B., Agans, K. N., Thi, E. P., Lee, A. C., Fenton, K. A., et al. (2016). Passive immunotherapy: Assessment of convalescent serum against Ebola virus Makona infection in nonhuman primates. *The Journal of Infectious Diseases, 214*(S3), S367–S374. https://doi.org/10.1093/infdis/jiw333.
148. Borisevich, I. V., Chemikova, N. K., Markov, V. I., Krasnianskiy, V. P., Borisevich, S. V., & Rozhdestvenskiy, E. V. (2017). An experience in the clinical use of specific immunoglobulin from horse blood serum for prophylaxis of Ebola haemorrhagic fever. *Voprosy Virusologii, 62* (1), 25–29.

149. Mendoza, E. J., Racine, T., & Kobinger, G. P. (2017). The ongoing evolution of antibody-based treatments for Ebola virus infection. *Immunotherapy, 9*(5), 435–450. https://doi.org/10.2217/imt-2017-0010.
150. Qiu, X., Audet, J., Wong, G., Pillet, S., Bello, A., Cabral, T., et al. (2012). Successful treatment of Ebola virus-infected cynomolgus macaques with monoclonal antibodies. *Science Translational Medicine, 4*(138), 138ra81. https://doi.org/10.1126/scitranslmed.3003876.
151. Moekotte, A. L., Huson, M. A., van der Ende, A. J., Agnandji, S. T., Huizenga, E., Goorhuis, A., et al. (2016). Monoclonal antibodies for the treatment of Ebola virus disease. *Expert Opinion on Investigational Drugs, 25*(11), 1325–1335. https://doi.org/10.1080/13543784.2016.1240785.
152. Gaudinski, M. R., Coates, E. E., Novik, L., Widge, A., Houser, K. V., Burch, E., Holman, L. A., Gordon, I. J., Chen, G. L., Carter, C., & Nason, M. (2019). Safety, tolerability, pharmacokinetics, and immunogenicity of the therapeutic monoclonal antibody mAb114 targeting Ebola virus glycoprotein (VRC 608): An open-label phase 1 study. *The Lancet, 393*, 889–898.
153. Holtsberg, F. W., Shulenin, S., Vu, H., Howell, K. A., Patel, S. J., Gunn, B., et al. (2016). Pan ebolavirus and pan-filovirus mouse monoclonal antibodies: Protection against Ebola and Sudan viruses. *Journal of Virology, 90*(1), 266–278. https://doi.org/10.1128/JVI.02171-15.
154. Kanapathipillai, R., Henao Restrepo, A. M., Fast, P., Wood, D., Dye, C., Kieny, M.-P., et al. (2014). Ebola vaccine – an urgent international priority. *New England Journal of Medicine, 371*, 2249–2251. https://doi.org/10.1056/NEJMp1412166.
155. Hayden, F. G., Friede, M., & Bausch, D. G. (2017). Experimental therapies for Ebola virus disease: What have we learned? *The Journal of Infectious Diseases, 215*, 167–170. https://doi.org/10.1093/infdis/jiw496.
156. Howell, K. A., Brannan, J. M., Bryan, C., McNeal, A., Davidson, E., Turner, H. L., et al. (2017). Cooperativity enables non-neutralizing antibodies to neutralize ebolavirus. *Cell Reports, 19*(2), 413–424. https://doi.org/10.1016/j.celrep.2017.03.049.
157. Wec, A. Z., Herbert, A. S., Murin, C. D., Nyakatura, E. K., Abelson, D. M., Fels, J. M., et al. (2017). Antibodies from a human survivor define sites of vulnerability for broad protection against Ebola viruses. *Cell, 169*(5), 878.e–890.e. https://doi.org/10.1016/j.cell.2017.04.037.
158. Howell, K. A., Qiu, X., Brannan, J. M., Bryan, C., Davidson, E., Holtsberg, F. W., et al. (2016). Antibody treatment of Ebola and Sudan virus infection via a uniquely exposed epitope within the glycoprotein receptor-binding site. *Cell Reports, 15*(7), 1514–1526. https://doi.org/10.1016/j.celrep.2016.04.026.
159. Lee, J. E., Fusco, M. L., Hessell, A. J., Oswald, W. B., Burton, D. R., & Saphire, E. O. (2008). Structure of the Ebola virus glycoprotein bound to an antibody from a human survivor. *Nature, 454*(7201), 177–182. https://doi.org/10.1038/nature07082.
160. Maruyama, T., Rodriguez, L. L., Jahrling, P. B., Sanchez, A., Khan, A. S., Nichol, S. T., et al. (1999). Ebola virus can be effectively neutralized by antibody produced in natural human infection. *Journal of Virology, 73*(7), 6024–6030. 186.
161. Parren, P. W. H. I., Geisbert, T. W., Maruyama, T., Jahrling, P. B., & Burton, D. R. (2002). Pre- and postexposure prophylaxis of Ebola virus infection in an animal model by passive transfer of a neutralizing human antibody. *Journal of Virology, 76*(12), 6408–6412. https://doi.org/10.1128/JVI.76.12.6408-6412.2002.
162. Teimoori, S., Seesuay, W., Jittavisutthikul, S., Chaisri, U., Sookrung, N., Densumite, J., et al. (2016). Human transbodies to VP40 inhibit cellular egress of Ebola virus-like particles. *Biochemical and Biophysical Research Communications, 479*, 245–252. https://doi.org/10.1016/j.bbrc.2016.09.052.
163. Seesuay, W., Jittavisutthikul, S., Sae-Lim, N., Sookrung, N., Sakolvaree, Y., & Chaicumpa, W. (2018). Human transbodies that interfere with the functions of Ebola virus VP35 protein in genome replication and transcription and innate immune antagonism. *Emerging Microbes & Infections, 7*(1), 41.

164. Wec, A. Z., Nyakatura, E. K., Herbert, A. S., Howell, K. A., Holtsberg, F. W., Bakken, R. R., et al. (2016). A "Trojan horse" bispecific antibody strategy for broad protection against ebolaviruses. *Science, 354*, 350–354. https://doi.org/10.1126/science.aag3267.
165. González-González, E., Alvarez, M. M., Márquez-Ipiña, A. R., Trujillode Santiago, G., Rodríguez-Martínez, L. M., Annabi, N., et al. (2017). Anti-Ebola therapies based on monoclonal antibodies: Current state and challenges ahead. *Critical Reviews in Biotechnology, 37* (1), 53–68. https://doi.org/10.3109/07388551.2015.1114465.
166. Furuyama, W., Marzi, A., Nanbo, A., Haddock, E., Maruyama, J., Miyamoto, H., et al. (2016). Discovery of an antibody for pan-ebolavirus therapy. *Scientific Reports, 6*, 20514. https://doi.org/10.1038/srep20514.
167. Flyak, A. I., Shen, X., Murin, C. D., Turner, H. L., David, J. A., Fusco, M. L., et al. (2016). Crossreactive and potent neutralizing antibody responses in human survivors of natural Ebola virus infection. *Cell, 164*(3), 392–405. https://doi.org/10.1016/j.cell.2015.12.022.
168. Hernandez, H., Marceau, C., Halliday, H., Callison, J., Borisevich, V., Escaffre, O., et al. (2015). Development and characterization of broadly cross-reactive monoclonal antibodies against all known ebolavirus species. *The Journal of Infectious Diseases, 212*(Suppl 2), S410–S413. https://doi.org/10.1093/infdis/jiv209.
169. Bornholdt, Z. A., Herbert, A. S., Mire, C. E., He, S., Cross, R. W., Wec, A. Z., Abelson, D. M., Geisbert, J. B., James, R. M., Rahim, M. N., & Zhu, W. (2019). A two-antibody pan-ebolavirus cocktail confers broad therapeutic protection in ferrets and nonhuman primates. *Cell Host & Microbe, 25*(1), 49–58.
170. Bornholdt, Z. A., Turner, H. L., Murin, C. D., Li, W., Sok, D., Souders, C. A., et al. (2016). Isolation of potent neutralizing antibodies from a survivor of the 2014 Ebola virus outbreak. *Science, 351*(6277), 1078–1083. https://doi.org/10.1126/science.aad5788.
171. Zhang, Q., Gui, M., Niu, X., He, S., Wang, R., Feng, Y., et al. (2016). Potent neutralizing monoclonal antibodies against Ebola virus infection. *Scientific Reports, 6*, 25856. https://doi.org/10.1038/srep25856.
172. Nguyen, V. K., & Hernandez-Vargas, E. A. (2017). Windows of opportunity for Ebola virus infection treatment and vaccination. *Scientific Reports, 7*, 8975. https://doi.org/10.1038/s41598-017-08884-0.
173. van Lieshout, L. P., Soule, G., Sorensen, D., Frost, K. L., He, S., Tierney, K., et al. (2018). Intramuscular adeno-associated virus-mediated expression of monoclonal antibodies provides 100% protection against Ebola virus infection in mice. *The Journal of Infectious Diseases, 217* (6), 916–925. https://doi.org/10.1093/infdis/jix644.
174. Jain, S., & Baranwal, M. (2019). Computational analysis in designing T cell epitopes enriched peptides of Ebola glycoprotein exhibiting strong binding interaction with HLA molecules. *Journal of Theoretical Biology, 465*, 34–44.
175. Picazo, E., & Giordanetto, F. (2015). Small molecule inhibitors of Ebola virus infection. *Drug Discovery Today, 20*, 277–286. https://doi.org/10.1016/j.drudis.2014.12.010.
176. Thi, E. P., Mire, C. E., Lee, A. C., Geisbert, J. B., Zhou, J. Z., Agans, K. N., Snead, N. M., Deer, D. J., Barnard, T. R., Fenton, K. A., MacLachlan, I., & Geisbert, T. W. (2015). Lipid nanoparticle siRNA treatment of Ebola-virus-Makona-infected nonhuman primates. *Nature, 521*, 362–365. https://doi.org/10.1038/nature14442.
177. Panchal, R. G., & Bavari, S. (2012). Discovery and early development of AVI-7537 and AVI-7288 for the treatment of Ebola virus and Marburg virus infections. *Viruses, 4*, 2806–2830. https://doi.org/10.3390/v4112806.
178. Warren, T. K., Warfield, K. L., Wells, J., Swenson, D. L., Donner, K. S., Van Tongeren, S. A., Garza, N. L., Dong, L., Mourich, D. V., Crumley, S., Nichols, D. K., Iversen, P. L., & Bavari, S. (2010). Advanced antisense therapies for postexposure protection against lethal filovirus infections. *Nature Medicine, 16*, 991–994. https://doi.org/10.1038/nm.2202.
179. Heald, A. E., Iversen, P. L., Saoud, J. B., Sazani, P., Charleston, J. S., Axtelle, T., Wong, M., Smith, W. B., Vutikullird, A., & Kaye, E. (2014). Safety and pharmacokinetic profiles of phosphorodiamidate morpholino oligomers with activity against Ebola virus and Marburg virus: Results of two single ascending- dose studies. *Antimicrobial Agents and Chemotherapy, 58*, 6639–6647. https://doi.org/10.1128/AAC.03442-14.

180. Sweiti, H., Ekwunife, O., Jaschinski, T., & Lhachimi, S. K. (2017). Repurposed therapeutic agents targeting the Ebola virus: A systematic review. *Current Therapeutic Research, 84*, 10–21. https://doi.org/10.1016/j.curtheres.2017.01.007.
181. Madariaga, M. G. (2015). Ebola virus disease: A perspective for the United States. *The American Journal of Medicine, 128*(7), 682–691. https://doi.org/10.1016/j.amjmed.2015.01.035.
182. Johnson, J. C., Martinez, O., Honko, A. N., Hensley, L. E., Olinger, G. G., & Basler, C. F. (2014). Pyridinyl imidazole inhibitors of p38 MAP kinase impair viral dendritic cells. *Antiviral Research, 107*, 102–109. https://doi.org/10.1016/j.antiviral.2014.04.014.
183. Dunning, J., Kennedy, S. B., Antierens, A., Whitehead, J., Ciglenecki, I., Carson, G., et al. (2016). Experimental treatment of Ebola virus disease with brincidofovir. *PLoS One, 11*, e0162199. https://doi.org/10.1371/journal.pone.0162199.
184. Yang, S., Xu, M., Lee, E. M., Gorshkov, K., Shiryaev, S. A., He, S., et al. (2018). Emetine inhibits Zika and Ebola virus infections through two molecular mechanisms: Inhibiting viral replication and decreasing viral entry. *Cell Discovery, 4*, 31. https://doi.org/10.1038/s41421-018-0034-1.
185. Kuehne, A., Tiffany, A., Lasry, E., Janssens, M., Besse, C., Okonta, C., et al. (2016). Impact and lessons learned from mass drug administrations of malaria chemoprevention during the Ebola outbreak in Monrovia, Liberia, 2014. *PLoS One, 11*, e0161311. https://doi.org/10.1371/journal.pone.0161311.
186. Geisbert, T. W., Hensley, L. E., Jahrling, P. B., Larsen, T., Geisbert, J. B., Paragas, J., et al. (2003). Treatment of Ebola virus infection with a recombinant inhibitor of factor VIIa/tissue factor: A study in rhesus monkeys. *Lancet, 362*, 1953–1958. https://doi.org/10.1016/S0140-6736(03)15012-X.
187. Olsen, N. J., Schleich, M. A., & Karp, D. R. (2013). Multifaceted effects of hydroxychloroquine in human disease. *Seminars in Arthritis and Rheumatism, 43*, 264–272. https://doi.org/10.1016/j.semarthrit.2013.01.001.
188. Dowall, S. D., Bosworth, A., Watson, R., Bewley, K., Taylor, I., Rayner, E., et al. (2015). Chloroquine inhibited Ebola virus replication in vitro but failed to protect against infection and disease in the in vivo Guinea pig model. *The Journal of General Virology, 96*(12), 3484–3492. https://doi.org/10.1099/jgv.0.000309.
189. Falzarano, D., Safronetz, D., Prescott, J., Marzi, A., Feldmann, F., & Feldmann, H. (2015). Lack of protection against Ebola virus from chloroquine in mice and hamsters. *Emerging Infectious Diseases, 21*(6), 1065–1067. https://doi.org/10.3201/eid2106.150176.
190. Long, J., Wright, E., Molesti, E., Temperton, N., & Barclay, W. (2015). Antiviral therapies against Ebola and other emerging viral diseases using existing medicines that block virus entry. Version 2. *F1000Res, 4*, 30. https://doi.org/10.12688/f1000research.6085.2.
191. Furuta, Y., Takahashi, K., Kuno-Maekawa, M., Sangawa, H., Uehara, S., Kozaki, K., Nomura, N., Egawa, H., & Shiraki, K. (2005). Mechanism of action of T-705 against influenza virus. *Antimicrobial Agents and Chemotherapy, 49*, 981–986. https://doi.org/10.1128/AAC.49.3.981-986.2005.
192. Smee, D. F., Hurst, B. L., Egawa, H., Takahashi, K., Kadota, T., & Furuta, Y. (2009). Intracellular metabolism of favipiravir (T-705) in uninfected and influenza A (H5N1) virus-infected cells. *The Journal of Antimicrobial Chemotherapy, 64*, 741–746. https://doi.org/10.1093/jac/dkp274.
193. Baranovich, T., Wong, S. S., Armstrong, J., Marjuki, H., Webby, R. J., Webster, R. G., & Govorkova, E. A. (2013). T-705 (favipiravir) induces lethal mutagenesis in influenza A H1N1 viruses in vitro. *Journal of Virology, 87*, 3741–3751. https://doi.org/10.1128/JVI.02346-12.
194. Smither, S. J., Eastaugh, L. S., Steward, J. A., Nelson, M., Lenk, R. P., & Lever, M. S. (2014). Post-exposure efficacy of oral T-705 (favipiravir) against inhalational Ebola virus infection in a mouse model. *Antiviral Research, 104*, 153–155. https://doi.org/10.1016/j.antiviral.2014.01.012.
195. Oestereich, L., Lüdtke, A., Wurr, S., Rieger, T., Muñoz-Fontela, C., & Günther, S. (2014). Successful treatment of advanced Ebola virus infection with T-705 (favipiravir) in a small animal model. *Antiviral Research, 105*, 17–21. https://doi.org/10.1016/j.antiviral.2014.02.014.

196. Henß, L., Beck, S., Weidner, T., Biedenkopf, N., Sliva, K., Weber, C., et al. (2016). Suramin is a potent inhibitor of Chikungunya and Ebola virus cell entry. *Virology Journal, 13*, 149. https://doi.org/10.1186/s12985-016-0607-2.
197. Kouznetsova, J., Sun, W., Martínez-Romero, C., Tawa, G., Shinn, P., Chen, C. Z., et al. (2014). Identification of 53 compounds that block Ebola virus-like particle entry via a repurposing screen of approved drugs. *Emerging Microbes Infections, 3*, e84. https://doi.org/10.1038/emi.2014.88.
198. Veljkovic, V., Loiseau, P. M., Figadere, B., Glisic, S., Veljkovic, N., Perovic, V. R., et al. (2015). Virtual screen for repurposing approved and experimental drugs for candidate inhibitors of EBOLA virus infection. *F1000Res, 4*, 34. https://doi.org/10.12688/f1000research.6110.1.
199. Zhao, Z., Martin, C., Fan, R., Bourne, P. E., & Xie, L. (2016). Drug repurposing to target Ebola virus replication and virulence using structural systems pharmacology. *BMC Bioinformatics, 17*, 90. https://doi.org/10.1186/s12859-016-0941-9.
200. Chopra, G., Kaushik, S., Elkin, P. L., & Samudrala, R. (2016). Combating Ebola with repurposed therapeutics using the CANDO platform. *Molecules, 21*(12), E1537. https://doi.org/10.3390/molecules21121537.
201. Paessler, S., Huang, C., Sencanski, M., Veljkovic, N., Perovic, V., Glisic, S., et al. (2018). Ibuprofen as a template molecule for drug design against Ebola virus. *Frontiers Bioscience (Landmark Ed), 23*, 947–953. https://doi.org/10.2741/4627.
202. Mupapa, K., Massamba, M., Kibadi, K., Kuvula, K., Bwaka, A., Kipasa, M., et al. (1999). Treatment of Ebola hemorrhagic fever with blood transfusions from convalescent patients. International Scientific and Technical Committee. *The Journal of Infectious Diseases, 179* (Suppl 1), S18–S23. https://doi.org/10.1086/514298.
203. Dye, J. M., Wu, H., Hooper, J. W., Khurana, S., Kuehne, A. I., Coyle, E. M., et al. (2016). Production of potent fully human polyclonal antibodies against Ebola Zaire virus in transchromosomal cattle. *Scientific Reports, 6*, 24897. https://doi.org/10.1038/srep24897.
204. Basu, A., Mills, D. M., Mitchell, D., Ndungo, E., Williams, J. D., Herbert, A. S., et al. (2015). Novel small molecule entry inhibitors of Ebola virus. *The Journal of Infectious Diseases, 212* (Suppl 2), S425–S434. https://doi.org/10.1093/infdis/jiv223.
205. Cheng, H., Lear-Rooney, C. M., Johansen, L., Varhegyi, E., Chen, Z. W., Olinger, G. G., et al. (2015). Inhibition of Ebola and Marburg virus entry by G protein-coupled receptor antagonists. *Journal of Virology, 89*(19), 9932–9938. https://doi.org/10.1128/JVI.01337-15.
206. McCarthy, S. D., Majchrzak-Kita, B., Racine, T., Kozlowski, H. N., Baker, D. P., Hoenen, T., et al. (2016). A rapid screening assay identifies monotherapy with interferon-ß and combination therapies with nucleoside analogs as effective inhibitors of Ebola virus. *PLoS Neglected Tropical Diseases, 10*(1), e0004364. https://doi.org/10.1371/journal.pntd.0004364.
207. Konde, M. K., Baker, D. P., Traore, F. A., Sow, M. S., Camara, A., Barry, A. A., et al. (2017). Interferon β-1a for the treatment of Ebola virus disease: A historically controlled, single-arm proof-of-concept trial. *PLoS One, 12*(2), e0169255. https://doi.org/10.1371/journal.pone.0169255.
208. Audet, J., Wong, G., Wang, H., Lu, G., Gao, G. F., Kobinger, G., et al. (2014). Molecular characterization of the monoclonal antibodies composing ZMAb: A protective cocktail against Ebola virus. *Scientific Reports, 4*, 6881. https://doi.org/10.1038/srep06881.
209. Duehr, J., Wohlbold, T. J., Oestereich, L., Chromikova, V., Amanat, F., Rajendran, M., et al. (2017). Novel cross-reactive monoclonal antibodies against ebolavirus glycoproteins show protection in a murine challenge model. *Journal of Virology, 91*(16), e00652. https://doi.org/10.1128/JVI.00652-17.
210. Modrof, J., Mühlberger, E., Klenk, H. D., & Becker, S. (2002). Phosphorylation of VP30 impairs Ebola virus transcription. *The Journal of Biological Chemistry, 277*(36), 33099–33104. https://doi.org/10.1074/jbc.M203775200.
211. Kolokoltsov, A. A., Adhikary, S., Garver, J., Johnson, L., Davey, R. A., & Vela, E. M. (2012). Inhibition of Lassa virus and Ebola virus infection in host cells treated with the kinase inhibitors genistein and tyrphostin. *Archives of Virology, 157*(1), 121–127. https://doi.org/10.1007/s00705-011-1115-8.

Prion Diseases: A Concern for Mankind

Syed Beenish Rufai, Ayush Gupta, and Sarman Singh

Abstract

Abnormal prion proteins are known to cause slowly progressive, invariably fatal neurodegenerative diseases in animals and humans. They are characterized by having prolonged incubation period and a clinically silent phase during which the normal cellular prion protein (PrP^c) is misfolded into an abnormal form (PrP^{sc}). This can occur sporadically, genetically or can be acquired by either oral or iatrogenic routes of transmission. Majority of human prion diseases are sporadic in occurrence and are termed as sporadic Creutzfeldt-Jakob disease (sCJD), whereas familial forms include familial or genetic CJD, fatal familial insomnia (FFI) and Gerstmann-Straussler-Scheinker syndrome (GSSS). About 5% of cases of human prion disease are infectious in origin, variant CJD (vCJD), iatrogenic CJD (iCJD) and a now extinct form "Kuru", being the manifestations. Though Prion diseases are rare in humans, being invariably fatal and difficult to diagnose especially in ageing population, they have become an area of active interest. This chapter will focus on various aspects of human prion diseases including epidemiology, clinical features, pathogenesis, treatment and prevention.

Keywords

Prions · Creutzfeldt-Jakob disease · Prion proteins · Transmissible spongiform encephalopathies

---

S. B. Rufai · A. Gupta · S. Singh (✉)
Department of Microbiology, All India Institute of Medical Sciences, Bhopal, India
e-mail: Sarman.singh@gmail.com

S. Hameed, Z. Fatima (eds.), *Pathogenicity and Drug Resistance of Human Pathogens*,
https://doi.org/10.1007/978-981-32-9449-3_14

Introduction

Prions are unique mysterious pathogens responsible for causing fatal neurodegenerative diseases by completely unique mechanisms in humans and animals. Prions seem to behave like other infectious prokaryotic organisms; however, they don't have any of fundamental characteristics and lack (deoxyribonucleic acid) DNA or (ribonucleic acid) RNA as genetic material [1].

During the course of time, the understanding of Prions was proposed as "proteinaceous infectious particles" which causes an abundant normal cellular protein (PrP^{c}, prion-related protein, in which C stands for the cellular form of the protein) in brain to change its shape to the misfolded, abnormal form (PrP^{sc}, in which SC stands for Scrapie, a prion disease of sheep and goats) of proteins with β-pleated sheet structure [2]. Prion diseases manifest as transmissible spongiform encephalopathies (TSEs), which are distinguished by variable incubation periods, typical spongiform changes in the Scrapie related with neuronal loss and a failure to persuade inflammatory response. Prion diseases are relatively sporadic and difficult to transmit; however, the nature of diseases for being progressive with no cure or treatment makes it alarming and concerning for mankind [3].

Discovery of Prions

Description of proteins as discrete biological molecules and their role in cellular processes were documented in the eighteenth century [4]. It was then a century later that other descriptions related to protein were brought explaining protein structure, protein folding and its significance of primary importance [5]. In 1920, Hans Gerhard Creutzfeldt and Alfons Maria Jakob first time explicated human neurological disorder of unknown origin aggravating scientific community and this condition was known as Creutzfeldt-Jakob disease (CJD) [6, 7]. Moreover, in 1938 after shepherds described irrelevant reflections affecting sheep Merino, Cuille and Chelle hypothesized "slow virus" as a causative agent of sheep disease due to its longer incubation period. In 1959, another human neurological disorder among the fore tribe in Papua New Guinea named as Kuru that presented symptoms closely related to CJD was identified [8]. Moreover, inoculation of cerebral spinal fluid (CSF) from Scrapie-infected sheep into healthy revealed the fact that disease is transmissible and diseases Kuru, Scrapie as well as CJD were distinct forms of the same neuropathy [8, 9]. Inactivation of the causative agent of Scrapie by formalin increased the curiosity among the scientific community further to scrutinize the firmness of the "slow virus" [10]. Further, the inactivation of the disease agent was failed using several stringent experiments like using extreme heat, high pressure, isolation from formalin-fixed tissue and high aggregates of ultraviolet radiation (UV) that concluded that the agent might be replicating without the presence of any form of nucleic acid [1]. In 1967, in an open debate the central dogma was challenged by explanation of J.S. Griffith by three different mechanisms which explained how a protein could be infectious and how its spontaneous occurrence can be controlled genetically [11]. The unusual and fascinating concept of protein particles as infectious agents increased curiosity of several research scientists that continued to suggest the infectious nature of protein particles in Scrapie disease [12]. It was in 1982, Prusiner et al. strengthened the "protein as infectious agent" hypothesis by segregating the proteinaceous infectious agents and devastating them by same methods used for protein inactivation [13].

For the discovery of human disease Kuru and showing its similarity to sheep disease Scrapie, in 1976, Carleton Gajdusek shared the Nobel Prize for his contribution. For developing a prion model, in 1997 Stanley Prusiner was awarded Nobel Prize. After the discovery of prion model source of the prion protein was investigated at genetic level to understand which gene codes the messages that is responsible for its translation into proteins which led to the identification of mRNA transcript from the protein sequence than encoded prion protein (PrP) [14, 15].

Prion Proteins

The human PRNP (PRioN) gene stipulates directions for production of protein called prion protein (PrP^c), which is functional in the brain and several other tissues. Prion protein has been suggested to have a role in transportation of copper into cells, neuroprotection and synapses from cell-to-cell communication [16]. Human PRNP gene is positioned on the short arm of chromosome 20, approximately 20 kbp upstream of a gene, which encodes a biochemically and structurally similar protein to the one encoded by this gene [17]. The protein coded by this gene is a membrane glycosyl phosphatidyl inositol-anchored glycoprotein that inclines to aggregate into rod-like structures. The encoded protein contains a highly unstable region of five tandem octapeptide repeats. Mutations present in this gene region as well as elsewhere in this gene have been associated with Creutzfeldt-Jakob disease, fatal familial insomnia (FFI), Gerstmann-Straussler disease (GSS), Huntington disease-like 1, and Kuru.

In all incidents, the pathogenic protein is the host-encoded PrP^C protein with a reformed conformation, called PrP^{sc}. In contrast, propagation of prions, and, more specifically, of PrP^{sc} prions, involves reproduction not of their primary, but of their secondary, tertiary and quaternary arrangements [18] (Fig. 1).

Normal structure of PrP^C protein has large alpha-helical with slight beta-strand content. However, it has been revealed that the PrP^{sc} protein contains more beta-strand content than the normal protein and is underglycosylated. Differences between glycosylation patterns of the PrP^c and PrP^{sc} have been observed, and this variability is believed to contribute to the pathogenesis of prion disease [19]. Both PrP^c and PrP^{sc} have the same group of >50 bi-antennary, tri-antennary and tetra-antennary N-linked oligosaccharides; however, PrP^{sc} exhibits fewer glycans with bisecting GlcNAc residues and more tri-antennary and tetra-antennary oligosaccharides. It is hypothesized that glycosylation increases the stability of proteins, thereby preventing the pathological conformational transition to PrP^{sc} [20]. A glycoform-selective prion formation pathway causing variably protease-sensitive prionopathy (VPSPr) has been reported highlighting the importance of glycosylation in the development, assortment and strain characterization of prions [21]. PrP^{sc} protein structure discloses that it consists of two tangled fibrils, which

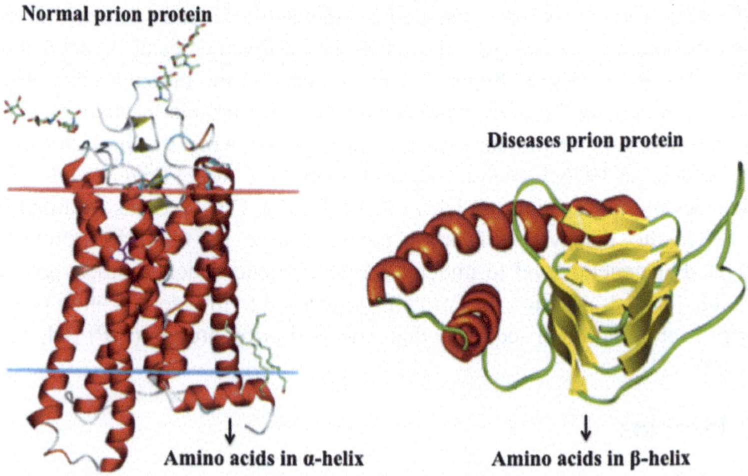

Fig. 1 Showing normal (PrP) and abnormal forms (PrPsc) of prion protein. Normal prion protein has alpha-helix in majority and beta-helix in less than 5% content; however, when the misfolding of prion proteins takes place, more than 50% of alpha-helix structure is converted into beta-helix

have a series of repeated beta-strands (beta-solenoid). The upper and lower steps of beta-solenoids are assumed as start outlets for hydrogen bonding through newfangled PrP^C molecules in many proteins with beta-solenoids; they are blocked to prevent propagation of beta-sheets. Once they are added to the fibrils, these ends would function to enlist additional proteins resulting in increasing length of the chain, and this explains little about the biology of unusual pathogen [22] (Fig. 2).

Pathogenesis of Human Prion Diseases

Pathogenesis of prions is significantly related due to the resistance to protease digestion, which supports the infectious agent to survive its way through the digestive tract and finally reaching to the central nervous system (CNS). Passage of prions from the intestinal mucosa is not yet clear; however, M cells, which are antigen-transporting epithelial cells, are somehow responsible for facilitating prion passage [23]. Thus, entry of the infectious protein particle through oral route penetrates across M cells and Peyer's patches from the mucosa. Dependent on host, prions can also replicate and accumulate in lymph nodes, spleen where they reach through lymphoreticular system (LRS) and has been observed as vCJD in man [24]. However, B cells play an important role for maturation of follicular dendritic cells for amplification and PrP^{sc} accumulation. Other cell types from the periphery are also reported for neuroinvasion in place of follicular dendritic cells in which amplification of prion and PrP^{sc} occur. Prion proteins reach the brain through the vagus nerve or spinal cord either along the peripheral nervous system through LRS and other sites (Fig. 3). There are very low evidences that prions were detected in

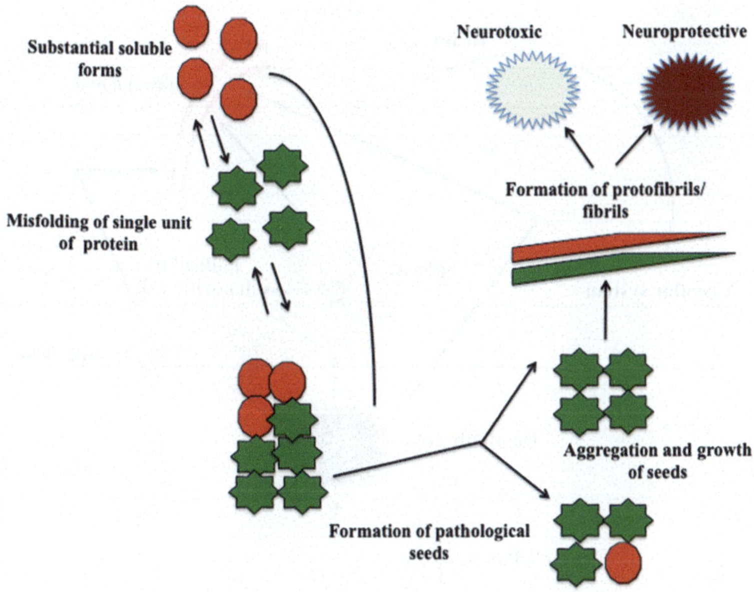

Fig. 2 Mechanism of protein folding in prion diseases

blood of Scrapie-infected hamsters and 1 of 19 sheep that were experimentally used for transfusion of blood developed prion disease [24].

Cell-to-Cell Spread of Prion Proteins

Exact mechanism of cell-to-cell spread of prions within central nervous system (CNS) has remained elusive to delineate. However, experiments involving permissive cell lines have shown it to be mediated by cell-to-cell contact through trans-synaptic spread or tunnelling nanotubes (TNTs) and/or through extracellular compartment involving extracellular vesicles (EVs) [25].

Several studies have shown that prion seeds are actively transported intra-axonally in neurons [26]; however, studies where axonal transport was impaired in vivo suggest that PrP^{sc} can also spread in the brain independent of this route [27]. Current evidence points towards spread of prions through this route though other mechanisms are also involved. Trans-synaptic spread of prions is a plausible mechanism as there is sequential development of pathology in brain in prion diseases. Tunnelling nanotubes (TNTs) are transient cytoplasmic extensions that bridge two distant cells together and have been shown to have a role in the spread of multiple pathogens including bacteria and viruses [28, 29]. These are open-ended channels, composed of F-actin and typically 50–200 nm in diameter, through which transfer of small molecules as well as whole organelles such as mitochondria, lysosomes and different types of vesicles can take place [30]. Both PrP^{c} and PrP^{sc}

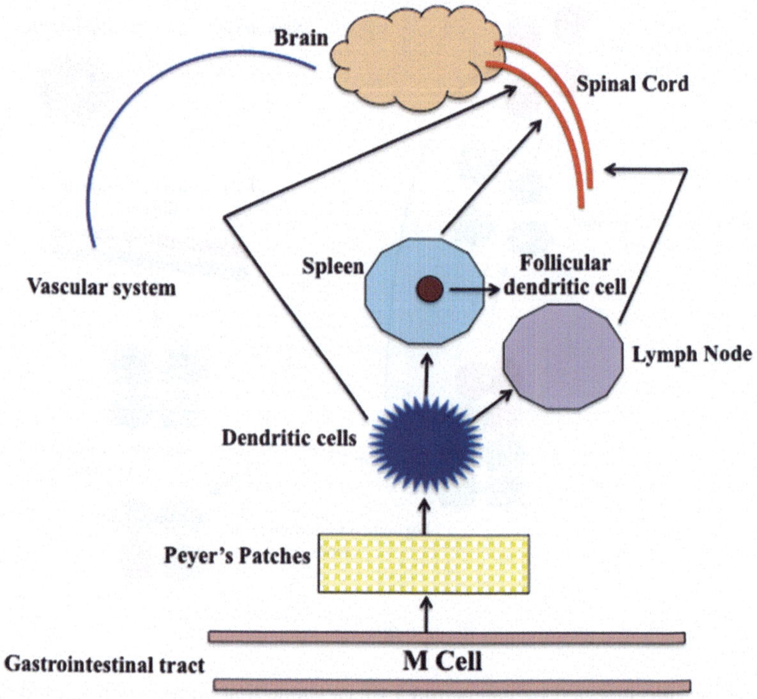

Fig. 3 Probable paths of circulation of ingested prions, from oral route it may penetrate through M cell and reach the enteric nervous system. Moreover, different routes may also be taken by prions depending on the host, which may be through transportation by dendritic cells to the LRS and finally to the brain

have been shown to be transferred through vesicles using TNTs [31]. Furthermore, co-culture experiments involving simultaneous culture of prion-infected cells and non-infected cells have shown that neurons co-cultured with infected bone marrow-derived dendritic cells are infected through TNTs.

Extracellular vesicles (EVs) could represent an important mechanism for extracellular spread of infective prions. Furthermore, the authors isolated 50–90 nm vesicles in the extracellular compartment containing prion proteins and subsequently showed them transmitting prion disease upon inoculation into mice, if they contained infective prion proteins [32]. Even though there is no direct evidence that prions detected in the culture media of both astrocytes and brain cortical neurons are also associated with EVs, both cell lines have been shown to propagate prions as well as release EVs separately [33].

Toxicity of Prion Protein Aggregates

Considerably less is known about the cellular pathways by which prion aggregates cause neurotoxicity. Insight in this aspect will be key to develop therapeutic targets not only against prionopathies but also against other neurodegenerative disorders such as Parkinson's disease, Alzheimer's disease and many others. A core synaptotoxic signalling pathway involving the activation of NMDA and AMPA receptors leading to stimulation of p38 MAPK phosphorylation and collapse of the actin cytoskeleton in dendritic spines has been suggested as a key pathway in causation of neurotoxicity due to prion aggregates [34].

Transmission of Prion Proteins

"Natural" Transmissions

Although prion diseases are not transmitted through air and are thus non-contagious, i.e. doesn't spread through direct contact, transmission through parenteral and oral route has been reported [23]. The skin normally acts as an obstruction against transmission of prions; however, certain research studies have shown that skin lesions characterize a proficient route by which prion infection can be proven [35]. Similarly, lesions of oral and nasal mucosa are known to enhance prion disease susceptibility and increase the efficiency of prion uptake across epithelial surfaces. Kuru disease transmission is recognized to Fore cannibalistic practiced rituals of mourning in which the brain of the dead was eaten, especially by women and children, and is believed to have originated from a case of sCJD [36]. Contaminated foodstuff through injuries in oral cavity from teething and loss of tooth in adolescence has been observed as a cause of vCJD. No evidence of vertical transmission of human prion disease from mother to child has yet been elucidated unlike animal prion diseases [37]. Studies have proven that colostrum and milk from Scrapie-affected ewes have similarly been shown to contain infectious prions, which demonstrates the possible risk of prion; BSE may also be transmitted to humans through consumption of cattle milk or milk products [37]. Several studies have shown natural and man-made materials that are more frequently found in the environments of wild and imprisoned animals can fix prions and may act as vectors for disease transmission. Previous findings indicated that prions could bind various types of soils and remain infectious even after many years attached to soil [36]. Route of transmission of human prion diseases is shown in Table 1.

Iatrogenic/Accidental Transmission (iCJD)

Iatrogenic form of CJD arises from contagious tissue from an infected person generally as a result of a medical procedure and contaminated neurological devices. These invasive procedures include brain biopsy and neurosurgery for the removal of

Table 1 Route of prion disease transmission in humans

Prion disease	Transmission route
Iatrogenic Creutzfeldt-Jakob disease (CJD	Accidental medical exposure to CJD-contaminated tissues or tissue products
Sporadic Creutzfeldt-Jakob disease	Unknown. Theories include somatic mutation or spontaneous conversion of PrPc to PrPsc
Variant Creutzfeldt-Jakob disease	Ingestion of BSE-contaminated food or transfusion of blood or blood products from variant CJD-donor
Kuru	Ritualistic cannibalism

meningioma or cortical undercut, implantation of inefficiently sterilized electrodes in the brain, grafts of dura mater and infected pituitary hormones from prion-infected patient, and unknowingly through corneal transplantation [17].

Geographical Distribution of Human Prion Disease

Typical CJD has been reported from most parts of the world since 1920. The most usual form, sporadic Creutzfeldt-Jakob disease (sCJD), has a global mortality rate of approximately 1 case per million people each year and is accountable for about 85% of all CJD cases globally. The cause of sporadic CJD remains unknown. Familial Creutzfeldt-Jakob disease (fCJD) is known as a consequence of inherited mutations and calculates for 10–15% of the cases from the countries from where it has been reported [38]. The remaining cases are iatrogenic and variant Creutzfeldt-Jakob disease (vCJD). Around 90% of the patients are known to die within the 1st year of symptom onset.

CJD has been reported from the United States at 0.5–1 per million populations with mean age group of 55–75 years. Approximately 90% of cases were of sCJD, followed by 10–15% cases of vCJD and 5% iatrogenic CJD. CJD surveillance studies from the United Kingdom (January 1990–2017) have reported 2359 cases of sCJD, of which 21 cases were alive till December 2017, and 123 definite cases of vCJD with mortality rates of 1.46 from Wales, 1.19 from Scotland, 1.16 from England and 0.85 from Northern Ireland (per million per year) [39]. Japan has also reported CJD cases from year 1998 to 2019 which includes overall 1685 suspected cases, of which 1222 are definite prion cases [40]. Canada reported a total of 974 sCJD cases during the previous decade 1998–2019, whereas China has shown estimates for 1585 cases during the year 2010–2015. With strengthening of the prion surveillance systems in various numbers of countries, reports of CJD cases are updated in database, which is essential for future research and understanding transmission of the unusual disease. Sporadic CJD cases (definite as well as probable) from surveillance reports of different countries have been statistically shown from 1993 to 2013 in Fig. 4.

Occurrence of iCJD has been noted after transplantation of infected cornea [41], dura mater grafts [42], after the use of dura mater material in radiographic embolization procedures [43], use of contaminated neurosurgical instruments [44], use of

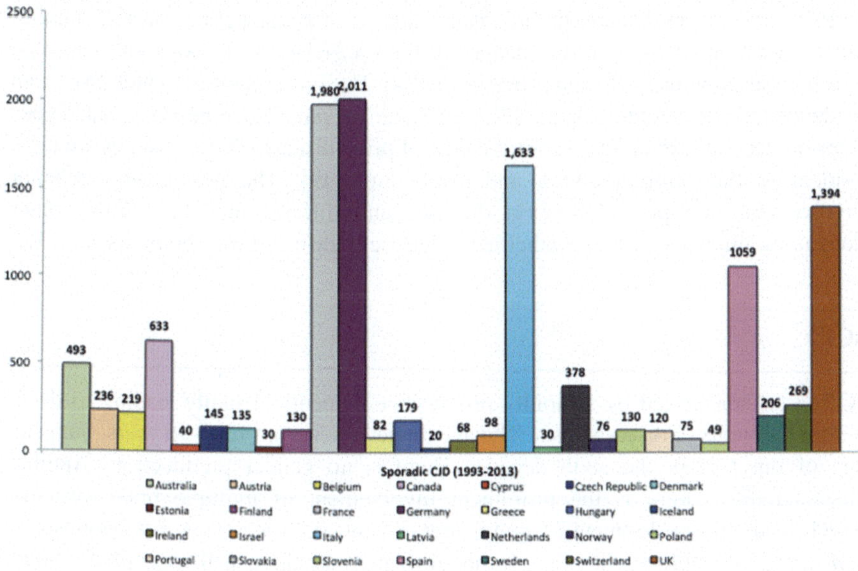

Fig. 4 Total number of sCJD cases from surveillance report (1993–2013). Estimates from Creutzfeldt-Jakob Disease International Surveillance Network formerly known as *Euro CJD*

stereotactic depth electrodes, use of pituitary growth hormones [45] and gonadotropins as well as after transfusion of whole blood and plasma concentrates [21]. Till date, more than 200 cases each of iCJD have been reported after transplantation of dura mater graft and administration of cadaveric human pituitary hormone. Cases due to dura mater graft have been reported mainly from Japan, France and Germany whereas due to growth hormone are from France, the United Kingdom and the United States. Only few cases (≤5) are reported, till date, which are due to corneal transplantation, administration of gonadotrophins, neurosurgical instruments and blood transfusion [46]. The neurodegenerative disease Kuru is thought to have originated in the initial part of this century from remote linguistic tribal groups of Eastern Highlands population of Papua New Guinea populated by 35,000 persons in 169 villages. The total population affected was 17,000 (Fore plus the neighbouring tribes), of which 11,000 were the Fore. After imposing restrictions on cannibalism, decline in Kuru-related disease was observed. Only two Kuru-related deaths were reported from 2003 to 2008, indicating that the epidemic is approaching its end [47].

Prion Diseases: Clinical Features and Molecular Diagnosis

Human prion diseases are differentiated clinically on the basis of age of onset, clinical symptoms, rapidity of their development and duration of symptoms. A definitive diagnosis is made on pathological findings in brain tissue by haematoxylin-eosin (HE) staining and immunohistochemistry that detects pattern of PrP^{sc} deposition. Classic pathological findings of prionopathies on HE staining are vacuolation or spongiform changes in the upper layers of cortex and neuronal loss without inflammation and astrogliosis [48]. Various patterns in which PrP^{sc} can be deposited are synaptic, plaque-like, perivacuolar, punctate or perineuronal. Types of prion deposits are indicative for the type of prion disease. These changes are most evident in the cerebral cortex and cerebellum [49]. The prominent cerebellar involvement is typical for prion diseases and differentiates them from other dementing illnesses such as Alzheimer's disease or diffuse Lewy body disease [31].

sCJD

sCJD is characterized by a rapidly progressive dementia, usually leading to death within 6 months of disease onset. The age range varies from 12 to 96 years, but peak age of onset is in the sixth decade. There is no gender predilection. Another characteristic feature is the prominent involvement of multiple brain systems, which leads to development of motor impairment, such as ataxia, bradykinesia or spasticity, myoclonus and extrapyramidal symptoms along with cognitive deficits and visual loss. The mean survival in sCJD cases is around 6 months, and most patients die within the 1st year from onset. As it affects many areas of the brain, its presentations are protean with a wide variety of symptoms, and in early stages it can mimic many other neurologic or psychiatric conditions [3].

Molecular classification of sCJD is on the basis of genetic polymorphism at codon 129 in the host's PRNP gene and the electrophoretic mobility of PrP^{sc} extracted from the affected brain, after cleavage by protease K and run on a Western blot. Codon 129 can code for either methionine (M) or valine (V). Cleavage of type 1 prions generates a 21-kDA fragment whereas of type 2, generates a 19-kDa fragment. Thus, six molecular subtypes of sJCD can occur: MM1, MV1, VV1, MM2, MV2 and VV2. Each subtype has different clinical and pathologic presentations with varying sensitivity to diagnostic tests. MM1 and MV1 are very similar clinically and pathologically and thus grouped together as a single subtype MM1/MV1. This molecular type is the commonest type found in sCJD, in around 40% cases. Other types are VV2 (15%) presenting as a rapidly progressive ataxia; MV2 (8%), which is slowly progressive than the prior two; and the MM2, which has cortical and thalamic subtypes. The MM2-thalamic subtype has overlapping features with familial fatal insomnia and is referred as "sporadic fatal insomnia". About 6% of sJCD cases have both prion types (1 and 2) and present along a spectrum, the phenotype of which depends on their relative ratio [50].

Establishing diagnosis of sCJD is challenging and is based on a combination of neuropathologic findings and ancillary tests, namely, electroencephalogram (EEG), CSF protein profiling, brain magnetic resonance imaging (MRI) and a new assay which measures the activity of PrP^{sc} aggregates triggered by PrP^{sc} seeds in the presence of recombinant PrP [45]. Characteristic EEG finding is the presence of periodic sharp-wave (often biphasic or triphasic) complexes. They are found in about two-thirds of cases, but appear in advanced clinical stage. In CSF, various biomarkers reflecting neuronal injury can be found in sCJD cases notably 14-3-3, S100β, neuron-specific enolase (NSE) and t-tau protein. Of these, 14-3-3 is the most commonly utilized assay and was reported to have 92% sensitivity and 80% specificity when tested on 1849 patients [52].

On brain MRI, most sensitive findings are diffusion restriction signal in the cerebral cortex or deep grey matter nuclei. The cortical abnormal hyperintensity is also commonly known as "cortical ribboning" [53]. For diagnosing sJCD, brain MRI has a sensitivity ranging from 92% to 96% and a specificity of about 94% [54]. MRI findings also vary according to the molecular subtypes. Basal ganglia hyperintensities are more commonly seen in the MV2, VV2 and MM1 subtypes, whereas abnormal "cortical ribboning" signal is more common in the VV1, MM2 and MV1 subtypes [55]. Diffusion-weighted images are more sensitive than the fluid-attenuated inversion recovery (FLAIR) sequence in detecting the hyperintensity abnormalities in sJCD [54]. A definite diagnosis for sJCD relies on a neuropathologic diagnosis along with identification of PrP^{sc} via immunohistochemistry or Western blot [48]; however, it is commonly available after postmortem. Thus several clinic-diagnostic criteria have been proposed for antemortem diagnosis, of which notable are University of California, San Francisco (UCSF), and European MRI-CJD Consortium criteria. Both criteria have the same clinical findings, which include the following: myoclonus, visual disturbance, cerebellar signs, pyramidal/extrapyramidal signs and akinetic mutism. In addition, UCSF criteria also include presence of focal cortical signs (e.g. neglect, aphasia, acalculia, apraxia). Diagnostic tests, which are common in both criteria, are presence of either typical EEG findings (periodic sharp-wave complexes) or suggestive MRI findings. In European consortium criteria, in addition, elevated levels of 14-3-3 proteins in CSF and positive real-time quaking-induced conversion assay in CSF or other tissues are also included in diagnostic criteria [56].

In recent years, a relatively new test, the real-time quaking-induced conversion assay (RT-QuIC) has been described. In this assay, sample-containing PrP^{sc} is incubated with a substrate containing normal PrP^{c}. With continuous shaking, the PrP^{sc} in samples come in contact with PrP^{C}, converting it into PrP^{sc}, which aggregate into amyloid fibrils. The PrP^{sc} amyloid fibrils can be detected using thioflavin T, which binds amyloid and emits fluorescent signal [51]. Various samples, which are suitable for this technique, are the brain, CSF, olfactory mucosa and skin. The sensitivity of RT-QuIC in sJCD in CSF samples generally ranges around 80% with a specificity of about 98%. Thus, a negative test does not exclude disease, but a positive test in the appropriate clinical context has great diagnostic value [57].

A novel form of sporadic Creutzfeldt-Jakob disease was described in 2008, called variably protease-sensitive prionopathy (VPSPr). Till date, 39 cases have been reported [58, 59]. It is characterized clinically by presence of psychiatric, dysphasic or cognitive symptoms initially followed by progressive parkinsonism or ataxia. The age of onset is usually in the late 60s–70s, with median disease duration of 2 years. The lack of gene mutations establishes VPSPr as a sporadic form of human prion disease. Like sCJD, VPSPr affects patients harbouring any of the three genotypes, MM, MV and VV, at the codon 129, with VPSPr VV accounting for 65% of all

VPSPr cases. EEG and CSF biomarkers are often non-diagnostic, with MRI just showing atrophy. Neuropathology comprises of moderate spongiform degeneration, PrP amyloid mini-plaques and a target-like or plaque-like PrP deposition [60].

gCJD

gCJDs are due to autosomal dominant mutations in the PRNP gene, usually with high, but not always 100%, penetrance. Most common type of mutations are missense mutations, although some are insertions or stop codons [61]. Many patients, who are found to have genetic prion disease, have no known family history of prion disease [62]. There are three clinical phenotypes of genetic prion diseases described: hereditary CJD, GSS and FFI. Diagnosis of genetic prion disease can be difficult as there are variations in the range of age of onset, duration of symptoms, type of symptoms and neuropathologic features.

fCJD is similar to sporadic CJD as it presents as a rapidly progressive dementia with ataxia and other motor features with similar MRI and neuropathologic features however, in a younger age group, i.e. between 30 and 55 years [3]. GSS typically presents as a slowly progressive ataxia or motor disorder (e.g. Parkinson's disease) with late-onset dementia. Median age of onset is in the 50s; however, age spectrum may increase on both ends. Neuropathologic findings of GSS are typical of other prion diseases except that patients have widespread amyloid plaques composed of densely aggregated PrPsc, mainly in the cerebellum. FFI is a rare form associated with a single PRNP point mutation, at codon 129. Onset is typically in the late 40s, and the average survival is about 18 months. Patients usually present with severe progressive insomnia over several months, which is followed by feature of dysautonomia such as tachycardia, hyperhidrosis and hyperpyrexia. Motor and cognitive manifestations tend to occur late in the course of the disease. On neuropathologic examination, disease is characterized by predominant involvement of the thalamus [63]. Families with fatal insomnia have been reported from Italy, the United Kingdom, Finland, the United States, China and Japan [64].

Acquired CJD

Prion diseases can also be acquired orally by ingestion; intravenously by pituitary extracts, transfusion of blood and blood products; and by implantation of tissues such as the cornea and dura mater that contain abnormal PrP^{sc} protein. Such infectiously transmitted cases constitute only about 1% of total reported cases. First reports of human-to-human transmission through oral route were known in a now-obsolete manifestation termed as "Kuru". Evidence point towards ingestion of meat from cattle suffering from bovine spongiform encephalopathy (BSE), a prion disease of ruminants.

Kuru

Kuru is remembered as the largest known outbreak of human-to-human transmission of a prion disease. It typically manifested insidiously as a prodrome of headache and aching limbs, which lasted for several months. This prodrome was followed by development of progressive neurological disease with cerebellar ataxia, tremors and involuntary movements (choreoathetosis, coarse fasciculations and myoclonic jerks) as cardinal features. This was followed by death due to malnutrition and intercurrent pneumonia [65].

Neuropathologic examination of brains of patients suffering from Kuru showed neuronal loss, astrogliosis and accumulation of PrP^{sc}. The pathological hallmark of the disease was the presence of amyloid plaques, predominantly involving cerebellar tissue [66]. People who suffered from Kuru also showed a high incidence of MM homozygosity at polymorphic codon 129 of the PRNP gene, whereas the presence of MV heterozygosity conferred a protective effect.

vCJD

It is the only form of human prionopathies known to be transmitted directly from animals to humans. vCJD can be differentiated from other prionopathies in its epidemiological, clinical and pathological features. Patients with this form are usually younger (mean age of onset is 26 years), have longer course of illness (about 14 months) and usually present with sensory disturbances such as pain in limbs, cold sensation and paraesthesia involving face and limbs and psychiatric manifestations which frequently include dysphoria, withdrawal, anxiety, insomnia and irritability. As the disease progresses, dysarthria and ataxia become the most prominent symptoms [67]. The median duration of disease is also longer about 14 months.

The neuropathological examination is also characteristic in having predominant cerebellar involvement and the presence of "florid" amyloid plaques. vCJD cases also show typical spongiform changes, neuronal loss and astrogliosis in the cortex, basal ganglia and thalamus along with the cerebellum. PrP^{sc} deposition follows a characteristic pattern, infrequently seen in other human prionopathies, which shows dense eosinophilic centre and pale periphery. These are surrounded by vacuoles in the neuropil arranged around the plaque-like petals on a flower, hence termed as "florid plaques" [68]. Another unique feature of vCJD is the presence of PrP^{sc} in lymphoreticular system, including the spleen, lymph nodes, tonsils and appendix [42]. As PrP^{sc} can be detected in these tissues by immunohistochemistry and Western blot, tonsillar biopsy has been suggested as a diagnostic method [69].

On EEG, periodic sharp-wave complexes are rarely found, and CSF biomarkers are less sensitive, unlike sCJD. MRI often shows thalamic involvement, and characteristically, the pulvinar is seen brighter than the anterior putamen on T2-weighted and DWI sequence of images. This is termed as "pulvinar sign" and is present in about 75% cases of vCJD [42]. However, this may be detected infrequently in sporadic form also. Patients with this form also have a distinct mobility pattern of PrP^{sc} on electrophoresis [70]. Almost all patients of variant CJD have been found homozygous for methionine at polymorphic codon 129. However, it is observed that M/V heterozygosity at codon 129 may delay the appearance of clinical symptoms. This may lead to a second wave of cases that are heterozygous at codon 129 [71].

The epidemic of vCJD has been successfully abated, in part due to cessation of the BSE epidemic among cattle. However, vCJD PrP^{sc} has been identified by immunohistochemistry in appendices of residents of the United Kingdom [72]. Thus, carriers for vCJD are present in UK population; however, it is not known whether they will eventually develop the disease or be capable of transmitting it [46].

iCJD

Incubation period of *i*CJD is more than 10 years after the transplantation or administration of contaminated product. Presenting clinical symptoms of iatrogenic cases indicate cerebellar involvement, i.e. ataxia, tremors and myoclonus [73], except in those due to transfusion of blood and blood products. In such cases, clinical presentation is that of psychiatric illness resembling vCJD [21]. Neuropathologically, cases due to dura mater present as both plaque type and non-plaque type resembling sCJD [74], whereas those due to growth hormone show predominantly plaque type of deposition [75]. Overall, cases due to iatrogenic transmission are very rare; however, due control measures should be taken especially for instruments which are used in neurosurgery.

Treatment

All prion diseases are incurable leading to fatality in all cases. As prion diseases are rare, there are multiple issues for conducting well-designed clinical studies, namely, lack of adequate number of patients for randomization, low funding opportunities, limited sample sizes leading to publication as case reports or series and publication bias towards studies with successful results [38]. In humans, till date no therapeutic trial has shown beneficial results. Various drugs which have been studied are quinacrine [3, 76], amantadine [77], acyclovir [78], doxycycline and flupirtine [79]. Of these, only flupirtine has shown beneficial effect on decline in cognitive function without improving the survival time.

Various approaches have been tried which involve inhibitors preventing conversion of PrP^{c} to PrP^{sc}; cellular pathways engaged in control of misfolded proteins; molecules which promote lysosomal degradation and autophagy; active and passive immunization against prions; and targeting peripheral replication, thereby preventing neuroinvasion [38]. Majority of these have been studied only in vitro against prions propagating either in cell lines or in animal models. Several of these approaches have abolished Scrapie infection in cell lines or delayed the onset of clinical illness in mice when administered prior to inoculation with prion proteins. Thus, in the absence of effective therapy, medical care for patients is essentially supportive and palliative.

Vaccination

The main obstacle for development of efficient vaccine against prionopathies is tolerance to T cells. Due to the expression of PrP^c in host cells, the host tolerates it, which results in deterrence of host cell-mediated and humoral immune responses to both PrP^c and PrP^{sc} proteins [80]. Several studies have also reported generation of anti-PrP antibodies towards the clinical phase of disease [81]. Recently, concept of cell-based immunotherapy by sensitizing CD4+ T cells can overwhelm host tolerance towards PrP wild-type expressing hosts [80]. DNA-based vaccines have also been verified incorporating immunization with cDNA encoding for heterologous (human) PrP fused to either a stimulatory T-cell epitope [82]. In silico methods have also been used to select a non-mammalian epitope bacterial succinylarginine dihydrolase with related arrangement to that predictable by the anti-PrP monoclonal antibody 6H4. Immunized mice with succinylarginine dihydrolase elicited anti-prion Abs, and attenuated prion disease in mice considerably delayed survival times (~7 to 10%) after intraperitoneal prion exposure [83]. These sophisticated advances acknowledge the cellular strategies, which affect the induction of defensive anti-PrP antibody responses to be studied. However, such approaches are improbable to be translated into clinical applications, as they are presently too overpriced and precisely challenging for widespread use.

Infection Control and Prevention

Concerns have been raised regarding the reuse of neurosurgical instruments or those used in performing brain biopsy on patients who have been subsequently found to have CJD. To address it, the Joint Commission has issued a sentinel event alert concerning this risk (Joint Commission 2001). Thus, medical institutions are advised to have a tracking system that will simplify recall of critical or semi-critical devices used on high-risk tissue and high-risk patients along with the date of use, procedure performed and surgeon's name [84]. Dentists working on patients that may have exposure to higher infectivity tissues like maxillectomy, cranial nerve exposure and oral surgical cancer ways also relating a neurosurgical approach would result in exposure to tissue of higher infectivity (potential brain tissue, central nerve exposure), and such work must be practised at a facility with healthcare worker with proper knowledge of CJD-specific infection prevention and control procedures. Though body fluids, body secretions and small amounts of blood are known to have low possibility for CJD transmission, large volumes of transfusion blood are known for transmission of vCJD [85]. Thus, the policy was established in the United Sates for the people not to donate their blood if they had resided for more than couple of months in a country where BSE is common.

However, recommendations are that CSF specimens should be considered potentially infectious for human prion disease and thus wearing of gloves and lab coats is highly recommended while working on such samples. Due to the resistance of prion proteins to inactivation by aldehydes and alcohols, brain specimens are recommended to be fixed in 4% formaldehyde solution (10% formal saline), followed by immersion in formic acid (>96%) for 1 h. For transportation of funerals post death of a CJD patient, the body is recommended to be packed in a leak-proof pack lined with absorbent material to absorb any leakage, and decontamination guidelines should be followed if any leakage is suspected [84].

Though lot of research has been done in the past decade on prion diseases, researchers are still enthusiastically working to decipher novel targets for treatment and vaccination. Moreover, uncovering prions with distinct novel properties and roles in the research frontier still lies ahead.

References

1. Prusiner, S. B. (1998). Prions. *Proceedings of the National Academy of Sciences of the United States of America, 95*(23), 13363–13383.
2. Das, A. S., & Zou, W.-Q. (2016). Prions: Beyond a single protein. *Clinical Microbiology Reviews, 29*(3), 633–658.
3. Geschwind, M. D. (2015). Prion diseases. *Continuum (Minneap Minn), 21*(6 neuroinfectious dISEASE), 1612–1638.
4. Moore, R. A., Faris, R., & Priola, S. A. (2015). Proteomics applications in prion biology and structure. *Expert Review of Proteomics, 12*(2), 171–184.
5. Zabel, M. D., & Reid, C. (2015). A brief history of prions. *Pathogens and Disease* [Internet]. cited 2019 April 12];*73*(9). Available from: https://www.ncbi.nlm.nih.gov/pmc/articles/PMC4626585/.
6. Creutzfeldt, H. G. (1920). Über eine eigenartige herdförmige erkrankung des zentralnervensystems (Vorläufige mitteilung). *Zeitschrift für die gesamte Neurologie und Psychiatrie, 57*(1), 1–18.
7. Jakob, A. (1921). Über eigenartige erkrankungen des zentralnervensystems mit bemerkenswertem anatomischen befunde. *Zeitschrift für die gesamte Neurologie und Psychiatrie, 64*(1), 147–228.
8. Gajdusek, D. C., & Zigas, V. (1959). Kuru; clinical, pathological and epidemiological study of an acute progressive degenerative disease of the central nervous system among natives of the Eastern Highlands of New Guinea. *The American Journal of Medicine, 26*(3), 442–469.
9. Plummer, P. J. (1946). Scrapie-a disease of sheep: A review of the literature. *Canadian Journal of Comparative Medicine and Veterinary Science, 10*(2), 49–54.
10. Gordon, W. S. (1946). Advances in veterinary research. *The Veterinary Record, 58*(47), 516–525.
11. Griffith, J. S. (1967). Self-replication and Scrapie. *Nature, 215*(5105), 1043–1044.
12. Hunter, G. D., & Millson, G. C. (1964). Studies on the heat stability and chromatographic behaviour of the Scrapie agent. *Journal of General Microbiology, 37*, 251–258.
13. Bolton, D. C., McKinley, M. P., & Prusiner, S. B. (1982). Identification of a protein that purifies with the Scrapie prion. *Science, 218*(4579), 1309–1311.
14. Chesebro, B., Race, R., Wehrly, K., Nishio, J., Bloom, M., Lechner, D., et al. (1985). Identification of Scrapie prion protein-specific mRNA in Scrapie-infected and uninfected brain. *Nature, 315*(6017), 331–333.
15. Locht, C., Chesebro, B., Race, R., & Keith, J. M. (1986). Molecular cloning and complete sequence of prion protein cDNA from mouse brain infected with the Scrapie agent. *Proceedings of the National Academy of Sciences of the United States of America, 83*(17), 6372–6376.
16. Kupfer, L., Hinrichs, W., & Groschup, M. H. (2009). Prion protein misfolding. *Current Molecular Medicine, 9*(7), 826.

17. Wille, H., & Requena, J. R.. (2018). The structure of PrPSc prions. *Pathogens* [Internet]. [cited 2019 April 29];7(1). Available from: https://www.ncbi.nlm.nih.gov/pmc/articles/PMC5874746/.
18. Bongianni, M., Orrù, C., Groveman, B. R., Sacchetto, L., Fiorini, M., Tonoli, G., et al. (2017). Diagnosis of human prion disease using real-time quaking-induced conversion testing of olfactory mucosa and cerebrospinal fluid samples. *JAMA Neurology, 74*(2), 155–162.
19. Xanthopoulos, K., Polymenidou, M., Bellworthy, S. J., Benestad, S. L., & Sklaviadis, T. (2009). Species and strain glycosylation patterns of PrPSc. *PLoS One, 4*(5), e5633.
20. Rudd, P. M., Endo, T., Colominas, C., Groth, D., Wheeler, S. F., Harvey, D. J., et al. (1999). Glycosylation differences between the normal and pathogenic prion protein isoforms. *Proceedings of the National Academy of Sciences of the United States of America, 96*(23), 13044–13049.
21. Zou, S., Fang, C. T., & Schonberger, L. B. (2008). Transfusion transmission of human prion diseases. *Transfusion Medicine Reviews, 22*(1), 58–69.
22. Vázquez-Fernández, E., Vos, M. R., Afanasyev, P., Cebey, L., Sevillano, A. M., Vidal, E., et al. (2016). The structural architecture of an infectious mammalian prion using electron cryomicroscopy.. Supattapone S, editor. *PLOS Pathogens, 12*(9), e1005835.
23. Weissmann, C., Enari, M., Klöhn, P.-C., Rossi, D., & Flechsig, E. (2002). Transmission of prions. *Proceedings of the National Academy of Sciences of the United States of America, 99* (Suppl 4), 16378–16383.
24. Prinz, M., Montrasio, F., Klein, M. A., Schwarz, P., Priller, J., Odermatt, B., et al. (2002). Lymph nodal prion replication and neuroinvasion in mice devoid of follicular dendritic cells. *PNAS, 99*(2), 919–924.
25. Vilette, D., Courte, J., Peyrin, J. M., Coudert, L., Schaeffer, L., Andréoletti, O., et al. (2018). Cellular mechanisms responsible for cell-to-cell spreading of prions. *Cellular and Molecular Life Sciences, 75*(14), 2557–2574.
26. Magalhães, A. C., Baron, G. S., Lee, K. S., Steele-Mortimer, O., Dorward, D., Prado, M. A. M., et al. (2005). Uptake and neuritic transport of Scrapie prion protein coincident with infection of neuronal cells. *The Journal of Neuroscience, 25*(21), 5207–5216.
27. Kratzel, C., Krüger, D., & Beekes, M. (2007). Prion propagation in a nerve conduit model containing segments devoid of axons. *The Journal of General Virology, 88*(Pt 12), 3479–3485.
28. Onfelt, B., Nedvetzki, S., Benninger, R. K. P., Purbhoo, M. A., Sowinski, S., Hume, A. N., et al. (2006). Structurally distinct membrane nanotubes between human macrophages support long-distance vesicular traffic or surfing of bacteria. *Journal of Immunology, 177*(12), 8476–8483.
29. Sowinski, S., Jolly, C., Berninghausen, O., Purbhoo, M. A., Chauveau, A., Köhler, K., et al. (2008). Membrane nanotubes physically connect T cells over long distances presenting a novel route for HIV-1 transmission. *Nature Cell Biology, 10*(2), 211–219.
30. Austefjord, M. W., Gerdes, H.-H., & Wang, X. (2014). Tunneling nanotubes: Diversity in morphology and structure. *Communicative & Integrative Biology, 7*(1), e27934.
31. Schoch, G., Seeger, H., Bogousslavsky, J., Tolnay, M., Janzer, R. C., Aguzzi, A., et al. (2006). Analysis of prion strains by PrPSc profiling in sporadic Creutzfeldt-Jakob disease. *PLoS Medicine, 3*(2), e14.
32. Castro-Seoane, R., Hummerich, H., Sweeting, T., Tattum, M. H., Linehan, J. M., Fernandez de Marco, M., et al. (2012). Plasmacytoid dendritic cells sequester high prion titres at early stages of prion infection. *PLoS Pathogens, 8*(2), e1002538.
33. Hajj, G. N. M., Arantes, C. P., Dias, M. V. S., Roffé, M., Costa-Silva, B., Lopes, M. H., et al. (2013). The unconventional secretion of stress-inducible protein 1 by a heterogeneous population of extracellular vesicles. *Cellular and Molecular Life Sciences, 70*(17), 3211–3227.
34. Fang, C., Wu, B., Le, N. T. T., Imberdis, T., Mercer, R. C. C., & Harris, D. A. (2018). Prions activate a p38 MAPK synaptotoxic signaling pathway. *PLoS Pathogens, 14*(9), e1007283.
35. Bartz, J. C., Kincaid, A. E., & Bessen, R. A. (2003). Rapid prion neuroinvasion following tongue infection. *Journal of Virology, 77*(1), 583–591.

36. Pritzkow, S., Morales, R., Moda, F., Khan, U., Telling, G. C., Hoover, E., et al. (2015). Grass plants bind, retain, uptake, and transport infectious prions. *Cell Reports, 11*(8), 1168–1175.
37. Xiao, X., Miravalle, L., Yuan, J., McGeehan, J., Dong, Z., Wyza, R., et al. (2009). Failure to detect the presence of prions in the uterine and gestational tissues from a Gravida with Creutzfeldt-Jakob disease. *The American Journal of Pathology, 174*(5), 1602–1608.
38. Aguzzi, A., Lakkaraju, A. K. K., & Frontzek, K. (2018). Toward therapy of human prion diseases. *Annual Review of Pharmacology and Toxicology, 58*, 331–351.
39. report26.pdf [Internet]. [cited 2019 Apr 14]. Available from: https://www.cjd.ed.ac.uk/sites/default/files/report26.pdf.
40. Nozaki, I., Hamaguchi, T., Sanjo, N., Noguchi-Shinohara, M., Sakai, K., Nakamura, Y., et al. (2010). Prospective 10-year surveillance of human prion diseases in Japan. *Brain, 133*(10), 3043–3057.
41. Brown, K., & Mastrianni, J. A. (2010). The prion diseases. *Journal of Geriatric Psychiatry and Neurology, 23*(4), 277–298.
42. Heath, C. A., Barker, R. A., Esmonde, T. F. G., Harvey, P., Roberts, R., Trend, P., et al. (2006). Dura mater-associated Creutzfeldt-Jakob disease: Experience from surveillance in the UK. *Journal of Neurology, Neurosurgery, and Psychiatry, 77*(7), 880–882.
43. Antoine, J. C., Michel, D., Bertholon, P., Mosnier, J. F., Laplanche, J. L., Beaudry, P., et al. (1997). Creutzfeldt-Jakob disease after extracranial dura mater embolization for a nasopharyngeal angiofibroma. *Neurology, 48*(5), 1451–1453.
44. Will, R. G., & Matthews, W. B. (1982). Evidence for case-to-case transmission of Creutzfeldt-Jakob disease. *Journal of Neurology, Neurosurgery, and Psychiatry, 45*(3), 235–238.
45. Billette de Villemeur, T., Deslys, J. P., Pradel, A., Soubrié, C., Alpérovitch, A., Tardieu, M., et al. (1996). Creutzfeldt-Jakob disease from contaminated growth hormone extracts in France. *Neurology, 47*(3), 690–695.
46. Tee, B. L., Longoria Ibarrola, E. M., & Geschwind, M. D. (2018). Prion diseases. *Neurologic Clinics, 36*(4), 865–897.
47. McGuire, L. I., Poleggi, A., Poggiolini, I., Suardi, S., Grznarova, K., Shi, S., et al. (2016). Cerebrospinal fluid real-time quaking-induced conversion is a robust and reliable test for sporadic creutzfeldt-jakob disease: An international study. *Annals of Neurology, 80*(1), 160–165.
48. Kretzschmar, H. A., Ironside, J. W., DeArmond, S. J., & Tateishi, J. (1996). Diagnostic criteria for sporadic Creutzfeldt-Jakob disease. *Archives of Neurology, 53*(9), 913–920.
49. Glatzel, M., & Aguzzi, A. (n.d.). Transmissible spongiform encephalopathies. In *Manual of clinical microbiology* (11th ed., pp. 1859–66). ASM Press.
50. Puoti, G., Bizzi, A., Forloni, G., Safar, J. G., Tagliavini, F., & Gambetti, P. (2012). Sporadic human prion diseases: Molecular insights and diagnosis. *Lancet Neurology, 11*(7), 618–628.
51. Atarashi, R., Wilham, J. M., Christensen, L., Hughson, A. G., Moore, R. A., Johnson, L. M., et al. (2008). Simplified ultrasensitive prion detection by recombinant PrP conversion with shaking. *Nature Methods, 5*(3), 211–212.
52. Muayqil, T., Gronseth, G., & Camicioli, R. (2012). Evidence-based guideline: Diagnostic accuracy of CSF 14-3-3 protein in sporadic Creutzfeldt-Jakob disease: Report of the guideline development subcommittee of the American Academy of Neurology. *Neurology, 79*(14), 1499–1506.
53. Bahn, M. M., & Parchi, P. (1999). Abnormal diffusion-weighted magnetic resonance images in Creutzfeldt-Jakob disease. *Archives of Neurology, 56*(5), 577–583.
54. Shiga, Y., Miyazawa, K., Sato, S., Fukushima, R., Shibuya, S., Sato, Y., et al. (2004). Diffusion-weighted MRI abnormalities as an early diagnostic marker for Creutzfeldt-Jakob disease. *Neurology, 63*(3), 443–449.
55. Meissner, B., Kallenberg, K., Sanchez-Juan, P., Collie, D., Summers, D. M., Almonti, S., et al. (2009). MRI lesion profiles in sporadic Creutzfeldt-Jakob disease. *Neurology, 72*(23), 1994–2001.

56. Zerr, I., Kallenberg, K., Summers, D. M., Romero, C., Taratuto, A., Heinemann, U., et al. (2009). Updated clinical diagnostic criteria for sporadic Creutzfeldt-Jakob disease. *Brain, 132* (Pt 10), 2659–2668.
57. Franceschini, A., Baiardi, S., Hughson, A. G., McKenzie, N., Moda, F., Rossi, M., et al. (2017). High diagnostic value of second generation CSF RT-QuIC across the wide spectrum of CJD prions. *Scientific Reports, 7*(1), 10655.
58. Aizpurua, M., Selvackadunco, S., Yull, H., Kipps, C. M., Ironside, J. W., & Bodi, I. (2019). Variably protease-sensitive prionopathy mimicking frontotemporal dementia. *Neuropathology, 39*(2), 135–140.
59. Vicente-Pascual, M., Rossi, M., Gámez, J., Lladó, A., Valls, J., Grau-Rivera, O., et al. (2018). Variably protease-sensitive prionopathy presenting within ALS/FTD spectrum. *Annals of Clinical Translational Neurology, 5*(10), 1297–1302.
60. Notari, S., Appleby, B. S., & Gambetti, P. (2018). Variably protease-sensitive prionopathy. *Handbook of Clinical Neurology, 153*, 175–190.
61. Takada, L. T., Kim, M.-O., Cleveland, R. W., Wong, K., Forner, S. A., Gala, I. I., et al. (2017). Genetic prion disease: Experience of a rapidly progressive dementia center in the United States and a review of the literature. *American Journal of Medical Genetics. Part B, Neuropsychiatric Genetics, 174*(1), 36–69.
62. Kovács, G. G., Puopolo, M., Ladogana, A., Pocchiari, M., Budka, H., van Duijn, C., et al. (2005). Genetic prion disease: The EUROCJD experience. *Human Genetics, 118*(2), 166–174.
63. Manetto, V., Medori, R., Cortelli, P., Montagna, P., Tinuper, P., Baruzzi, A., et al. (1992). Fatal familial insomnia: Clinical and pathologic study of five new cases. *Neurology, 42*(2), 312–319.
64. Montagna, P., Gambetti, P., Cortelli, P., & Lugaresi, E. (2003). Familial and sporadic fatal insomnia. *Lancet Neurology, 2*(3), 167–176.
65. Liberski, P. P., Gajos, A., Sikorska, B., & Lindenbaum, S. (2019). Kuru, the first human prion disease. *Viruses, 11*(3), 232.
66. McLean, C. A. (2008). Review. The neuropathology of Kuru and variant Creutzfeldt-Jakob disease. *Philosophical Transactions of the Royal Society B Biological Science, 363*(1510), 3685–3687.
67. Spencer, M. D., Knight, R. S. G., & Will, R. G. (2002). First hundred cases of variant Creutzfeldt-Jakob disease: Retrospective case note review of early psychiatric and neurological features. *BMJ, 324*(7352), 1479–1482.
68. Sikorska, B., Liberski, P. P., Sobów, T., Budka, H., & Ironside, J. W. (2009). Ultrastructural study of florid plaques in variant Creutzfeldt-Jakob disease: A comparison with amyloid plaques in Kuru, sporadic Creutzfeldt-Jakob disease and Gerstmann-Sträussler-Scheinker disease. *Neuropathology and Applied Neurobiology, 35*(1), 46–59.
69. Hill, A. F., Joiner, S., Wadsworth, J. D. F., Sidle, K. C. L., Bell, J. E., Budka, H., et al. (2003). Molecular classification of sporadic Creutzfeldt-Jakob disease. *Brain, 126*(Pt 6), 1333–1346.
70. Collinge, J., Whitfield, J., McKintosh, E., Beck, J., Mead, S., Thomas, D. J., et al. (2006). Kuru in the 21st century--an acquired human prion disease with very long incubation periods. *Lancet, 367*(9528), 2068–2074.
71. Garske, T., & Ghani, A. C. (2010). Uncertainty in the tail of the variant Creutzfeldt-Jakob disease epidemic in the UK. *PLoS One, 5*(12), e15626.
72. Hilton, D. A., Ghani, A. C., Conyers, L., Edwards, P., McCardle, L., Ritchie, D., et al. (2004). Prevalence of lymphoreticular prion protein accumulation in UK tissue samples. *The Journal of Pathology, 203*(3), 733–739.
73. Kobayashi, A., Kitamoto, T., & Mizusawa, H. (2018). Iatrogenic Creutzfeldt-Jakob disease. *Handbook of Clinical Neurology, 153*, 207–218.
74. Shimizu, S., Hoshi, K., Muramoto, T., Homma, M., Ironside, J. W., Kuzuhara, S., et al. (1999). Creutzfeldt-Jakob disease with florid-type plaques after cadaveric dura mater grafting. *Archives of Neurology, 56*(3), 357–362.

75. Delisle, M. B., Fabre, N., Rochiccioli, P., Doerr-Schott, J., Rumeau, J. L., & Bes, A. (1993). Creutzfeldt-Jakob disease after treatment with human extracted growth hormone.A clinico-pathological study. *Revue Neurologique (Paris), 149*(10), 524–527.
76. Love, R. (2001). Old drugs to treat new variant Creutzfeldt-Jakob disease. *Lancet, 358*(9281), 563.
77. Terzano, M. G., Montanari, E., Calzetti, S., Mancia, D., & Lechi, A. (1983). The effect of amantadine on arousal and EEG patterns in Creutzfeldt-Jakob disease. *Archives of Neurology, 40*(9), 555–559.
78. David, A. S., Grant, R., & Ballantyne, J. P. (1984). Unsuccessful treatment of Creutzfeldt-Jakob disease with acyclovir. *Lancet, 1*(8375), 512–513.
79. Otto, M., Cepek, L., Ratzka, P., Doehlinger, S., Boekhoff, I., Wiltfang, J., et al. (2004). Efficacy of flupirtine on cognitive function in patients with CJD: A double-blind study. *Neurology, 62* (5), 714–718.
80. Mabbott, N. A. (2015). Prospects for safe and effective vaccines against prion diseases. *Expert Review of Vaccines, 14*(1), 1–4.
81. Sassa, Y., Kataoka, N., Inoshima, Y., & Ishiguro, N. (2010). Anti-PrP antibodies detected at terminal stage of prion-affected mouse. *Cellular Immunology, 263*(2), 212–218.
82. McGuire, L. I., Peden, A. H., Orrú, C. D., Wilham, J. M., Appleford, N. E., Mallinson, G., et al. (2012). Real time quaking-induced conversion analysis of cerebrospinal fluid in sporadic Creutzfeldt-Jakob disease. *Annals of Neurology, 72*(2), 278–285.
83. Ishibashi, D., Yamanaka, H., Mori, T., Yamaguchi, N., Yamaguchi, Y., Nishida, N., et al. (2011). Antigenic mimicry-mediated anti-prion effects induced by bacterial enzyme succinylarginine dihydrolase in mice. *Vaccine, 29*(50), 9321–9328.
84. Infection Control | Creutzfeldt-Jakob Disease, Classic (CJD) | Prion Disease | CDC [Internet]. [cited 2019 Apr 29]. Available from: https://www.cdc.gov/prions/cjd/infection-control.html.
85. Brown, P., Brandel, J.-P., Sato, T., Nakamura, Y., MacKenzie, J., Will, R. G., et al. (2012). Iatrogenic Creutzfeldt-Jakob Disease, Final Assessment. *Emerging Infectious Diseases, 18*(6), 901–907.

Part V

Translational Research in Human Microbes

Harnessing Microbiota as Anti-infectives

Malik M. Ahmad, Danishuddin, Rabiya Basri, Salman Ahmad, and Mohammad Haris Siddiqui

Abstract

Probiotics are living bacteria that are present in yogurt and other fermented food products including pills. They promote and enhance health benefits especially human digestive system. Normally the human body has trillions upon trillions of bacteria within the digestive track, and even they are ingested every time we swallow. Many of these swallowed bacteria may be advantageous, while most are simply inactive and does not cause any problem. Most important human probiotic microorganisms include *Lactobacillus* spp., *Bifidobacterium* spp., *Lactococcus* spp., and *Streptococcus*. Probiotics are used to recover the internal organ health and to activate the system. They are used to improve diarrhea, lactose intolerance, and calcium absorption from the gut, act against carcinogens, enhance immune response, prevent allergies and atopic diseases in infants, and improve cholesterol levels in blood and in turn helpful to heart disorders. In this chapter, we had discussed the benefits of probiotics on digestive track and overall human health.

M. M. Ahmad · S. Ahmad
Department of Agriculture, Integral Institute of Agricultural Science and Technology (IIAST), Integral University, Lucknow, India

Danishuddin
School of Computation and Integrated Sciences, Jawaharlal Nehru University, New Delhi, India

R. Basri
Department of Plant Protection, Aligarh Muslim University, Aligarh, India

Department of Bioengineering, Faculty of Engineering, Integral University, Lucknow, India

M. H. Siddiqui (✉)
Department of Agriculture, Integral Institute of Agricultural Science and Technology (IIAST), Integral University, Lucknow, India

Department of Bioengineering, Faculty of Engineering, Integral University, Lucknow, India
e-mail: mohdharis.siddiqui@gmail.com

S. Hameed, Z. Fatima (eds.), *Pathogenicity and Drug Resistance of Human Pathogens*, https://doi.org/10.1007/978-981-32-9449-3_15

Keywords
Probiotics · Gut microorganisms · GI tract · Lactobacillus · Health benefits

Introduction

Probiotics are either culture of bacteria or the living organisms, which stimulate and magnify human health benefits on its ingestion in a definite quantity. Some kind of food taken by us contains probiotics, viz., yogurt, cheese, and milk, both fermented and non-fermented. Extensively used probiotic bacteria are generally the strains of genera *Bifidobacterium, Lactobacillus, and Saccharomyces*. Typically, probiotics have been allied with the gastrointestinal tract. However, in recent investigations, it is also found associated with the oral health. More than 400 m^2 of surface area of gastrointestinal tract contains microflora of over 500 bacterial species, out of which some perform to sustain human health. It implies close association of human beings with huge range of microorganisms present in the oral aperture and gastrointestinal tract and on the skin. The greatest accumulations of these organisms are found in GI tract. These microflora are acquired promptly after birth and remains in adequate amount for the whole life, and it is essential for regulating homeostasis [12]. The proof for the impact of gut microorganisms on human healthiness is fast increasing, and numerous host-microbial interactions, positive as well as negative, have been reported [11, 16]. The large intestine of humans is known to be one of the metabolically most active organs, and the microbial ecosystem present in it is exceptionally complex and diverse in nature [2]. It has been proved that the nutritional adjustments of the composition and activity of the microbiota have been achieved by the use of syn-, pre-, and probiotics. They can be divided into amendment of immune responses of host and interference with the colonic microbial ecosystem, consequential change in the metabolism of colonic bacteria [3].

Prebiotics can be described as "a carefully fermented component" that permits certain precise changes, both in composition and/or activity in GI microflora which provides benefits upon host wellbeing and health [10, 23]. They include nondigestible food components such as lactulose, inulin, and fructooligosaccharides which affect the colon of the host's body by specifically stimulating the growth and/or increase in the activity of probiotic bacteria. Also, these prebiotic carbohydrates do not undergo any further changes while cooking which fascinates the food scientists. Synbiotics refer to the food items or nutritional supplements combining both pro- and prebiotics in a form of synergism and hence synbiotics [12]. This perception was first pioneered as a blend of probiotics and prebiotics which can help the host by advancing the continued existence and implantation of live microbial dietary supplements in the colon of the intestine, by selectively stimulating the growth and/or by activating the metabolism of one or a limited number of health-promoting bacteria, thus improving host welfare [11]. Probiotics are the live bacterial cultures which upon ingestion in a particular amount, promotes and enhances health. A few conventional foods that contain probiotics are fermented as well as non-fermented milk, yogurt, cheese, etc., and in these, mostly

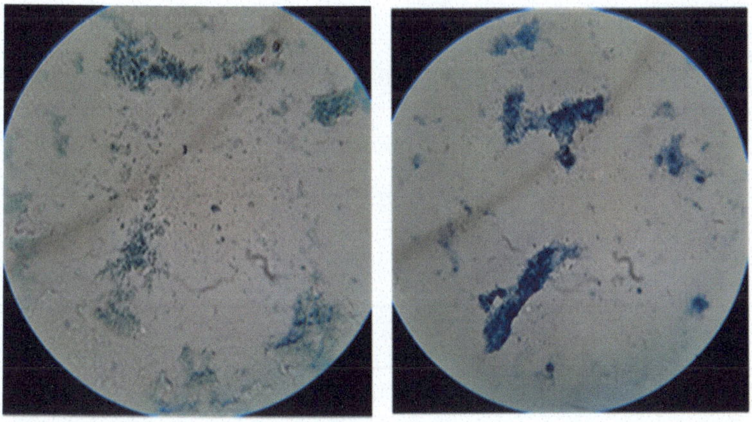

Fig. 1 *Lactobacillus* colony seen under microscope

nonpathogenic strains of genera *Bifidobacterium* and *Lactobacillus* are widely used as probiotic bacteria ([1, 24, 25]; Fig. 1). Conventionally, probiotics have been associated with the GI tract; however, in recent studies, a number of investigators have recommended the use of probiotics for oral health.

With the discovery of penicillin in 1928 by Alexander Fleming, the golden era of microbiology began [5]. Introduction of antibiotics revolutionized the field of medicine, and increased life expectancy, with improvement in the quality of human life due to a decrease in the mortality rate. However, the main disadvantage with antibiotics is that besides killing harmful bacteria, they also kill good bacteria, thus disturbing the ecosystem of the body, and can, in turn, result in superinfection and drug resistance. The word probiotic is derived from the Latin word "pro" meaning "for," while "biotic" is a Greek word which means "life" [20]. The Food and Agriculture Organization (FAO) and World Health Organization (WHO) have defined probiotics as live microorganisms which when taken in sufficient amount boost the health on the host [1, 19].

Probiotic Microorganisms

Bifidobacterium, *Lactococcus*, *Lactobacillus*, and *Streptococcus* are some of the human probiotic microorganisms. In addition, certain G+ve bacteria of genus *Bacillus* and yeast strains of genus *Saccharomyces* are also frequently used in probiotic products [6]. Probiotics are subjected to laws contained within the general food law; in step with that, they ought to be safe for human as well as animal health. In the USA, the Food and Drug Administration (FDA) regulates such laws wherein it says microorganisms used for consumption functions ought to regard as generally regarded as safe (GRAS), while the European Food Safety Authority (EFSA) in Europe launched a new term "QPS (qualified presumption of safety)" that involves some extra criteria of the security assessment of microorganism supplements, together with the history of safe usage and absence of the danger of nonheritable

Table 1 Probiotic utility in human nutrition

Genus	Type	Pharmaceutical product	Food additives
Lactobacillus	*L. acidophilus*	Yes	No
	L. amylovorus	No	Yes
	L. casei	Yes	Yes
	L. gasseri	Yes	No
	L. helveticus	Yes	No
	L. johnsonii	No	Yes
	L. pentosus	No	Yes
	L. plantarum	No	Yes
	L. reuteri	Yes	No
	L. rhamnosus	Yes	Yes
Bifidobacterium	*B. adolescentis*	Yes	No
	B. animalis	Yes	No
	B. bifidum	Yes	No
	B. breve	No	Yes
	B. infantis	Yes	No
	B. longum	Yes	No
Others	*Enterococcus faecium*	Yes	No
	Bacillus clausii	Yes	No
	Streptococcus thermophilus	Yes	No

resistance to antibiotics [13]. There are basically two types of microbes used for preparing probiotics, viz., (i) lactic acid-producing bacilli and (ii) non-lactic acid-producing bacterial species (Table 1).

Lactic acid-producing bacilli include:

(i) *Lactobacillus*: *L. acidophilus*, *L. sporogenes*, *L. rhamnosus*, *L. reuteri*, *L. fermentum*, *L. lactus*, *L. brevis*, *L. paracasei*, *L. gasseri*, *L. salivarius*, and *L. casei*
(ii) *Bifidobacterium*: *B. bifidum* and *B. lactis*
(iii) *Streptococcus*: *S. lactis*, *S. salivarius*, and *S. thermophilus*.

Non-lactic acid-producing bacterial species

(i) *Bacillus propionibacterium*
(ii) Nonpathogenic yeasts: *Saccharomyces*

Selection Criteria and Characteristics for Ideal Probiotic Microorganisms

Probiotics are preparations of live microorganisms that beneficially have an effect on the host by up the properties of the autochthonal microbes. Since the human microorganism plays a very significant role in health and sickness of man, probiotics

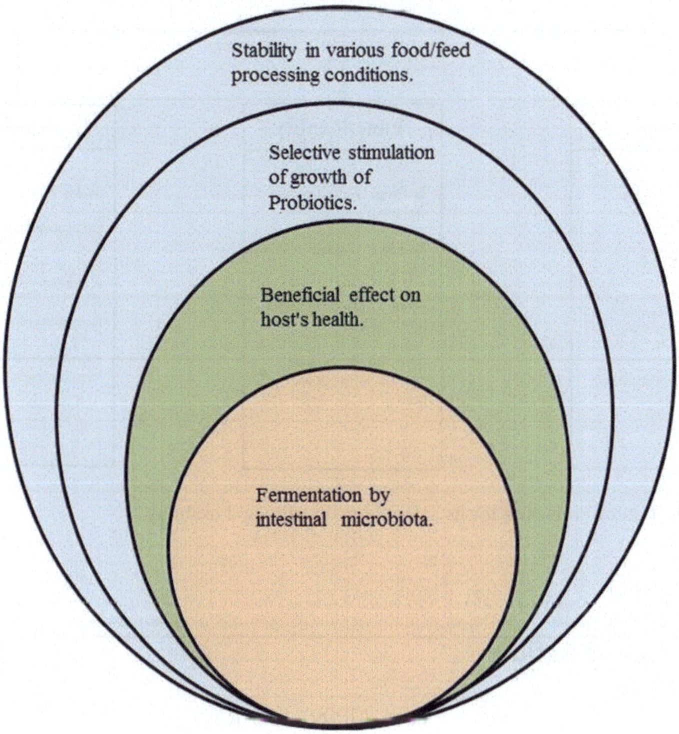

Fig. 2 Characterstics of ideal probiotic microorganism

are accustomed to improve internal organ health and to stimulate the system. The microbes normally used as probiotics for human consumption are the carboxylic acid microorganisms. In early studies the strains used for chemical change in milk merchandise for human consumption were often times used as probiotics, but later on, it had been proposed that if the strains originated from the human internal organ tract, they may additionally be used with carboxylic acid microorganisms. It was normally concurred that the strain must be of host birthplace, all around represented, and able to endure the trials of the abdomen connected tract and maybe colonize, organically dynamic against the target even as to be steady and amiable to business creation and appropriation (Figs. 2 and 3).

Fuller [7] stated that the traits of a good probiotic were as follows:

- It should be competent of applying a beneficial effect on the host, for example, increased growth or resistance to disease.
- It should be nonpathogenic and nontoxic.
- It should be present as viable cells, preferably in large numbers.
- It should be able to survive and metabolize in the gut environment, for example, resistance to low pH, organic acids, and bile.

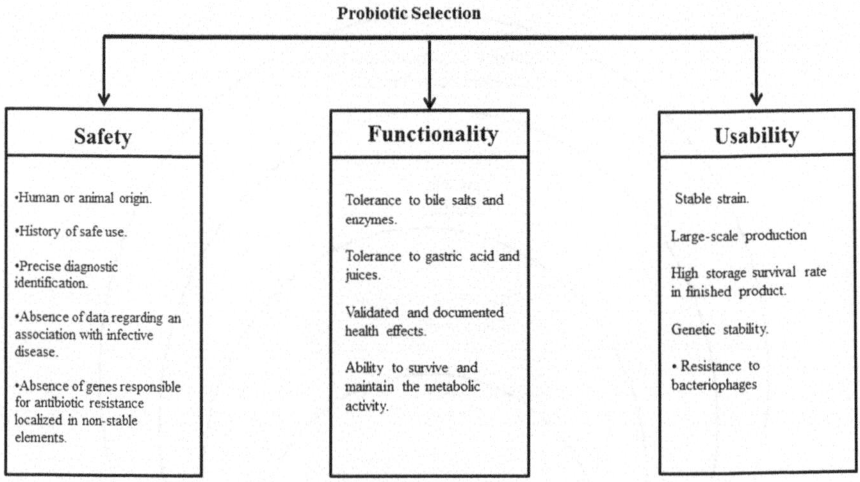

Fig. 3 Selection of Probiotics for its safety, functionality and usability

Mechanism of Action

Gupta and Garg [12], Stamatova et al. (2009), and Rastogi et al. [17] had proposed several mechanisms to explain the action of probiotics:

- Normalization of the intestinal microflora of the host.
- Nonspecific immunity can be stimulated, and the humoral and cellular immune response can be adapted.
- Modification of the metabolic activity and composition of host microbiota at the precise location.
- Secretion of organic acids, hydrogen peroxide, and bacteriocins (antimicrobial substances).
- Competition with pathogenic agents for adhesion sites on mucosa.
- Modulating the pH and adapting the surrounding environment by oxidation-reduction potential.

Vehicles

Probiotics can be provided in products in any of the following ways [19].

- Added to food or beverage as culture concentrate
- Inoculated into prebiotic fibers
- Inoculated into dairy products or milk-based food (milk, milk drinks, yogurt, cheese, kefir, biodrink), asparagus, and soybeans
- Dried and concentrated cells packaged and as dietary supplements (nondairy products such as powder, capsule, and gelatin tablets)

Health Benefits Through Probiotics

Conventionally, probiotics have been associated with health of digestive track and mainly used in the prevention or treatment of gastrointestinal infections and diseases; however, during the past decade, clinical interest has focused on an increasing number of established and proposed health effects of probiotic bacteria [15].

Prevention and/or Reduction of Diarrhea

Probiotic bacteria such as *Lactobacillus* spp., *Lactobacillus casei*, *B. bifidum*, and *Streptococcus thermophilus* have been shown to conserve the intestinal integrity and arbitrate the effects of irritable bowel syndrome, inflammatory bowel diseases, colitis, and even alcoholic liver disease [18]. The intestinal effects of probiotics are to encourage recovery from diarrhea (rotavirus, travelers', and antibiotic-induced), alleviate symptoms of lactose intolerance and malabsorption, produce lactase, relieve constipation, treat colitis, and stimulate gastrointestinal immunity [11]. Diarrhea is a major cause of infant death throughout the world and can be devastating in adults; hence, the use of probiotics can be a significant, nonpersistent means in its prevention and treatment, particularly in developing countries like India.

Lactose Intolerance

S. thermophilus, *Lactobacillus delbrueckii* subsp. *bulgaricus*, and other lactobacilli in fermented milk products provide bacterial lactase to the stomach and intestine and can improve lactose intolerance symptoms [11]. People with lactose intolerance are able to tolerate the lactose in yogurt instead of lactose in milk. It is assumed that the occurrence of lactase-producing bacteria in the yogurt, especially *L. acidophilus*, contributed to the digestion as well as absorption of lactose. Probiotics reduce hydrogen exhalation and improves lactose digestion in comparison with milk, but they are not as much efficient in the unfermented form [4].

Calcium Absorption

Milk is considered to be an abundant source of calcium apart from other dietary sources. Low (acidic) pH favors the absorption of calcium in the gut. The consumption of probiotics in lactose-intolerant patients results in hydrolysis of milk lactose by the probiotic strains, thus favoring calcium absorption [11]. A study on *Lactobacillus casei*, *Lactobacillus reuteri*, and *Lactobacillus gasseri* reported higher absorption of calcium in growing rats and 35% bone weight among the sets that are fed by probiotic as compared to the control sets [9].

Reduction of Concentration of Cancer by Promoting Enzymes in the Gut

It has been suggested that probiotics also own a defensive effect against colon carcinogenesis. This is attributed to the production of short-chain fatty acids through fermentation by gut microflora, and gene expressions in tumor cells are altered. Gut flora, especially after the intake of resistant starch, persuades the chemopreventive enzyme glutathione transferase in the colon. Altogether, these factors guide to a reduced load of genotoxic agents in the gut and to augment the production of butyrate, which deactivate toxic components [22].

Immune Response

Probiotics can augment the immune responses that are specific and nonspecific by triggering macrophages, growing levels of cytokines, increasing natural killer cell activity, and/or boosting the levels of immunoglobulins. This can prevent growth of pathogens and translocation and decrease the chance of infection from frequent pathogens such as *Shigella* and *Salmonella* [11]. Studies also revealed that probiotics can regulate the immune response in human beings by producing an exclusive material that affects the expression of particular genes that are connected to the immune regulation. The effects of gastric microbiota such as *L. fermentum* and *L. plantarum* are significant examples to reveal the regulation of the immune system. In fact, the studies indicate that gastric mucosal inflammation can be avoided by pretreatment of *L. plantarum* ZDY 2013 induced by *H. pylori*. In addition, *L. fermentum* UCO-979C obtained from human gut showed increased growth in *M. gerbil*. It can discontinue stimulation to produce cytokines like IL8 by *H. pylori* in gastric adenocarcinoma in human cells (AGS) [8].

Prevention or Alleviation of Allergies and Atopic Diseases in Infants

Numerous in vitro studies had shown that probiotics have an active role in treating allergic disorders as well as atopic diseases. Probiotics may put forth useful effect on the allergic reaction by improving mucosal barrier function. In addition, probiotic utilization by children may beneficially influence the development of the immune system [11]. Human trials had established a restricted advantage for the utilization of probiotic microorganisms in atopic dermatitis in both preventive and therapeutic capacities [14].

Cancer Prevention

Clinical studies on the development of cancer in animal models had shown defensive role of probiotics against cancer. Studies reported that some of the probiotic strains could bring down the incidence of postoperative tenderness in the patients having cancer. Radiotherapy- and chemotherapy-related diarrhea problems can be eased in patients who were on probiotics. The role of probiotics includes inactivation of agent compounds, suppression of the modification of procarcinogens into operational carcinogens, formation of antimutagenic compounds, suppression of growth of procarcinogenic microorganism, diminution of the assimilation of carcinogens, enhancement of immune response, and influence on salt concentrations of the body. Furthermore, some strains of probiotics can be used as an adjuvant for cancer prevention or/and treatment by modulating intestinal microbiota and immune response [27].

Prevention of Heart Diseases/Influence on Blood Cholesterol Levels

In a recent finding, it was suggested that the pathogenesis of cardiovascular diseases (CVD) is increased by imbalance of microbes in the gut. Probiotic microbes aid in reducing the low definition lipids (LDL-cholesterol) and thus meliorate the LDL/HDL ratio, as well as bring down blood pressure, inflammatory mediators, blood sugar levels, and body mass index (BMI) [26]. Besides, probiotic microorganisms produce acids that can counteract cholesterol production. Probiotic bacteria absorb fibers from the intestines to produce acids, for example, *Propionibacterium freudenreichii* produces proprionic acid which reduces production of cholesterol by the liver, while some useful bacteria have shown to break down cholesterol for its energy requirement [21].

Conclusion

Consumption of probiotics as food sources is strongly recommended in the developing countries like India because of the affirmed health benefits which include reduction in acute diarrheal diseases, increase in natural resistance to infectious disease in gastrointestinal tract and acquired immunity diseases (like HIV/AIDS), improvement in intolerance of lactose, reduction in serum cholesterol levels, improved nutrition, allergic treatments, and serving as an adjuvant to vaccine. There is a demand of awareness regarding the health benefits of probiotics in developing countries, and standards are needed to ensure that quality and safety standards are met. In the absence of regulations, there could be manufacturing of erratic products containing devalued levels of the beneficial strains without benefits. Hence, funding in the area of probiotics in developing countries is very important to establish that they can help to improve the nutritional and health status of the people. Every country should develop a strategy concerned to the needs of the people, medical facilities, economic issues, increased awareness among consumers, and their confidence which will ultimately lead to high-level adoption of probiotic products in the society. They can be incorporated in the medical facilities aided by the government because it is proved that some strains of probiotics can be used as an adjuvant for cancer prevention, help in lowering cholesterol and LDL levels, and can even be used to raise specific and nonspecific immune responses. In conclusions, as compared to normal drugs or therapeutic agents, probiotic strains are cheaper to produce and easily available, stored, and delivered which would overall be advantageous for developing countries.

References

1. Agarwal, E., Bajaj, P., Guruprasad, C. N., Naik, S., & Pradeep, A. R. (2011). Probiotics: A novel step towards oral health. *Archaeology and Oral Science Research, 1*, 108–115.
2. Cummings, J. H., Antoine, J. M., Azpiroz, F., Bourdet- Sicard, R., et al. (2004). PASSCLAIM – gut health and immunity. *European Journal of Nutrition, 43*, II118–II173.
3. De Preter, V., Hamer, H. M., Windey, K., & Verbeke, K. (2011). The impact of pre- and/or probiotics on human colonic metabolism: Does it affect human health? *Molecular Nutrition and Food Research, 55*, 46–57.
4. de Vrese, M., Stegelmann, A., Richter, B., Fenselau, S., Laue, C., & Schrezenmeir, J. (2001). Probiotics—compensation for lactase insufficiency. *The American Journal of Clinical Nutrition, 73*(suppl), 421S–429S.
5. Figueiredo, N. R., Dinkar, A. D., & Khorate, M. M. (2018). Probiotics in human health. *International Journal of Orofacial Biology, 2*, 1–5.
6. Fijan, S. (2014). Microorganisms with claimed probiotic properties: An overview of recent literature. *International Journal of Environmental Research and Public Health, 11*, 4745–4767.
7. Fuller, R. (1989). Probiotics in man and animals. *The Journal of Applied Bacteriology, 66*, 365–378.
8. Ganjbakhsh, S. E., & Rezaee, P. (2017). The effect of probiotics on immune system. *Journal of Bacteriology and Mycology, 5*, 319–320.
9. Ghanem, K. Z., Badawy, I. H., & Abdel-Salam, A. M. (2004). Influence of yoghurt and probiotic yoghurt on the absorption of calcium, magnesium, iron and bone mineralization in rats. *Milchwissenschaft, 59*, 472–475.
10. Gibson, G. R., & Roberfroid, M. B. (1995). Dietary modulation of the human colonic microbiota: Introducing the concept of prebiotics. *The Journal of Nutrition, 125*, 1401–1412.
11. Gill, S. R., Pop, M., Deboy, R. T., Eckburg, P. B., et al. (2006). Metagenomic analysis of the human distal gut microbiome. *Science, 312*, 1355–1359.
12. Gupta, V., & Garg, R. (2009). Probiotics. *Indian Journal of Medical Microbiology, 27*, 202–209.
13. Levin, B. R., & Rozen, D. E. (2006). Non-inherited antibiotic resistance. *Nature Reviews. Microbiology, 4*(7), 556–562.
14. Michail, S. (2009). The role of probiotics in allergic diseases. *Allergy, Asthma and Clinical Immunology, 5*(1), 5.

15. Ng, S. C., Hart, A. L., Kamm, M. A., Stagg, A. J., & Knight, S. C. (2009). Mechanisms of action of probiotics: Recent advances. *Inflammatory Bowel Diseases, 15*, 300–310.
16. Nicholson, J. K., Holmes, E., & Wilson, I. D. (2005). Gut microorganisms, mammalian metabolism and personalized health care. *Nature Reviews. Microbiology, 3*, 431–438.
17. Rastogi, P., Saini, H., Dixit, J., & Singhal, R. (2011). Probiotics and oral health. *National Journal of Maxillofacial Surgery, 2*, 6–9.
18. Reddy, P., & Srivastava, R. (2011). Probiotics-the next saviour in oral diseases. *International Journal of Dental Clinics, 3*, 54–57.
19. Reddy, V. N., Rao, A. P., Mohan, G., & Kumar, R. R. (2011). Probiotic lactobacilli and oral health. *Annals and Essences of Dentistry, 3*, 100–103.
20. Reid, G., Jass, J., Sebulsky, M. T., & McCormick, J. K. (2003). Potential uses of probiotics in clinical practice. *Clinical Microbiology Reviews, 16*, 658–672.
21. Saini, R., Saini, S., & Sharma, S. (2010). Potential of probiotics in controlling cardiovascular diseases. *Journal of Cardiovascular Disease Research, 1*, 213–214.
22. Saraf, K., Shashikanth, M. C., Priya, T., Sultana, N., & Chaitanya, N. C. (2010). Probiotics-do they have a role in medicine and dentistry? *The Journal of the Association of Physicians of India, 58*, 488–492.
23. Schrezenmeir, J., & de Vrese, M. (2001). Probiotics, prebiotics, and synbiotics-approaching a definition. *The American Journal of Clinical Nutrition, 73*, 361S–364S.
24. Singh, K., Kallali, B., Kumar, A., & Thaker, V. (2011). Probiotics: A review. *Asian Pacific Journal of Tropical Biomedicine, 1*, S287–S290.
25. Stamatova, I., & Meurman, J. H. (2009). Probiotics: Health benefits in the mouth. *American Journal of Dentistry, 22*, 329–338.
26. Thushara, R. M., Gangadaran, S., Solati, Z., & Moghadasian, M. H. (2016). Cardiovascular benefits of probiotics: A review of experimental and clinical studies. *Food & Function, 7*, 632–642.
27. Yu, A. Q., & Li, L. (2016). The potential role of probiotics in cancer prevention and treatment. *Nutrition and Cancer, 68*(4), 535–544.

Human Microbiome and Malignancy: Principles, Mechanisms, and Challenges

Munindra Ruwali and Rahul Shukla

Abstract

Cancer is a disease caused by several factors, and there are interindividual differences in susceptibility toward cancer. Recent advances in cancer studies have failed to explain the reason as to why out of people exposed to carcinogens or genetically predisposed, only selective individuals develop cancer, while others are not affected. Among the factors that have received attention as a causal factor for cancer development in recent times is the human microbiome. The human microbiome is the genetic material of all the microbes that are present in the human body. A typical human gut microbiome is made up of trillions of organisms, which closes matches with the number of cells in the human body. Also, the genome of these organisms is made up of around 3 million genes, which is much more than the number of genes in the human genome. Microbes induce about 20% of all fatal cancers in humans and therefore provide options for interventions for cancer prevention. With the advent of modern techniques for sequencing, data storage, and analysis, a more detailed analysis of human microbiome is becoming a reality which is paving the way for establishing the relationship between microbiome and cancer. A better understanding of this relationship would equip us to develop new therapeutic and preventive strategies against cancer.

Keywords

Microbiome · Cancer · Microbiota · Microbe

M. Ruwali (✉)
Amity Institute of Biotechnology, Amity University Haryana, Manesar, Gurugram, India
e-mail: munindraruwali@gmail.com; mruwali@ggn.amity.edu

R. Shukla
Department of Pharmaceutics, National Institute of Pharmaceutical Education and Research (NIPER-Raebareli), Lucknow, Uttar Pradesh, India

S. Hameed, Z. Fatima (eds.), *Pathogenicity and Drug Resistance of Human Pathogens*,
https://doi.org/10.1007/978-981-32-9449-3_16

Introduction

Human Microbiome

The human microbiome can be defined as the genetic material of all the microbes that live in the human body [66]. With rapid advances in DNA sequencing tools, the microbes can be identified and quantified with more accuracy compared to the traditional methods of bacterial culture. Worldwide research initiatives have found millions of protein-coding genes in human microbes, and interestingly, the number of these genes is much higher compared to that present in human genome. These studies have explored the genetic basis of the useful effects of these microbes on human health [86]. It is estimated that out of an approximate number of 100 trillion, the majority of human microbes are present in the gut. The basic difference between the host genome and microbiome is stability, as the genome of the host is relatively constant, while the microbiome exhibits a dynamic nature. The dynamic nature of microbiome is revealed when it changes in response to early development, environmental factors, and especially in response to disease [71]. Infancy and early childhood are the stages in which the most dramatic changes in composition of microbiome take place [80]. The microbiome of the gut of an infant is affected by gestational age, methods of childbirth, type of feed, food intake of mother, and use of antibiotics [75]. Two factors, namely, composition and the adaptability of microbes to changes in environment of the infant microbiota during initial developmental stages, help in maintaining the balance with the host immune responses and influence the health as the age progresses [17].

The human gut microbiome is mainly composed of bacteria belonging to four phyla of the *Firmicutes*, *Bacteroidetes*, *Proteobacteria*, and *Actinobacteria*. The *Firmicutes* and *Bacteroidetes* are more abundant (90%), while the *Proteobacteria* and *Actinobacteria* are comparatively less abundant constituents [62]. Among these, the Gram-positive are the *Firmicutes* (e.g., anaerobic clostridia, streptococci, and enterococci) and *Actinobacteria* (e.g., bifidobacteria) which have high GC content, while the Gram-negative are the *Bacteroidetes* (e.g., *Bacteroides thetaiotaomicron*, *Bacteroides fragilis*) which can digest complex carbohydrates and the *Proteobacteria* (e.g., *Gammaproteobacteria*, *Escherichia coli*, and *Klebsiella* species). With information about the microbiome composition in healthy people, a relationship can be established between the changes in microbiome and disease susceptibilities. Among the microbes which inhabit our gut, the majority of studies have been conducted on bacteria compared to viruses, fungi, and archaea.

The Human Microbiome: Dynamic Interactions, Diversity, and Role in Digestion and Nutrition

The gastrointestinal (GI) tract of a newly born human infant is a perfect site for colonization by the microbes whose population largely depends on the method of childbirth [23]. This can be corroborated by the fact that there are differences in the microbiota of infants delivered through the vagina of mother compared to the infants delivered via Cesarean section. Immediately after birth, similarities have been observed in the microbial composition of infants and vagina of the mother (in case of vaginal delivery) or skin of the mother (in case of Cesarean section delivery) [24]. With the increase in age of the infant, the microbial composition exhibits marked increase in numbers of microbes coupled with an enhanced diversity in a linear manner based on crucial factors such as diet composition and antibiotic treatment. Surprisingly, each change in dietary pattern also modifies the microbial composition as well as the microbial genome as seen in the case of genes associated with synthesis of vitamins and digestion of polysaccharides (Koenig et al. 2011).

Studies have also shown that there is an interaction of microbes with environment and the microbes can move from one surface to another. A study reported that microbes can move from the tips of the fingers of humans to surfaces in contact such as keyboards of computers thereby acting as signatures to differentiate individuals with high levels of accuracy [32]. Investigators have also explored the composition and stability of microbes in a person over a period of time. A long time series study revealed that communities at different body sites were readily distinguishable from one another [14], and the diversity is different in different sites such as oral cavity and gut exhibiting the highest levels of microbe diversities [21]. Studies also provide evidence that people have a similar genetic composition of microbes when compared to the nature of microbes. A study conducted by Turnbaugh et al. [108] characterized the microbes present in feces of twins and their mothers and observed that microbes in the gut are shared among the members of a family though differences are also present. It was further observed that obesity can have an influence on the microbial composition, diversity, and metabolism. The results demonstrated that deviations from the "core microbiome" can take place under different physiological states such as obesity and leanness.

Efforts were also made to establish the relationship between host's microbiota, digestion, and metabolism. In order to explore the evolutionary relationship of mammals and their resident microbes, Ley et al. [69] conducted a study based on the gene sequence of 16S ribosomal RNA of bacteria present in the feces of human and 59 other species of mammals. The results from the study indicated that dietary factors and phylogeny change bacterial diversity, bacterial diversity increases as the food habits change, microbes exhibited increased diversity as their hosts, and the gut microbes of present humans resemble omnivorous primates. Studies have also shown that a fascinating relationship exists between gut microbes and obese status. Bäckhed et al. (2004) reported that if normal microbes collected from the distal part of the intestine of animals are introduced into adult germ-free mice, it results in a significant increase in fat content and insulin resistance in a short span of 14 days though the food consumption goes down resulting from increased monosaccharide absorption and lipogenesis. Obesity may also influence the diversity of the gut microbes as seen in the case of obese and lean mice [68]. Perhaps, the most interesting observation with regard to the correlation between microbes and obesity was seen in a study of weight loss by Ley et al. [70]. They observed that *Bacteroidetes* are present in lesser proportion in obese people compared to the nonobese people though the number increases with a reduction in weight promoted by a low fat or low carbohydrate diet. These findings may provide therapeutic options in obese patients for a reduction in body weight by modulation of the patient's microbiota. Another surprising observation is the introduction of new genes to our microbiome by the microbes that we ingest with a particular food which then assist in digesting that food. This was observed in a study in which a hydrolase enzyme was added in human system as a result of consumption of porphyran [50] in Japanese populations which commonly use seaweed in their diet but not in the American populations.

The Human Microbiome and Disease

Several studies have investigated the possibilities of a relationship between the variations in the composition of microbiome and disease development. In inflammatory bowel disease (IBD), the diversity of gut microbiota exhibits differences between healthy people and patients. Studies have shown that in patients with IBD, there is reduced prevalence of *Firmicutes* and *Bacteroidetes*, while *Proteobacteria* exhibit increased presence and there is also a lowering of bacterial diversity [73]. IBD pathogenesis results from interplay between environmental factors and host genetic composition, and this interplay is controlled by commensal microbiota resulting in activation of immune responses which can be protective or deleterious [95]. Another disease called necrotizing enterocolitis (NEC) has a multifactorial pathogenesis with gut microbiota serving as an important pathogenesis factor corroborated by the fact that there is an increased prevalence of *Proteobacteria* in preterm infants with NEC as opposed to the healthy ones [44].

The incidence rates of atopic diseases have seen an increase in recent times. This can be attributed to the fact that if an infant is not exposed to antigens of the microbes in initial stages, it results in changes of the gut microbiota leading to a disruption of the immune system [110]. *Bacteroides fragilis* modulates the pathways of immune receptor signaling which creates immunological tolerance. An exposure to pets and siblings resistant to asthma and allergies leads to changes in infant gut microbiota [26]. Factors such as type of birth, feeding of powdered milk, and antibiotic exposure can increase chances of atopic diseases in infants [88]. Diabetes mellitus (DM) is a type of metabolic disorder characterized by elevated blood sugar levels. Studies have shown that for type 1 diabetes, the gut microbes control metabolic-immune axis [94] as shown by the protective effect of bifidobacteria and causative effect of *Proteobacteria*. A modification of gut microbes by factors such as birth by Cesarean section and diet can make one prone to develop type 1 diabetes (Charbonneau et al. 2016). Similar to influencing the metabolic-immune axis, gut microbes can also influence gut-brain axis leading to changes in the human behavior, a condition called as autistic spectrum disorder (ASD). The effect is mediated via the hormonal, immune, and neural signaling pathways. Children with ASD have been shown to have varied composition of gut microbes compared to the normal children. It has been shown experimentally that changes in microbes in the gut may result in symptoms which are consistent with that of autism (Ding et al. 2017).

Human Microbiome and Cancer

Cancer is a condition in which there is a loss of the regulation of multiplication of cells resulting in a mass of cells called as tumor. The etiology of cancer is multifactorial involving both genetic and environmental factors. An exposure to causative factors of cancer such as carcinogens, radiations, or microbes follows a very diverse pattern [99]. Currently, we cannot clearly demarcate individuals exposed to environmental factors such as smoking or those having hepatitis B infections to develop cancer. This brings into focus other environmental factors [111] having equally important roles in cancer development as that of other established causative factors. Among these factors, human microbiome has garnered immense interest owing to the fact that nobody expected microbes to play an important role in cancer pathogenesis. It has been estimated that microbes induce about 20% of all fatal cancers in humans [81], signifying the importance of microbes in cancer progression. Cancer is caused by many infectious agents, but the microbes that have been believed to be the major causative factors are human papillomaviruses (HPV), *Helicobacter pylori*, and hepatitis B and C viruses. Table 1 lists some of the major microbes associated with cancers and possible mechanisms.

Oncogenic Microbes: Classification, Principles, and Mechanisms of Microbial Oncogenesis

Microbes can be classified into two important classes based on the oncogenic interactions that may induce cancers. Class A microbes, such as human T-cell lymphotropic virus type 1 and HIV, target immune cells and may promote lymphomas and immunosuppression that can lead to other cancers caused by microbes [41, 51, 113]. Class B microbes work by interacting with the parenchymal cells such as epithelial, endothelial, or mesenchymal cells. The host responds to these microbe-parenchyma interactions which promote malignancies by metaplasia and dysplasia. Typical examples of this class of microbes are hepatitis viruses, *H. pylori*, and helminths such as *Clonorchis sinensis* and schistosomes [109].

Though there is a great diversity of the microbes and the malignancies caused by them, yet some principles are common to most of these interactions. First principle relates to the persistence of the microbes in the host body wherein a part of oncogenic microbes persists in the host for a long duration, sometimes years, even after their removal from the body. The host body tries to eliminate these microbes, and in this process there is tissue damage that may promote malignancy [11]. The second principle deals with the variations observed in the oncogenic potential of the microbes. This can be exemplified by taking the cases of HPV and for hepatitis viruses and *H. pylori* (few types are more carcinogenic). Microbial load in cases of hepatitis viruses and schistosomes, interactions between microbial genotype and load, interaction of microbial genotype load with host genotypes and phenotypes, and microbial prevalence in prior generations are other important principles which are common to most of the malignancies promoted by the microbes.

Table 1 Some microbes associated with cancers and possible mechanisms

Cancer	Pathogen	Mechanism	References
Colorectal cancer	*Fusobacterium*	Increase in cytokines and NF-κB signaling	Rubinstein et al. [92]
GI tract adenocarcinomas	*Helicobacter*	Increase in cytokines and NF-κB signaling through CagA	Peek and Blaser [82]
Breast cancer	*Alistipes*	Microbial dysbiosis	Xuan et al. [112]
Esophageal cancer	*Streptococcus*	Lowered gastric acidity	Fischbach et al. [33]
Head and neck cancer	*Fusobacterium*	Different compositions of bacterial community	Gong et al. [43]
Prostate cancer	*Propionibacterium*	Inflammation	Cohen et al. [19]
Pancreatic cancer	*Enterobacteriaceae*	Tumor resistance to the chemotherapeutic drug gemcitabine	Geller et al. [38]
Cervical cancer	Human papillomavirus (HPV)	Oncogenic proteins	[107]
Kaposi sarcoma, non-Hodgkin's lymphoma, Hodgkin's lymphoma, cancers of the cervix, anus, and conjunctiva	Human immunodeficiency virus (HIV)	Immunosuppression	IARC [54]
Hepatocellular carcinoma (HCC)	Hepatitis B virus (HBV)	Viral proteins, inflammation, and genetic instability	Beasley et al. [7]
Leukemia	Human T-cell leukemia retrovirus (HTLV-1)	Basic leucine zipper factor (HBZ) gene	Satou et al. [96]
Liver cancer	*Aspergillus flavus*	Aflatoxin toxicity	Ross et al. [90]
			Qian et al. [85]
Hepatocellular carcinoma (HCC)	Hepatitis C virus (HCV)	Chronic immune inflammatory response	Choo [18]

The mechanisms of microbial oncogenesis are myriad but can be grouped into four important types. The first process involves alteration of signaling in host cells. The microbes try to increase the longevity of the host cells by several mechanisms such as by action on p53 [79], upregulation of cell cyclins [16], or effects on stroma [84]. The second process involves chronic inflammation which can promote malignancy through activation of the immune system, an increase in host cell turnover that may increase incidences of DNA damage either by mutagenic events or by reactive oxygen and nitrogen species [10]. The third process involves alterations in the physiology of the host which occurs from cellular to organ level and can result in events such as compromised immune system, hormonal imbalances, and perturbed organ circulation [34]. The fourth process involves the effect of microbes on other microbes to promote malignancy as is seen in case of gastric *H. pylori*.

Oncogenic Microbes: Nonviral Oncogenesis

Parasites, Fungus, and Bacteria

Some parasites have been found to promote malignancy. Parasitic flukes such as *Schistosoma haematobium* (*S. haematobium*) and *Opisthorchis viverrini* (*O. viverrini*) have been implicated in bladder cancer and bile duct cancer (known as cholangiocarcinoma), respectively. *S. haematobium* is mainly found in African and Asian regions and gains entry to human body through skin [55]. Ferguson [29] worked in Egypt and found that a majority of males had parasites, with bladder cancer patients having fluke ova in a blood vessel that carries blood from the gastrointestinal tract, gallbladder, pancreas, and spleen to the liver vein, bladder, or cancer cells. *Opisthorchis viverrini* is spread by eating uncooked fish and is predominantly found in Thai and other East Asian populations. Stewart [101] suspected *O. viverrini* as a carcinogen in 1931 which was confirmed by the work of Bhamarapravati and Virranuvatti [8].

The role of fungus in promoting cancer was confirmed as a result of several reports. Groundnut-based diet was found to be the causative factor of acute hepatic necrosis in turkey poults in England and liver tumors without cirrhosis in rats after extended feeding [1, 65]. The diet contained aflatoxin B1 as the carcinogenic factor, produced by the fungus *Aspergillus flavus* and can cause liver tumors in mammals. However, after these findings, it was relatively difficult to establish aflatoxin as a carcinogen in human though high rates of liver cancer are observed in areas which are hot and humid, and these areas also have higher risks of aflatoxin contamination of stored food items. Such observation is also inconsistent as heterogeneous reports exist establishing the relationship between aflatoxin contamination and HBV infection [46, 56]. A pioneering study to investigate the relationship between aflatoxin and hepatic carcinogenesis was conducted in Shanghai, China. The study involved collection of blood and urine samples from a huge sample size, and then a prospective follow-up was done for 4 years [85, 90]. After follow-up, it was observed that there were higher chances of aflatoxin metabolites in blood or urine samples of these cases with liver cancer compared to the controls.

The role of bacteria in promoting malignancy was established by the pioneering work of Robin Warren who was a clinical pathologist in Perth, Western Australia. In the studies, gastritis and ulcers were found to be associated with the presence of spiral or curved bacilli [74]. In 1981, Barry Marshall collaborated with Warren, and they worked together to establish the role of these unknown bacteria with ulcers. However, it was challenging to convince their colleagues about the role of these bacteria in gastritis as it was largely believed that acidic environment of the stomach restricts the growth of bacteria and the main reasons for ulcers are stress and hyperacidity. *Helicobacter pylori* was found to protect itself from acidic environment by inhabiting gastric mucosa and by producing ammonia [2, 106]. The year 1994 became an important year for *H. pylori* when the NIH report made it mandatory to test for and treat *H. pylori* in ulcer disease [77], and in the same year, World Health Organization's (WHO) International Agency for Research on Cancer [97] declared *H. pylori* as a class 1 carcinogen. A significant amount of research has been conducted after that which has confirmed the role of *H. pylori* as an important agent in gastric carcinogenesis. Some of the important findings are discovery of the *H. pylori* bacterial oncoprotein gene product CagA, *H. pylori*-induced cancer animal models, and randomized trials [35]. Studies have also shown that a high rate of seropositivity for *H. pylori* is present in gastric cancer cases compared to the normal persons. About twofold increase in risk for gastric cancer was found for the presence of *H. pylori* [53].

Oncogenic Microbes: Viral Oncogenesis

The first human virus that was identified in 1900 was the one that causes yellow fever [87]. Following the discovery of yellow fever virus, Shimkin (1977) put forward the hypothesis that cancer can be caused by viral agents which was confirmed by the studies wherein it was found that an extract which is devoid of cells can induce cancer [27, 91]. Several other discoveries prompted scientists to think about viral role in human cancers. Among these, the discovery of mouse and chicken leukemia virus as well as feline lymphoma virus played an important role in identifying similar viral agents in human cancers. In 1980s, rapid strides were made in deciphering viral oncogenesis through some prominent studies which helped in establishing the role of viruses in the development of human cancers. Following are some prominent examples:

Human T-Cell Leukemia Virus

Gallo and co-workers were motivated by the findings of a retrovirus being the causative factor for leukemia in nonhuman primates and in cattle [59, 60] and discovered human T-cell leukemia retrovirus (HTLV-1) as a causative agent of T-cell lymphoma [36]. HTLV-1 inserts in the genome of the host and encodes a protein termed Tax which induces transformation [89] though about 60% of ATL cases lose its expression [20, 105]. This was quite confusing, which led the investigators to search for other possible mechanisms. The mechanism became clear when it was discovered that T-cell proliferation was promoted by HBZ gene which was experimentally validated in transgenic mice which expressed HBZ gene [96].

Human Immunodeficiency Virus (HIV) and Hepatitis B Virus (HBV)/Hepatitis C Virus (HCV)

In 1981, acquired immunodeficiency syndrome (AIDS) was identified in the USA which was characterized by abnormally high prevalence of life-threatening lung infection [42] and exceptionally high rates of Kaposi's sarcoma (KS). For identifying the cause of AIDS, Barre-Sinoussi (1983) studied the biopsy sample of the lymph node of an AIDS patient and subsequently discovered the cause of AIDS to be a retrovirus. For this outstanding work, the duo was awarded the Nobel Prize in Medicine on October 6, 2008. HIV does not directly lead to KS or non-Hodgkin's lymphoma, as these cancer cells do not have the infection of HIV. An infection of T-cells and macrophages with HIV releases HIV-specific protein Tat that promotes angiogenesis and carcinogenesis in the cells which take up this protein [58], and furthermore, since KS and lymphomas are frequent in transplant patients having a compromised immune system, it can be concluded that a suppressed immune system caused by HIV may be a factor in the development of KS and lymphomas. HIV was accepted as a causative factor for several carcinomas by IARC in 2012 [54].

Robert Beasley performed pioneering studies in establishing the role of HBV in the development of hepatocellular carcinoma (HCC). He performed a large study involving approximately 22,000 healthy males for studying HBV infection, and the study group had a follow-up of 3 years [7]. HCC resulted in the death of about 41 individuals out of which 40 were tested positive for surface antigen of HBV [114]. HBV can lead to the development of HCC via several mechanisms. Viral proteins were implicated as one of the causative factors for the development of liver cancer as it was discovered that transgenic mice which express viral proteins HBsAg or HBx develop liver cancer [61]. Chronic inflammation, as a result of HBV infection, leads to cirrhosis that ultimately progresses into HCC [9]. Genome instability resulting from the insertion of the HBV into the genome of the host leads to the activation of the processes that ultimately result in HCC.

The cause of transfusion-associated hepatitis was discovered to be neither hepatitis A virus nor hepatitis B virus. A new virus was believed to cause transfusion-associated hepatitis and subsequently HCC, though the identification of hepatitis C virus (HCV) was extremely difficult [40]. After several attempts failed to identify HCV in subsequent years, a large scale investigation was conducted by Houghton [52] along with Daniel Bradley. It took about 7 years of hard work which led to the identification of HCV having a genetic material made up of RNA (about 10 kb) and was named as HCV [18]. HCC caused by HCV needs a long duration of several years and progresses through intermediate stages of chronic hepatitis and cirrhosis. A meta-analysis conducted among HCV-positive cirrhosis patients revealed that about 5 years is needed for the development of HCC in 7–30% of cases [28]. Another long-term study in patients found that risk of HCC was higher in cases which had higher RNA content of HCV [67]. Of the possible mechanisms, chronic immune inflammatory response has been recognized as the major contributor in the development of HCC mediated by HCV [48].

Human Papillomavirus (HPV)

Harald zur Hausen was a pioneer in establishing the role of human papillomavirus in the development of cervical cancer. After initial attempts to establish this relationship failed due to limitations in technology, zur Hausen's group could successfully sequence HPV by using the recombinant DNA technology in the mid-1970s. They identified different types of HPV such as HPV-16 and HPV-18 in cervical cancer biopsies [12, 25] which produce oncogenic proteins [107] such as E6 and E7 which

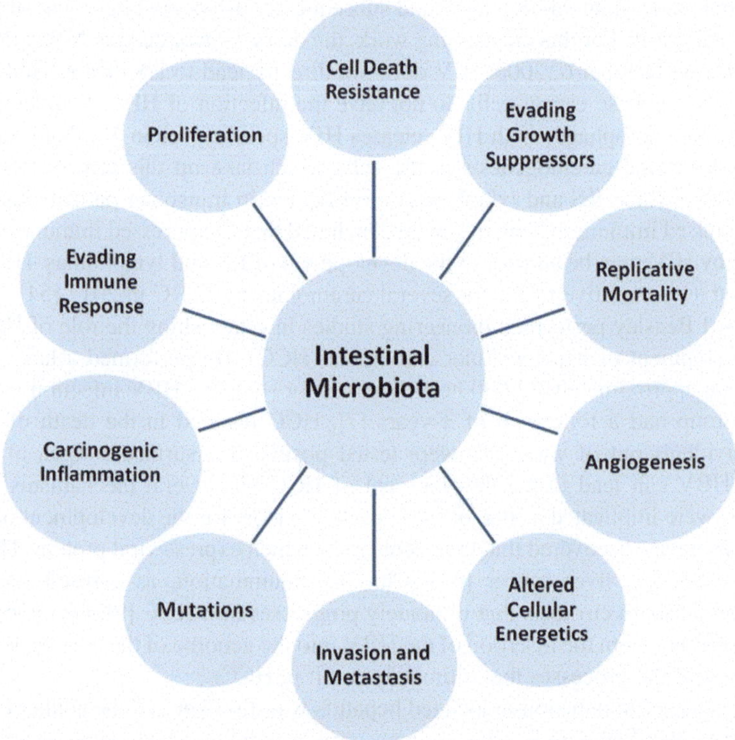

Fig. 1 Factors governed by intestinal microbiota in carcinogenesis

induce chromosomal instability. E6 binds to p53 and proteins, which play a key role in regulating advancement of the cell division cycle from the G1 to S-phases, leading to their lysis and a perturbed function. HPV was not found to integrate into the host's cell genome in a study [83] though the virus usually integrates into the host cell's genome. The precise mechanism of integration is not known though the most likely mechanism could be the points where the DNA breaks under the influence of E6 and E7. Furthermore, it was also found that though the inserted HPV is smaller in size, there is no effect on the E6 and E7. On the contrary, both have higher transcription levels and a higher stability [57]. It has also been observed that invasive cancers are characterized by higher viral insertion compared to the premalignant cancers [3].

Association of Microbiome with Cancer: Mechanistic Studies

The colon is the region of the body that has the highest concentration of microbes as a result of which has been the target of studies on carcinogenesis in experimental models. Some of the key factors governed by intestinal microbiota in carcinogenesis have been summarized in Fig. 1. Studies have demonstrated that in colon cancer, the microbial composition has a function in tumorigenesis via modulation of inflammation [45]. Treatment with an antibiotic had a reducing effect on tumor as is observed when the IL-23 receptor is knocked off. Zackular et al. [115] showed that colorectal tumors can be developed in a healthy mouse by transfer of microbiota from a mouse with tumor to healthy mouse. Interestingly, resident microbiome is an important factor for the survival of some cancers by evading the immune system. Treatment with antibiotics results in damage to the microbiome which leads to the treatment of the cancer. Studies have observed this phenomenon in the treatment of some lymphomas [30, 31, 102] and pseudomyxoma peritonei [39] which is a rare malignant growth marked by the progressive accumulation of mucus-secreting tumor cells within the abdomen and pelvis. Some cancers develop as a result of genotoxic and other toxins generated by specific microbes such as calobactin that can lead to breaks in both the strands of DNA [78]. To separate the effect of inflammation and effect of specific *E. coli*-borne genotoxin in cancer development, experiments were conducted in which mice prone to colorectal cancer were infected with either *E. coli* containing the pks gene or lacking it. The results were quite interesting as former mice exhibited increased number of tumors while the latter mice did not have such effect [4]. Furthermore, this association was further confirmed by the discovery of the presence of pks gene in the microbial genome present in colon tissue in majority of colorectal cancer patients compared to the controls [4].

The role of *Fusobacterium* species as a causative factor for colorectal cancer (CRC) has been investigated, and several studies have observed presence of *Fusobacterium* species in CRC samples [15, 64]. Moreover, *Fusobacterium* was also found present in colorectal liver metastases, thereby suggesting a conducive niche [13]. One of the mechanisms *Fusobacterium* employs to protect itself and also promote cancer is by reducing the capability of immune system to target itself. Myeloid-derived suppressor cells promote cancer and lead to poor prognosis in cancer patients [76]. The other mechanisms include the blocking of NK cell-mediated destruction of cancer cells [47] and host cell invasion [92]. As seen in mouse models, increased presence of *Fusobacterium* in tumor tissues leads to increased levels of NF-κB mRNA [64, 92] and decreased CD3-positive T-cells suggesting that it inhabits the tumors that are not targeted by the immune system.

Bacterial role in promoting carcinogenesis has been studied in detail in *Helicobacter* which is a causative factor for gastric MALT lymphoma and gastric adenocarcinoma [82]. The mechanisms employed by *H. pylori* for promoting carcinogenesis include host colonization, manipulation of the immune system, inducing key steps of carcinogenesis including inflammation, inhibiting immune response against tumors, and an increased proliferation. *H. pylori* achieves colonization by moving across the thick gastric mucous layer via flagella and employing adhesins such as SabA and BabA to adhere to the gastric epithelial cells [72, 103]. After attachment to the epithelium, it deposits CagA and other virulence factors which in turn dysregulates SHP2, leading to increased cellular proliferation [49]. The immune system responds to an *H. pylori* infection by promoting inflammatory mechanisms mediated by neutrophils, macrophages, inflammatory cytokines, and NF-κB which lead to carcinogenesis. There is also DNA damage in host cells caused by the ROS while *H. pylori* is protected by ROS with the help of some proteins [93]. *H. pylori* employs a fascinating mechanism to not only protect itself from the immune response but also to evade screening by the immune system. The T-cell proliferation is compromised by using the virulence factor VacA and γ-glutamyltranspeptidase (GGT) of *H. pylori* [37, 98, 104] which diverts Th1 and Th17 immune responses against *H. pylori* toward regulatory T-cell responses.

Conclusions

Microbial oncogenesis is a complex process mediated by several mechanisms which range from derailing the immune system to production of oncoproteins and other metabolites and toxins.

The mechanism of carcinogenesis of few pathogens is reasonably well understood, but a lot needs to be done to elucidate the pathways or mechanisms of a still large group of pathogens. Having an in-depth knowledge of microbial oncogenesis paves the way for interventions to control the process and prevent cancer. As the field grows, new and better clinical diagnostics are being invented to identify more microbial genotypes given the dynamic nature of microbial oncogenesis. As opposed to viral infections, bacterial infections are curable and offer the possibility of antibiotic treatments to cure cancers. Vaccination against cancer-causing pathogens offers an attractive opportunity to prevent infection and thus eliminate the risk of cancer. Pharmacomicrobiomics has recently emerged as a new branch in which the role of genetic variations of human microbiome in personalized medicine is being investigated. Studies have shown that response to a particular chemotherapeutic agent is dependent upon the metabolic activities of a specific gut microbiome. A better knowledge of microbiome composition and metabolic activities will help to understand the microbes and their targets that will pave way for developing new clinical tools for preventing and treating cancers.

Acknowledgments The financial assistance provided by Science and Engineering Research Board, Department of Science and Technology, Government of India, under Teachers Associateship for Research Excellence (TARE) scheme is gratefully acknowledged.

References

1. Allcroft, R. (1969). Chapter IX. Aflatoxicosis in farm animals. In L. A. Goldblatt (Ed.), *Aflatoxin, scientific background, control, and implications* (pp. 237–264). New York: Academic.
2. Amieva, M. R., & El-Omar, E. M. (2008). Host-bacterial interactions in *Helicobacter pylori* infection. *Gastroenterology, 134*, 306–323.
3. Arias-Pulido, H., Peyton, C. L., Joste, N. E., Vargas, H., & Wheeler, C. M. (2006). Human papillomavirus type 16 integration in cervical carcinoma in situ and in invasive cervical cancer. *Journal of Clinical Microbiology, 44*, 1755–1762.

4. Arthur, J. C., Gharaibeh, R. Z., Mühlbauer, M., Perez-Chanona, E., Uronis, J. M., McCafferty, J., Fodor, A. A., & Jobin, C. (2014). Microbial genomic analysis reveals the essential role of inflammation in bacteria-induced colorectal cancer. *Nature Communications, 5*, 4724.
5. Backhed, F., Ding, H., Wang, T., Hooper, L. V., Koh, G. Y., Nagy, A., Semenkovich, C. F., & Gordon, J. I. (2004). The gut microbiota as an environmental factor that regulates fat storage. *Proceedings of the National Academy of Sciences of the United States of America, 101*, 15718–15723.
6. Barré-Sinoussi, F., Chermann, J. C., Rey, F., Nugeyre, M. T., Chamaret, S., Gruest, J., Dauguet, C., Axler-Blin, C., Vézinet-Brun, F., Rouzioux, C., Rozenbaum, W., & Montagnier, L. (1983). Isolation of a T-lymphotropic retrovirus from a patient at risk for acquired immune deficiency syndrome (AIDS). *Science, 220*, 868–871.
7. Beasley, R. P., Hwang, L. Y., Lin, C. C., & Chien, C. S. (1981). Hepatocellular carcinoma and hepatitis B virus. A prospective study of 22 707 men in Taiwan. *Lancet, 2*, 1129–1133.
8. Bhamarapravati, N., & Virranuvatti, V. (1966). Liver diseases in Thailand. An analysis of liver biopsies. *The American Journal of Gastroenterology, 45*, 267–275.
9. Blackadar, C. B. (2013). Systematic review of hepatocellular carcinoma mortality rates among hepatitis B virus-infected renal transplant recipients, with supplemental analyses of liver failure and all-cause mortality. *International Journal of Infectious Diseases, 17*, e24–e36.
10. Blaser, M. J., & Atherton, J. (2004). *Helicobacter pylori* persistence: Biology and disease. *The Journal of Clinical Investigation, 113*, 321–333.
11. Blaser, M. J., & Kirschner, D. (2007). The equilibria that permit bacterial persistence in human hosts. *Nature, 449*, 843–849.
12. Boshart, M., Gissmann, L., Ikenberg, H., Kleinheinz, A., Scheurlen, W., & zur Hausen, H. (1984). A new type of papillomavirus DNA, its presence in genital cancer biopsies and in cell lines derived from cervical cancer. *The EMBO Journal, 3*, 1151–1157.
13. Bullman, S., Pedamallu, C. S., Sicinska, E., Clancy, T. E., Zhang, X., Cai, D., Neuberg, D., Huang, K., Guevara, F., Nelson, T., Chipashvili, O., Hagan, T., Walker, M., Ramachandran, A., Diosdado, B., Serna, G., Mulet, N., Landolfi, S., Ramon, Y., Cajal, S., Fasani, R., Aguirre, A. J., Ng, K., Élez, E., Ogino, S., Tabernero, J., Fuchs, C. S., Hahn, W. C., & Nuciforo, P. (2017). Meyerson M3. Analysis of *Fusobacterium* persistence and antibiotic response in colorectal cancer. *Science, 358*, 1443–1448.
14. Caporaso, J. G., Lauber, C. L., Costello, E. K., Berg-Lyons, D., Gonzalez, A., Stombaugh, J., Knights, D., Gajer, P., Ravel, J., Fierer, N., Gordon, J. I., & Knight, R. (2011). Moving pictures of the human microbiome. *Genome Biology, 12*, R50.
15. Castellarin, M., Warren, R. L., Freeman, J. D., Dreolini, L., Krzywinski, M., Strauss, J., Barnes, R., Watson, P., Allen-Vercoe, E., Moore, R. A., & Holt, R. A. (2012). *Fusobacterium nucleatum* infection is prevalent in human colorectal carcinoma. *Genome Research, 22*, 299–306.
16. Chang, Y. J., Wu, M. S., Lin, J. T., Peslell, R. G., Blaser, M. J., & Chen, C. C. (2006). Mechanisms in *Helicobacter pylori* CagA-induced cyclin D1 expression that affect cell cycle. *Cellular Microbiology, 8*, 1740–1752.
17. Charbonneau, M. R., Blanton, L. V., DiGiulio, D. B., Relman, D. A., Lebrilla, C. B., Mills, D. A., & Gordon, J. I. (2016). A microbial perspective of human developmental biology. *Nature, 535*, 48–55.
18. Choo, Q. L., Kuo, G., Weiner, A. J., Overby, L. R., Bradley, D. W., & Houghton, M. (1989). Isolation of a cDNA clone derived from a blood-borne non-A, non-B viral hepatitis genome. *Science, 244*, 359–362.
19. Cohen, R. J., Shannon, B. A., McNeal, J. E., Shannon, T., & Garrett, K. L. (2005). *Propionibacterium acnes* associated with inflammation in radical prostatectomy specimens: A possible link to cancer evolution? *The Journal of Urology, 173*, 1969–1974.
20. Cook, L. B., Melamed, A., Niederer, H., Valganon, M., Laydon, D., Foroni, L., Taylor, G. P., Matsuoka, M., & Bangham, C. R. (2014). The role of HTLV-1 clonality, proviral structure, and genomic integration site in adult T-cell leukemia/lymphoma. *Blood, 123*, 3925–3931.

21. Costello, E. K., Lauber, C. L., Hamady, M., Fierer, N., Gordon, J. I., & Knight, R. (2009). Bacterial community variation in human body habitats across space and time. *Science, 326*, 1694–1697.
22. Ding, H. T., Taur, Y., & Walkup, J. T. (2017). Gut microbiota and autism: Key concepts and findings. *Journal of Autism and Developmental Disorders, 47*, 480–489.
23. Dominguez-Bello, M. G., Blaser, M. J., Ley, R. E., & Knight, R. (2011). Development of the human gastrointestinal microbiota and insights from high-throughput sequencing. *Gastroenterology, 140*, 1713–1719.
24. Dominguez-Bello, M. G., Costello, E. K., Contreras, M., Magris, M., Hidalgo, G., Fierer, N., & Knight, R. (2010). Delivery mode shapes the acquisition and structure of the initial microbiota across multiple body habitats in newborns. *Proceedings of the National Academy of Sciences of the United States of America, 107*, 11971–11975.
25. Dürst, M., Gissmann, L., Ikenberg, H., & zur Hausen, H. (1983). A papillomavirus DNA from a cervical carcinoma and its prevalence in cancer biopsy samples from different geographic regions. *Proceedings of the National Academy of Sciences of the United States of America, 80*, 3812–3815.
26. Ege, M. J., Mayer, M., Normand, A. C., Genuneit, J., Cookson, W. O., Braun-Fahrländer, C., Heederik, D., Piarroux, R., von Mutius, E., & GABRIELA Transregio 22 Study Group. (2011). Exposure to environmental microorganisms and childhood asthma. *The New England Journal of Medicine, 364*, 701–709.
27. Ellermann, V., & Bang, O. (1909). Experimental leukemia in chickens II (from German). *Zeitschrift fur Hygiene und Infektionskrakheiten, 63*, 231–273.
28. Fattovich, G., Stroffolini, T., Zagni, I., & Donato, F. (2004). Hepatocellular carcinoma in cirrhosis: Incidence and risk factors. *Gastroenterology, 127*, S35–S50.
29. Ferguson, A. R. (1911). Associated bilharziosis and primary malignant disease of the urinary bladder, with observations on a series of forty cases. *The Journal of Pathology and Bacteriology, 16*, 76–98.
30. Ferreri, A. J., Govi, S., Pasini, E., Mappa, S., Bertoni, F., Zaja, F., Montalbán, C., Stelitano, C., Cabrera, M. E., Giordano Resti, A., Politi, L. S., Doglioni, C., Cavalli, F., Zucca, E., Ponzoni, M., & Dolcetti, R. (2012). Chlamydophila psittaci eradication with doxycycline as first-line targeted therapy for ocular adnexae lymphoma: Final results of an international phase II trial. *JCO, 30*, 2988–2994.
31. Ferreri, A. J., Ponzoni, M., Guidoboni, M., De Conciliis, C., Resti, A. G., Mazzi, B., Lettini, A. A., Demeter, J., Dell'Oro, S., Doglioni, C., Villa, E., Boiocchi, M., & Dolcetti, R. (2005). Regression of ocular adnexal lymphoma after Chlamydia psittaci-eradicating antibiotic therapy. *JCO, 23*, 5067–5073.
32. Fierer, N., Lauber, C. L., Zhou, N., McDonald, D., Costello, E. K., & Knight, R. (2010). Forensic identification using skin bacterial communities. *Proceedings of the National Academy of Sciences of the United States of America, 107*, 6477–6481.
33. Fischbach, L. A., Graham, D. Y., Kramer, J. R., Rugge, M., Verstovsek, G., Parente, P., Alsarraj, A., Fitzgerald, S., Shaib, Y., Abraham, N. S., Kolpachi, A., Gupta, S., Vela, M. F., Velez, M., Cole, R., Anand, B., & El Serag, H. B. (2014). Association between *Helicobacter pylori* and Barrett's esophagus: A case-control study. *The American Journal of Gastroenterology, 109*, 357–368.
34. Francois, F., Roper, J., Goodman, A. J., Pei, Z., Ghumman, M., Mourad, M., de Perez, A. Z., Perez-Perez, G. I., Tseng, C. H., & Blaser, M. J. (2008). The association of gastric leptin with esophageal inflammation and metaplasia. *Gut, 57*, 16–24.
35. Fuccio, L., Zagari, R. M., Eusebi, L. H., Laterza, L., Cennamo, V., Ceroni, L., Grilli, D., & Bazzoli, F. (2009). Meta-analysis: Can *Helicobacter pylori* eradication treatment reduce the risk for gastric cancer? *Annals of Internal Medicine, 151*, 121–128.
36. Gallo, R. C. (2005). History of the discoveries of the first human retroviruses: HTLV-1 and HTLV-2. *Oncogene, 24*, 5926–5930.

37. Gerhard, M., Schmees, C., Voland, P., Endres, N., Sander, M., Reindl, W., Rad, R., Oelsner, M., Decker, T., Mempel, M., Hengst, L., & Prinz, C. (2005). A secreted low-molecular weight protein from *Helicobacter pylori* induces cell-cycle arrest of T cells. *Gastroenterology, 128*, 1327–1339.
38. Geller, L. T., Barzily-Rokni, M., Danino, T., Jonas, O. H., Shental, N., Nejman, D., Gavert, N., Zwang, Y., Cooper, Z. A., Shee, K., Thaiss, C. A., Reuben, A., Livny, J., Avraham, R., Frederick, D. T., Ligorio, M., Chatman, K., Johnston, S. E., Mosher, C. M., Brandis, A., Fuks, G., Gurbatri, C., Gopalakrishnan, V., Kim, M., Hurd, M. W., Katz, M., Fleming, J., Maitra, A., Smith, D. A., Skalak, M., Bu, J., Michaud, M., Trauger, S. A., Barshack, I., Golan, T., Sandbank, J., Flaherty, K. T., Mandinova, A., Garrett, W., Thayer, S. P., Ferrone, C. R., Huttenhower, C., Bhatia, S. N., Gevers, D., Wargo, J. A., Golub, T. R., & Straussman, R. (2017). Potential role of intratumor bacteria in mediating tumor resistance to the chemotherapeutic drug gemcitabine. *Science, 357*, 1156–1160.
39. Gilbreath, J. J., Semino-Mora, C., Friedline, C. J., Liu, H., Bodi, K. L., McAvoy, T. J., Francis, J., Nieroda, C., Sardi, A., Dubois, A., Lazinski, D. W., Camilli, A., Testerman, T. L., & Merrell, D. S. (2013). A core microbiome associated with the peritoneal tumors of pseudomyxoma peritonei. *Orphanet Journal of Rare Diseases, 8*, 105.
40. Gilliam, J. H., Geisinger, K. R., & Richter, J. E. (1984). Primary hepatocellular carcinoma after chronic non-A, non-B post-transfusion hepatitis. *Annals of Internal Medicine, 101*, 794–795.
41. Goedert, J. J., Coté, T. R., Virgo, P., Scoppa, S. M., Kingma, D. W., Gail, M. H., Jaffe, E. S., & Biggar, R. J. (1998). Spectrum of AIDS-associated malignant disorders. *Lancet, 351*, 1833–1839.
42. Gottlieb, M. S., Schanker, H. M., Fan, P. T., Saxon, A., Weisman, J. D., & Pozalski, I. (1981). Pneumocystis pneumonia--Los Angeles. *MMWR. Morbidity and Mortality Weekly Report, 30*, 250–252.
43. Gong, H. L., Shi, Y., Zhou, L., Wu, C. P., Cao, P. Y., Tao, L., Xu, C., Hou, D. S., & Wang, Y. Z. (2013). The composition of microbiome in larynx and the throat biodiversity between laryngeal squamous cell carcinoma patients and control population. *PLoS One, 8*, e66476.
44. Grishin, A., Papillon, S., Bell, B., Wang, J., & Ford, H. R. (2013). The role of the intestinal microbiota in the pathogenesis of necrotizing enterocolitis. *Seminars in Pediatric Surgery, 22*, 69–75.
45. Grivennikov, S. I., Wang, K., Mucida, D., Stewart, C. A., Schnabl, B., Jauch, D., Taniguchi, K., Yu, G. Y., Osterreicher, C. H., Hung, K. E., Datz, C., Feng, Y., Fearon, E. R., Oukka, M., Tessarollo, L., Coppola, V., Yarovinsky, F., Cheroutre, H., Eckmann, L., Trinchieri, G., & Karin, M. (2012). Adenoma-linked barrier defects and microbial products drive IL-23/IL-17-mediated tumour growth. *Nature, 491*, 254–258.
46. Groopman, J. D., Wang, J. S., & Scholl, P. (1996). Molecular biomarkers for aflatoxins: From adducts to gene mutations to human liver cancer. *Canadian Journal of Physiology and Pharmacology, 74*, 203–209.
47. Gur, C., Ibrahim, Y., Isaacson, B., Yamin, R., Abed, J., Gamliel, M., Enk, J., Bar-On, Y., Stanietsky-Kaynan, N., Coppenhagen-Glazer, S., Shussman, N., Almogy, G., Cuapio, A., Hofer, E., Mevorach, D., Tabib, A., Ortenberg, R., Markel, G., Miklić, K., Jonjic, S., Brennan, C. A., Garrett, W. S., Bachrach, G., & Mandelboim, O. (2015). Binding of the Fap2 protein of *Fusobacterium nucleatum* to human inhibitory receptor TIGIT protects tumors from immune cell attack. *Immunity, 42*, 344–355.
48. Hajarizadeh, B., Grebely, J., & Dore, G. J. (2013). Epidemiology and natural history of HCV infection. *Nature Reviews. Gastroenterology & Hepatology, 10*, 53–562.
49. Hatakeyama, M. (2004). Oncogenic mechanisms of the *Helicobacter pylori* CagA protein. *Nature Reviews. Cancer, 4*, 688–694.
50. Hehemann, J., Correc, G., Barbeyron, T., Helbert, W., Czjzek, M., & Michel, G. (2010). Transfer of carbohydrate reactive enzymes from marine bacteria to Japanese gut microbiota. *Nature, 464*, 908–912.

51. Hinuma, Y., Nagata, K., Hanaoka, M., Nakai, M., Matsumoto, T., Kinoshita, K. I., Shirakawa, S., & Miyoshi, I. (1981). Adult T-cell leukemia: Antigen in an ATL cell line and detection of antibodies to the antigen in human sera. *Proceedings of the National Academy of Sciences of the United States of America, 78*, 6476–6480.
52. Houghton, M. (2009). The long and winding road leading to the identification of the hepatitis C virus. *Journal of Hepatology, 51*, 939–948.
53. Huang, J. Q., Sridhar, S., Chen, Y., & Hunt, R. H. (1998). Meta-analysis of the relationship between *Helicobacter pylori* seropositivity and gastric cancer. *Gastroenterology, 114*, 1169–1179.
54. IARC (International Agency for Research on Cancer). (2012). Human immunodeficiency virus-1. In *IARC monographs on the evaluation of carcinogenic risk of chemicals to man volume 100B: A review of human carcinogens: Biological agents* (pp. 215–253). Lyon: International Agency for Research on Cancer.
55. IARC (International Agency for Research on Cancer). (1994). Infection with schistosomes (Schistosoma haematobium, S. mansoni and S. japonicum). In *IARC monographs on the evaluation of carcinogenic risk of chemicals to man*. Volume 61, Schistosomes, liver flukes and *Helicobacter pylori* (pp. 45–120). Lyon: International Agency for Research on Cancer, esp 45–59.
56. Jackson, P. E., & Groopman, J. D. (1999). Aflatoxin and liver cancer. *Baillière's Best Practice & Research Clinical Gastroenterology, 13*, 545–555.
57. Jeon, S., & Lambert, P. F. (1995). Integration of human papillomavirus type 16 DNA into the human genome leads to increased stability of E6 and E7 mRNAs: Implications for cervical carcinogenesis. *Proceedings of the National Academy of Sciences of the United States of America, 92*, 1654–1658.
58. Johri, M. K., Mishra, R., Chhatbar, C., Unni, S. K., & Singh, S. K. (2011). Tits and bits of HIV Tat protein. *Expert Opinion on Biological Therapy, 11*, 269–283.
59. Kawakami, T. G., Huff, S. D., Buckley, P. M., Dungworth, D. L., Synder, S. P., & Gilden, R. V. (1972). C-type virus associated with gibbon lymphosarcoma. *Nature: New Biology, 235*, 170–171.
60. Kettmann, R., Mammerickx, M., Dekegel, D., Ghysdael, J., Portetelle, D., & Burny, A. (1975). Biochemical approach to bovine leukemia. *Acta Haematologica, 54*, 201–209.
61. Kim, C. M., Koike, K., Saito, I., Miyamura, T., & Jay, G. (1991). HBx gene of hepatitis B virus induces liver cancer in transgenic mice. *Nature, 351*, 317–320.
62. Kim, S., Covington, A., & Pamer, E. G. (2017). The intestinal microbiota: Antibiotics, colonization resistance, and enteric pathogens. *Immunological Reviews, 279*, 90–105.
63. Koenig, J. E., Spor, A., Scalfone, N., Fricker, A. D., Stombaugh, J., Knight, R., Angenent, L. T., & Ley, R. E. (2011). Succession of microbial consortia in the developing infant gut microbiome. *Proceedings of the National Academy of Sciences of the United States of America, 108*, 4578–4585.
64. Kostic, A. D., Chun, E., Robertson, L., Glickman, J. N., Gallini, C. A., Michaud, M., Clancy, T. E., Chung, D. C., Lochhead, P., Hold, G. L., El-Omar, E. M., Brenner, D., Fuchs, C. S., Meyerson, M., & Garrett, W. S. (2013). *Fusobacterium nucleatum* potentiates intestinal tumorigenesis and modulates the tumor-immune microenvironment. *Cell Host & Microbe, 14*, 207–215.
65. Lancaster, M. C., Jenkins, F. P., & Philip, J. M. (1961). Toxicity associated with certain samples of groundnuts. *Nature, 192*, 1095–1096.
66. Lederberg, J., & McCray, A. T. (2001). 'Ome sweet 'omics a genealogical treasury of words. *Scientist, 17*, 8.
67. Lee, M. H., Yang, H. I., Lu, S. N., Jen, C. L., Yeh, S. H., Liu, C. J., Chen, P. J., You, S. L., Wang, L. Y., Chen, W. J., & Chen, C. J. (2010). Hepatitis C virus seromarkers and subsequent risk of hepatocellular carcinoma: Long-term predictors from a community-based cohort study. *Journal of Clinical Oncology, 28*, 4587–4593.
68. Ley, R. E., Backhed, F., Turnbaugh, P., Lozupone, C. A., Knight, R. D., & Gordon, J. I. (2005). Obesity alters gut microbial ecology. *Proceedings of the National Academy of Sciences of the United States of America, 102*, 11070–11075.

69. Ley, R. E., Hamady, M., Lozupone, C., Turnbaugh, P. J., Ramey, R. R., Bircher, J. S., Schlegel, M. L., Tucker, T. A., Schrenzel, M. D., Knight, R., & Gordon, J. I. (2008). Evolution of mammals and their gut microbes. *Science, 320*, 1647–1651.
70. Ley, R. E., Turnbaugh, P. J., Klein, S., & Gordon, J. I. (2006). Microbial ecology: Human gut microbes associated with obesity. *Nature, 444*, 1022–1023.
71. Lozupone, C. A., Stombaugh, J. I., Gordon, J. I., Jansson, J. K., & Knight, R. (2012). Diversity, stability and resilience of the human gut microbiota. *Nature, 489*, 220–230.
72. Mahdavi, J. (2002). *Helicobacter pylori* SabA adhesin in persistent infection and chronic inflammation. *Science, 297*, 573–578.
73. Manichanh, C., Rigottier-Gois, L., Bonnaud, E., Gloux, K., Pelletier, E., Frangeul, L., Nalin, R., Jarrin, C., Chardon, P., Marteau, P., Roca, J., & Dore, J. (2006). Reduced diversity of faecal microbiota in Crohn's disease revealed by a metagenomic approach. *Gut, 55*, 205–211.
74. Marshall, B. J., & Warren, J. R. (1984). Unidentified curved bacilli in the stomach of patients with gastritis and peptic ulceration. *Lancet, 1*, 1311–1315.
75. Meropol, S. B., & Edwards, A. (2015). Development of the infant intestinal microbiome: A bird's eye view of a complex process. *Birth Defects Research. Part C, Embryo Today, 105*, 228–239.
76. Montero, A. J., Diaz-Montero, C. M., Kyriakopoulos, C. E., Bronte, V., & Mandruzzato, S. (2012). Myeloid-derived suppressor cells in cancer patients: A clinical perspective. *Journal of Immunotherapy, 35*, 107–115.
77. NIH Consensus Conference. (1994). *Helicobacter pylori* in peptic ulcer disease. NIH Consensus development panel on *Helicobacter pylori* in peptic ulcer disease. *JAMA, 272*, 65–69.
78. Nougayrede, J. P. (2006). Escherichia coli induces DNA double-strand breaks in eukaryotic cells. *Science, 313*, 848–851.
79. Oliviera, D. E. (2007). DNA viruses in human cancer: An integrated overview on fundamental mechanisms of viral carcinogenesis. *Cancer Letters, 247*, 182–196.
80. Palmer, C., Bik, E. M., DiGiulio, D. B., Relman, D. A., & Brown, P. O. (2007). Development of the human infant intestinal microbiota. *PLoS Biology, 5*, e177.
81. Parkin, D. M. (2006). The global health burden of infection associated cancers in the year 2002. *International Journal of Cancer, 118*, 3030–3044.
82. Peek, R. M., & Blaser, M. J. (2002). *Helicobacter pylori* and gastrointestinal tract adenocarcinomas. *Nature Reviews. Cancer, 2*, 28–37.
83. Pett, M., & Coleman, N. (2007). Integration of high-risk human papillomavirus: A key event in cervical carcinogenesis? *The Journal of Pathology, 212*, 356–367.
84. Pillinger, M. H., Marjanovic, N., Kim, S. Y., Lee, Y. C., Scher, J. U., Roper, J., Abeles, A. M., Izmirly, P. I., Axelrod, M., Pillinger, M. Y., Tolani, S., Dinsell, V., Abramson, S. B., & Blaser, M. J. (2007). *Helicobacter pylori* stimulates gastric epithelial cell MMP-1 secretion via CagA-dependent and independent ERK activation. *The Journal of Biological Chemistry, 282*, 18722–18731.
85. Qian, G. S., Ross, R. K., Yu, M. C., Yuan, J. M., Gao, Y. T., Henderson, B. E., Wogan, G. N., & Groopman, J. D. (1994). A follow-up study of urinary markers of aflatoxin exposure and liver cancer risk in Shanghai, People's republic of China. *Cancer Epidemiology, Biomarkers & Prevention, 3*, 3–10.
86. Qin, J., Li, R., Raes, J., Arumugam, M., Burgdorf, K. S., Manichanh, C., Nielsen, T., Pons, N., Levenez, F., Yamada, T., Mende, D. R., Li, J., Xu, J., Li, S., Li, D., Cao, J., Wang, B., Liang, H., Zheng, H., Xie, Y., Tap, J., Lepage, P., Bertalan, M., Batto, J. M., Hansen, T., Le Paslier, D., Linneberg, A., Nielsen, H. B., Pelletier, E., Renault, P., Sicheritz-Ponten, T., Turner, K., Zhu, H., Yu, C., Li, S., Jian, M., Zhou, Y., Li, Y., Zhang, X., Li, S., Qin, N., Yang, H., Wang, J., Brunak, S., Doré, J., Guarner, F., Kristiansen, K., Pedersen, O., Parkhill, J., Weissenbach, J., MetaHIT Consortium, B. P., Ehrlich, S. D., & Wang, J. (2010). A human gut microbial gene catalogue established by metagenomic sequencing. *Nature, 464*, 59–65.
87. Reed, W., Carroll, J., Agramonte, A., & Lazear, J. W. (1900). The etiology of yellow fever-a preliminary note. *Public Health Papers and Reports, 26*, 37–53.

88. Riiser, A. (2015). The human microbiome, asthma, and allergy. *Allergy, Asthma and Clinical Immunology, 11*, 35.
89. Romanelli, M. G., Diani, E., Bergamo, E., Casoli, C., Ciminale, V., Bex, F., & Bertazzoni, U. (2013). Highlights on distinctive structural and functional properties of HTLV tax proteins. *Frontiers in Microbiology, 4*, 271.
90. Ross, R. K., Yuan, J. M., Yu, M. C., Wogan, G. N., Qian, G. S., Tu, J. T., Groopman, J. D., Gao, Y. T., & Henderson, B. E. (1992). Urinary aflatoxin biomarkers and risk of hepatocellular carcinoma. *Lancet, 339*, 943–946.
91. Rous, P. (1911). Transmission of a malignant new growth by means of a cell-free filtrate. *JAMA, 56*, 198.
92. Rubinstein, M. R., Wang, X., Liu, W., Hao, Y., Cai, G., & Han, Y. W. (2013). *Fusobacterium nucleatum* promotes colorectal carcinogenesis by modulating E-cadherin/b-catenin signaling via its FadA adhesin. *Cell Host & Microbe, 14*, 195–206.
93. Salama, N. R., Hartung, M. L., & Müller, A. (2013). Life in the human stomach: Persistence strategies of the bacterial pathogen *Helicobacter pylori. Nature Reviews. Microbiology, 11*, 385–399.
94. Sanz, Y., Olivares, M., Moya-Pérez, Á., & Agostoni, C. (2015). Understanding the role of gut microbiome in metabolic disease risk. *Pediatric Research, 77*, 236–244.
95. Sartor, R. B. (2006). Mechanisms of disease: Pathogenesis of Crohn's disease and ulcerative colitis. *Nature Clinical Practice. Gastroenterology & Hepatology, 3*, 390–407.
96. Satou, Y., Yasunaga, J., Zhao, T., Yoshida, M., Miyazato, P., Takai, K., Shimizu, K., Ohshima, K., Green, P. L., Ohkura, N., Yamaguchi, T., Ono, M., Sakaguchi, S., & Matsuoka, M. (2011). HTLV-1 bZIP factor induces T-cell lymphoma and systemic inflammation in vivo. *PLoS Pathogens, 7*, e1001274.
97. Schistosomes, Liver Flukes and *Helicobacter pylori*. IARC Working Group on the Evaluation of Carcinogenic Risks to Humans. Lyon, 7–14 June 1994. IARC Monogr Eval Carcinog Risks Hum 1994; 61:1–241.
98. Schmees, C., Prinz, C., Treptau, T., Rad, R., Hengst, L., Voland, P., Bauer, S., Brenner, L., Schmid, R. M., & Gerhard, M. (2007). Inhibition of T-cell proliferation by *Helicobacter pylori* γ-glutamyl transpeptidase. *Gastroenterology, 132*, 1820–1833.
99. Schottenfeld, D., & Fraumeni, J. F. (2006). *Cancer epidemiology and prevention* (Third ed.). New York: Oxford University Press.
100. Shimkin, M. B. (1977). Virus hyspothesis: Borrel and Oberling. In *Contrary to nature: Being an illustrated commentary on some persons and events of historical importance in the development of knowledge concerning cancer* (NIH Publication No. 76–720) (pp. 213–214). Washington, DC: US Department of Health, Education and Welfare.
101. Stewart, M. J. (1931). Precancerous lesions of the alimentary tract. *Lancet, 218*, 669–675.
102. Stolte, M., Bayerdörffer, E., Morgner, A., Alpen, B., Wündisch, T., Thiede, C., & Neubauer, A. (2002). Helicobacter and gastric MALT lymphoma. *Gut, 50*(suppl 3), III19–III24.
103. Styer, C. M., Hansen, L. M., Cooke, C. L., Gundersen, A. M., Choi, S. S., Berg, D. E., Benghezal, M., Marshall, B. J., Peek, R. M., Jr., Borén, T., & Solnick, J. V. (2010). Expression of the BabA adhesin during experimental infection with *Helicobacter pylori. Infection and Immunity, 78*, 1593–1600.
104. Sundrud, M. S., Torres, V. J., Unutmaz, D., & Cover, T. L. (2004). Inhibition of primary human T cell proliferation by *Helicobacter pylori* vacuolating toxin (VacA) is independent of VacA effects on IL-2 secretion. *Proceedings of the National Academy of Sciences of the United States of America, 101*, 7727–7732.
105. Takeda, S., Maeda, M., Morikawa, S., Taniguchi, Y., Yasunaga, J., Nosaka, K., Tanaka, Y., & Matsuoka, M. (2004). Genetic and epigenetic inactivation of tax gene in adult T-cell leukemia cells. *International Journal of Cancer, 109*, 559–567.
106. Thomsen, L. L., Gavin, J. B., & Tasman-Jones, C. (1990). Relation of *Helicobacter pylori* to the human gastric mucosa in chronic gastritis of the antrum. *Gut, 31*, 1230–1236.

107. Tommasino, M. (2014). The human papillomavirus family and its role in carcinogenesis. *Seminars in Cancer Biology, 26*, 13–21.
108. Turnbaugh, P. J., Hamady, M., Yatsunenko, T., Cantarel, B. L., Duncan, A., Ley, R. E., Sogin, M. L., Jones, W. J., Roe, B. A., Affourtit, J. P., Egholm, M., Henrissat, B., Heath, A. C., Knight, R., & Gordon, J. I. (2009). A core gut microbiome in obese and lean twins. *Nature, 457*, 480–484.
109. Watanapa, P., & Watanapa, W. B. (2002). Liver fluke-associated cholangiocarcinoma. *The British Journal of Surgery, 89*, 962–970.
110. Wold, A. E. (1998). The hygiene hypothesis revised: Is the rising frequency of allergy due to changes in the intestinal flora? *Allergy, 53*, 20–25.
111. Wynder, E. L., & Gori, G. B. (1977). Contribution of the environment to cancer incidence: An epidemiologic exercise. *Journal of the National Cancer Institute, 58*, 825–832.
112. Xuan, C., Shamonki, J. M., Chung, A., Dinome, M. L., Chung, M., Sieling, P. A., & Lee, D. J. (2014). Microbial dysbiosis is associated with human breast cancer. *PLoS One, 9*, e83744.
113. Yoshida, M., Miyoshi, I., & Hinuma, Y. (1982). Isolation and characterization of retrovirus from cell lines of human T-cell leukemia and its implication in the disease. *Proceedings of the National Academy of Sciences of the United States of America, 79*, 2031–2035.
114. Yoshida, M., Seiki, M., Yamaguchi, K., & Takatsuki, K. (1984). Monoclonal integration of human T-cell leukemia provirus in all primary tumors of adult T-cell leukemia suggests causative role of human T-cell leukemia virus in the disease. *Proceedings of the National Academy of Sciences of the United States of America, 81*, 2534–2537.
115. Zackular, J. P., Baxter, N. T., Iverson, K. D., Sadler, W. D., Petrosino, J. F., Chen, G. Y., & Schloss, P. D. (2013). The gut microbiome modulates colon tumorigenesis. *MBio, 4*, e00692–e00613.

Nanobiotechnology: Current and Future Perspectives in Combating Microbial Pathogenesis

Indu Singh, Hemant K. Gautam, and Gagan Dhawan

Abstract

Nanobiotechnology is the bridge between biology and chemistry interface, with disparate biomedical as well as microbiological applications. Nanomaterials, nanoconjugates and nanowires have extensively been used for the detection of diverse pathological conditions as well as in the chemotherapy of the diagnosed disorders. Targeted drug and gene delivery has been shown to produce encouraging results. In current scenario, nosocomial infections have been affecting developing countries with a high frequency. Eradication of these infections may be achieved by introduction of novel nanodrugs effective for longer duration of time as well as with fewer side effects. Some peculiar properties of nanostructures such as cost-effectiveness, biocompatibility, mammalian cell compatibility and less toxicity to the environment make these nanoparticles as major candidates for various therapeutic purposes. In agriculture too, nanoparticles synthesized from marine sources or several bacteria, fungi, algae, actinomycetes and biofungicides have been shown to possess the potential to prevent the crops from pests. Nanobiotechnology provides a platform for designing and developing nanomaterials with promising effects that can be delivered at specific target sites. Combining nanoscience with biotechnology provides a broad term for

I. Singh
Department of Biomedical Science, Acharya Narendra Dev College, University of Delhi, New Delhi, India

Microbial Biotechnology Laboratory, CSIR-Institute of Genomics and Integrative Biology, New Delhi, India

H. K. Gautam
Microbial Biotechnology Laboratory, CSIR-Institute of Genomics and Integrative Biology, New Delhi, India

G. Dhawan (✉)
Department of Biomedical Science, Acharya Narendra Dev College, University of Delhi, New Delhi, India
e-mail: gagandhawan@andc.du.ac.in

S. Hameed, Z. Fatima (eds.), *Pathogenicity and Drug Resistance of Human Pathogens*,
https://doi.org/10.1007/978-981-32-9449-3_17

exploring the design and synthesis of novel molecules which can further be inculcated in various studies. At present, microbial infections are playing a major havoc due to improper use of antibiotics in hospitals, improper use of pesticides in fields, poor sanitation as well as lack of awareness among population. In this chapter, we mainly focus on the areas affected by nanobiotechnology, such as how microbial population can be affected, current trends in microbial infection inflation rate, various nanomaterials used to combat microbial infections as well as their future aspects.

Keywords
Nanobiotechnology · Nanomedicines · Bacteriophages · Bacterial magnetic nanoparticles (BMPs) · Biosensors

Introduction

Nanobiotechnology is the most burgeoning field in the current scenario. It has offered an opportunity to make significant changes in various properties through expansion of the uses. It reduces the need of energy and raw material through the development of environment-friendly processes. This field offers various advantages in medical perspective such as protein chips [1], quantum dots [2], drug delivery and gene delivery [3, 4], liposomes for gene therapy [5], biosensors [6] and development of biopharmaceuticals [7]. Nanotechnology has created platform in the knowledge domain of science, which has resulted in the advancements in the field of electronics, food, medical and agriculture. A variety of nanoformulations has been explored for different biomedical applications such as anticancer therapy, biosensing and molecular imaging [8]. This field is anticipated to produce innovations and plays an important role in different biomedical applications such as developing and designing biomarkers, molecular imaging and biosensors. The fundamental role of this area is to create targeted drug delivery methods and gene therapy [9]. Fusion of biotechnology and nanotechnology with microtechnology including molecular and biological aspects collectively is termed as nanobiotechnology. Nanobiotechnology is a modern form of life science which makes use of knowledge to develop products by interfering cellular, molecular and genetic processes [10]. Scientists around the globe are working on potential aspects of nanomedicines and nanobiotechnology. Targeting approach in both drug and gene delivery has attracted several scientific groups as it has become a major issue of concern in numerous treatment strategies such as enhancing pharmacodynamic as well as pharmacokinetic properties of a drug [11]. Nanotechnology promises to create novel nanomaterials with well-defined properties/morphology. Nanotechnological approach provides methods for manipulations in food polymers and dyes which help in improving food quality and safety.

Nanotechnology has been explored to an extent of developing nanoparticles (NPs) that serve as carrier capable of penetrating themselves through tight junction of cells present in the brain, for delivering therapeutic agents for the cure of brain-related tumours via detecting nucleic acids using beneficial biological weapon [4]. Use of nanomaterials in daily practice such as cosmetics, food packaging, controlling pest, water treatment and disease treatment strategies has become a routine fashion in current era. They have many new properties that do not accompany their counterparts at the macro scale. Nanotechnology is the art in which structure of nanoparticles can be manipulated depending on the physicochemical characteristics (size, surface charge, self-fabrication, polydispersity index and count rate) as well as biological applications. Size of nanoparticles determines their dispersibility and stability in a particular solvent. Simultaneously, surface charge reflects stability and extent by which drug binds to nucleic acid for targeted drug delivery [12, 13]. A large number of nanoparticles are available in variable sizes with disparate shapes. They are being used in various applications, which include automobiles, aerospace, electronics, household, hygiene and medicine. The synthesis of nanoparticles follows three different methods: biochemical methods (synthesis of NPs using biological active molecule present in microorganisms); herbal approach (synthesis of NPs using active ingredients in plant material); and chemical mediated (formation of nanomaterial using chemicals). Nanoparticles are divided in subgroups depending on their physical and chemical properties: (i) inorganic NPs, (ii) organic NPs, (iii) bionanopolymeric NPs, (iv) peptide/protein NPs, (v) carbon nanotubes, (vi) nanopolymers, etc. Quantum dots help in detection of in vitro as well as in vivo biological systems [14].

Nanobiotechnology in Microbiology

Nanomaterials are well known for antimicrobial activity especially silver, silicon, carbon, zinc, cobalt, etc. Plant-mediated synthesis of nanomaterials is the recent approach in nanoscience. These nanoparticles (NPs) possess antibacterial activity against variety of clinically harmful pathogens. As far as healthcare sector is concerned, these NPs may reduce chances of developing resistance against drugs due to abrupt use of antibiotics. Mechanistically, these nanoparticles hinder growth via enzyme deactivation, cell membrane disruption and osmotic imbalance. Antimicrobial enzyme attachment with nanoparticles increases retention time in bacteria which directly inhibit their growth. This approach may prove to be good alternative for antibiotic [15]. Gold nanoparticles serve as better NPs that can be bound to different ligands such as antibodies, disaccharides, peptides and other biomolecules for targeting to specific cells. They can detect nucleic acids specifically [16]. Peptide NPs cross the blood-brain barrier more effectively than other drugs particularly antibiotics. Hence, these NPs are considered for preventing and cure of brain infections. Nowadays, nanomachines are being used capable of sensing bacterial behaviour followed by hindering bacterial quorum sensing process and finally preventing infection [15].

Nanobiotechnology in Agriculture

The concept of nanoparticle synthesis using herbal approach has become a benchmark in plant treatment and parasitic control. Nanobiotechnology plays a key role in agriculture as pest control agents, treating plant infection with targeting diseased organ, use of nanoengineered material for targeted delivery and controlled release of bioactive constituents [17]. Nanoforms such as carbon, silica and silver are being used for preventing plant diseases [18].

Agricultural microbiology is the connecting bridge liable for propagation of newer techniques from basic microbiological aspects to ecological microbiology, leading to development of advanced methods for better crop production and agro-practices [19]. In current scenario, a wide variety of microorganisms have been incorporated for assuring and enhancing crop quality and production methods [20]. Microbial biotechnology persisting in soil promotes optimum nutrient absorption by plant. Environment-friendly microbes and plant materials promote 'nutrient recycling'. The non-pathogenic environment-friendly microorganisms support plant for uptake of essential nutrients and energy sources. Biodegradable waste produced by plant is further utilized for survival of microbes. Scientists use biocompatible microorganisms to foster the production of biofertilizers.

Recently, the concept of using microorganisms for synthesizing nanomaterials is of much use in combating pests, herbs, insects and nematodes. Different forms of carbon nanotubes such as single and multiwall nanotubes are in current fashion for studying effects on plants. Titanium dioxide NPs are used in several industries as these do not possess any harmful effects. Photocatalytic potential of TiO_2 contributes in pathogen disinfection. These NPs are used in preventing plant from harmful pathogens without interrupting food chain [21]. Plant pathogens such as fungus, virus and bacteria are exploited to form newer nanoforms that are used to detect early pathogenic state in plant and prevent them from harmful effects [22].

Nanobiotechnology in Environmental Activity

Emerging trends in biological sciences are contributing towards environmental toxicity. Repercussion of increased usage of plastics, non-biodegradable substances and environmental toxicants allows limited practice of bioremediation. As far as nanoscience is concerned, nanoparticles are known for their promising benefits due to which these NPs are integrated with bioremediation methods for waste management and environmental toxicant eradication.

Nanoclusters of silver are used to reduce effluent from various chemical industries [23]. Development of solar cells using silicone wires has been undertaken by several researchers. Entrapment of silicone wires within the polymeric covering enhances efficiency and reduces cost of solar cells [24]. Use of iron NPs for purifying underground water bodies has currently attracted scientists around the globe. Iron nanoparticles cleanse organic impurities from groundwater [25]. Efforts by various researchers have been made in the direction to create nanoengineered materials with promising effects and less toxicity to environment.

Nanobiotechnology in Medical Microbiology

Antibiotics are well known for treating microbial infections. They are called 'wonder drugs', but when used irrationally, they convert themselves to 'superbug' like gram-positive *Staphylococcus aureus* strain. From several decades, the difference between the discovery of antibiotics and development of resistance is increasing rapidly. Due to lack of awareness among patients and communication gap between healthcare personnel, the problem has reached to its peak. Higher rate of antimicrobial resistance (AMR) has surfaced in developed countries in comparison to developing countries. Repetitive use of antibiotics and leaving doses in between regimen enforce natural selection process in microbes which directly trigger AMR. Inculcation of nanoscience has opened an avenue to synthesize a newer class of modified conjugates, which would help in disruption of bacterial or viral synthesis process via several inhibitory mechanisms that further lead to eruption of combating strategy for AMR.

Advancements in the field of microbial nanobiotechnology have played a vital role in the production of antibiotics and antimicrobial agents. Herbal and chemically synthesized microbial growth inhibitors may eradicate human, animal and plant infections. Various nanomolecular techniques have allowed production of vaccines that help in protecting the subject with fewer risks of causing infection. Barman et al. mentioned potential microorganisms for nanomaterial synthesis containing several intrinsic properties as well as cost-effective, biocompatible and less toxic [26]. Peptide nanofibres are considered to be outstanding candidates for vaccine adjuvants for developing better immune responses [27].

Nanobiotechnology in Food Microbiology

Nanotechnology focuses on manipulation of molecular as well as biological techniques to create newer and effective nanocomposites. Food industries are looking towards development of nanoformulations useful for food during storage, packaging and its processing. Nanobiotechnology supports development of intelligent packaging techniques that make consumers to modify food intake with proper nutritional values [28]. The application of nanobiotechnological techniques is the current concept of using genetically transformed microorganisms for enhancing productivity. Products of microbial metabolism such as enzymes, dyes and other compounds have been utilized for enhancing organoleptic properties and inflating nutritional functions to certain food products. Microbial floras are well known for fermentation process, and they are used in production of food ingredients [29].

Synthesis of nanobarcode devices helps in identifying food pathogens [30]. Presence of *Escherichia coli* in food product can be detected using reflective interferometric analysis in which label-free identification of biomolecules in complex mixture is carried out [31]. Synthesis of nanostructures from combination of biocompatible polymers, peptides and lipids is observed with prominent microbial cell lethal effects on *Escherichia coli* and gram-positive bacteria *Staphylococcus aureus* [32]. Food safety as well as food quality assurance was a major issue of concern that could be overcome using nanocomposite structures. Liposomes are the novel nanomaterials which comprise of characteristic physicochemical properties in comparison to other NPs. These novel materials have been utilized for improving flavours and nutritional values of food product. Liposomes have capacity to encapsulate natural metabolites that may prevent spoiling and degrading of food products. Liposomes are commercially being used as medicinal vehicles, signal enhancers, solubilizing agent for disparate food products and penetrating agents in variety of cosmetic products [33]. Biofilm formation is the most common process involving microbial pathogens. Gkana et al. have stated that organosilane nanostructures possess anti-biofilm and anti-adhesive activity against food-borne microbes [34].

Impact of Different Forms of Nanofabricated Structures on Microbial Growth

Anti-biofilm Agents

Biofilms are colonies of different microbes grown together via cellular signalling called quorum sensing. This film forms a uniform and heterogeneous layer on substrate and secretes extracellular polymeric substances (EPS). EPS protect internal environment of biofilm from external environment such as antimicrobial agents and limited nutrition within cells [35]. Ability to adhere of bacterial cells on any substrate enhances mature biofilm development which further leads to form double layer and detachment of bacterial cells. This condition gives rise to transmission of infection. Nowadays, medical devices such as catheter, nebulizer tubes, insertion tubes, diagnostic tools and implants contaminated with microbial biofilm are directly causing hazards to human health. Mature biofilms become resistant to external environment and not affected by existing drugs. Various research groups are working in these areas for preventing biofilm formation within the tubular structure. Nanoparticle-coated surfaces inhibit the formation of biofilms. Nanoparticles are small enough to penetrate bacterial cell membrane and disrupting mitochondria. Nanoparticles have large surface to volume ratio, and this property inflates chemical reactivity and bioactivity [36, 37]. Metal nanoparticles act through mechanisms such as production of reactive oxygen species within the bacterial cell, release of nitric oxide, altering respiratory chain and disrupting EPS network [38]. Polymeric nanoconjugates act through biocidal effects on cell membrane and alter protein absorption with interruption in internal cellular mechanism [39].

Bacteriophage Template for Nanostructures

Bacteriophages are known as infecting agents for pathogenic microorganisms and affecting host for replicating its genetic make-up. Filamentous phage and M13 are the best known for their promising biomedical perspectives. These phages possess nanoscaled fibrous filaments that serve as carrier for propagating genetic information among cells. Phage virions allow self-fabrication that only allows capsid protein to retain genetic information without interfering other proteins. Genetic flexibility of these phages allows development of disparate binding sites, so that these alterations serve as building block for nanostructures [40]. Recently, M13 phage NPs have attracted researchers for the purpose of designing and synthesizing nanoforms for neural tissue construct [41, 42]. Several groups are working on phage targeting to develop optimum therapeutic agent. As far as phage targeting is concerned, M13 is considered as reference as it contains chain of polypeptide on its surface. Novel tailored nanostructure containing fusion of libraries of amino acids and peptides can be easily matched with standard [43–45].

Biosensors

Biosensors have its own importance as diagnostic tool as well as in identifying drug target site. On the other hand, nanobiotechnology plays a key role in designing and development of biosensors using protein molecules coupled with nanoelectronic devices for developing nanobioelements and nanobiotransistors. Nanobiotechnology exploits existing knowledge for nanoscience for creating novel and efficient tools which directly analyse structural and biological phenomenon of nanomaterials. Antigens, antibodies and peptides are meant for the development of biosensors for detecting several pathogenic strains. Monoclonal antibodies have higher specificity as compared to biosensors, while polyclonal antibodies identify different epitopes on the same pathogen. Antibodies containing nanobiosensors have high thermal stability, refolding capacity, small size and profound solubility [46].

Quantum Dots

It is reported in the literature that colloidal quantum dots are semiconductor nanocrystals, conjugated with molecule that label bacteria specifically. The binding tendency depends on two factors, viz. metabolism rate and environmental conditions. Quantum dots facilitate in the detection of strain among diverse community of bacteria, i.e. biofilm (a delicate film formed by different bacteria through quorum sensing). Diagnosing mycobacterium using magnetic beads coupled with monoclonal or polyclonal antibodies leads to formation of complexes, which further conjugate with cadmium selenide QDs and bind to bacterial surface antigen followed by emission of fluorescent signals [47].

Bacterial strains as well as microbial monitoring can be done using quantum dots as fluorescent labels in an immunoassay [48]. Recent study reveals that curcumin quantum dots help in biodegradation of bacterial biofilm via targeting susceptible bacterial protein, delta toxin and curli proteins [49].

Bacterial Magnetic Nanoparticles (BMPs)

Recently, an interesting application of nanotechnology with microbiology has erupted for the development of bacterial magnetic nanoparticles. BMPs are mostly magnetite and greigite which form single domain crystals. The surface of these nanocrystals exhibits biological bilayer with phospholipids and surface proteins. During biomineralization process, the membrane layer helps in determining characteristic morphology of BMPs. Biologically isolated nanomolecules chemically synthesize them with high dispersity, crystallinity, high percentage yield surrounded with lipid bilayer membrane, high chemical purity, narrow size range and crystal morphology with specificity for particular species. *Magnetospirillum gryphiswaldense MSR-1* and *Magnetospirillum magneticum AMB-1* are well-known species for extracting magnetic nanoparticles [50]. Other species are under process for extracting out magnetic nanoparticles. These magnetic nanoparticles help in treatment of tumours as well as other pathogenic states. Mechanistically, magnetic hyperthermia is the process which inculcates heat exposure to kill the tumour cells [51] (Figs. 1 and 2).

Nanobiotechnological Techniques for Enzyme Hybridization

Nanoparticle-biomolecule hybrid system shows combined effects such as biocatalysis, recognition with optical, catalytic and acoustic effects of biological nanoparticles. Hybrid methods include fabrication of nanostructured patterns on magnetic as well as metallic nanoparticles. Recently, gold and silver nanoparticles

Fig. 1 Biological membrane of bacterial magnetic nanoparticles and their arrangement of surface proteins

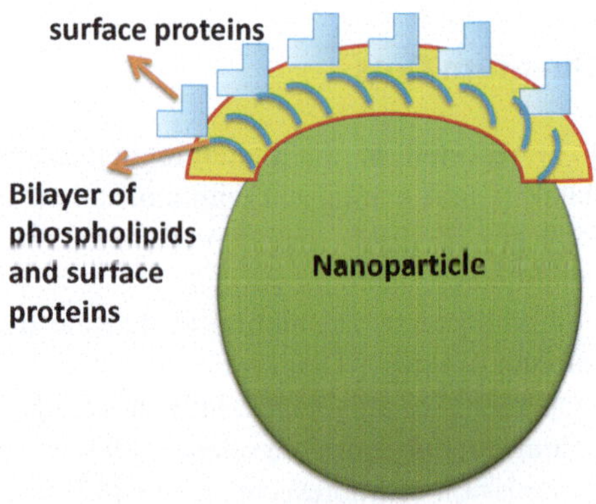

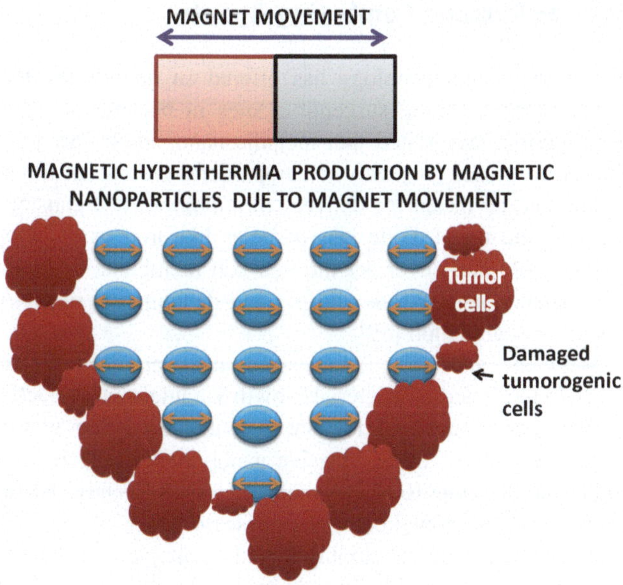

Fig. 2 Effect of magnetic hyperthermia produced by bacterial magnetic nanoparticles on tumour cell necrosis

have been used for development of biosensors to detect glucose levels. Biocatalytic growth property of metallic nanomaterial promotes fabrication of nanowires on surface. Semiconductor NPs, connected with electrodes, work as photoactivate bio-electrocatalytic cascades with photoelectric current generation [52].

Nanorobots

Use of nanorobots is a thrust area of research in the field of nanoscience and nanomedicines. Designing and development of nanorobots inculcate hybrid cell membrane of RBCs and platelets, coated with gold nanowires. These nanorobots perform different tasks in one shot, by binding to pathogens like methicillin-resistant *Staphylococcus aureus* (MRSA). Erythrocytes used in these nanorobots help in releasing toxins secreted by bacterial strain and detoxifying the surroundings. Gold nanorobots have been reported with higher extent mobility which helps in releasing toxins produced by pathogenic bacteria [53]. Nanorobots have potentially used as phagocytic agents. These nanorobots termed as microbivores bear various binding sites specifically for antigens and pathogenic strains. Hence, they eradicate infection within limited period of time without affecting the host [54, 55]. Microbots and nanobots are coupled with ligated bacterial strain carrying therapeutic agents for targeted delivery. Simultaneously, customizing nanorobots with some probes act as sensors for detecting pathogenic state and collecting information [56].

Nanoparticles as Microbe Combating Agents

In a current scenario, nanotechnology has offered an opportunity for exploiting microbial diversity via targeting susceptible sites in bacteria or other microbes. Some literature reports have stated that metallic nanoparticles are an outstanding tool for killing HIV viruses [57]. Silver nanoparticles are one of the most known bactericidal nanomaterial meant for delivery of antibiotics and causing synergistic effects [58]. Metal and metal oxide nanoparticles like iron oxide, copper, copper oxide, zinc oxide, gold and silver act through cell membrane disruption, enzyme deactivation, imbalance of electrons and pH and denaturation of DNA and RNA subunit followed by degradation [59].

Some biopolymers and their synthesized nanoparticles possess adhesive properties that help to hinder the bacterial growth via interrupting bacterial survival mechanisms. Polymeric nanoparticles are recently used for gluing gels and biological tissues. Somehow, these adhesive nanoparticles help in wound healing via connecting tissues together [60]. Polymeric nanoparticles have become a benchmark in treating several microbial pathogenic disease states.

Several research groups are developing metallic nanoparticles based on algae in which algae proteins are responsible for reduction and stabilization of nanoparticles. Polysaccharides present in algae cell wall participate in reduction process. Fucoxanthin (hydroxyl group rich element), present in algae, helps in the reduction of gold into gold nanoparticles. These pigments have reductive as well as capping property and enhance maintenance of surface integrity of nanoparticles [61, 62]. Marine source for reduction of metal salts into nanoparticles results in less toxic nanomaterials as compared to chemical approach. Fungal-, actinomycetes-, plant- and bacterial-mediated nanoparticles have different aspects for production of nanoparticles with different morphology and stability [63]. Bio-fabrication of nanoparticles like silver (Ag) and copper (Cu) is used as novel antimicrobials for inhibiting fungal and conidial growth in crops, which has become a major issue in agriculture that can be taken care of by using biofabricated gold and silver nanoparticles. Enzymatic reduction of metal to nanoparticles is a unique approach. Well-known proteins encapsulins that help to encapsulate cargo proteins are generally found in bacteria. Nanobiotechnology emphasizes on increasing efficiency of detecting pathogens present in crops in one shot as well as development of cost-effective biosensors and phage proteins for eradication of crop microbes [64].

Future Aspect of Nanobiotechnology Use in Combating Microbial Diverse Population

Nanobiotechnology is the most fascinating field of science. Development of disparate nanolayers, nanochips and nanomaterials leads to combating microbial pathogens. From several decades, incorporation of nanoforms reduces the frequency of disease-causing agent. Presently, enormous inculcation of nanobiotechnology is meant for detection and treatment of pathogenic disease with lower side effects

Nanomaterials are conjugated to form biofungicides via covalent bonding. This peculiar property of nanoparticle manages to break pathogen pest chain. This practice leads to save agriculture crop by reducing pest effect [65]. Recent literature mention that newer and novel drugs are not able to overcome microbial pathogen susceptibility, but nanoforms are the only way to reduce these diseases causing diverse microbial community via anti-biofilm activity by quorum sensing, osmotic imbalance among cell membranes and reactive oxygen species eruption from stressed condition [59]. In current state of art, peptide nanoparticles are becoming a most attracting topic for research in nanobiotechnology. Bactericidal and bacteriostatic potency of these nanoforms increase its interest in major research groups. Fwu-Long Mi and co-workers have suggested that chitosan-conjugated peptidal nanoparticles help in degrading harmful bacterial growth. Hybrid forms of nanoparticles are observed with synergistic activity against resistant as well as clinical pathogenic microbes [66]. Combination therapy of peptide and metallic nanoparticles has shown promising results.

Conclusion

Nanobiotechnology is the burgeoning field in current era of nanomedicine and nanoscience. Development of nanocomposites, nanofibres, nanowires, nanotubes, nanoforms, nanorobots, polymeric nanoconjugates, inorganic NPs, magnetic NPs and bacteriophage nanostructures has received a great attention in the field of microbiology for eradicating microbial infections. Nowadays, increasing use of antibiotics and poor sanitation facilities have inflated chances of getting discovery void and genetic mutation in patients receiving antibiotics due to development of resistance. These conditions have led to opening up of a new avenue, i.e. synthesis of above-mentioned nanomaterials. This chapter has shown the efficacy of variety of nanomaterials in sensing microbial growth and inhibiting their survival as well as current and future perspectives of microbial aspect of nanomedicine.

References

1. Carrara, S. (2010). Nano-bio-technology and sensing chips: New systems for detection in personalized therapies and cell biology. *Sensors, 10*(1), 526–543.
2. Fakruddin, M., Hossain, Z., & Afroz, H. (2012). Prospects and applications of nanobiotechnology: A medical perspective. *Journal of Nanobiotechnology, 10*(1), 31.
3. Khan, I., Khan, M., Umar, M. N., & Oh, D.-H. (2015). Nanobiotechnology and its applications in drug delivery system: A review. *IET Nanobiotechnology, 9*(6), 396–400.
4. Davis, S. (1997). Biomedical applications of nanotechnology—implications for drug targeting and gene therapy. *Trends in Biotechnology, 15*(6), 217–224.
5. Hart, S. L. (2005). Lipid carriers for gene therapy. *Current Drug Delivery, 2*(4), 423–428.
6. Touhami, A. (2014). Biosensors and nanobiosensors: Design and applications. *Nanomedicine, 15*, 374–403.
7. Maine, E., Thomas, V., Bliemel, M., Murira, A., & Utterback, J. (2014). The emergence of the nanobiotechnology industry. *Nature Nanotechnology, 9*(1), 2.

8. Morais, M. G. D., Martins, V. G., Steffens, D., Pranke, P., & da Costa, J. A. V. (2014). Biological applications of nanobiotechnology. *Journal of Nanoscience and Nanotechnology, 14* (1), 1007–1017.
9. Sahoo, S. K., & Labhasetwar, V. (2003). Nanotech approaches to drug delivery and imaging. *Drug Discovery Today, 8*(24), 1112–1120.
10. Sahoo, S., Parveen, S., & Panda, J. (2007). The present and future of nanotechnology in human health care. *Nanomedicine: Nanotechnology, Biology and Medicine, 3*(1), 20–31.
11. Crommelin, D. J., Storm, G., Jiskoot, W., Stenekes, R., Mastrobattista, E., & Hennink, W. E. (2003). Nanotechnological approaches for the delivery of macromolecules. *Journal of Controlled Release, 87*(1–3), 81–88.
12. Kumar, N., & Kumbhat, S. (2016). *Essentials in nanoscience and nanotechnology*. Hoboken: Wiley.
13. Mahato, M., Sharma, A. K., & Kumar, P. (2017). Nanoparticles for DNA delivery. *ACS Applied Materials & Interfaces, 9*, 11546–11556.
14. Watson, A., Wu, X., & Bruchez, M. (2003). Lighting up cells with quantum dots. *BioTechniques, 34*(2), 296–303.
15. C D. Nanobiotechnology: Relevance in Microbiology 2019.
16. Daniel, M.-C., & Astruc, D. (2004). Gold nanoparticles: Assembly, supramolecular chemistry, quantum-size-related properties, and applications toward biology, catalysis, and nanotechnology. *Chemical Reviews, 104*(1), 293–346.
17. Resham, S., Khalid, M., & Kazi, A. G. (2015). *Nanobiotechnology in agricultural development. PlantOmics: The omics of plant science* (pp. 683–698). Berlin: Springer.
18. Singh, S., Singh, B. K., Yadav, S., & Gupta, A. (2015). Applications of nanotechnology in agricultural and their role in disease management. *Journal of Nanoscience and Nanotechnology, 5*(1), 1–5.
19. Singh, J. S., Pandey, V. C., & Singh, D. (2011). Efficient soil microorganisms: A new dimension for sustainable agriculture and environmental development. *Agriculture, Ecosystems & Environment, 140*(3–4), 339–353.
20. Arnold, A. E., Lamit, L. J., Gehring, C. A., Bidartondo, M. I., & Callahan, H. (2010). Interwoven branches of the plant and fungal trees of life. *New Phytologist, 185*(4), 874–878.
21. Yao, K. S., Li, S., Tzeng, K., Cheng, T. C., Chang, C. Y., Chiu, C., et al. (2009). Fluorescence silica nanoprobe as a biomarker for rapid detection of plant pathogens. In: *Advanced materials research*. Trans Tech Publ.
22. Young, M., Debbie, W., Uchida, M., & Douglas, T. (2008). Plant viruses as biotemplates for materials and their use in nanotechnology. *Annual Review of Phytopathology, 46*, 361–384.
23. Kühr, S., Schneider, S., Meisterjahn, B., Schlich, K., Hund-Rinke, K., & Schlechtriem, C. (2018). Silver nanoparticles in sewage treatment plant effluents: Chronic effects and accumulation of silver in the freshwater amphipod Hyalella azteca. *Environmental Sciences Europe, 30*(1), 7.
24. Wang, Y. Y. (2015). *Graphene based transistors and supported lipid bilayer*. Irvine: University of California-Irvine Irvine United States.
25. Zhao, X., Liu, W., Cai, Z., Han, B., Qian, T., & Zhao, D. (2016). An overview of preparation and applications of stabilized zero-valent iron nanoparticles for soil and groundwater remediation. *Water Research, 100*, 245–266.
26. Prasad, R., Pandey, R., & Barman, I. (2016). Engineering tailored nanoparticles with microbes: Quo vadis? *Wiley Interdisciplinary Reviews: Nanomedicine and Nanobiotechnology, 8*(2), 316–330.
27. Yang, C., Shi, F., Li, C., Wang, Y., Wang, L., & Yang, Z. (2017). Single dose of protein vaccine with peptide nanofibers as adjuvants elicits long-lasting antibody titer. *ACS Biomaterials Science Engineering, 4*(6), 2000–2006.
28. Neethirajan, S., & Jayas, D. S. (2011). Nanotechnology for the food and bioprocessing industries. *Food and Bioprocess Technology, 4*(1), 39–47.

29. Salmerón, I. (2017). Fermented cereal beverages: From probiotic, prebiotic and synbiotic towards nanoscience designed healthy drinks. *Letters in Applied Microbiology, 65*(2), 114–124.
30. Alvarado, K., Bolaños, M., Camacho, C., Quesada, E., & Vega-Baudrit, J. (2019). Nanobiotechnology in agricultural sector: Overview and novel applications. *Journal of Biomaterials and Nanobiotechnology, 10*(02), 120.
31. Horner, S. R., Mace, C. R., Rothberg, L. J., & Miller, B. L. (2006). A proteomic biosensor for enteropathogenic *E. coli*. *Biosensors and Bioelectronics, 21*(8), 1659–1663.
32. Carrasco, L. D. M., Bertolucci, R., Jr., Ribeiro, R. T., Sampaio, J. L., & Carmona-Ribeiro, A. M. (2016). Cationic nanostructures against foodborne pathogens. *Frontiers in Microbiology, 7*, 1804.
33. Shukla, S., Haldorai, Y., Hwang, S. K., Bajpai, V. K., Huh, Y. S., & Han, Y.-K. (2017). Current demands for food-approved liposome nanoparticles in food and safety sector. *Frontiers in Microbiology, 8*, 2398.
34. Gkana, E. N., Doulgeraki, A. I., Chorianopoulos, N. G., & Nychas, G.-J. E. (2017). Anti-adhesion and anti-biofilm potential of organosilane nanoparticles against foodborne pathogens. *Frontiers in Microbiology, 8*, 1295.
35. Whitchurch, C. B., Tolker-Nielsen, T., Ragas, P. C., & Mattick, J. S. (2002). Extracellular DNA required for bacterial biofilm formation. *Science, 295*(5559), 1487.
36. Taylor, E., & Webster, T. J. (2011). Reducing infections through nanotechnology and nanoparticles. *International Journal of Nanomedicine, 6*, 1463.
37. Suci, P. A., Berglund, D. L., Liepold, L., Brumfield, S., Pitts, B., Davison, W., et al. (2007). High-density targeting of a viral multifunctional nanoplatform to a pathogenic, biofilm-forming bacterium. *Chemistry & Biology, 14*(4), 387–398.
38. Taheri, S., Vasilev, K., & Majewski, P. (2015). Silver nanoparticles: Synthesis, antimicrobial coatings, and applications for medical devices. *Recent Patents on Materials Science, 8*(2), 166–175.
39. Ramasamy, M., & Lee, J. (2016). Recent nanotechnology approaches for prevention and treatment of biofilm-associated infections on medical devices. *BioMed Research International, 2016*.
40. Bakhshinejad, B., & Sadeghizadeh, M. (2014). Bacteriophages and development of nanomaterials for neural regeneration. *Neural Regeneration Research, 9*(22), 1955.
41. Chung, W.-J., Merzlyak, A., Yoo, S. Y., & Lee, S.-W. (2010). Genetically engineered liquid-crystalline viral films for directing neural cell growth. *Langmuir, 26*(12), 9885–9890.
42. Zhu, W., O'brien, C., O'brien, J. R., & Zhang, L. G. (2014). 3D nano/microfabrication techniques and nanobiomaterials for neural tissue regeneration. *Nanomedicine, 9*(6), 859–875.
43. Kehoe, J. W., & Kay, B. K. (2005). Filamentous phage display in the new millennium. *Chemical Reviews, 105*(11), 4056–4072.
44. Gray, B. P., & Brown, K. C. (2013). Combinatorial peptide libraries: Mining for cell-binding peptides. *Chemical Reviews, 114*(2), 1020–1081.
45. Su, M.-T., Venkatesh, T. V., & Bodmer, R. (1998). Large-and small-scale preparation of bacteriophage λ lysate and DNA. *BioTechniques, 25*(1), 44–46.
46. Dover, J. E., Hwang, G. M., Mullen, E. H., Prorok, B. C., & Suh, S.-J. (2009). Recent advances in peptide probe-based biosensors for detection of infectious agents. *Journal of Microbiological Methods, 78*(1), 10–19.
47. Liandris, E., Gazouli, M., Andreadou, M., Sechi, L. A., Rosu, V., & Ikonomopoulos, J. (2011). Detection of pathogenic mycobacteria based on functionalized quantum dots coupled with immunomagnetic separation. *PLoS One, 6*(5), e20026.
48. Yaohua, H., Chengcheng, W., Bing, B., Mintong, L., Wang, R., & Li, Y. (2014). Detection of *Staphylococcus aureus* using quantum dots as fluorescence labels. *International Journal of Agricultural and Biological Engineering, 7*(1), 77–83.
49. Singh, A. K., Prakash, P., Singh, R., Nandy, N., Firdaus, Z., Bansal, M., et al. (2017). Curcumin quantum dots mediated degradation of bacterial biofilms. *Frontiers in Microbiology, 8*, 1517.

50. Chen, C., Wang, P., & Li, L. (2016). Applications of bacterial magnetic nanoparticles in nanobiotechnology. *Journal of Nanoscience and Nanotechnology, 16*(3), 2164–2171.
51. Hildebrandt, B., Wust, P., Ahlers, O., Dieing, A., Sreenivasa, G., Kerner, T., et al. (2002). The cellular and molecular basis of hyperthermia. *Critical reviews in Oncology/Hematology, 43*(1), 33–56.
52. Willner, I., Basnar, B., & Willner, B. (2007). Nanoparticle–enzyme hybrid systems for nanobiotechnology. *The FEBS Journal, 274*(2), 302–309.
53. Li, J., Angsantikul, P., Liu, W., Esteban-Fernández de Ávila, B., Chang, X., Sandraz, E., et al. (2018). Biomimetic platelet-camouflaged nanorobots for binding and isolation of biological threats. *Advanced Materials, 30*(2), 1704800.
54. Sandhiya, S., Dkhar, S. A., & Surendiran, A. (2009). Emerging trends of nanomedicine–an overview. *Fundamental & Clinical Pharmacology, 23*(3), 263–269.
55. Freitas, R. A., Jr. (2005). Microbivores: Artificial mechanical phagocytes using digest and discharge protocol. *Journal of Evolution and Technology, 14*, 1–52.
56. Saadeh, Y., & Vyas, D. (2014). Nanorobotic applications in medicine: Current proposals and designs. *American Journal of Robotic Surgery, 1*(1), 4–11.
57. Elechiguerra, J. L., Burt, J. L., Morones, J. R., Camacho-Bragado, A., Gao, X., Lara, H. H., et al. (2005). Interaction of silver nanoparticles with HIV-1. *Journal of Nanobiotechnology, 3* (1), 6.
58. Franci, G., Falanga, A., Galdiero, S., Palomba, L., Rai, M., Morelli, G., et al. (2015). Silver nanoparticles as potential antibacterial agents. *Molecules, 20*(5), 8856–8874.
59. Zielińska-Górska, M. K., Sawosz, E., Górski, K., & Chwalibog, A. (2017). Does nanobiotechnology create new tools to combat microorganisms? *Nanotechnology Reviews, 6* (2), 171–189.
60. Rose, S., Prevoteau, A., Elzière, P., Hourdet, D., Marcellan, A., & Leibler, L. (2014). Nanoparticle solutions as adhesives for gels and biological tissues. *Nature, 505*(7483), 382.
61. Venkatpurwar, V., & Pokharkar, V. (2011). Green synthesis of silver nanoparticles using marine polysaccharide: Study of in-vitro antibacterial activity. *Materials Letters, 65*(6), 999–1002.
62. Singh, M., Kalaivani, R., Manikandan, S., Sangeetha, N., & Kumaraguru, A. (2013). Facile green synthesis of variable metallic gold nanoparticle using *Padina gymnospora*, a brown marine macroalga. *Applied Nanoscience, 3*(2), 145–151.
63. Alenazi, N. M. K. (2018). Green nanoparticles: Biogenerators; mechanistic aspects of biosynthesis; potential applications and future prospective.
64. Kaushal, M. (2018). Role of microbes in plant protection using intersection of nanotechnology and biology. In *Nanobiotechnology applications in plant protection* (pp. 111–135). Cham: Springer.
65. Bhattacharyya, A., Duraisamy, P., Govindarajan, M., Buhroo, A. A., & Prasad, R. (2016). Nano-biofungicides: Emerging trend in insect pest control. In *Advances and applications through fungal nanobiotechnology* (pp. 307–319). Cham: Springer.
66. Tamara, F., Lin, C., Mi, F.-L., & Ho, Y.-C. (2018). Antibacterial effects of chitosan/cationic peptide nanoparticles. *Nanomaterials, 8*(2), 88.

Microbes in the Reproductive Tract Spectrum: Inferences from the Microbial World

Saika Manzoor, Sabiha Majid, and Fouzia Rashid

Abstract

Reproductive tract infections are the most venerable diseases which affect not only the reproductive tract organs but overall the whole body of an organism. RTIs can involve infection of multiple organs of the human host like the skin, brain, eyes, lymph, and urogenital, perivascular, endocrine, and reproductive systems, respectively. RTIs affect both the sexes, but the burden of the diseases is found higher in case of females as compared to males because female are more susceptible to these infections due to various life processes. RTIs are known to degrade the quality of life as these infections affect the reproductive efficiency and can even lead to infertility in both the sexes. One of the major routes of transmission of the reproductive tract infections is oral-genital sexual contact with an infected person. Other modes of STIs are bodily fluids like blood, mother's milk, mucous secretions, exudates, transplacental secretions, etc. There is an interrelationship of microorganisms with human hosts since their existence and the slam contact of the two cannot be denied. A balanced microflora is present in the human body as a part of the system, but the disequilibrium can result in various types of RTIs. Moreover, exogenous factors can also be responsible for transmission of other types of RTIs. RTIs are most prevalent in low-income and developing countries due to following reasons: lack of awareness regarding the diseases, poor hygiene, social stigmas, lack of medical facilities, poverty, shyness, inaccessibility to condoms, sanitary pads, hand washes, hand sanitizers, traditional child delivery methods, etc. This chapter will highlight the microbes involved in RTIs, their management, prevention, and drug resistance scenario.

S. Manzoor · F. Rashid (✉)
Department of Clinical Biochemistry/Biochemistry, University of Kashmir, Srinagar, J&K, India
e-mail: rashid.fouzia@gmail.com

S. Majid
Department of Biochemistry, Government Medical College, Srinagar, J&K, India

S. Hameed, Z. Fatima (eds.), *Pathogenicity and Drug Resistance of Human Pathogens*,
https://doi.org/10.1007/978-981-32-9449-3_18

Keywords
Balanced microflora · Genital tract infections · Vaginal and urethral discharges · Iatrogenic infections · HPV · Gonorrhea

Introduction

RTI is an infection of the reproductive tract caused by organisms normally present in our reproductive tract or by those introduced externally during sexual intercourse or medical intervention procedures. Reproductive-aged women are at higher risk of RTI due to associated life process events like menstrual cycles, pregnancy, and childbirth [35, 71]. Chronic RTI can lead to various complications like pelvic inflammatory diseases (PID), infertility, ectopic pregnancy, miscarriages, pelvic pain, and cervical cancers [92]. A strong correlation has been reported regarding microbial flora and various genital tract infections in humans. Genital tract infections caused by microbes like bacteria, viruses, protozoa, and fungi interfere in the reproductive functions of both the sexes, i.e., male and female. A general understanding of the normal composition of microbiota in the genital tract provides insight regarding the susceptibility toward various diseases associated with the reproductive tract [80].

Occurrence RTIs occur worldwide, and there are various factors that govern it like economic, social, behavioral, and biological factors. Its occurrence is not uniform as it varies from region to region and community to community, e.g., STIs are commonly reported in regions where commercial sex networks are active; iatrogenic infections are mostly reported in regions where STIs are common, and health-care providers are not fully trained to cope with it; endogenous infections are influenced by hygienic, hormonal, and environmental factors (WHO guidelines).

Symptoms of RTI in general are:

- Abnormal vaginal discharge
- Pain in the lower abdomen and back
  - Genital ulcers
  - Vulval itching
  - Inguinal swelling
  - Feeling of burning sensation during urination
  - Blood stain discharges

Overview of the Reproductive System

The reproductive system being one of the most important systems of the human body comprises of various internal and external organs. Reproductive organs are either primary or secondary structures. The primary reproductive organs are mainly responsible for the production of gametes, i.e., sperm and egg, and are called gonads. The secondary reproductive organs are responsible for other reproductive functions like development and maturation of gametes.

Male Reproductive Organs

The male urogenital system (merging of the urinary and reproductive system) consists of many parts that include the testes, epididymis, vas deferens, ejaculatory ducts, urethra, penis, prostate, and accessory glands. The predominant function of the male reproductive system is to provide sperm for the fertilization of the egg.

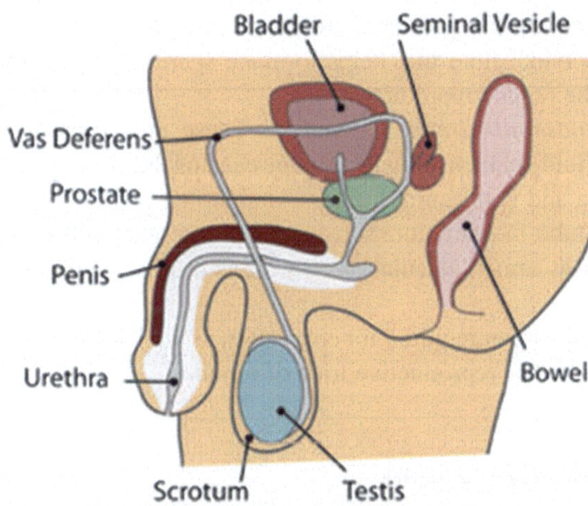

Testes – are oval-shaped paired structures present in the scrotum and are considered as a primary reproductive organ as it produces sperm and male sex hormone called testosterone. Tunica albuginea (dense connective tissue layer) covers the testis and also invaginates to form septa that divide the testis into 300–400 structures called lobules. The sperm within these lobules further develops into seminiferous tubules. Sperm moves from the lumens of the seminiferous tubules into tubuli recti and then enters into the rete testis and finally enters into the epididymis via the efferent ductules.

Epididymis (plural = epididymides) – a coiled tube almost 6 m (20 ft) long is attached to the testis, and the immature sperm enters through the head of the epididymis that matures during its movement along the length of the epididymis tube. All the matured sperms are then collected and stored in the epididymis tail and are ready to ejaculate.

Duct system – A thick muscular tube that is attached to the tail of the epididymis in the scrotum is called ductus deferens or vas deferens that extends upward into the abdominal cavity via the inguinal canal and then ends into the bladder where it dilates into the ampulla. The sperm moves from the epididymis tail to the vas deferens with the help of smooth muscle contraction.

Seminal vesicles – Seminal vesicles are accessory glands that produce fluid containing fructose that is required to generate ATP for the movement of sperm through the female reproductive tract. Their secretions contribute almost 60% of semen. Sperm mixes with seminal vesicle secretion after being ejaculated at the ampulla of ductus deferens. The ampulla of ductus deferens joins *with the seminal vesicle duct* to form the ejaculatory ducts and transports seminal fluid (sperm as well as semen) into the prostate gland.

Prostate gland – is a glandular tissue that is situated just below the urinary bladder and surrounds the prostatic urethra. It secretes an alkaline fluid that is mixed with the seminal fluid that helps to make sperm sticky and thus helps it to retain longer in the female reproductive tract.

Cowper's glands/bulbourethral glands – These are accessory glands which release alkaline fluid that mixes with semen causing lubrication of the urethra as well as the vagina and is released during sexual arousal.

Urethra – is a tube-like structure surrounded by the shaft of the penis. It is dual in function as it causes semen ejaculation as well as urine excretion from the bladder through the penis.

Penis – The male organ required for copulation as well as for urination. It helps to penetrate semen into the reproductive tract of females.

Female Reproductive Organs

The female reproductive system includes internal as well as external structures. External structures like mons pubis (mons veneris), clitoris, labia majora, labia minora, vestibule of the vagina, bulb of the vestibule, and greater vestibular glands are present outside the vagina and hymen, while internal structures include the vagina, uterus, fallopian tubes, and ovaries. External reproductive structures of female help the sperm to enter into the female reproductive tract and protects the internal organs from susceptible infections from external environment.

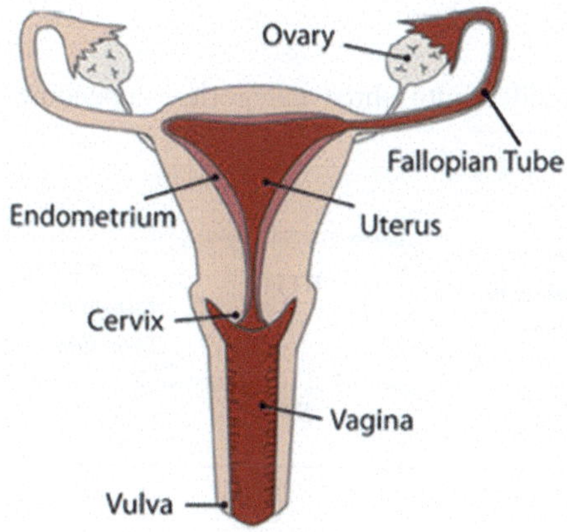

Vagina – It is a canal that extends from the cervix (lower end of uterus) to the vestibule of the vagina (cleft present between labia minora), forming 90° with the uterus. It is located in front of the rectum and anal canal but lies behind the urinary bladder and urethra. The orifice of the vagina is guarded by the hymen – a thick mucous membrane. There is absence of mucus glands in the vaginal lining, and the glands of the cervix secrete the mucus present. It receives sperm via the penis during copulation and is called birth canal because the baby leaves through it during birth.

Uterus – It is an inverted pear-shaped hollow muscular structure with glandular lining called the endometrium. The lower narrow end of the uterus is the cervix that ends into the vagina. The opening of the two opposite tubes called fallopian tubes takes place into it and the part of uterus above it called fundus and below it called the body. It receives the ovum sent by the ovary via the fallopian tube into it. The main body of the uterus called the corpus extends in size to hold and nurture the developing embryo until birth. The innermost layer of tissues present in the uterus is called the endometrium that gets shed off during menstrual cycles.

Ovaries – The paired / two almond shaped hollow structures present on opposite sides of the uterus are called ovaries, each suspended by the mesentery – a membranous ligament of the uterus which keeps both the ovaries in proper position. They are considered as primary sex organs in the female reproduction system as they contain eggs and produce various hormones – estrogens and progesterone.

Fallopian tubes – The narrow tubes that act as a passage for the movement of egg from the ovary to the uterus are called fallopian tubes. It contains three main outer parts: infundibulum (flask-shaped) projected with finger-like projections called fimbrae, ampulla the middle part, and isthmus (cord-like) that opens into the uterus. Fertilization of egg by sperm takes place at the ampulla of the fallopian tubes and then descends into the uterus for embryo implantation.

Types of RTIs

RTIs are broadly divided into three categories depending upon the mode of infection [99].

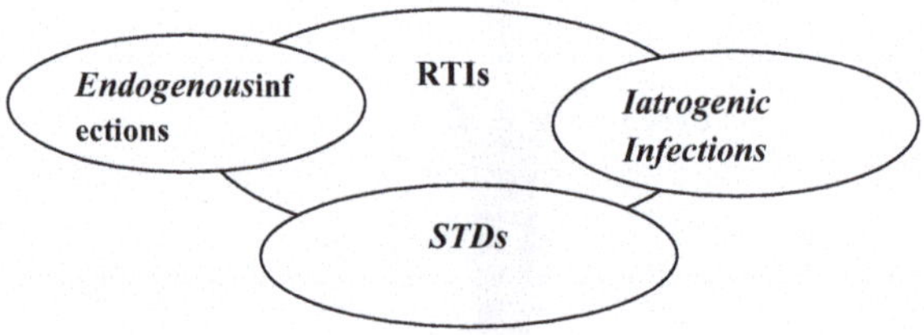

Sexually Transmitted Diseases

STDs are infections that are transmitted from one infected person to another by vaginal, oral, or anal sexual contact. Microorganisms like bacteria, virus, yeast, and protozoan are agents that are responsible for different types of STIs. STIs can cause genital neoplasia, neonatal damages, and infertility. According to WHO, STIs can be classified according to one of the five presenting features for which patients seek medical help:

- Oral, genital, anal, or perianal ulcers
- Vaginal and urethral discharges
- Genital warts
- HIV
- Hepatitis C infection (not typically sexually transmitted disease)

STDs affect both men and women and can also be transmitted from mother to child during pregnancy and childbirth. Various microorganisms like bacteria, protozoa, fungus, and viruses cause STDs which are described below.

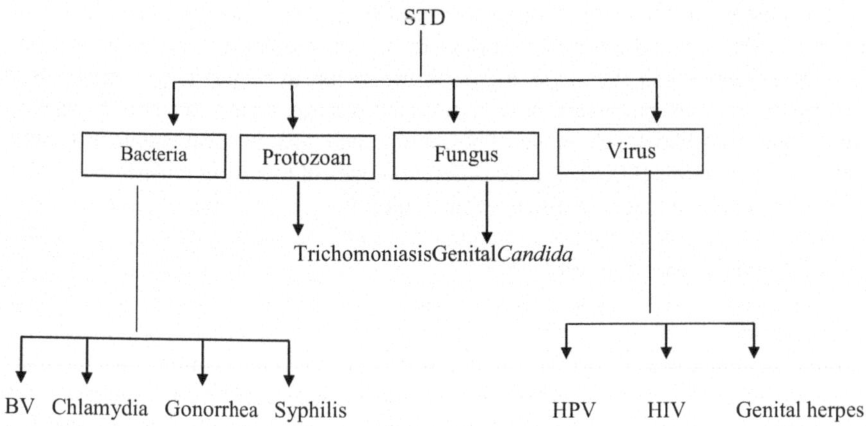

STDs Caused by Bacteria

Bacterial Vaginosis

The presence of microbial flora in the lower genital tract of the female reproductive system gives protection against the pathogens and predominantly consists of *Lactobacillus* spp., *L. crispatus*, *L. jensenii*, and *L. iners* [69, 85, 105]. Vaginal lactobacilli maintain the normal pH (<4.5), metabolizing glycogen secreted by the vaginal epithelia into lactic acid. Hence, the acidic environment created by lactobacilli prevents the growth of other harmful pathogens that cannot survive the acidic environment of the healthy vagina [5, 22, 42]. BV involves replacement of lactobacilli with other microorganisms like *Gardnerella vaginalis*, *Prevotella*, *Peptostreptococcus*, and *Bacteroides* spp. [31, 47, 51, 88]. Hence, the drop-down level of lactobacilli and the disequilibrium of normally present bacterial flora in the reproductive system result in a condition called bacterial vaginosis. It is the most common infection among reproductive-aged women and is considered as sexually transmitted disease [34, 97]. Many researchers point out that BV is not merely an infection but an imbalance in the normally present microbiota of the reproductive tract system [38, 42, 81]. The diagnosis of BV depends upon the presence of three out of four given criteria [4]:

1. Elevated vaginal pH (>4.5)
2. Presence of white adherent discharge
3. Presence of exfoliated epithelial cells with bacteria (gram-variable polymorphic rods) attached to their surface (clue cells)
4. Fishy odor (whiff test)

BV possess different microorganisms which secrete various toxins which can cross the placenta and pose various risks for the developing embryo ranging from brain injury to various other neurological disorders like hyperactivity, cerebral palsy, and periventricular leukomalacia [37]. BV is also a risk factor for various gynecologic conditions like development of postpartum and post-abortion endometritis and pelvic infection [50, 100]. There are different modes of treatment regarding BV like probiotic treatment that replenishes vaginal microflora, application of antimicrobials produced by vaginal lactobacilli that target and kill the specific pathogens, and the use of immunomodulators [96].

Chlamydia

Chlamydia trachomatis – gram-negative bacterium that causes infection of genital sites like the cervix, urethra, and rectum and non-genital sites like lungs and eyes and is the most common cause of sexually transmitted disease [104]. It is responsible for male infertility as it causes non-gonococcal urethritis (NGU), epididymitis-orchitis, sperm tract obstruction, and prostatitis. Chlamydia infections affect both the sexes, but it is more prevalent in females as compared to males, i.e., almost three times higher than males [7]. It causes cervicitis, urethritis, endometritis, salpingitis, and perihepatitis that can result in pelvic inflammatory disease, infertility, and ectopic pregnancy in women [54, 59]. People with *C. trachomatis infection are* asymptomatic; however abnormal vaginal discharge and vaginal bleeding (bleeding after intercourse) are symptoms in women, while penile discharge and pruritus are symptoms among men. Dysuria is a common symptom among both males and females [79]. Antibiotics that are considered useful in the treatment of chlamydia are azithromycin and doxycycline, and screening is recommended to all women below 25 years of age; however, screening is not recommended to males.

Gonorrhea

Neisseria gonorrhoeae, a gram-negative bacterium, causes sexually transmitted disease called gonorrhea. Gonorrhea infection can lead to various complications if left untreated like gonorrheal endocarditis, gonorrheal meningitis, and gonorrheal arthritis that can affect both males and females. Gonorrhea can result in male infertility by causing infection to the testes and urethritis. The primary site of infection in women is endocervix that causes PID (pelvic inflammatory diseases) which can manifest in endometritis, salpingitis, and tubo-ovarian abscess, which ultimately lead to scarring, ectopic pregnancy, sterility, and chronic pelvic pain [73]. *Neisseria gonorrhoeae* is well adapted to survive within the human host because it can hijack the immune system by not eliciting the immunologic memory of the host with the help of various virulence factors like Opa (opacity-associated outer membrane protein), LOS (Lipooligosaccharide), and pilus. This organism gets adhere to genital areas by means of pilus and moves upward from the vagina to the cervix of the uterus by climbing epithelial cells. Binding of pilus to CR3 present in ecto- and endocervix activates the compliment system which leads the formation of inactivated C3b (iC3b); it results in tight association between iC3b and porin with CR3 on the cervical cell surface by binding to the I-domain of this receptor. This whole process of CR3 engagement triggers gonococcal internalization within macropinosomes. Secreted gonococcal proteins, which ensure CR3 recruitment to the cervical cell surface, mediate these processes. The sexual transmission of gonococci mostly takes place from male to their partner as sialylated LOS; however presence or absence of sialic acid on LOS does not affect its interaction with cervical epithelial cells in the female reproductive tract. Moreover, vaginal microflora can remove sialic acid on LOS, thus modifying the gonococcus, which can lead to enhanced disease transmission to men [27]. Infection with *N. gonorrhoeae* results in the production of various pro-inflammatory cytokines like IL-1a, IL-1b, and TNF-α by tubal epithelial cells that cause genesis of gonococcus-induced infertility [53].

Syphilis

Syphilis is a sexually transmitted disease caused by the bacteria *Treponema pallidum*, and humans are considered as their predominant hosts [9]. Syphilis is spread by contact with ulcers present on the genital areas, anus, rectum, mouth, and lips. Orogenital route is one of the main routes of transmission of the bacteria (*Treponema pallidum)* into the human host, and other modes of its transmission are blood transfusion and transplacenta, i.e., from mother to fetus [14].The exact pathogenesis of the bacteria is yet to be elucidated; however, it has been found that entry of bacteria into host occurs via primary and secondary lesions that are present in the skin or mucus membranes, during sexual activities, and travel into lymph nodes within no time [30, 52, 72]. The attack of bacteria is considered to be virulent as it penetrates deep and adhere to the intact endothelial cell membranes [8, 93]. Humoral cell immunity of the host defense mechanism is known to play a minor role against bacteria; however the cell-mediated immunity is known to play a key role against the pathogenicity of the bacteria. The disease has been divided into four stages, namely, primary, secondary, latent, and tertiary [66]. The first three stages of this disease are venerable, i.e., primary, secondary, and latent.

Primary stage – It is characterized by ulceration and lymphadenopathy. The presence of painless lesions lasts only for a few weeks and then heals up, but the disease progresses to second stage of infection. Moreover, inflammatory response in the nearby dermis cells and perivascular area has also been observed in this stage of syphilis.

Secondary stage – It is characterized by the presence of specific pattern of changes like parakeratosis, spongiosis, acanthosis, and exocytosis. Granulomatous infiltration represents the latter stage of the secondary stage [1, 28].

Latent stage – It is a kind of dormancy stage in which no sign of infection is visible, but still the infection prevails inside the body of the organism. It can be divided further into two stages: early latent stage <1 year of infection and late latent stage >1 year of infection.

Tertiary stage – It is the most destructive stage among all the four stages of syphilis in terms of the complications implicated by this stage. It involves the heart, blood vascular system, nervous system, and multiple organs system which can eventually lead to death of an infected individual [86].

Neurosyphilis and ocular syphilis are two conditions associated with the disease prognosis at any of given stages, if left untreated. Penicillin is considered as primary recommended antibiotic for the treatment of syphilis infection, but the dosage depends upon the stage of infection. However, penicillin-allergic patients are provided with other options of drug treatment like erythromycin, tetracycline, and doxycycline [14, 47].

STDs Caused by Protozoa

Trichomoniasis

It is the most common sexually transmitted non-viral infection which is caused by a protozoan parasite called *Trichomonas vaginalis* (anaerobic and flagellated parasite) and affects both males and females [101]. Most of the infected persons are asymptomatic; however symptoms may develop after a few weeks, months, or years [77]. Infection with *Trichomonas vaginalis* results in the inflammation of the reproductive parts like the vagina, cervix, and urethra; thus the symptoms associated with this condition appear as mentioned below [91, 103].

- Abnormal volume of vaginal discharge with color variation from white clear fluid to yellowish-green
- Vaginal discharge odor
- Discomfort during urination
- Pain with intercourse

Symptoms associated with *Trichomonas vaginalis* in males include (CDC):

- Burning sensation after urination or ejaculation
- Irritation in the penis
- Abnormal discharge from the penis

Upon the infection of the vagina or prostate, parasite multiples by binary fission in the urogenital tract of human host and gets adhered to squamous epithelial cells resulting in inflammation and eventually their destruction. The menstrual cycle provides an ideal environment for microbes to flourish in the vagina of the female reproductive tract and worsens the already existing condition. Although the immunologic response against this pathogen is not known yet, response of various cytokines has been found like IL-8 and leukotreine B4 (LTB4) [50, 63, 64, 75 84]. Antimicrobial drugs like nitroimidazoles (metronidazole or tinidazole) are recommended as drug of choice in trichomoniasis treatment. However, if one of the two partners is still infected or without treatment, it can lead to recurrence of infection [104].

STDs Caused by Fungus

Genital Candida

Candida albicans is naturally present in the GI tract, the mouth, and the vagina of a healthy individual, but its overgrowth leads to the various infections in the human body [2]. Various factors like *stress change in local environment* (e.g., shifts in pH or nutritional content) and disruption of normal microflora (e.g., *Lactobacillus is known to keep the balance in the amount of Candida in the genital tract) can result in overgrowth of the candida that can lead to yeast genital infections. Candida albicans* are predominantly colonized in the male urethra and possess an inhibitory effect on human sperm motility. It impairs the ultrastructure of human spermatozoa that results in male infertility [94] and in females causes vaginitis and cervicitis [81]. *The* genitourinary candidiasis infection includes balanitis and balanoposthitis in men and vulvovaginal candidiasis (VVC) in women, while candiduria is common to both men and women [2]. Symptoms of a genital candida infection can include:

- A burning feeling while having sex or while urinating
- An itchy or painful feeling in or around the vagina
- Redness, irritation, or swelling around the vagina
- Abnormal vaginal discharge that can be either watery or thick and white
- A rash around the vagina
- A rash on the penis

STDs Caused by Virus

Human Papilloma Virus or Genital Warts (HPV)

HPV is a double-stranded non-enveloped DNA virus. HPV is divided into three main genera, α, β, and γ species. The genus α includes high-risk as well as low-risk HPV that can cause various types of cancers and skin warts, respectively. The genus β is associated with various types of lesions like malignant-associated lesions, cutaneous lesions, and lesions in immunosuppressed patients. The genus γ is associated with cutaneous lesions only [21]. HPV initiates infection at the basal membrane of epithelial cells and enters into the cell by a process of endocytosis; however the detailed mechanism of viral entry and receptors associated is not clear yet. Virus forms an episome within basal cell nuclei and uses the machinery of host cell for its DNA transcription. Then the infected basal cells starts shedding off from the basement membrane and piles up to join stratum spinosum. Next epithelial layer, usually formed post-mitotically, reaches up till top cornified layer of epithelial called as stratum corneumand this is how viruse scapes the cell cycle. Finally, the viral genome gets packed at this stage and is released in the extracellular environment along shedding skin cells in such a manner so as not to illicit immune response of the host [11].

Although the immune system can detect the virus and the infection can get cleared within 2 years, papilloma virus can hijack the immune response of the host cell in different ways. The virus can alter codon usage that can result in reduced expression of viral protein and change in its expression that prevents its detection by the immune system. Moreover, downregulation of various host associated immune responses like immune histocompatibility responses and reduced signaling through T cell receptor 9 also affects [6, 10, 18, 39, 56, 104]. HPV infection is the most common sexually transmitted disease; hence sexual contact with an infected partner is the major risk factor for its transmission. HPV is mainly found in genital and anal areas and can cause infection in the vagina, vulva, penis, cervix, rectum, and anus.

The most common associated risk factors with this infection are genital warts/ condylomata acuminata (painless bumps, sometimes itchy and bleeding, mostly found in the genital areas and groins and treated by topical medications) and cervical cancers. Persistent sexual contact with an infected partner can lead to constancy of this infection for longer durations, and that can result in various types of cancers like throat, cervical, vulval, anal, and penile cancers [103]. HPV-16 and 18 are in particular associated with cervical cancers while vaccines namely 2vHPV, 4vHPV and 9vHPV have also been developed to get protection against them [67].

Human Immunodeficiency Virus (HIV)

HIV is a retrovirus that lives on bodily fluids like blood, sexual fluids, and breast milk. It involves infection of two retroviruses HIV-1 and HIV-2 or both of them. Once the virus enters the body, it adheres to $CD4^+$ molecule of T-helper cells and chemokine receptors of host cells. The virus forms proviral DNA with the help of reverse transcriptase and gets incorporated in the host DNA by using various enzymes like integrase; hence proviral DNA is duplicated along the host DNA during a process of cell cycle. Moreover, the virus undergoes transcription that results in formation of RNA and viral proteins within the host cell membrane. Subsequently, the viral proteins get budded off from the cell membrane as immature virions and finally form infectious virions by HIV protease. The infected lymphocytes produce large numbers of virions. HIV infection causes adverse effect on the host immune response by causing $CD4^+$ lymphocyte depletion as well as immune suppression and affects both the cell-mediated and humoral immunity. The normal $CD4^+$ count is about 750/μL, but when it drops to a level below 200/μL, the risk of developing AIDS increases [57]. AIDS is characterized by decreased or absence of host $CD4^+$ count; it's when the immune response of the body almost stops against pathogenicity of various attacking agents. The transmission is possible through sexual contact with an infected person, use of infected needles, blood transfusion, breast-feeding, exudates from wound or cut containing virions, childbirth, etc. Acute HIV infection includes flu like symptoms, but long term infection of almost 10 years will result in the following symptoms: diarrhea, fever, loss of appetite, weight loss, fatigue, lymphadenopathy, and sweating [82]. There is not any vaccine available against HIV infection till date; however, significant efforts have been made regarding the management of various HIV-associated conditions and prevention of its progression toward the disease progress, i.e., development of AIDS.

Genital Herpes

Genital herpes is caused by herpes simplex virus (HSV) which is a linear, enveloped, double-stranded virus with humans as its only known host [102]. Two variants of this virus, namely, HSV-1 and HSV-2, cause oral and genital infections, respectively. However both of these viruses can cause infection at any site on the skin. Virus can enter the body via the skin or mucous membranes by sexual contact with an infected person and then multiplies in the epithelial layer. The virus descends along the sensory nerve roots to dorsal ganglion root and undergoes a period of latency. Upon reactivation, the virus retrieve from the dorsal ganglion root to the nerve root and causes mucocutaneous outbreaks, thus HSV infection becomes symptomatic or even sometimes, being asymptomatic [102]. The primary genital HSV is characterized by localized pain and burning and tingling sensation that last for few hours (i.e., 2–24 h) with associated symptoms like fever, inguinal lymphadenopathy, anorexia, malaise, and headaches. With the progression of diseases, small papules, vesicles on erythematous base, and erosions appear that can persist over days, and lesions can crust and heal without scarring. In females, ulcers are found mainly on the labia, urethral meatus, perineum, and introitus, while in males, it is found mainly on the shaft of the penis. However, ulcers or lesions are found on the thighs, buttocks, and perianal region in both males and females [49]. Oral antiviral drugs are available to suppress the symptoms associated with genital herpes, but there is no long-term treatment available that can cure the disease altogether [83]. The prevention of secondary infection due to outbreak of ulcerative lesion is necessary to avoid complications associated with the diseases. Moreover, strong interaction between HIV and HSV has been found that results in the efficient transmission of HIV during HSV reactivation [78].

The diagnostic tests which are used for clinical evaluation of STDs are highlighted in Table 1.

Endogenous Infections

These infections are caused by the overgrowth of normally present microbial flora in the genital tract of a woman that includes bacterial vaginosis and vulvovaginal candidiasis (as described earlier). Endogenous infections are very common compared to sexually transmitted infections with a prevalence of 17.8% for BV and 8.5% for candidiasis, while STDs have prevalence of 2.8%. Some of the endogenous infections are sexually transmitted, while others may not be sexually transmitted [68]. There are certain individuals who are at higher risk of endogenous infections like pregnant women and those who are on oral contraceptives because of change in vaginal pH. Even the use of certain medications like steroids taken for treatment of diabetes and long-term use of antibiotic can result in the overgrowth of yeasts, etc. and also pose treatment for yeast infections [105].

Table 1 Techniques involved in the diagnosis of STDs

Infection	Diagnostic test
Bacterial vaginosis	Saline wet mount, gram-stained smear, pH test, KOH test
Chlamydia trachomatis	Antigen detection
	(1) Direct fluorescent antibody technique (DFA)
	(2) ELISA, culture, nucleic acid amplification, dark-field microscopy, RPR test (qualitative as well as quantitative), and VDRL (qualitative)
Neisseria gonorrhoeae	Gram-stained smear, culture, identification of organism, β-lactamase test, antimicrobial susceptibility test, nucleic acid amplification
Herpes simplex virus-1 and 2	Tzanck Giemsa-stained smear
	Antigen detection
	(1) Immunofluorescent staining
	(2) ELISA
	Antibody detection
	Anti-HSV IgM (type-specific glycoprotein) G[IgG] culture and nucleic acid amplification
Candida spp.	Potassium hydroxide wet mount, gram-stained smear, culture, speciation, and AST
Human papilloma viruses	Nucleic acid amplification
Hepatitis B virus	Immunochromatography, HBs Ag detection, ELISA for HBs Ag detection, and nucleic acid amplification tests
HIV	Antibody screening: ELISA, Western blot, qualitative and quantitative PCR
Syphilis	Dark-field microscopy, PCR, and direct fluorescent antibody testing for *T. pallidum* and serological tests, i.e., treponemal (e.g., the FTA-ABS) and non-treponemal tests (e.g., RPR or VDRL)
Trichomoniasis	Microscopy, vaginal swab culture, antibody testing (ELISA), and nucleic acid detection tests

Iatrogenic Infections

Infections due to unhygienic medical procedures like unsafe abortion, insertion of intrauterine devices (IUD), poor delivery practices, postoperative infections, and contaminated needles or instrument usage actually introduce pathogenic agents into the reproductive tract or simply develop already existing lower reproductive tract infection into upper reproductive tract infection [60]. Sexually transmitted infections in men are much more common than iatrogenic or endogenous infections.

Insertion of intrauterine device (IUD) can increase the risk of PID following period of insertion; however, IUD insertion is not recommended for women who are at greater risk of chlamydia and gonorrhea [105].The various practices carried out by clinicians during labor and childbirth can pose mother as well as newborn under the threat of various infections like sepsis. Newborn babies with EONS – early-onset sepsis present within 72 h after birth – can occur either from maternal genital tract or during labor. EONS are most prevalent in poor countries, which are associated with poor birth hygiene practices and maternal infections. Another type is LONS – late-onset sepsis which occurs after 72 h after birth – which occurs usually due to pathogen encounter of newborns during hospital stay or after discharge and is mainly linked with birth practices [33]. *E. coli*, *Klebsiella*, *Staphylococcus aureus*, and group B streptococcus (GBS) are the main organisms responsible for hospital-based infections [102]. Various events during labor can predispose the woman to chorioamnionitis like premature rupture of membrane (>18 h before birth), prolonged labor (>24 h), more than three vaginal examinations before labor, and unclean examination of the vagina during labor [26, 95]. Chorioamnionitis is an infection of the fetal membrane and amniotic fluid linked with maternal infection caused by bacteria. It can lead to prolonged rupture of the membrane and long labor and, if left untreated, can lead to severe infections to the uterus and is proven fatal to both the mother and the newborn.

Operative site infections are the most common infections associated with pelvic surgeries, which include caesarian deliveries, termination of pregnancy, and vaginal and abdominal hysterectomy. Pelvic cellulitis and endometritis are considered the most common complications of operative site infections. Pelvic cellulitis in one-third of women is found associated with vaginal hysterectomy, and endometritis is found associated with preexisting infection of the lower genital tract (BV and GBS), prolonged labor and ruptured membranes, invasive fetal monitoring, and multiple vaginal examinations [23]. However, surgical wound infections occur in 3% of patients, while its frequency is less than 1% in women who undergo interval sterilization, postpartum sterilization, and other laparoscopic procedures. The organisms which are responsible for wound infections include aerobic microorganism like staphylococci, streptococci, *E. coli*, *K. pneumoniae*, and *Proteus* species and various other anaerobic organisms as well [36]; the two forms of wound infection are incisional abscess and wound cellulitis.

General Concerns Related to RTIs

Most of the RTIs affect both males and females; however its consequences are found to be detrimental in the case of females as it carries higher risk of morbidity and mortality. Endogenous or sexually transmitted infections result in infertility (both in males and in females), pelvic inflammatory diseases (PID), ectopic pregnancy, abortions, preterm deliveries, stillbirth, pre-labor rupture of membranes, and other complications like various kind of disabilities, blindness, and death of newborn babies [105].

Uterine infection and puerperal sepsis are among the RTIs that are common causes of maternal morbidity and mortality. Infertility can be defined as inability to conceive after 12 months or longer with unprotected sex. *Salpingitis (inflammation of epithelial fallopian tube) and pelvic peritoneal adhesions are considered as the common causes of tubal factor infertility (TFI) resulting from persistent infections like Chlamydia trachomatis* and *Neisseria gonorrhoeae* [46]. *Moreover, bacterial movement from the cervix to endometrium and then finally to the fallopian tubes develops PID which ultimately changes 15% of women with PID to TFI* [74, 101].

In low- and middle-income countries, infertility leads to stigmatization, social isolation, embarrassment, and cost burden of treatment [19, 29]. RTIs can result in poor reproductive health that can lead to domestic abuse, emotional trauma, and unhealthy life span. Furthermore, RTIs can even lead to chronic pain, genital ulcers, adverse pregnancy outcome, and HIV which can end up with AIDs if left untreated and various types of cancers like cervical cancer which is mostly associated with HPV.

Management and Preventive Measures for RTI

Management

Although RTIs are not curable, they can be prevented. The main aim of primary preventive measures is to stop occurrence of disease, that is, when the pathogen has not interacted with the host body. Secondary prevention measure is to reduce the complications associated with the infection. The primary prevention is possible only by avoiding pathogenic transmission either by avoiding direct sexual contact or by avoiding other means of transmission with an infected person [32]. *However,* there are certain impediments regarding the management or treatment of RTIs and are broadly divided into three categories (Nagarkar and Mhaskar2015):

- Social and cultural factors: feeling of shame or embarrassment, shyness, and social stigma
- Economic factors: poor socioeconomic culture, poverty, and high cost of treatment
- Environmental factors: lack of knowledge or education, inaccessibility, and ignorance
- Health-care facility factors: lack of privacy, absence of female doctors, lack of care, indifferent attitude of medical staff, and lack of well-trained staff

Epidemiological studies have revealed that STIs are more common and mismanaged in low-income countries due to several reasons like migration in search of livelihood, poverty, inappropriate sexual behavior, stigmas associated with them, incompetent health-care providers, inadequate supply of drugs, ignorance, and lack of awareness, which can put them at higher risk of various infections [45, 98, 105]. *Lack of awareness and screening in itself is a big milestone that needs to be achieved in order to manage various sexually transmitted infections by organizing various programs, awareness camps, conferences,* etc.

Another reason for mismanagement of these infections is asymptomatic nature of various STIs (chlamydia, gonorrhea) as they remain unrecognized for longer period of time; hence the long-term consequences of the infection put them at greater risk of other health consequences [100]. *Moreover, expensive diagnostic tests required for the identification of the various STIs also provide a great hindrance to the management of these infections* [105].

Drug Resistance

Resistance of various microorganisms against a wide range of antibiotics is one of the major problems that the world is facing in controlling RTIs. Since antibiotics have crucial role to play in control of various RTIs, all other medical procedure like caesarian deliveries, surgeries, post-operative infections, operative site infections and other medical emergencies are safe and convenient only under the coverage of antimicrobial actions. The examples of various resistant reproductive tract infections include:

- *Neisseria gonorrhoeae that shows a remarkable ability of resistance against various antibiotics like* sulfonamides, penicillin, fluoroquinolones, tetracycline, and extended-spectrum cephalosporins (ESC).
- *Treponema pallidum which causes syphilis* shows resistance against macrolides, the main alternative to penicillin, azithromycin, erythromycin, clindamycin, rifampin, and tetracyclines.
- *C. trachomatis* shows resistance against the following drugs like sulfamethoxazole, doxycycline, erythromycin, ofloxacin, tetracycline, azithromycin, and clindamycin.
- *Trichomoniasis is the most common STI that shows resistance against* metronidazole.

  However, nontraditional approaches are being exploited for treating antimicrobial-resistant infections which include antimicrobial peptides (AMPs), efflux pump inhibitors, bacteriophage therapy, metallo-antibiotics, phytochemicals, etc. [44].

  Furthermore, self-medication being one of the major problems needs to be addressed because a patient taking medication on their own puts them at the risk of drug resistance and then it becomes more difficult to provide treatment in such cases [13]. *Although, patients might get relief from the pain and other symptoms associated with the diseases but is not effective in long term, as it cannot cure the disease. Mostly, incompetent persons in this field who lack knowledge about the disease, medicine, dosage, etc. provide those medications to them.*

Addressing all these problems can help a lot to reduce the burden of RTIs on all the spheres wherever it affects badly and can be done by various means.

Prevention

Prevention for STIs

<u>Male circumcision</u> – *Studies have shown that this practice is very beneficial for the prevention of various STIs both for males and for female partners as it is found to prevent acquisition of HIV, genital HPV, genital herpes, and trichomoniasis in males and in reduction of incidences of bacterial vaginosis, trichomoniasis, and genital ulcers in female partners* [76].

Use of condoms – Use of condoms in males prevent acquisition of large number of STIs like HIV, HPV, chlamydia, syphilis, genital herpes, and lesion of the penis associated with HPV. Use of male condoms has been found beneficial for women as well as it prevents the acquisition of the same infection.

Diaphragm (birth control) – Its role in males is unknown yet; however it may prevent acquisition of gonorrhea in females.

Vaccination – Preexposure of certain vaccines like hepatitis B and HPV is recommended in order to prevent transmission of these two infections [20, 55].

Counseling – Patient-centered counseling has been proved beneficial for prevention of STIs especially HIV in cases of sexually active adolescents and adults [12, 50].

Rescreening for prevention of recurrent infections – Those who have been previously diagnosed with infections like *N. gonorrhoeae*, *C. trachomatis*, and chlamydia need rescreening 3 months after diagnosis as a prevention method [70].

Prevention for Endogenous Infections

Genital hygiene – General genital hygiene practices, intravaginal modified practices, vaginal washing after sexual contact and douching, etc. need to be evaluated so as to assess and reduce the risk of various infections like BV, HIV, and other STIs associated with them [40, 61]. *Menstrual unhygienic practices like use of cloth during period are one of the factors that cause a greater risk of endogenous infections* [17].

Other hygienic practices like the use of douching, detergents, disinfectant, and drying agents need to be avoided as their use can also change the normal microbial flora of the vagina and can pose a greater risk of endogenous infections. Use of water and mild soap is sufficient for healthier genital hygiene [105].

Prevention for Iatrogenic Infections

Greater number of children can be a risk factor of iatrogenic infections for women as each childbirth increases the number of deliveries and the gynecological procedures. Use of contraceptive devices and gynecological surgeries are among the risk factors of iatrogenic infections in women [90]. *There is increased risk of iatrogenic infections through transcervical procedures like* manual vacuum aspiration, endometrial biopsy, and insertion of an intrauterine device (IUD); hence prevention practices are needed to be followed to avoid the risk of transcervical infections which include proper handwashing, use of gloves throughout the procedure, use of all instruments required only after proper decontamination, disinfecting all instruments like specula, tenacula, forceps, etc. through boiling for at least 20 min in a container with closed lid, use of antiseptic in washing the vagina and cervix, etc. *Moreover, frequent insertion and removal of IUDs should not be done so as to avoid unnecessary infections. The provider should give proper information regarding the expiration and effectiveness regarding the device inserted. Although women can develop infection like* PID, purulent cervicitis, chlamydial infection, or gonorrhea with IUD insertion, still there is no need to remove IUDs during treatment procedures [105].

Most of the infections transmitted in health-care practices are mainly due to improper handwashing of the workers putting the patient at risk for various infections and even leading to death [3]. Therefore, use of alcohol-based hand sanitizer before and after contact with each woman, newborn, blood, and bodily fluids should be imposed to avoid iatrogenic infections. Use of partograph (a graphical tool that examines the process of labor) is encouraged for delivery practices as it reduces vaginal examinations to once every 4 h hence reduces the chances of entry of various pathogens into the vagina and cervix, and thus prevents various labor-linked infections.

Moreover, patients that undergo various surgical procedures should receive prophylactic antibiotics in order to reduce the chances of postoperative infections. Prophylactic antibiotics have three modes of action: decrease the size of bacterial inoculums at the surgical site, alter the environment of operative site so as to make it unfavorable for the growth of bacteria, and enhance the phagocytosis of bacteria by concentrating in WBCs [16, 24, 25, 43, 87]. Furthermore, combination antibiotic therapy also plays a key role in treatment of operative site infections like pelvic cellulitis and endometritis which makes drug resistance a more serious issue for scientists to be tackled well in time.

Future Perspective

1. *Raising knowledge or awareness regarding the disease symptoms and signs through various programs and conferences can help a lot in management of the diseases and its consequences. Various programs have been launched in developed as well as developing countries so as to create awareness in general masses like the* Reproductive and Child Health (RCH) program *in India and* International Conference on Population and Development held at Cairo [76].
2. *Syndromic case management is a useful tool in the management of RTIs that means diagnosis and management of these infections based upon the signs and symptoms of the syndrome or diseases observed during the clinical diagnosis as well as reported by the patients and then prescribing a recommended treatment to the patients* [105].
3. *Better services need to be provided to patients in terms of medical facilities like trained and well-behaved staff who can address their problems very well, and thus the self-medication will be discouraged by itself* [41].

Conclusion

There is a considerable socioeconomic burden of RTIs on all the continents of this world as most of the RTIs are endemic and communicable; therefore there is a demand for proper strategy to control and prevent most of these infections. Various guidelines have been put forward by national as well as international agencies regarding the preventive measures of RTIs. A number of vaccines have been developed by different countries globally which are being prescribed as a part of routine vaccination panel. Although, most of the RTIs like HIV, etc. cannot be cured, most of their symptoms are treatable with proper medications, and hence their widespread prevalence can also be prevented. An extensive research is needed in this field to understand the disease process, transmission, prevention, and control measures, and most importantly mass awareness is much needed regarding this concern so as to reduce mortality and morbidity. Furthermore, RTI occurrence, transmission, and management spectrum becomes highly challenging for the clinicians and lab professionals/scientists owing to the greater incidences of drug resistance. As rapid growth of drug resistance against reproductive tract infections has emerged in last few years; hence the reduced treatment options for these infections. However, if the infections are left untreated, it can lead to long-term risk of various pathogenicity. Hence, proper follow-ups of various guidelines need to be followed to avoid the overuse or misuse of antibiotic drugs, and newer methods of treatment need to be explored to treat various antibiotic-resistant reproductive tract infections.

References

1. Abell, E., Marks, R., & Jones, E. W. (1975). Secondary syphilis: A clinico-pathological review. *The British Journal of Dermatology, 93*(1), 53–61.
2. Achkar, J. M., & Fries, B. C. (2010). Candida infections of the genitourinary tract. *Clinmicrobiol Reviews, 23*(2), 253–273.
3. Allegrenzi, B., & Pittet, D. (2009). Role of hand hygiene in healthcare-associated infection prevention. *Journal of Hospital Infection, 73*(4), 305–315.
4. Amsel, R., Totten, P. A., Spiegel, C. A., Chen, K. C., Eschenbach, D., & Holmes, K. K. (1983). Nonspecific vaginitis: Diagnostic criteria and microbial and epidemiologic associations. *The American Journal of Medicine, 74*(1), 14–22.
5. Aroutcheva, A., Gariti, D., Simon, M., Shott, S., Faro, J., Simoes, J. A., & Faro, S. (2001). Defense factors of vaginal lactobacilli. *American Journal of Obstetrics and Gynecology, 185* (2), 375–379.
6. Ashrafi, G. H., Tsirimonaki, E., Marchetti, B., O'Brien, P. M., Sibbet, G. J., Andrew, L., & Campo, M. S. (2002). Down-regulation of MHC class I by bovine papillomavirus E5 oncoproteins. *Oncogene, 21*(2), 248–259.
7. Atlanta, G. A. (2009). Centers for Disease Control and Prevention 2009.
8. Baseman, J. B., & Hayes, E. C. (1980). Molecular characterization of receptor binding proteins and immunogens of virulent Treponemapallidum. *The Journal of Experimental Medicine, 151*(3), 573–586.
9. Baughn, R. E., & Musher, D. M. (2005). Secondary syphilitic lesions. *Clinical Microbiology Reviews, 18*, 205–216.
10. Brady, C. S., Bartholomew, J. S., Burt, D. J., Duggan-Keen, M. F., Glenville, S., Telford, N., Little, A. M., Davidson, J. A., Jimenez, P., Ruiz-Cabello, F., Garrido, F., & Stern, P. L. (2000). Multiple mechanisms underlie HLA dysregulation in cervical cancer. *Tissue Antigens, 55*(5), 401–411.
11. Brendle, S. A., Bywaters, S. M., & Christensen, N. D. (2014). Pathogenesis of infection by human papillomavirus. *Human Papillomavirus, 45*, 47–57. Karger Publishers.
12. Calonge, N., Petitti, D. B., DeWitt, T. G., Dietrich, A. J., Gordis, L., Gregory, K. D., & Loveland-Cherry, C. (2008). Behavioral counseling to prevent sexually transmitted infections:

US Preventive Services Task Force recommendation statement. *Annals of Internal Medicine, 149*(7), 491–496.
13. Cars, O., Olsson-Liljequist, B., & Lundh, K. (2007). Short summary of Swedres 2005, a report on Swedish antibiotic utilisation and resistance. *Eurosurveillance, 12*(3), 9–10.
14. Centers for Disease Control & Prevention (US). (1998). Epidemiology Program Office. MMWR. Recommendations and Reports (Vol. 47). US Department of Health and Human Services, Public Health Service, Centers for Disease Control, Epidemiology Program Office.
15. Chelmow, D., Ruehl, M. S., & Huang, E. (2001). Prophylactic use of antibiotics for non laboring patients undergoing cesarean delivery with intact membranes: A meta-analysis. *American Journal of Obstetrics and Gynecology, 184*(4), 656–661.
16. Costello Daly, C., Francos, L., Chilongozi, D. A. T., & Dallabetta, G. (1998). A cost comparison of approaches to sexually transmitted disease treatment in Malawi. *Health Policy and Planning, 13*(1), 87–93.
17. D'Costa, Z. J., Jolly, C., Androphy, E. J., Mercer, A., Matthews, C. M., & Hibma, M. H. (2012). Transcriptional repression of E-cadherin by human papillomavirus type 16 E6. *PLoS One, 7*(11), e48954.
18. Daar, A. S., & Merali, Z. (2002). Infertility and social suffering: The case of ART in developing countries. *Current Practices and Controversies in Assisted Reproduction,* 15–21.
19. Daniels, D., Grytdal, S., & Wasley, A. (2009). Surveillance for acute viral hepatitis—United States, 2007. *Morbidity and Mortality Weekly Report. Surveillance Summaries, 58*(3), 1–27.
20. De Villiers, E. M., Fauquet, C., Broker, T. R., Bernard, H. U., & zur Hausen, H. (2004). Classification of papillomaviruses. *Virology, 324*(1), 17–27.
21. Donati, L., Di, V. A., Nucci, M., Quagliozzi, L., Spagnuolo, T., Labianca, A., Bracaglia, M., Ianniello, F., Caruso, A., & Paradisi, G. (2010). Vaginal microbial flora and outcome of pregnancy. *Archives of Gynecology and Obstetrics, 281*, 589–600.
22. Duff, P. (1986). Pathophysiology and management of post cesarean endo myometritis. *Obstetrics and Gynecology, 67*(2), 269–276.
23. Duff, P. (1987). Prophylactic antibiotics for cesarean delivery: A simple cost-effective strategy for prevention of postoperative morbidity. *American Journal of Obstetrics and Gynecology, 157*(4), 794–798.
24. Duff, P., & Park, R. C. (1980). Antibiotic prophylaxis in vaginal hysterectomy: A review. *Obstetrics and Gynecology, 55*(5 Suppl), 193–202.
25. Dutta, S., Reddy, R., Sheikh, S., Kalra, J., Ray, P., & Narang, A. (2010). Intrapartum antibiotics and risk factors for early onset sepsis. *Archives of Disease in Childhood. Fetal and Neonatal Edition, 95*(2), F99–F103.
26. Edwards, J. L., & Apicella, M. A. (2004). The molecular mechanisms used by Neisseria gonorrhoeae to initiate infection differ between men and women. *Clinical Microbiology Reviews, 17*(4), 965–981.
27. Engelkens, H. J. H., Ten Kate, F. J. W., Vuzevski, V. D., Van der Sluis, J. J., & Stolz, E. (1991). Primary and secondary syphilis: A histopathological study. *Indian Journal of Sexually Transmitted Diseases Acquired Immunodeficiency Syndrome, 2*(4), 280–284.
28. Ezeh, A., Bankole, A., Cleland, J., García-Moreno, C., Temmerman, M., & Ziraba, A. K. (2016). *Burden of reproductive Ill- Health.* Washington, DC: The World Bank.
29. Fitzgerald, T. J. (1981). Pathogenesis and immunology of *Treponema pallidum. Annual Review of Microbiology, 35*(1), 29–54.
30. Forsum, U., Holst, E., Larsson, P. G., Vasquez, A., Jakobsson, T., & Mattsby-Baltzer, I. (2005). Bacterial vaginosis—a microbiological and immunological enigma. *APMIS, 113*, 81–90.
31. Friis, R. H., & Sellers, T. A. (2014). *Measures of morbidity and mortality. Epidemiology for public health practice* (pp. 107–155). Burlington: Jones & Bartlett Learning.
32. Ganatra, H. A., Stoll, B. J., & Zaidi, A. K. (2010). International perspective on early-onset neonatal sepsis. *Clinics in Perinatology, 37*(2), 501–523.

33. Gardner, H. L. (1980). Haemophilusvaginalisvaginitis after twenty-five years. *American Journal of Obstetrics and Gynecology, 137*(3), 385–391.
34. Geetha, M. (2014). Prevalence of reproductive tract infections among rural married women in Tamil Nadu, India: A community based study. *The Journal of Pioneering Medical Sciences, 4*(1), 18–24.
35. Gibbs, R. S., Blanco, J. D., & St. Clair, P. J. (1983). A case-control study of wound abscess after cesarean delivery. *Obstetrics and Gynecology, 62*(4), 498–501.
36. Grether, J. K., & Nelson, K. B. (2000). Possible decrease in prevalence of cerebral palsy in premature infants. *The Journal of Pediatrics, 136*(1), 133.
37. Guise, J. M., Mahon, S. M., Aickin, M., Helfand, M., Peipert, J. F., & Westhoff, C. (2001). Screening for bacterial vaginosis in pregnancy. *American Journal of Preventive Medicine, 20*, 62–72.
38. Hasan, U. A., Bates, E., Takeshita, F., Biliato, A., Accardi, R., Bouvard, V., Mansour, M., Vincent, I., Gissmann, L., Iftner, T., Sideri, M., Stubenrauch, F., & Tommasino, M. (2007). TLR9 expression and function is abolished by the cervical cancer-associated human papillomavirus type 16. *Journal of Immunology, 178*(5), 3186–3197.
39. Hassan, W. M., Lavreys, L., Chohan, V., Richardson, B. A., Mandaliya, K., Ndinya-Achola, J. O., & McClelland, R. S. (2007). Associations between intravaginal practices and bacterial vaginosis in Kenyan female sex workers without symptoms of vaginal infections. *Sexually Transmitted Diseases, 34*(6), 384–388.
40. Hawkes, S., Nayyar, A., Van Dam, J., O'Reilly, K., Deperthes, B., & Agarwal, D. (2001). *Reproductive tract infections: A guide for Programme. Managers*. New York: UNFPA and Population Council.
41. Hay, P. (2005). Life in the littoral zone: Lactobacilli losing the plot. *Sexually Transmitted Infections, 81*, 100–102.
42. Jamie, W., & Duff, P. (2003). Preventing infections during elective C/S and abdominal hysterectomy (C/S hysterectomy infections). *Contemporary Obstetrics and Gynecology, 48*(1), 60–69.
43. Krupp, K., & Madhivanan, P. (2015). Antibiotic resistance in prevalent bacterial and protozoan sexually transmitted infections. *Indian Journal of Sexually Transmitted Diseases AIDS, 36*(1), 3–8.
44. Khandwalla, H. E., Luby, S., & Rahman, S. (2000). Knowledge, attitudes, and practices regarding sexually transmitted infections among general practitioners and medical specialists in Karachi, Pakistan. *Sexually Transmitted Infections, 76*(5), 383–385.
45. Khanum, S., Ahmed, J. U., Rahim, M. A., Sultana, N., & Begum, R. (2014). Evidence based diagnostic approach to tubal factor infertility. *BIRDEM Medical Journal, 4*(1), 33–37.
46. Laboratory Centre for Disease Control (LCDC) Expert Working Group on Canadian Guidelines for Sexually Transmitted Disease, & Laboratory Centre for Disease Control (Canada).Division of STD Prevention. (1998). Canadian STD guidelines. Canadian Government Publishing.
47. Larsson, P. G., & Forsum, U. (2005). Bacterial vaginosis–A disturbed bacterial flora and treatment enigma. *APMIS, 113*(5), 305–316.
48. Lautenschlager, S., & Eichmann, A. (2001). The heterogeneous clinical spectrum of genital herpes. *Dermatology, 202*(3), 211–219.
49. Lin, J. S., Whitlock, E., O'Connor, E., & Bauer, V. (2008). Behavioral counseling to prevent sexually transmitted infections: A systematic review for the U.S. preventive services task force. *Annals of Internal Medicine, 149*(7), 497–508.
50. Livengood, C. H., III. (2009). Bacterial vaginosis: An overview for 2009 rev. *Obstetrics and Gynecology, 2*(1), 28–37.
51. Mahoney, J. F., & Bryant, K. K. (1933). Contact infection of rabbits in experimental syphilis. *American Journal of Syphilis, 17*, 188–193.
52. Maisey, K., Nardocci, G., Imarai, M., Cardenas, H., Rios, M., Croxatto, H. B., & Velasquez, L. A. (2003). Expression of proinflammatory cytokines and receptors by human fallopian

tubes in organ culture following challenge with Neisseria gonorrhoeae. *Infection and Immunity, 71*(1), 527–532.

53. Manavi, K. (2006). A review on infection with chlamydia trachomatis. Best practice & research. *Clinical Obstetrics & Gynaecology, 20*(6), 941–951.
54. Markowitz, L. E., Dunne, E. F., Saraiya, M., Lawson, H. W., Chesson, H., & Unger, E. R. (2007). Quadrivalent human papillomavirus vaccine: Recommendations of the advisory committee on immunization practices (ACIP). *MMWR – Recommendations and Reports, 56*, 1–24.
55. Matthews, K., Leong, C. M., Baxter, L., Inglis, E., Yun, K., Bäckström, B. T., & Hibma, M. (2003). Depletion of Langerhans cells in human papillomavirus type 16-infected skin is associated with E6-mediated down regulation of E-cadherin. *Journal of Virology, 77*(15), 8378–8385.
56. May, M., Gompels, M., Delpech, V., Porter, K., Post, F., Johnson, M., & Hill, T. (2011). Impact of late diagnosis and treatment on life expectancy in people with HIV-1: UK Collaborative HIV Cohort (UK CHIC) Study. *BMJ, 343*, d6016.
57. Motrich, R. D., Cuffini, C., Oberti, J. P. M., Maccioni, M., & Rivero, V. E. (2006). Chlamydia trachomatis occurrence and its impact on sperm quality in chronic prostatitis patients. *The Journal of Infection, 53*(3), 175–183.
58. Muual, D., & Geubbels, E. (2006). Epidemiology of reproductive tract infections (RTIs) in Malawi. *Malawi Medical, J18*(4), 176–190.
59. Myer, L., Kuhn, L., Stein, Z. A., Wright, T. C., Jr., & Denny, L. (2005). Intravaginal practices, bacterial vaginosis, and women's susceptibility to HIV infection: Epidemiological evidence and biological mechanisms. *The Lancet Infectious Diseases, 5*(12), 786–794.
60. Nam, Y. H., Min, D., Kim, H. P., Song, K. J., Kim, K. A., Lee, Y. A., & Shin, M. H. (2011). Leukotriene B4 receptor BLT-mediated phosphorylation of NF-κB and CREB is involved in IL-8 production in human mast cells induced by Trichomonasvaginalis-derived secretory products. *Microbes and Infection, 13*(14–15), 1211–1220.
61. Nam, Y. H., Min, D., Park, S. J., Kim, K. A., Lee, Y. A., & Shin, M. H. (2011). NF-κB and CREB are involved in IL-8 production of human neutrophils induced by Trichomonasvaginalis-derived secretory products. *The Korean Journal of Parasitology, 49*(3), 291.
62. Oriel, J. D. (2012). *The scars of venus: A history of venereology*. London: Springer Science & Business Media.
63. Park, I. U., Introcaso, C., & Dunne, E. F. (2015). Human papillomavirus and genital warts: A review of the evidence for the 2015 centers for disease control and prevention sexually transmitted diseases treatment guidelines. *Clinical Infectious Diseases, 61*(suppl_8), S849–S855.
64. Patel, V., Weiss, H. A., Mabey, D., West, B., D'souza, S., Patil, V., & Kirkwood, B. R. (2006). The burden and determinants of reproductive tract infections in India: A population based study of women in Goa, India. *Sexually Transmitted Infections, 82*(3), 243–249.
65. Pavlova, S. I., Kilic, A. O., Kilic, S. S., So, J. S., Nader-Macias, M. E., Simoes, J. A., & Tao, L. (2002). Genetic diversity of vaginal lactobacilli from women in different countries based on 16S rRNA gene sequences. *Journal of Applied Microbiology, 92*(3), 451–459.
66. Peterman, T. A., Tian, L. H., Metcalf, C. A., Satterwhite, C. L., Malotte, C. K., DeAugustine, N., & Douglas, J. M. (2006). High incidence of new sexually transmitted infections in the year following a sexually transmitted infection: A case for rescreening. *Annals of Internal Medicine, 145*(8), 564–572.
67. Puthuchira Ravi, R., & AthimulamKulasekaran, R. (2014). Care seeking behaviour and barriers to accessing services for sexual health problems among women in rural areas of Tamil nadu state in India. *Journal of Sexually Transmitted Diseases, 2014*, 292157.
68. Raiziss, G. W., & Severac, M. (1937). Rapidity with which Spirochaetapallida invades the blood stream. *Archives of Dermatology and Syphilology, 35*(6), 1101–1109.

69. Rosemond, A., Lanotte, P., Watt, S., Sauget, A. S., Guerif, F., Royere, D., & Mereghetti, L. (2006). Systematic screening tests for Chlamydia trachomatis, Mycoplasma hominis and Ureaplasma urealyticum in urogenital specimens of infertile couples. *Pathologie-Biologie, 54*(3), 125–129.
70. Ross, J. D. (2014). Pelvic inflammatory disease. *Medicine, 42*(6), 333–337.
71. Ryu, J. S., Kang, J. H., Jung, S. Y., Shin, M. H., Kim, J. M., Park, H., & Min, D. Y. (2004). Production of interleukin-8 by human neutrophils stimulated with Trichomonasvaginalis. *Infection and Immunity, 72*(3), 1326–1332.
72. Sabarwal, S., & Santhya, K. G. (2012). Treatment-seeking for symptoms of reproductive tract infections among young women in India. *International Perspectives on Sexual and Reproductive Health, 38*, 90–98.
73. Satterwhite, C. L., Torrone, E., Meites, E., Dunne, E. F., Mahajan, R., Ocfemia, M. C. B., & Weinstock, H. (2013). Sexually transmitted infections among US women and men: Prevalence and incidence estimates, 2008. *Sexually Transmitted Diseases, 40*(3), 187–193.
74. Schacker, T. (2001). The role of HSV in the transmission and progression of HIV. *Herpes: The Journal of the IHMF, 8*(2), 46–49.
75. Schillinger, J. A., Dunne, E. F., Chapin, J. B., Ellen, J. M., Gaydos, C. A., Willard, N. J., & Markowitz, L. E. (2005). Prevalence of Chlamydia trachomatis infection among men screened in 4 US cities. *Sexually Transmitted Diseases, 32*(2), 74–77.
76. Schwebke, J. R., Richey, C. M., & Weiss, H. L. (1999). Correlation of behaviors with microbiological changes in vaginal flora. *The Journal of Infectious Diseases, 180*(5), 1632–1636.
77. Schwiertz, A., Taras, D., Rusch, K., & Rusch, V. (2006). Throwing the dice for the diagnosis of vaginal complaints? *Annals of Clinical Microbiology and Antimicrobials, 5*(1), 4.
78. Selik, R. M., Mokotoff, E. D., Branson, B., Owen, S. M., Whitmore, S., & Hall, H. I. (2014). Revised surveillance case definition for HIV infection—United States. *Morbidity and Mortality Weekly Report. Recommendations and Reports, 63*(3), 1–10.
79. Centers for Disease Control and Prevention. (2002). Sexually transmitted diseases treatment guidelines 2002. *MMWR – Recommendations and Reports, 51*, 1–78.
80. Shi, Y., Chen, L., Tong, J., & Xu, C. (2009). Preliminary characterization of vaginal microbiota in healthy Chinese women using cultivation-independent methods. *The Journal of Obstetrics and Gynaecology Research, 35*, 525–532.
81. Singh, A. E., & Romanowski, B. (1999). Syphilis: Review with emphasis on clinical, epidemiologic, and some biologic features. *Clinicla Microbiology Reviews, 12*(2), 187–209.
82. Smaill, F., & Hofmeyr, G. J. (2002). Antibiotic prophylaxis for cesarean section. *The Cochrane Database of Systematic Reviews, 3*, CD000933–CD000933.
83. Srinivasan., S., & Fredricks, D. N. (2008). The human vaginal bacterial biota and bacterial vaginosis. *Interdisciplinary Perspectives on Infectious Diseases, 2008*, 750479.
84. Sutcliffe, S., Taha, T. E., Kumwenda, N. I., Taylor, E., & Liomba, G. N. (2002). HIV-1 prevalence and herpes simplex 2, hepatitis C virus and hepatitis B virus infections among male workers at a sugar estate in Malawi. *Journal of Acquired Immune Deficiency Syndromes (1999), 31*(1), 90–97.
85. Swygard, H., Sena, A. C., Hobbs, M. M., & Cohen, M. S. (2004). Trichomoniasis: Clinical manifestations, diagnosis and management. *Sexually Transmitted Infections, 80*(2), 91–95.
86. Thekdi, K. P., Patel, N. K., Patel, K. G., & Thekdi, P. I. (2017). Health seeking attitude of women regarding reproductive tract infections in a rural area of Surendranagar district. *IJRMS, 1*(4), 552–556.
87. Thomas, D. D., Navab, M., Haake, D. A., Fogelman, A. M., Miller, J. N., & Lovett, M. A. (1988). Treponemapallidum invades intercellular junctions of endothelial cell monolayers. *Proceedings of the National Academy of Sciences of the United States of America, 85*(10), 3608–3612.

88. Tian, Y. H., Xiong, J. W., Hu, L., Huang, D. H., & Xiong, C. L. (2007). Candida albicans and filtrates interfere with human spermatozoal motility and alter the ultrastructure of spermatozoa: An in vitro study. *International Journal of Andrology, 30*(5), 421–429.
89. Tita, A. T., & Andrews, W. W. (2010). Diagnosis and management of clinical chorioamnionitis. *Clinics in Perinatology, 37*(2), 339–354.
90. Turovskiy, Y., Noll, K. S., & Chikindas, M. L. (2011). Theaetiology of bacterial vaginosis.J. *Applied Microbiology, 110*(5), 1105–1128.
91. Verstraelen, H. (2008). Bacterial vaginosis: A sexually enhanced disease. *International Journal of STD & AIDS, 19*(8), 575–576.
92. Voeten, H. A., Otido, J. M., O'hara, H. B., Kuperus, A. G., Borsboom, G. J., Ndinya-Achola, J. O., & Habbema, J. D. F. (2001). Quality of sexually transmitted disease case management in Nairobi, Kenya: A comparison among different types of healthcare facilities. *Sexually Transmitted Diseases, 28*(11), 633–642.
93. Wasserheit, J. N., & Holmes, K. K. (1992). Reproductive tract infections: Challenges for international health policy, programs, and research. In *Reproductive tract infections* (pp. 7–33). Boston: Springer.
94. Watts, D. H., Krohn, M. A., Hillier, S. L., & Eschenbach, D. A. (1990). Bacterial vaginosis as a risk factor for post-cesarean endometritis. *Obstetrics and Gynecology, 75*(1), 52–58.
95. Westrom, L. (1995). Effect of pelvic inflammatory disease on fertility. *Venereology: Official Publication of the National Venereology Council of Australia, 8*(4), 219–222.
96. Whitley, R. J., Kimberlin, D. W., & Roizman, B. (1998). Herpes simplex viruses. *Clinical Infectious Diseases, 26*, 541–553.
97. Wølner-Hanssen, P., Krieger, J. N., Stevens, C. E., Kiviat, N. B., Koutsky, L., Critchlow, C., & Holmes, K. K. (1989). Clinical manifestations of vaginal trichomoniasis. *JAMA, 261*(4), 571–576.
98. Workowski, K. A., & Bolan, G. A. (2015). Sexually transmitted diseases treatment guidelines, 2015. MMWR. Recommendations and reports: Morbidity and mortality weekly report. *Recommendations and Reports, 64*(RR-03), 1–110.
99. World Health Organization. (2005). *Sexually transmitted and other reproductive tract infections: A guide to essential practice*. Geneva: WHO.
100. World Health Organization. (2006). *Prevention and control of sexually transmitted infections: Draft global strategy: Report by the Secretariat*. Geneva: WHO. In Prevention and control of sexually transmitted infections: Draft global strategy: Report by the Secretariat.
101. World Health Organization. (2011). *Prevalence and incidence of selected sexually transmitted infections, Chlamydia trachomatis, Neisseria gonorrhoeae, syphilis and Trichomonas vaginalis: Methods and results used by WHO to generate 2005 estimates (no. chlamydia infections)*. Geneva: World Health Organization.
102. Zaidi, A. K., Thaver, D., Ali, S. A., & Khan, T. A. (2009). Pathogens associated with sepsis in newborns and young infants in developing countries. *The Pediatric Infectious Disease Journal, 28*(1), S10–S18.
103. Zandberg, D. P., Bhargava, R., Badin, S., & Cullen, K. J. (2013). The role of human papillomavirus in nongenital cancers. *CA: a Cancer Journal for Clinicians, 63*(1), 57–81.
104. Zhao, K. N., & Chen, J. (2011). Codon usage roles in human papillomavirus. *Reviews in Medical Virology, 21*, 397–411.
105. Zhou, X., Bent, S. J., Schneider, M. G., Davis, C. C., Islam, M. R., & Forney, L. J. (2004). Characterization of vaginal microbial communities in adult healthy women using cultivation-independent methods. *Microbiology, 150*, 2565–2573.

Discovery of Novel Drug Targets in Microbial Pathogens Among Hypothetical Proteins: Methods and Significance

Ahmad Abu Turab Naqvi, Taj Mohammad, and Md. Imtaiyaz Hassan

Abstract

Bacteria are ubiquitously found in our surroundings. Everything that we come across in our routine life witnesses presence of at least some kind of bacteria. Even human bodies are inhabited by several species of bacteria. Some of these bacteria are symbiotic in nature, and others cause diseases in human beings. Diseases caused by these bacteria are moderate to lethal if remain untreated. Fortunately, effective drugs have been developed for treating the diseases caused by these bacteria. Still there is some scope of further developments in the drug discovery against such diseases. Advancements in the genome sequence methods have produced huge amount of proteomic data for these bacteria. Some of the proteins belonging to these organisms are well characterized, and the rest of these proteins fall in the category of uncharacterized and hypothetical proteins (HPs). This chapter discusses some state-of-the-art methods of function prediction of HPs using bioinformatics techniques. The pipeline developed for this purpose uses prediction servers, tools and protein databases for successfully predicting the functions of the HPs. The functional characterization of HPs facilitates the identification of potential therapeutic targets and biomarkers of disease-causing bacteria.

Keywords

Bacterial pathogens · Microbial disease · Hypothetical proteins · Function prediction · Therapeutic targets · Drug discovery

A. A. T. Naqvi · T. Mohammad · Md. Imtaiyaz Hassan (✉)
Centre for Interdisciplinary Research in Basic Sciences, Jamia Millia Islamia, New Delhi, India
e-mail: mihassan@jmi.ac.in

S. Hameed, Z. Fatima (eds.), *Pathogenicity and Drug Resistance of Human Pathogens*,
https://doi.org/10.1007/978-981-32-9449-3_19

Introduction

Bacteria are surrounding us, obviously all around, on objects and typically found in and on our bodies. When they are on our body and does not cause any disease, they are so-called a colonizer. But people can get infected from the pathogenic bacteria called bacterial pathogen. A bacterial pathogen is the bacterium that is able to cause infectious disease in the host [39]. Only a few (fewer than 100) bacteria can cause diseases in humans, while most bacteria are inoffensive or often beneficial for human health and industrial purpose [52]. Pathogenic bacteria such as *Staphylococcus, Vibrio cholerae and Mycobacterium tuberculosis* (see Table 1) do cause infection and disease in the host body, and they are different from normal bacteria [2]. These pathogens use various strategies for causing infection: producing toxins and spreading through different parasites or adhesion for causing inflammation [6].

Associated Diseases

Factually, bacterial pathogens have been the reason for several lethal diseases such as tuberculosis, typhoid, cholera, typhus, plague, diphtheria, and pneumonia [55]. At the beginning of the twentieth century, these diseases were the leading causes of epidemic deaths worldwide. Apart from these diseases, a number of maladies such as toxic shock syndrome, Lyme disease, gastric and duodenal ulcers, salmonellosis, atherosclerosis and nosocomial infections are the result of bacterial infections [10].

Current Status

Since early 1950s, health communities are facing an emergent persistence of infectious diseases where emergence of new bacterial pathogens is a major public health threat [52]. In the emergence of infections from the pathogenic bacteria, several man-made attempts and services are responsible such as developing new analytical and diagnostics tools and bacterial culture; developing molecular practices and other microbiology research experimentation; increasing human exposure to pathogenic bacteria resulting in several environmental changes; and emergence and development of more virulent bacterial strains [52]. Now, the medical field requires a comprehensive integration of bacteriological, medical and epidemiologic aspects and the use of experimental models in order to escape the rise of such emerging infectious diseases. During the last four decades, a number of pathogenic bacteria have been identified known to cause different severe diseases and showing distinct clinical topographies which require more specific diagnosis and treatments [52].

Today, in the twenty-first century, water refinement, vaccination and antibiotic treatment have reduced instances of bacterial diseases [3]. Despite this, a safe and novel treatment is also needed to treat such bacterial infections with lesser complications. In modern biological science, genomics and proteomics possess enormous possibilities in order to treat pathogenic infections where structure-function analysis of the bacterial proteins can be carried out identify bacterial drug

Table 1 List of common Indian bacteria and associated diseases

S. no.	Bacterium	Disease
1.	*Acinetobacter baumannii*	Nosocomial infections
2.	*Bacillus cereus*	Food-borne illnesses: vomiting and diarrhoea
3.	*Bartonella henselae*	Cat scratch fever
4.	*Borrelia burgdorferi*	Lyme disease and complications
5.	*Brucella abortus*	Brucellosis
6.	*Campylobacter jejuni*	Gastroenteritis and campylobacteriosis
7.	*Chlamydia pneumoniae*	Atherosclerosis
8.	*Clostridium botulinum*	Sudden infant death syndrome (SIDS)
9.	*Clostridium difficile*	Diarrhoea; pseudomembranous colitis
10.	*Clostridium perfringens*	Gas gangrene
11.	*Clostridium tetani*	Tetanus
12.	*Corynebacterium amycolatum*	Hospital-acquired endocarditis
13.	*Corynebacterium diphtheriae*	Diphtheria
14.	*Ehrlichia chaffeensis*	Human ehrlichiosis
15.	*Enterococcus faecalis*	Nosocomial infections
16.	*Escherichia coli*	Haemorrhagic colitis; haemolytic uremic syndrome
17.	*Helicobacter pylori*	Gastric and duodenal ulcers
18.	*Klebsiella pneumoniae*	Bloodstream infections
19.	*Legionella pneumophila*	Legionnaires pneumonia
20.	*Listeria monocytogenes*	Listeriosis
21.	*Mycobacterium tuberculosis*	Tuberculosis
22.	*Mycoplasma pneumoniae*	Atypical pneumonia
23.	*Parachlamydia acanthamoebae*	Pneumonia
24.	*Rickettsia rickettsii*	Rickettsiosis, Rocky Mountain spotted fever
25.	*Salmonella enterica*	Salmonellosis
26.	*Salmonella typhi*	Typhoid
27.	*Shigella dysenteriae*	Severe dysentery
28.	*Staphylococcus aureus*	Pneumonia, otitis media, meningitis
29.	*Streptococcus pyogenes*	Streptococcal toxic shock syndrome
30.	*Treponema pallidum*	Syphilis
31.	*Vibrio cholerae*	Epidemic cholera
32.	*Vibrio vulnificus*	Cellulitis, wound and gastrointestinal disease
33.	*Yersinia enterocolitica*	Yersiniosis

targets as proteins play a vital role in the bacterial pathogenesis. Some of these bacterial proteins are well characterized and annotated which are involved in different biological functions but many proteins have not been classified, and their functions are not known which are referred to as hypothetical proteins (HPs).

Hypothetical Proteins

HPs are the proteins whose existence has been predicted, but doesn't have functional annotation because of lack of experimental evidence, and haven't been isolated and characterized yet [14]. Today, most of the hypothetical proteins result from the computational analysis of genomic data where bioinformatics tools are used for the identification of new gene (mainly ORF) lacking defined homologue in the protein databases [35]. However, not all HPs are completely uninformative, some of them have the theoretical description of their structural domains and functions predicted through various bioinformatics applications [8]. Now, using bioinformatics applications of similarity search methods, different HP genes are being identified in different species, progressively. Apart from that, various bioinformatics methods can also be used for the prediction of biological function of such HPs [9].

In Silico Function Prediction of Hypothetical Proteins

In most of the sequenced genomes, many proteins are uncharacterized, and the knowledge of their biological functions is not available [42]. Therefore, a detailed characterization, annotation and function prediction of these HPs using *in silico* methods is necessary which will be very useful in modern biological research. The function of HPs can often be predicted by comparing it with known proteins of similar folds [38]. A number of bioinformatics methods and tools are used for the functional annotation of HPs by domain homology search approach where conserved domains of HPs are compared with the known family of domains. By this approach, HPs can be classified into particular family of the proteins. The function of HPs can also be predicted by comparative/homology modeling, where HP is compared using seqeunce alignment with annotated proteins whose three-dimensional structures have been solved and deposited in the Protein Data Bank [22]. Further approach of functional annotation of HPs includes computational prediction of three-dimensional structure by different modeling approaches such as threading/fold recognition and *ab initio* methods [22]. The functions of the HPs can also be predicted by integrating different protein classification systems, motif discovery tools as well as methods that are based on characteristics of the protein sequence and metabolic pathways [42]. *In silico* methods may lead to the discovery of new structures with novel functions, which will be helpful in detailed analysis of undiscovered protein pathways present in the organisms [32]. This chapter deals with the function prediction of HPs from common Indian human pathogenic bacteria using *in silico* approach.

Common Indian Bacterial Pathogens

This section deals with the common Indian pathogenic bacteria. Detailed information of some major bacterial pathogens, their pathogenesis and associated diseases is reviewed here. A list of common Indian disease-causing bacteria along with their associated diseases can be seen in Table 1.

Mycobacterium tuberculosis

Mycobacterium tuberculosis is a nonmotile, obligate aerobe, acid-fast pathogenic bacterium belonging to the family *Mycobacteriaceae* that causes tuberculosis [16]. Tuberculosis (TB) is one of the oldest known human diseases and still is one of the major causes of human deaths [37]. Primarily, *M. tuberculosis* is a pathogen of the mammalian respiratory system which infects the lungs [40]. *M. tuberculosis* was first discovered by Robert Koch in 1882 [5]. It is transmitted through the air and affects the lungs, bone, central nervous system, liver, spleen, lymph nodes, genito-urinary tract and many other organ systems [18].

Mycoplasma pneumoniae

Mycoplasma pneumoniae is a simplest self-replicating bacterium belonging to the family of *Mycoplasmataceae*. It has been found to cause acute respiratory tract infection known as acquired pneumonia and various types of neurological, hepatic and haemolytic anaemia, cardiac diseases and polyarthritis [36]. *M. pneumoniae* accounts for more than 40% of community-acquired pneumonia cases in children [11]. *M. pneumoniae* has also been found to cause several chronic diseases such as juvenile idiopathic arthritis, rheumatoid arthritis, asthma and Crohn's disease [4]. It is a surface pathogen which develops adherence to mucosal epithelium of respiratory and urogenital tracts of the host by using various surface proteins. The important adherence proteins of *M. pneumoniae* are P1 adhesins [20] which are responsible for the primary cause of its virulence [26].

Treponema pallidum

Treponema pallidum ssp. *pallidum* is a Gram-negative bacterium, a member of family *Spirochaetaceae*, and is an obligate human parasite which is found to cause syphilis, a globally occurring sexually transmitted disease (STD) [28, 34]. The syphilis infection is often spread through sexual contacts, which is a main epidemic reason of this disease [19]. The primary effect of the syphilis infection is skin lesions on the site of infection [34], while the secondary and tertiary stages are assumed to be deadly for the host [48]. The syphilis infection is severe in nature as 12 million new cases of this disease were reported by the World Health Organization (WHO) in 1999 with most of the cases coming from the developing countries [34].

Chlamydia trachomatis

Chlamydia trachomatis is a Gram-negative bacterium and an intracellular obligate pathogen of humans that is responsible for causing trachoma and chlamydia, a STD in humans. Chlamydial genital infections are the most common among all infectious diseases. According to the WHO, every year 92 million new cases of chlamydia infections are reported worldwide [34]. Chlamydia has been found to increase HIV shedding in the genital tract [13]. Apart from the spread of STDs in humans, it is also found to cause female infertility and pelvic inflammatory disease (PID) worldwide [27].

Helicobacter pylori

Helicobacter pylori is a Gram-negative slow-growing microaerophilic bacterium that causes the most common bacterial infections including several gastric problems in humans [47]. It has a spiral shape flagellated body which helps in locomotion and colonization in the host environment. It is capable of living in the acidic gastric environment of the stomach with the help of acid-adaptive genes for survival [49]. Continued infection of *H. pylori* can be transformed into a chronic infection that causes severe gastric diseases such as duodenal ulcer, gastric ulcer, gastric lymphoma and cancer [7, 54].

Mycobacterium leprae

Mycobacterium leprae is an intracellular Gram-positive bacillus that belongs to the family *Mycobacteriaceace* [17]. *M. leprae* causes leprosy in humans which was first discovered by Armauer Hansen in 1873. Thus, the disease is also known as Hansen's disease or Hanseniasis. It leads to the destruction of peripheral nerves and skin deformation. It is an obligate parasite that requires a huge amount of nutrients from the host [21].

Haemophilus influenzae

Haemophilus influenza is a nonmotile Gram-negative rod-shaped pathogenic bacterium that belongs to the family *Pasteurellaceae* [41]. *H. influenzae* causes serious invasive diseases such as bacteremia, severe pneumonia and acute bacterial meningitis especially in young children and newborns. The most virulent strain of *H. influenzae* is type b (Hib) which is spread through the respiratory tract from diseased person to another one. It is also found to cause severe inflammatory infections of the face, mouth, epiglottis, blood, joints, heart, bones, etc. [41].

Neisseria gonorrhoeae

Neisseria gonorrhoeae is a Gram-negative diplococcus obligate pathogenic bacterium, mainly found to cause a common STD known as gonorrhoea [51]. It forms colonies on the mucosal surface of urogenital tract and causes infection in the urethra of men and endocervix of women. This infection leads to severe problems such as epididymitis and pelvic inflammatory disease. *N. gonorrhoeae* infection mainly occurs in the eyes of newborn baby during their passage in the birth canal of the infected mother which may result into blindness [51].

Hypothetical Proteins

What Is Hypothetical Protein?

HP is a protein whose existence has been predicted, but it lacks experimental evidence, and hasn't been isolated and characterized yet [14]. Usually, they are predicted from nucleic acid sequences and whose existence is not reported with experimental evidence [53]. It means that the functional expression of these proteins is not shown in experimental studies so far. However, not all the HPs are completely uninformative, some of them have theoretical description of their structural domains and functions predicted through various bioinformatics applications [8]. Generally, prediction of these proteins is made by gene identification methods during the genome analysis. Nowadays, using a number of bioinformatics applications of similarity search methods, function prediction of HPs is being carried out [9].

Relevance and Importance of Hypothetical Proteins

A number of HPs deposited in the databases is unavoidable. According to the statistics, half of the genes in most of the genomes are HP candidate genes [15]. Therefore, this group contains ample amount of genomic and proteomic information. Further, identification of HPs among the genomic pool may pave the way for unraveling of novel structural and functional features among the proteins. With the identification of novel HPs along with functional and structural information, it is likely to unearth some unknown protein pathways. Such information will also help identify the protein-protein interaction cascades among the HPs. Besides, all this information may lead to identification of potential biomarkers and drug targets. Consequently, there will be more and more information about predicted genes and those that are still unknown.

In Silico Function Prediction of Hypothetical Proteins

HPs represent a subsequently large part of the proteomes. Much of these data are lying untouched and awaiting annotations. These annotations either functional or structural are possible state-of-the-art bioinformatics pipelines that have been and are being developed by the people active in functional and structural genomic research. In this section of the chapter, we will describe some methods for functional annotation of the HPs that are in trend and that we have applied (Fig. 1) for functional

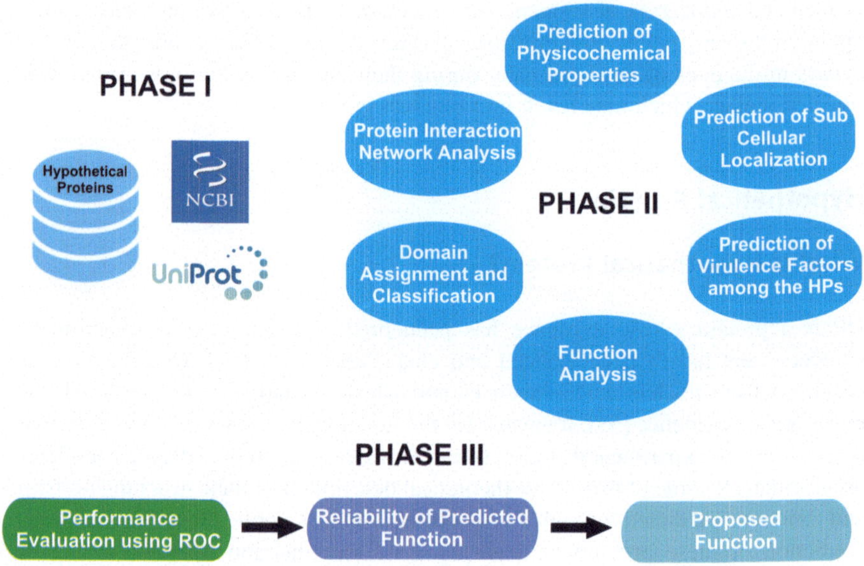

Fig. 1 Protocol used for sequence-based function prediction of HPs [46]

annotation of the HPs of common Indian bacterial pathogens during the recent years [23–25, 29–31, 43–46, 50]. Basically, based on the preliminary information available, there are two main approaches to the problem. First, sequence-based function prediction of the proteins uses amino acid sequences of the HPs to assign functions to them using various tools and databases that will be discussed in detail later on. The second approach is structural-based function prediction that following structure to function relationship paradigm uses protein structure information and matches it with predefined functional classes of the proteins with known structures and assigns them functions accordingly. Both these approaches have their advantages and limitations that are important to be considered. Although there are a number of detailed reviews that shed light on all the methods of *in silico* function prediction of unknown proteins, we may like to give a brief overview of these methods with more attention to these approaches used for functional annotation of HPs from the bacterial pathogens. (See Table 2 for the list of some commonly used tools and databases with their web addresses).

Sequence-Based Function Prediction of the HPs

Amino acid sequence of a protein contains all the needed 'information' to define its function. It also contains the code for structural organization of the protein. *In silico* methods of function prediction use amino acid sequence as a tool for assigning function to unknown proteins using the information from proteins of known function.

Table 2 List of tools and databases for functional annotation of hypothetical proteins

S. No.	Tool/web-server	URL
	Sequence similarity search tool	
1.	BLAST: basic local alignment search tool	http://blast.ncbi.nlm.nih.gov/Blast.cgi
	ClustalW2	https://www.ebi.ac.uk/Tools/msa/clustalw2/
	HHpred	ftp://toolkit.genzentrum.lmu.de/pub/HH-suite/
	PRALINE (PRofile ALIgNment)	http://ibivu.cs.vu.nl/programs/pralinewww/
2.	Biophysical and chemical characterization	
	ProtParam	http://web.expasy.org/protparam/
3.	Subcellular localization of the protein	
(i)	SOSUI	http://bp.nuap.nagoyau.ac.jp/sosui/sosui_submit.html
	TMHMM	http://www.cbs.dtu.dk/services/TMHMM/
	Psort II	http://psort.hgc.jp/form2.html
	SignalP	http://www.cbs.dtu.dk/services/SignalP/
	HMMTOP	http://www.enzim.hu/hmmtop/index.php
	SecretomeP	http://www.cbs.dtu.dk/services/SecretomeP/
	CELLO	http://cello.life.nctu.edu.tw
4.	Functional analysis tools	
	Conserved domain	http://www.ncbi.nlm.nih.gov/Structure/cdd/wrpsb.cgi
	InterProScan	http://www.ebi.ac.uk/Tools/pfa/iprscan/
	Interpro	http://www.ebi.ac.uk/interpro/
	ScanProsite	http://prosite.expasy.org/scanprosite/
	Panther	http://www.pantherdb.org/
	SMART	http://smart.embl-heidelberg.de/
	CATH	http://www.cathdb.info
	Pfam	http://pfam.sanger.ac.uk
	ProtoNet	http://www.protonet.cs.huji.ac.il
	SVMProt	http://bidd2.nus.edu.sg/cgi-bin/svmprot/svmprot.cgi
5.	Prediction of fold pattern	
	PFP-FunDSeqE	http://www.csbio.sjtu.edu.cn/bioinf/PFP-FunDSeqE/
	HHpred	http://toolkit.tuebingen.mpg.de/hhpred
	Dali server	http://ekhidna.biocenter.helsinki.fi/dali_server/start
6.	Virulence prediction	
	VICMpred	crdd.osdd.net/raghava/vicmpred/index.html
7.	Protein-protein interaction	
	STRING	http://string-db.org
	MINT	https://mint.bio.uniroma2.it/

Similarity Search Methods

In this approach, unknown protein is used as query to be searched against the database of knows proteins. BLAST [1], FASTA [33], HMMER [12], etc., are some prominent tools that are used for similarity search. Usually, classical protein BLAST tool is used for similarity search. Besides, some variants of BLAST such as PSI-BLAST and PHI-BLAST can also be used for more precise results. FASTA is a first-generation algorithm of heuristic similarity searching. It is also used for the similarity searching in the *in silico* function annotation of HPs. HMMER uses Hidden Markov Models (HMM) for the similarity search. In the *in silico* function prediction of HPs, any one of these tools can be used. Further, they can also be used in group to derive consensus results.

Classification Methods

This approach is somehow similar to the similarity search approach in concept, but in this approach, specific databases are used instead of any interface. The emergence of huge amount of protein information has made it necessary to organize the storage of proteins. Family-based classification of proteins has become the prominent tool in this regard. There are mainly two approaches used for protein classification, i.e. functional classification and structural classification. There are dedicated databases to hold the information of both types. Panther, CATH, SCOP, InterPro, Prints, PROSITE, ProDom and Pfam are some prominent protein family databases based on the structural and functional information. Apart from these, the information of conserved domains in the unknown proteins can be obtained using CDART tool and CDD database of the NCBI.

Protein-Protein Interaction Networks for Function Prediction

Proteins are very diverse macromolecules that show a wide range of molecular functions and are involved in a number of biological processes by interacting with other proteins regulating their function whether inhibiting or activating them. The phenomenon of protein-protein interaction lies in the heart of a lot of biological process guided by the individual function of a protein. Thus, knowledge of a protein's role in an interaction network may be helpful for assigning functions to unknown proteins. Fortunately, there is enough information on protein-protein interaction networks stored in the databases that can be exploited through the dedicated interfaces of these databases. Some of these databases that are widely used for analysis are SPRING, MINT, and Intact etc. The interfaces of these databases are used to retrieve information for unknown proteins, and consequently functions are assigned to them. The pipeline developed by our group uses SPRING for the prediction of protein-protein interaction maps of the HPs. Based on the information provided by the SPRING, a consensus-based function is assigned to each HP.

Physicochemical Properties and Subcellular Localization

The cellular localization of proteins tells a lot about its function in the cell. There is an evident distinction between the functional behaviour of cytoplasmic and membranous proteins. Membranous proteins are mostly involved in intercellular communication and transport. Therefore, using the information of proteins with known functions and definite subcellular localization, we can predict the similarity for unknown proteins and based on that functions to unknown proteins can also be assgined. There are a number of tools used for subcellular localization prediction such as Cello, PSORTb, and PSLpred etc. Some of these tools are dedicated to bacterial proteins only (PSLpred), and others can be used for both prokaryotic and eukaryotic organisms (PSORTb and CELLO). Presence of transmembrane helices is an indicative feature of membrane proteins. Tools like TMHMM and HMMTOP can be used to predict transmembrane helices in the HPs thus membrane proteins can be identified among the pool of HPs which is consequently helpful in function prediction. Besides, physicochemical properties of a protein may also tell a lot about its function. We use ProtParam server for the theoretical prediction of physicochemical properties such as molecular weight, theoretical PI, aliphatic index, stability, etc.

Prediction of Virulence Factors

Virulence factors are proteins that facilitate the pathogenicity of the bacteria by helping it in invasion, intrusion, locomotion and adherence to the host cell. There are a large number of known virulence factors of the pathogenic bacteria whose information is stored in the databases. Prediction servers such as VICMpred and Virulentpred are dedicated for the prediction of virulence factors among the unknown proteins. By querying the unknown proteins against the set of known virulence factors, we can identify the potential factors among the HPs along with their functions conforming their role in pathogenesis.

Structure-Based Function Prediction of the HPs

The structure-based function prediction of the proteins is based on the structure to function paradigm which says that the function of a protein is defined by specific organization in its structural elements. Therefore, structure prediction of unknown protein may be helpful for function prediction. There are mainly two approaches that are used such as homology modeling and *ab initio* modeling with lesser success stories of the later one. Detailed description of modeling methods is beyond the scope of this chapter. In the first approach of structure prediction through homology modeling is followed by assigning function to the HPs on the basis of structural elements. ProFunc server predicts the molecular function of a protein using its three-dimensional structure. Another way is by applying fold recognition methods. In this approach the predicted structure is searched against the library of protein folds to find out possible similarity between the query and known structures of the database. Based on the retrieved information, putative function to the unknown protein is assigned. Despite recent advancements in computational techniques for structure prediction, the *ab initio* modeling is still confined due to its inability to predict accurate or close to accurate structure. Besides, it is also very costly in respect to time and computational resources.

Consensus-Based Function Prediction

During the course of *in silico* function prediction of HPs, an array of tools and databases is used. The reason behind consulting more resources is to establish a consensus approach for function prediction. Using lesser number bioinformatics tools for annotation is more likely to give false positive results. To counter this problem, we use consensus-based function prediction approach in which a number of tools and databases are used for analysis. We then set a threshold value for the number of resources predicting the same function, for example, if we are using 12 tools and databases, then we consider the function which is predicted by four or more tools. To increase the sensitivity and reduce the number of false positives, this threshold value can be increased accordingly. Therefore, there are chances of getting function prediction with high accuracy.

Conclusions

In this chapter, so far, we describe some of the common Indian bacterial pathogens and the diseases associated with them. Along with that, we further talk about hypothetical proteins, their incidence in the genomic data, their importance and the methods to predict their functions using *in silico* techniques. It has been decades since Margaret Dayhoff started developing the first ever database of protein sequence. Since then, there is an astronomical increase in the production of protein sequence data. From primary to derived databases, all contain millions of entries. Some of them are experimentally characterized, and some are predicted using comparative genomics techniques. We look for the plethora of proteomic data featured as hypothetical in the databases of proteins. As discussed earlier, HPs don't have experimental level existence; rather these are predicted proteins. These proteins are likely to have similarity with other proteins that have known functions and structure. Therefore, we develop the pipeline incorporating a various bioinformatics prediction servers, protein family databases, domain and motif databases, physicochemical property prediction server, subcellular localization prediction and some virulence factor prediction tools.

The course of computational function prediction is bifurcated in usually two approaches based on the available information. The first approach is sequence-based function prediction where only sequence information, in most of cases for the HPs, is available in the databases. The second one is structure-based functional characterization, where the three-dimensional structure of the protein is used for function prediction. We choose either of the approaches based on preliminary data available. After extensive analysis of HPs using all the tools and databases, we get putative functions for them that are further validated using extensive literature survey or some statistical methods such as ROC. The HPs then are classified based on their molecular functions in the wide range of protein functional classes such as enzymes, transporter proteins, nucleic acid and protein-binding proteins, virulence factors, etc. Based on their function and their possible involvement in the pathogenicity of the bacteria, these HPs are marked potential drug targets and biomarkers that pave the way for further experimental experiments.

References

1. Altschul, S. F., Gish, W., Miller, W., Myers, E. W., & Lipman, D. J. (1990). Basic local alignment search tool. *Journal of Molecular Biology, 215*, 403–410. https://doi.org/10.1016/S0022-2836(05)80360-2S0022-2836(05)80360-2.
2. Andersen, J. L., et al. (2015). Multidrug efflux pumps from Enterobacteriaceae, Vibrio cholerae and *Staphylococcus aureus* bacterial food pathogens. *International Journal of Environmental Research and Public Health, 12*, 1487–1547.
3. Arias, C. A., & Murray, B. E. (2009). Antibiotic-resistant bugs in the 21st century – A clinical super-challenge. *New England Journal of Medicine, 360*, 439–443.
4. Atkinson, T. P., & Waites, K. B. (2014). Mycoplasma pneumoniae infections in childhood. *The Pediatric Infectious Disease Journal, 33*, 92–94.
5. Cambau, E., & Drancourt, M. (2014). Steps towards the discovery of *Mycobacterium tuberculosis* by Robert Koch, 1882. *Clinical Microbiology and Infection, 20*, 196–201.
6. Certad, G., Viscogliosi, E., Chabé, M., & Cacciò, S. M. (2017). Pathogenic mechanisms of Cryptosporidium and Giardia. *Trends in Parasitology, 33*, 561–576.
7. Cid, T. P., Fernández, M. C., Benito Martínez, S., & Jones, N. L. (2013). Pathogenesis of Helicobacter pylori infection. *Helicobacter, 18*, 12–17.
8. Doerks, T., Bairoch, A., & Bork, P. (1998). Protein annotation: Detective work for function prediction. *Trends in Genetics, 14*, 248–250.
9. Eisenstein, E., et al. (2000). Biological function made crystal clear—annotation of hypothetical proteins via structural genomics. *Current Opinion in Biotechnology, 11*, 25–30.
10. Falkow, S. (1998). The microbe's view of infection. *Annals of Internal Medicine, 129*, 247–248.
11. Ferwerda, A., Moll, H. A., & de Groot, R. (2001). Respiratory tract infections by Mycoplasma pneumoniae in children: A review of diagnostic and therapeutic measures. *European Journal of Pediatrics, 160*, 483–491.
12. Finn, R. D., Clements, J., & Eddy, S. R. (2011). HMMER web server: Interactive sequence similarity searching. *Nucleic Acids Research, 39*, W29–W37. https://doi.org/10.1093/nar/gkr367gkr367.
13. Fleming, D. T., & Wasserheit, J. N. (1999). From epidemiological synergy to public health policy and practice: The contribution of other sexually transmitted diseases to sexual transmission of HIV infection. *Sexually Transmitted Infections, 75*, 3–17.
14. Galperin, M. Y. (2001). Conserved 'hypothetical'proteins: New hints and new puzzles. *International Journal of Genomics, 2*, 14–18.
15. Galperin, M. Y., & Koonin, E. V. (2004). 'Conserved hypothetical' proteins: Prioritization of targets for experimental study. *Nucleic Acids Research, 32*, 5452–5463. https://doi.org/10.1093/nar/gkh885.
16. Gordon, S. V., & Parish, T. (2018). Microbe profile: *Mycobacterium tuberculosis*: Humanity's deadly microbial foe. *Microbiology (Reading, England), 164*, 437–439.
17. Gough, J., Karplus, K., Hughey, R., & Chothia, C. (2001). Assignment of homology to genome sequences using a library of hidden Markov models that represent all proteins of known structure1. *Journal of Molecular Biology, 313*, 903–919.
18. Harisinghani, M. G., McLoud, T. C., Shepard, J.-A. O., Ko, J. P., Shroff, M. M., & Mueller, P. R. (2000). Tuberculosis from head to toe 1: (CME available in print version and on RSNA link). *Radiographics, 20*, 449–470.

19. Hertel, M., Matter, D., Schmidt-Westhausen, A. M., & Bornstein, M. M. (2014). Oral syphilis: A series of 5 cases. *Journal of Oral and Maxillofacial Surgery, 72*, 338–345.
20. Hu, P. C., Collier, A. M., & Baseman, J. (1977). Surface parasitism by Mycoplasma pneumoniae of respiratory epithelium. *Journal of Experimental Medicine, 145*, 1328–1343.
21. Irgens, L. (2002). The discovery of the leprosy bacillus. *Paediatric and Perinatal Epidemiology, 16*, 290–296.
22. Kelley, L. A., & Sternberg, M. J. (2009). Protein structure prediction on the web: A case study using the Phyre server. *Nature Protocols, 4*, 363.
23. Khan, S., Shahbaaz, M., Bisetty, K., Ahmad, F., & Hassan, M. I. (2017). Classification and functional analyses of putative conserved proteins from Chlamydophila pneumoniae CWL029. *Interdisciplinary Sciences, 9*, 96–106. https://doi.org/10.1007/s12539-015-0134-710.1007/s12539-015-0134-7.
24. Kumar, K., Prakash, A., Anjum, F., Islam, A., Ahmad, F., & Hassan, M. I. (2015). Structure-based functional annotation of hypothetical proteins from *Candida dubliniensis*: A quest for potential drug targets 3. *Biotech, 5*, 561–576. https://doi.org/10.1007/s13205-014-0256-3.
25. Kumar, K., Prakash, A., Tasleem, M., Islam, A., Ahmad, F., & Hassan, M. I. (2014). Functional annotation of putative hypothetical proteins from *Candida dubliniensis*. *Gene, 543*, 93–100. https://doi.org/10.1016/j.gene.2014.03.060S0378-1119(14)00398-9.
26. Layh-Schmitt, G., & Herrmann, R. (1994). Spatial arrangement of gene products of the P1 operon in the membrane of Mycoplasma pneumoniae. *Infection and Immunity, 62*, 974–979.
27. Malhotra, M., Sood, S., Mukherjee, A., Muralidhar, S., & Bala, M. (2013). Genital Chlamydia trachomatis: An update. *The Indian Journal of Medical Research, 138*, 303.
28. McGill, M. A., Edmondson, D. G., Carroll, J. A., Cook, R. G., Orkiszewski, R. S., & Norris, S. J. (2010). Characterization and serologic analysis of the Treponema pallidum proteome. *Infection and Immunity, 78*, 2631–2643.
29. Naqvi, A. A., Ahmad, F., & Hassan, M. I. (2015a). Identification of functional candidates amongst hypothetical proteins of Mycobacterium leprae Br4923, a causative agent of leprosy. *Genome, 58*, 25–42. https://doi.org/10.1139/gen-2014-0178.
30. Naqvi, A. A., Anjum, F., Khan, F. I., Islam, A., Ahmad, F., & Hassan, M. I. (2016). Sequence analysis of hypothetical proteins from helicobacter pylori 26695 to identify potential virulence factors. *Genomics Informatics, 14*, 125–135. https://doi.org/10.5808/GI.2016.14.3.125.
31. Naqvi, A. A., Shahbaaz, M., Ahmad, F., & Hassan, M. I. (2015b). Identification of functional candidates amongst hypothetical proteins of Treponema pallidum ssp. pallidum. *PLoS One, 10*, e0124177. https://doi.org/10.1371/journal.pone.0124177PONE-D-14-36059.
32. Nimrod, G., Schushan, M., Steinberg, D. M., & Ben-Tal, N. (2008). Detection of functionally important regions in "hypothetical proteins" of known structure. *Structure, 16*, 1755–1763.
33. Pearson, W. R., & Lipman, D. J. (1988). Improved tools for biological sequence comparison. *Proceedings of the National Academy of Sciences of the United States of America, 85*, 2444–2448.
34. Peeling, R. W., & Hook, E. W., III. (2006). The pathogenesis of syphilis: The great mimicker, revisited. *The Journal of Pathology: A Journal of the Pathological Society of Great Britain and Ireland, 208*, 224–232.
35. Pellegrini, M., Marcotte, E. M., Thompson, M. J., Eisenberg, D., & Yeates, T. O. (1999). Assigning protein functions by comparative genome analysis: Protein phylogenetic profiles. *Proceedings of the National Academy of Sciences, 96*, 4285–4288.
36. Razin, S., Yogev, D., & Naot, Y. (1998). Molecular biology and pathogenicity of mycoplasmas. *Microbiology and Molecular Biology Reviews, 62*, 1094–1156.
37. Russell, D. G., Barry, C. E., & Flynn, J. L. (2010). Tuberculosis: What we don't know can, and does, hurt us. *Science, 328*, 852–856.
38. Russell, R. B., & Barton, G. J. (1994). Structural features can be unconserved in proteins with similar folds: An analysis of side-chain to side-chain contacts secondary structure and accessibility. *Journal of Molecular Biology, 244*, 332–350.
39. Ryan, K. J., & Ray, C. G. (2004). Medical microbiology. *McGraw Hill, 4*, 370.

40. Schnappinger, D., et al. (2003). Transcriptional adaptation of *Mycobacterium tuberculosis* within macrophages: Insights into the phagosomal environment. *Journal of Experimental Medicine, 198*, 693–704.
41. Sethi, S., & Murphy, T. F. (2001). Bacterial infection in chronic obstructive pulmonary disease in 2000: A state-of-the-art review. *Clinical Microbiology Reviews, 14*, 336–363.
42. Shahbaaz, M., Bisetty, K., Ahmad, F., & Hassan, M. I. (2015a). In silico approaches for the identification of virulence candidates amongst hypothetical proteins of Mycoplasma pneumoniae 309. *Computational Biology and Chemistry, 59*, 67–80.
43. Shahbaaz, M., Bisetty, K., Ahmad, F., & Hassan, M. I. (2015b). In silico approaches for the identification of virulence candidates amongst hypothetical proteins of Mycoplasma pneumoniae 309. *Computational Biology and Chemistry, 59*(Pt A), 67–80. https://doi.org/10.1016/j.compbiolchem.2015.09.007S1476-9271(15)30165-1.
44. Shahbaaz, M., Bisetty, K., Ahmad, F., & Hassan, M. I. (2015c). Towards new drug targets? Function prediction of putative proteins of Neisseria meningitidis MC58 and their virulence characterization. *OMICS, 19*, 416–434. https://doi.org/10.1089/omi.2015.0032.
45. Shahbaaz, M., Bisetty, K., Ahmad, F., & Hassan, M. I. (2016). Current advances in the identification and characterization of putative drug and vaccine targets in the bacterial genomes. *Current Topics in Medicinal Chemistry, 16*, 1040–1069.
46. Shahbaaz, M., Hassan, M. I., & Ahmad, F. (2013). Functional annotation of conserved hypothetical proteins from Haemophilus influenzae Rd KW20. *PLoS One, 8*, e84263. https://doi.org/10.1371/journal.pone.0084263PONE-D-13-40546.
47. Shiota, S., Suzuki, R., & Yamaoka, Y. (2013). The significance of virulence factors in Helicobacter pylori. *Journal of Digestive Diseases, 14*, 341–349.
48. Singh, A. E., & Romanowski, B. (1999). Syphilis: Review with emphasis on clinical, epidemiologic, and some biologic features. *Clinical Microbiology Reviews, 12*, 187–209.
49. Testerman, T. L., & Morris, J. (2014). Beyond the stomach: An updated view of helicobacter pylori pathogenesis, diagnosis, and treatment. *World journal of gastroenterology: WJG, 20*, 12781.
50. Turab Naqvi, A. A., et al. (2017). Genome analysis of Chlamydia trachomatis for functional characterization of hypothetical proteins to discover novel drug targets. *International Journal of Biological Macromolecules, 96*, 234–240. https://doi.org/10.1016/j.ijbiomac.2016.12.045.
51. Unemo, M., & Dillon, J.-A. R. (2011). Review and international recommendation of methods for typing *Neisseria gonorrhoeae* isolates and their implications for improved knowledge of gonococcal epidemiology, treatment, and biology. *Clinical Microbiology Reviews, 24*, 447–458.
52. Vouga, M., & Greub, G. (2016). Emerging bacterial pathogens: The past and beyond. *Clinical Microbiology and Infection, 22*, 12–21.
53. Watson, J. D., Laskowski, R. A., & Thornton, J. M. (2005). Predicting protein function from sequence and structural data. *Current Opinion in Structural Biology, 15*, 275–284.
54. Yamaoka, Y., & Graham, D. Y. (2014). Helicobacter pylori virulence and cancer pathogenesis. *Future Oncology, 10*, 1487–1500.
55. Yoshikawa, T. T. (2001). Epidemiology of aging and infectious diseases. In *Infectious disease in the aging* (pp. 3–6). Springer.

Microbial Natural Products: Exploiting Microbes Against Drug-Resistant Bugs

Suriya Rehman, Zainab Al Salem, Reem Al Jindan, and Saif Hameed

Abstract

The unsystematic and incoherent usage of antibiotics has led to an exceptional challenge for mankind, due to the development of resistance in pathogens toward the available drugs by a phenomenon known as multidrug resistance (MDR). By year 2050, the drug-resistant infections across the globe are predicted to reach an alarming number of ten million deaths. The development of MDR in pathogens to the known drugs could be attributed to a number of factors, like target modification, efflux pump, inactivation of antibiotics, etc. Therefore, with the current rate of acquiring drug resistance, there is a need to investigate all the promising sources of antimicrobials. One among them is the microbial flora associated with soil, water, and plants producing multitude of antimicrobials like terpenes, coumarins, alkaloids, etc., which could have immense and diverse potential in this field. Different antimicrobials recognized so far are known as per their mode of action, like inhibiting the cell wall synthesis, interfering with the protein synthesis, synthesis of nucleic acids, cell membrane lysis, etc. The present chapter emphasizes on few antimicrobials produced by microorganisms with their mechanisms of action, thereby urging the need to study microbial flora extensively as an effective strategy to fight the resistance of deadly superbugs.

S. Rehman (✉) · Z. Al Salem
Department of Epidemic Disease Research, Institute of Research and Medical Consultations (IRMC), Imam Abdulrahman Bin Faisal University, Dammam, Saudi Arabia
e-mail: surrehman@iau.edu.sa; suriyamir@gmail.com; http://www.researcherid.com/rid/D-7999-2017; https://orcid.org/0000-0003-3087-852X

R. Al Jindan
Department of Microbiology, College of Medicine, Imam Abdulrahman Bin Faisal University, Dammam, Saudi Arabia

S. Hameed
Amity Institute of Biotechnology, Amity University Haryana, Manesar, Gurugram, India

S. Hameed, Z. Fatima (eds.), *Pathogenicity and Drug Resistance of Human Pathogens*,
https://doi.org/10.1007/978-981-32-9449-3_20

Keywords
Antimicrobials · Multidrug resistance (MDR) · Microorganisms · Infectious diseases

Introduction

Over the millions of years, the growing human population with its pandemic effects has resulted in the development of infectious diseases [1, 2]. The search and quest for the weapons against these infections started just before the turn of the twentieth century. A major medical breakthrough came in 1928, when Fleming incidentally discovered penicillin, which was later considered as the miracle drug of the twentieth century [3]. During this century, the discovery and development of new antibiotics made it possible to treat deadly infectious diseases, hence resulted in improving the human health [4]. Although misuse and abuse of antibiotics have led to development of drug resistance, more and more bacteria with multiple resistances are being observed. It has now become evident that microorganisms are countering the impact of antimicrobial resistance at an alarming speed, and therefore demanding alternative approaches to fight microbial infections [5, 6].

A wide variety of pathogens have developed resistance to most of the antimicrobials available to date. Microbial resistance mechanisms involve a variety of strategies to protect microbes against the lethal effects of drugs (Fig. 1). These mechanisms include restricting the drug entry into the bacterial cell, changing the drug target site, and producing enzymes, which modify or degrade the antimicrobial compound, hence making it ineffective. All these resistance mechanisms developed by bacteria are either intrinsic or extrinsic. Intrinsic mechanisms include mutations in the genetic makeup of bacteria, thus changing the structure of proteins which are the target side of antimicrobials or help in their entry across bacterial membrane, while extrinsic mechanisms include borrowing of antibiotic resistance genes from other bacteria through a variety of horizontal gene transfer mechanisms such as conjugation, transformation, and transduction [7]. New mechanisms of resistance are described, and new genes and vectors responsible for resistance are being reported on regular basis [8]. This existing resistance to available antimicrobials is somehow attributed to inappropriate use of antibiotics, poor quality of drugs, exposure to suboptimal level of antibiotics, incorrect prescription without supervision, ineffective prevention and control of infections, and many more. Therefore, the search for new and more effective antimicrobial is the need of a day [9].

The sustenance and endurance of infectious microbes are characterized by multitude of factors, necessary for them to cause pathogenicity in the host [10]. Such factors include attachment of the microbes to host cell, colonization in the tissue, and invasion of microbes on host cells to cause virulence. Microbial cell wall components, toxins, biofilm formation, and spore formation are features of microbes that cause virulence and help the pathogens to persist and thrive in adverse environments [11, 12].

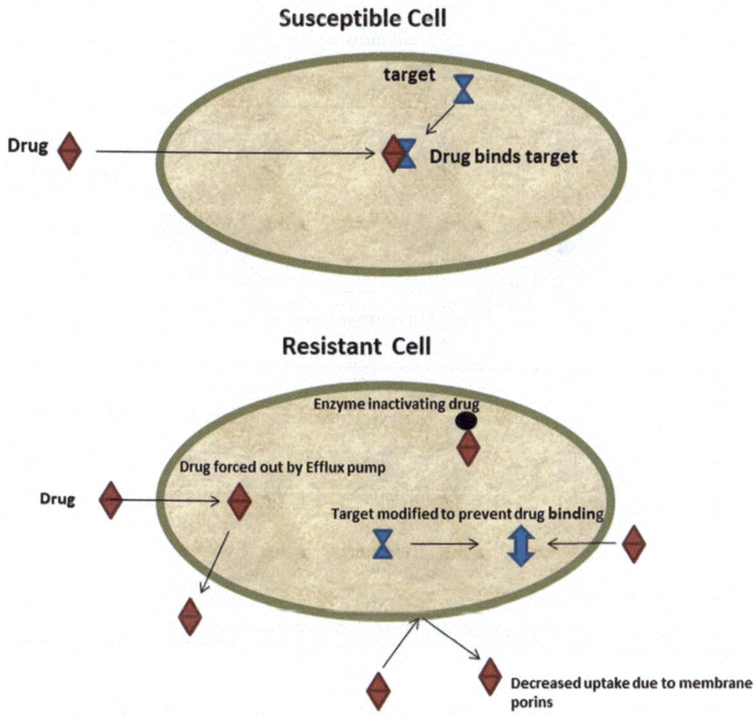

Fig. 1 Mechanism of drug resistance by microbes

Since time immemorial, natural products have played a crucial role in improvement and prosperity of mankind. A surplus of natural products has been used in food industry, pharmaceutical industry, and many more. Natural products can be obtained from different sources including plants, animals, and microorganisms (Fig. 2) [13–18]. Different countries have the folklore claims of the use of molds, soils, and plants to treat infections, for example, China, Greece, and Egypt in ancient times have been using moldy bread for the wounds to prevent infections [19]. Antimicrobial compounds produced by microorganisms have received much attention as a medium of novel antimicrobial substances for both prophylactic and therapeutic use for controlling the diseases [20, 21]. The bioactive compounds produced by microorganisms with antimicrobial activity have been previously reported [22–27]. A plethora of these compounds are mainly secondary metabolites produced at the late growth phase by microorganisms. The secondary metabolism is greatly influenced by environmental factors like interaction within ecosystem, nutrient exhaustion, or competition with the fellow microbes for survival. Formation of secondary metabolites by microbes like antimicrobials seems to have no direct role in their physiology and is mainly influenced by nutrients, metals, growth rate, enzyme inactivation, etc. The biggest advantage of microbial-derived antibiotics to be used as therapeutic is that they do not have any adverse effect on healthy cells,

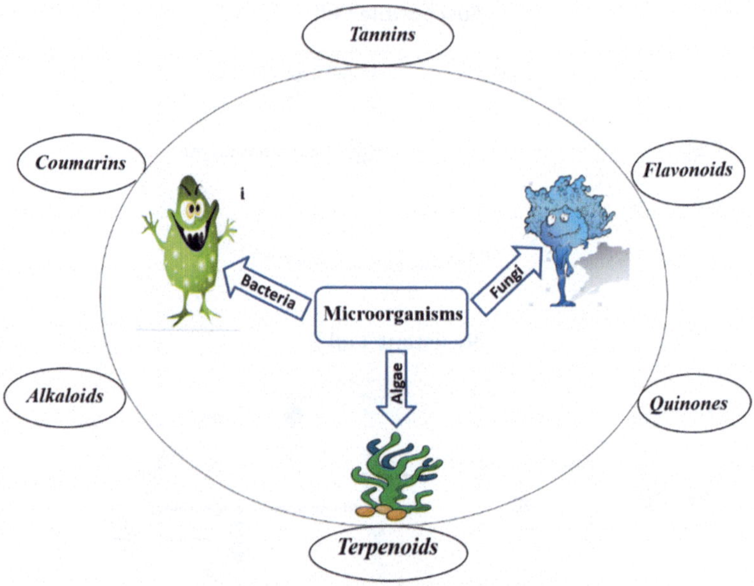

Fig. 2 Microorganisms involved in the synthesis of bioactive compounds

which is often associated with synthetic drugs, due to the presence of basic similarity between the enzyme systems in evolutionary development of host and producer. This is the fascinating feature of microbial antibiotics and has lot of implications for humans [28]. The advantages of microbial-derived antibiotics are their high potency, structural diversity, selectivity, and diverse mode of action. Above all the affordability of microbial compounds forms the scope for their wide use as effective antimicrobials and disinfectants in food industry, veterinary, and untapped source for new and rare antibiotics. The following sections review various antimicrobials from microorganisms and emphasize on the effect of microbial-derived compounds on virulence of pathogens.

Microbe-Derived Antimicrobials and Their Mode of Action

Here are the major groups of antimicrobials produced by microorganisms, which can be classified on the basis of their chemical structure which has an impact on their antimicrobial property. Table 1 provides few examples of antimicrobial compounds produced by various microorganisms.

Flavonoids: These are the low molecular weight compounds with a simple 15 carbon backbone. Flavonoids are pigmented secondary metabolites mainly found in fruits and flowers of plants [29, 30]. They have been also produced from bacteria and fungus such as *Escherichia coli*, *Streptomyces* species, *Saccharomyces cerevisiae*, etc. The bioactivity of flavonoids against a variety of bacterial [31, 32]

Table 1 List of some of the microorganisms producing various antimicrobials

Microorganisms	Source	Compound	Activity	References
Pseudomonas stutzeri, *Alcanivorax dieselolei*	Marine	Furan	Antibacterial	[72]
Vibrio owensii	Marine	2-pyrrolidinone, phenol	Antibacterial	[72]
Aspergillus sydowii	Marine	Unidentified	Antibacterial	[73]
Streptomyces sp.	Marine	Divergolides	Antibacterial	[74]
Chalara sp.	Terrestrial,	Isofusidienols	Antibacterial	[75]
A nonsporulating fungus	Terrestrial,	Spiro-mamakone	Antibacterial	[76]
Penicillium brocae	Marine	Spirobrocazines	Antibacterial	[77]
Pestalotiopsis fici	Terrestrial	Chloropupukeananin	Antibacterial, antiviral	[78]
Blennoria sp.	Terrestrial	*Carpobrotus edulis* blennolides	Antifungal, antialgal	[79]
Cryptosporiopsis sp.	Terrestrial	*Viburnum tinus* viburspiran	Antifungal	[80]
Cryptosporiopsis cf. *quercina*	Terrestrial	*Tripterygium wilfordii* cryptocin	Antifungal	[58]
Nocardiopsis sp.	Terrestrial	Sulfamethoxypyridazine	Antifungal	[81]
Fusarium sp.	Terrestrial	Podophyllotoxin	Antimicrobial	[82]
Streptomyces sp. BO-07	Terrestrial	3′-hydroxy-5-methoxy-3,4-methylenedioxybiphenyl 3′-hydroxy-5,5′-dimethoxy-3,4-methylenedioxybiphenyl	Antimicrobial	[83]
Cephalosporium acremonium	Terrestrial	*Trachelospermum jasminoides* cephalosol	Antimicrobial	[84]
Trichoderma koningiopsis, *Trichoderma lixii*, and *Aspergillus sydowii*	Terrestrial	Unknown compounds	Antimicrobial	[85]
Daldinia eschscholtzii	Terrestrial	*Paphiopedilum exul* daldionin	Antimicrobial	[86]
Rhizoctonia solani	Terrestrial	*Cyperus rotundus* solanioic acid	Antimicrobial	[87]
Trichoderma spp.	Marine	Harziandione and harzianone	Antimicrobial	[88]
Paecilomyces variotii	Marine	Varioxepine A	Antimicrobial	[89]
Pestalotiopsis sp.	Marine	Pestalotiopens	Antimicrobial	[90]
Pestalotiopsis fici	Terrestrial	Chloropestolide	Anti-HIV	[91]
Periconia sp.	Terrestrial	Pericoannosin	Anti-HIV	[92]

(continued)

Table 1 (continued)

Microorganisms	Source	Compound	Activity	References
Neosartorya	Marine	Neosartoryadins	Antiviral	[93]
Periconia sp.	Terrestrial	Periconiasins	Antiviral	[94, 95]
Aspergillus versicolor	Terrestrial	Aspergillines	Antiviral	[96]
Fusarium pallidoroseum	Terrestrial	Apicidins	Antiprotozoal	[97]
Actinoallomurus fulvus	Terrestrial	Actinoallolides	Anti-trypanosomal	[98]
Aspergillus sp.	Marine	Asperterpenoid	Antituberculosis	[99]

and fungal pathogens [33] is mainly on the cell membranes [34]. Its mode of action is interaction with membrane proteins of the cell wall, which leads to increased permeability and disintegration. The naturally occurring flavonoids depict a wide range of antiviral activity against a number of RNA (RSV, Pf-3, polio) and DNA (HSV-1) viruses inhibiting replication and infectivity [35]. Recently, it is also reported to have an antiviral activity against few more viral infections [36].

Quinones: The organic compounds which consist of aromatic rings and two ketone substitutions are quinones. Isoprenoid quinones are essential components in electron transport system like those present in the membrane of mitochondria or chloroplasts, and these quinones are present in most microorganisms like *Cyanobacteria, Chlorobium thiosulphatophilium, archaea, Rhodospirillum rubrum*, and *R. photometricum* [37]. The major action against the pathogenic cell includes cell wall polypeptides, proteins exposed to surface, and membrane-bound proteins [38]. Quinone is found to be bactericidal against *Burkholderia pseudomallei* and bacteriostatic against *Corynebacterium pseudodiphthericum, Bacillus anthracis*, and *Pseudomonas aeruginosa* [39]. The quinone-based compound library has also been found to be active against the parasites like *Trypanosoma cruzi* and *Leishmania*, the disease-causing agents of Chagas disease and cutaneous leishmaniasis, respectively [40].

Tannins: Tannins are the polymeric polyphenolic compounds mainly found in bark, leaves, fruits, and roots of plants [41]. Tannins are also found to be produced by microorganisms like *Penicillium* sp., such as *P. frequentans* [42, 43]. The antimicrobial activities of tannins are well documented and are active against bacteria and fungi as well [44]. The mechanism of action of tannins as an antimicrobial is to inhibit the activity of surface adhesins and transport proteins of cell envelope [45]. Tannins have also been reported to bind to the cell wall of bacteria like *Staphylococcus aureus* and *Salmonella typhimurium* and fungus like *Aspergillus niger* and *Saccharomyces cerevisiae*, therefore inhibiting the growth and protease activity [46].

Coumarins: Coumarins are the compounds that consist of fused benzene and an α-pyrone rings. Several bacteria and fungi have been reported to produce coumarin, e.g., *Streptomyces niveus, Streptomyces spheroids, Agaricus* sp., *Fomitopsis officinalis, Talaromyces flavus, Ganoderma lucidum*, and *Phellinus* sp. [47–51]. Coumarins and its derivatives have been found to exhibit pronounced antibacterial, antifungal, and antiviral activity [52–54].

Alkaloids: These are the group of heterocyclic nitrogenous compounds with profound antimicrobial activity. Some of the alkaloids obtained from microorganisms are pyrrolopyrazine, an alkaloid by *Acremonium lolii* [55], ergot alkaloids by genus *Neotyphodium* [56], phomopsichalasin by *Phomopsis* sp. [57], phomoenamide by *Phomopsis* sp., cryptocin by *Cryptosporiopsis quercina* [58], ammosamide D from *Streptomyces variabilis* [59], and aqabamycins A–G from *Vibrio* sp. [60]. The mode of action of these alkaloids is mostly the intercalating of DNA leading in impaired cell division and death [61]. Flindersine alkaloids are reported to have activity against *Staphylococcus epidermidis*, *Staphylococcus aureus*, *Bacillus subtilis*, *Enterococcus faecalis*, *Pseudomonas aeruginosa*, *Acinetobacter baumannii* and fungi *Trichophyton rubrum* 57, *Trichophyton mentagrophytes*, *Trichophyton simii*, *Magnaporthe grisea*, *Epidermophyton floccosum*, and *Candida albicans* [62]. In the last few decades, around 40 fumiquinazoline analogues have been reported from different fungal species. In 2016, Jähne et al. reported two new fumiquinazoline analogues, neosartoryadins A and B from the endophytic fungus *Neosartorya udagawae* HDN13-313 depicting anti-influenza virus A (H1N1) effects [63].

Terpenoids: They are one of the diverse groups of compounds with five carbon isoprene units. As found mainly in plants, these bioactive molecules have also been found to be produced by microorganisms. These terpenoids exhibit a broad range of bioactivities against malaria and a variety of viral and bacterial infectious diseases [64]. Sesquiterpenes, diterpenoids, and triterpenoids are among the major groups of terpenoids produced by microbes, e.g. trichodermin from *Trichoderma harzianum* [65] and phomenone from *Xylaria* sp. [66]. The antifungal drug as well as anticancer drug, paclitaxel (Taxol), was found to be produced by the fungus *Taxomyces andreanae* [67], and nordammarane triterpenoid was isolated from the yeast, *Pichia guilliermondii* [68]. Terpenoids named 5-dimethylallylindole-3-carboxylic acid and A80915G-8″-acid [69], napyradiomycins is isolated from *Streptomyces antimycoticus* NT17 [70]. Mainly sesquiterpenoids have been shown to possess bactericidal activity against mostly Gram-positive bacteria, including *M. tuberculosis* [71].

Conclusion

The widespread of drug-resistant microorganisms has triggered the search for finding novel antimicrobials. Despite the fact that more than half of the drugs available are based on natural products, the discovery of natural products has still been neglected. Nevertheless, there is a need to search for new antibiotics by exploring the natural and unusual habitats like microorganisms, lying deep below the earth's surface, deep oceans, thermal vents, glaciers, associated with plant diversity, in the skin of animals, insects, and so on. Such discoveries could be a renaissance of naturally derived antimicrobials and as a lead to combat drug resistance, particularly if novel microorganisms can be unveiled.

Conflict of Interest Authors declare no conflict of interest.

References

1. McMichael, A. J. (2004). Environmental and social influences on emerging infectious diseases: Past, present and future. *Philosophical Transactions of the Royal Society of London. Series B: Biological Sciences, 359*(1447), 1049–1058.
2. Bin, Z. S., et al. (2017). A review on antibiotic resistance: Alarm bells are ringing. *Cureus, 9*, 6.
3. Martens, E., & Demain, A. L. (2017). The antibiotic resistance crisis, with a focus on the United States. *Journal of Antibiotics, 70*(5), 520–526.
4. Aminov, R. I. (2010). A brief history of the antibiotic era: Lessons learned and challenges for the future. *Frontiers in Microbiology, 1*, 134.
5. Levy, S. B. (1994). Balancing the drug-resistance equation. *Trends in Microbiology, 2*(10), 341–342.
6. Livermore, D. M. (2003). Bacterial resistance: Origins, epidemiology, and impact. *Clinical Infectious Diseases, 36*(1), S11–S23.
7. Yelin, I., & Kishony, R. (2018). Antibiotic Resistance. *Cell, 172*(5), 1136–1136.e1.
8. Blair, J. M. A., et al. (2015). Molecular mechanisms of antibiotic resistance. *Nature Reviews Microbiology, 13*(1), 42–51.
9. Alkharsah, K. R., et al. (2018). Comparative and molecular analysis of MRSA isolates from infection sites and carrier colonization sites. *Annals of Clinical Microbiology and Antimicrobials, 17*(1), 7.
10. Falkow, S. (1991). *What is pathogen?* Washington, DC: American Society for Microbiology.
11. Taylor, C., & Roberts, I. S. (2005). Capsular polysaccharides and their role in virulence. *Contributions to Microbiology, 12*, 55–66.
12. Wu, H.-J., Wang, A. H. J., & Jennings, M. P. (2008). Discovery of virulence factors of pathogenic bacteria. *Current Opinion in Chemical Biology, 12*(1), 93–101.
13. Burt, S. (2004). Essential oils: their antibacterial properties and potential applications in foods—A review. *International Journal of Food Microbiology* (pp. 223–253). Amsterdam: Elsevier B.V.
14. Holley, R. A., & Patel, D. (2005). Improvement in shelf-life and safety of perishable foods by plant essential oils and smoke antimicrobials. *Food Microbiology, 22*(4), 273–292.
15. Nychas, G.-J., Skandamis, P., Tassou, C. (2003). Antimicrobials from herbs and spices. In *Natural antimicrobials for the minimal processing of foods* (pp. 176–200). USA: Woodhead Publishing.
16. Osbourn, A. E. (1996). Preformed antimicrobial compounds and plant defense against fungal attack. *The PlantCell, 8*(10), 1821–1831.
17. Antony, M. L., & Singh, S. V. (2011). Molecular mechanisms and targets of cancer chemoprevention by garlic-derived bioactive compound diallyl trisulfide. *Indian Journal of Experimental Biology, 49*(11), 805–816.
18. Nash, R. J., et al. (2011). Iminosugars as therapeutic agents: Recent advances and promising trends. *Future Medicinal Chemistry, 3*(12), 1513–1521.
19. Keyes, K., Lee, M., Maurer, J. (2008). Antibiotics: Mode of action, mechanisms of resistance, and transfer. In *Microbial food safety in animal agriculture: Current topics* (pp. 45–56). USA: Lowa State Press.
20. Bhagavathy, S., Sumathi, P., & Jancy Sherene, I. (2011). Bell, Green algae Chlorococcum humicola-a new source of bioactive compounds with antimicrobial activity. *Asian Pacific Journal ofTropical Biomedicine, 1*(1), S1–S7.

21. Ramesh, C., & Pattar, M. G. (2010). Antimicrobial properties, antioxidant activity and bioactive compounds from six wild edible mushrooms of western ghats of Karnataka, India. *Pharmacognosy Research, 2*(2), 107–112.
22. Demain, A. L., & Sanchez, S. (2009). Microbial drug discovery: 80 years of progress. *The Journal of Antibiotics, 62*(1), 5–16.
23. Imhoff, J. F., Labes, A., & Wiese, J. (2011). Bio-mining the microbial treasures of the ocean: New natural products. *Biotechnology Advances, 29*(5), 468–482.
24. Kusari, S., & Spiteller, M. (2011). Are we ready for industrial production of bioactive plant secondary metabolites utilizing endophytes? *Natural Product Reports, 28*(7), 1203–1207.
25. Crawford, J. M., & Clardy, J. (2011). Bacterial symbionts and natural products. *Chemical Communications, 47*(27), 7559–7566.
26. Yu, H., et al. (2010). Recent developments and future prospects of antimicrobial metabolites produced by endophytes. *Microbiological Research, 165*(6), 437–449.
27. Simmons, T. L., et al. (2008). Biosynthetic origin of natural products isolated from marine microorganism-invertebrate assemblages. *Proceedings of the National Academy of Sciences of the United States of America, 105*(12), 4587–4594.
28. Bérdy, J. (2012). Thoughts and facts about antibiotics: Where we are now and where we are heading. *Journal of Antibiotics, 65*(8), 385–395.
29. Cowan, M. M. (1999). Plant products as antimicrobial agents. *Clinical Microbiology Reviews, 12*(4), 564–582.
30. Luisa Helena, C., et al. (2008). Flavonoids: Prospective drug candidates. *Mini-Reviews in Medicinal Chemistry, 8*(13), 1429–1440.
31. Locher, C. P., et al. (1995). Anti-microbial activity and anti-complement activity of extracts obtained from selected Hawaiian medicinal plants. *Journal of Ethnopharmacology, 49*(1), 23–32.
32. Zeng, F., et al. (2010). Two prenylated and C-methylated flavonoids from Tripterygium wilfordii. *Planta Medica, 76*(14), 1596–1599.
33. Kramer, R. P., et al. (1984). Antifungal activity of soybean and chickpea isoflavones and their reduced derivatives. *Phytochemistry, 23*(10), 2203–2205.
34. Davidson, & Naidu. (2000). Phytophenols. In *Natural food antimicrobial systems* (pp. 265–293).
35. Kaul, N., Middleton, T. E., Jr., & Ogra, P. (1985). Antiviral effect of flavonoids on human viruses. *Journal of Medical Virology, 15*, 71–79.
36. Shibata, C., et al. (2014). The flavonoid apigenin inhibits hepatitis C virus replication by decreasing mature microRNA122 levels. *Virology, 462–463*, 42–48.
37. Perez-Mateos, M. (1990). In J. M. Bollag & G. Stotzky (Eds.), *Soil biochemistry* (Vol. 6, p. 584). New York: Marcel Dekker. $150 ISBN 0-8247-8232-1. Biochemical Education. 2010. 215-215.
38. Ciocan, D., & Ioan, B. (2007). Plant products as antimicrobial agents. *AnaleleStiintifice ale Universitatii "Al. I. Cuza" Din Iasi. (Serie Noua). Sectiunea 2.a. Genetica si Biologie Moleculara, 8*(1), 151–156.
39. Kazmi, M. H., et al. (1994). An anthraquinone derivative from Cassia italica. *Phytochemistry, 36*(3), 761–763.
40. Sullivan, J. A., et al. (2015). Unravelling the role of SNM1 in the DNA repair system of Trypanosoma brucei. *Molecular Microbiology, 96*(4), 827–838.
41. Scalbert, A. (1991). Antimicrobial properties of tannins. In *Phytochemistry* (pp. 3875–3883). Oxford: Elsevier Ltd.
42. Bhardwaj, A., et al. (2015). Antimicrobial and phytochemical screening of endophytic fungi isolated from spikes of Pinus roxburghii. *Archives of Clinical Microbiology, 6*, 1.
43. Devi, N. N., Prabakaran, J. J., & Wahab, F. (2012). Phytochemical analysis and enzyme analysis of endophytic fungi from *Centella asiatica. Asian Pacific Journal of Tropical Biomedicine, 2*(3), S1280–S1284.
44. Chung, K.-T., et al. (1998). Tannins and human health: A review. *Critical Reviews in Food Science and Nutrition, 38*(6), 421–464.

45. Knasmüller, S., et al. (2009). *Tannins: Bioavailability and mechanisms of action, in chemoprevention of cancer and DNA damage by dietary factors* (pp. 499–508). Weinheim: Wiley-VCH Verlag GmbH & KGaA.
46. Chandra, H., et al. (2017). Antimicrobial resistance and the alternative resources with special emphasis on plant-based antimicrobials-a review. *Plants, 6*(2), 16.
47. Steffensky, M., et al. (2000). Identification of the novobiocin biosynthetic gene cluster of Streptomyces spheroides NCIB 11891. *Antimicrobial Agents and Chemotherapy, 44*(5), 1214–1222.
48. Wang, Y., et al. (2007). Quantitation of Armillarisin A in human plasma by liquid chromatography–electrospray tandem mass spectrometry. *Journal of Pharmaceutical and Biomedical Analysis, 43*(5), 1860–1863.
49. Hwang, C. H., et al. (2013). Chlorinated coumarins from the polypore mushroom, fomitopsis officinalis, and their activity against *Mycobacterium tuberculosis. Journal of Natural Products, 76*(10), 1916–1922. https://doi.org/10.1021/np400497f.
50. He, J.-W., et al. (2014). Two new coumarins from talaromyces flavus. *Molecules, 19*(12), 20880–20887.
51. Islam, M., et al. (2015). Phytochemicals and antibacterial activity screening of three edible mushrooms Pleurotus ostreatus, Ganoderma lucidum and Lentinula edodes accessible in Bangladesh. *American Journal of Biology and Life Sciences, 3*, 31–35.
52. Cao, J. L., et al. (2013). ChemInform abstract: A catalyst-free one-pot construction of skeletons of 5-methoxyseselin and alloxanthoxyletin in water. *ChemInform, 44*(51), no–no.
53. Venugopala, K. N., Rashmi, V., & Odhav, B. (2013). Review on natural coumarin lead compounds for their pharmacological activity. *BioMed Research International, 2013*, 1–14.
54. Singh, L. R., et al. (2017). Coumarin-benzimidazole hybrids as a potent antimicrobial agent: Synthesis and biological elevation. *Journal of Antibiotics, 70*(9), 954–961.
55. Rowan, D. D. (1993). Lolitrems, peramine and paxilline: Mycotoxins of the ryegrass/endophyte interaction. *Agriculture, Ecosystems and Environment, 44*(1), 103–122.
56. Schardl, C. L. (2010). The epichloae, symbionts of the grass subfamily poödeae. *Annals of the Missouri Botanical Garden, 97*(4), 646–665.
57. Horn, W. S., et al. (1995). Phomopsichalasin, a novel antimicrobial agent from an endophytic Phomopsis sp. *Tetrahedron, 51*(14), 3969–3978.
58. Li, J. Y., et al. (2000). Cryptocin, a potent tetramic acid antimycotic from the endophytic fungus cryptosporiopsis cf. quercina. *Organic Letters, 2*(6), 767–770.
59. Pan, E., et al. (2012). Ammosamide D, an oxidatively ring opened ammosamide analog from a marine-derived Streptomyces variabilis. *Organic Letters, 14*(9), 2390–2393.
60. Al-Zereini, W., et al. (2010). Aqabamycins A-G: Novel nitro maleimides from a marine Vibrio species. I. Taxonomy, fermentation, isolation and biological activities. *The Journal of Antibiotics, 63*, 297.
61. Savoia, D. (2012). Plant-derived antimicrobial compounds: Alternatives to antibiotics. *Future Microbiology, 7*(8), 979–990.
62. Shang, X. F., et al. (2018). Biologically active quinoline and quinazoline alkaloids part I. *Medicinal Research Reviews, 38*(3), 775–828.
63. Jähne, E. A., et al. (2016). Pharmacokinetics and in vitro blood-brain barrier screening of the plant-derived alkaloid tryptanthrin. *Planta Medica, 82*, 1021–1029.
64. Wang, G., Tang, W., & Bidigare, R. R. (2005). Terpenoids as therapeutic drugs and pharmaceutical agents. In *Natural products* (pp. 197–227). Totowa: Humana Press.
65. Chen, L., et al. (2007). Identification and antifungal activity of the metabolite of endophytic fungi isolated from Llex cornuta. *Chinese Journal of Pesticide Science, 151*, 143–148.
66. Silva, D. H. S., et al. (2010). Sesquiterpenes from Xylaria sp., an endophytic fungus associated with *Piper aduncum* (Piperaceae). *Phytochemistry Letters, 3*(3), 164–167.
67. Stierle, A., et al. (1995). The search for a taxol-producing microorganism among the endophytic Fungi of the Pacific Yew, *Taxus brevifolia. Journal of Natural Products, 58*(9), 1315–1324.

68. Zhao, J., et al. (2010). Antimicrobial metabolites from the endophytic fungus pichia guilliermondii isolated from Paris polyphylla var. yunnanensis. *Molecules, 15*(11), 7961–7970.
69. Motohashi, K., et al. (2008). Studies on terpenoids produced by actinomycetes. *The Journal of Antibiotics, 61*(2), 75–80.
70. Motohashi, K., et al. (2008). Studies on terpenoids produced by actinomycetes: 5-dimethylallylindole-3- carboxylic acid and A80915G-8″-acid produced by marine-derived Streptomyces sp. MS239. *Journal of Antibiotics, 61*(2), 75–80.
71. Kurek, A., et al. (2011). New antibacterial therapeutics and strategies. *Polish Journal of Microbiology, 60*(1), 3–12.
72. Karthick, P., & Mohanraju, R. (2018). Antimicrobial potential of epiphytic bacteria associated with seaweeds of Little Andaman, India. *Frontiers in Microbiology, 9*, 611.
73. Trianto, A., Widyaningsih, S., Radjasa, O. K., & Pribadi, R. (2017). Symbiotic fungus of marine sponge Axinella sp. producing antibacterial agent. *IOP Conference Series: Earth and Environmental Science, 55*(1), 012005.
74. Ding, L., et al. (2011). Divergolides A-D from a mangrove endophyte reveal an unparalleled plasticity in ansa-macrolide biosynthesis. *Angewandte Chemie (International Ed. in English), 50*(7), 1630–1634.
75. Lösgen, S., et al. (2008). Isofusidienols: Novel chromone-3-oxepines produced by the endophytic fungus Chalara sp. *European Journal of Organic Chemistry, 2008*(4), 698–703.
76. van der Sar, S. A., Blunt, J. W., & Munro, M. H. G. (2006). Spiro-Mamakone a: A unique relative of the spirobisnaphthalene class of compounds. *Organic Letters, 8*(10), 2059–2061.
77. Meng, L.-H., et al. (2016). Three diketopiperazine alkaloids with spirocyclic skeletons and one bisthiodiketopiperazine derivative from the mangrove-derived endophytic fungus Penicillium brocae MA-231. *Organic Letters, 18*(20), 5304–5307.
78. Liu, L., et al. (2008). Chloropupukeananin, the first chlorinated Pupukeanane derivative, and its precursors from Pestalotiopsis fici. *Organic Letters, 10*(7), 1397–1400.
79. Zhang, W., et al. (2008). New mono- and dimeric members of the secalonic acid family: Blennolides A-G isolated from the fungus Blennoria sp. *Chemistry (Weinheim an der Bergstrasse, Germany), 14*(16), 4913.
80. Saleem, M., et al. (2011). Viburspiran, an antifungal member of the octadride class of maleic anhydride natural products. *European Journal of Organic Chemistry, 2011*(4), 808–812.
81. Bibi, F., et al. (2018). Microbial flora associated with the halophyte–Salsola imbricate and its biotechnical potential. *Frontiers in Microbiology, 9*, 65.
82. Tan, X., et al. (2018). Diversity and bioactive potential of culturable fungal endophytes of Dysosma versipellis; A rare medicinal plant endemic to China. *Sci Rep, 8*, 5929.
83. Taechowisan, T., Chaisaeng, S., & Phutdhawong, W. S. (2017). Antibacterial, antioxidant and anticancer activities of biphenyls from Streptomyces sp. BO-07: An endophyte in *Boesenbergia rotunda* (L.) Mansf A. *Article in Food and Agricultural Immunology, 28*, 1–17.
84. Zhang, H., et al. (2008). Cephalosol: An antimicrobial metabolite with an unprecedented skeleton from endophytic *Cephalosporium acremonium* IFB-E007. *A European Journal, 14*, 10670–10674.
85. Handayani, D., et al. (2018). Antimicrobial and cytotoxic activities of endophytic fungi isolated from mangrove plant *Sonneratia alba* Sm. *Applied Pharmaceutical Science, 8*, 49–53.
86. Barnes, E., et al. (2016). Daldionin, an unprecedented binaphthyl derivative, and diverse polyketide congeners from a fungal orchid endophyte. *Chemistry, 22*, 4551–4555.
87. Ratnaweera, P., et al. (2015). Solanioic acid, an antibacterial degraded steroid produced in culture by the fungus Rhizoctonia solani isolated from tubers of the medicinal plant *Cyperus rotundus*. *Organic Letters, 17*, 2074–2077.
88. Miao, F.-P., et al. (2012). Absolute configurations of unique harziane diterpenes from trichoderma species. *Organic Letters, 14*(15), 3815–3817.
89. Zhang, P., et al. (2014). Varioxepine a, a 3H-Oxepine-containing alkaloid with a new oxa-cage from the marine algal-derived endophytic fungus Paecilomyces variotii. *Organic Letters, 16* (18), 4834–4837.

90. Hemberger, Y., et al. (2014). ChemInform abstract: Pestalotiopens A (I) and B (II): stereochemically challenging flexible sesquiterpene-cyclopaldic acid hybrids from Pestalotiopsis sp. *Chemistry – A European Journal, 19*, 6571–6575.
91. Liu, L., et al. (2009). Chloropestolide a, an antitumor metabolite with an unprecedented spiroketal skeleton from Pestalotiopsis fici. *Organic Letters, 11*(13), 2836.283.
92. Zhang, D., et al. (2015). Pericoannosin a, a polyketide synthase–nonribosomal peptide synthetase hybrid metabolite with new carbon skeleton from the endophytic fungus Periconia sp. *Organic Letters, 17*(17), 4304–4307.
93. Yu, G., et al. (2016). Neosartoryadins A and B, fumiquinazoline alkaloids from a mangrove-derived fungus Neosartorya udagawae HDN13-313. *Organic Letters, 18*(2), 244–247.
94. Zhang, D., et al. (2016). Periconiasin G, a new cytochalasan with unprecedented 7/6/5 tricyclic ring system from the endophytic fungus Periconia sp. *Tetrahedron Letters, 57*(7), 796–799.
95. Liu, J., et al. (2016). Periconiasins I and J, two new cytochalasans from an endophytic fungus Periconia sp. *Tetrahedron Letters, 57*, 796–799.
96. Zhou, M., et al. (2014). Aspergillines A-E, hhighly oxygenated hexacyclic indole-tetrahydrofuran-tetramic acid derivatives from Aspergillus versicolor. *Organic Letters, 16*, 5016.
97. Singh, S. B., et al. (2001). Structure, histone deacetylase, and antiprotozoal activities of Apicidins B and C, congeners of Apicidin with proline and valine substitutions. *Organic Letters, 3*(18), 2815–2818.
98. Inahashi, Y., et al. (2015). Actinoallolides A–E, new anti-trypanosomal macrolides, produced by an endophytic Actinomycete, Actinoallomurus fulvus MK10-036. *Organic Letters, 17*(4), 864–867.
99. Huang, X., et al. (2013). Asperterpenoid a, a new Sesterterpenoid as an inhibitor of *Mycobacterium tuberculosis* protein tyrosine phosphatase B from the culture of Aspergillus sp. 16-5c. *Organic Letters, 15*(4), 721–723.

The manufacturer's authorised representative in the EU is Springer Nature Customer Service Centre GmbH, Europaplatz 3, 69115 Heidelberg, Germany. If you have any concerns regarding our products, please contact ProductSafety@springernature.com

Printed and bound by CPI Group (UK) Ltd, Croydon, CR0 4YY

25/09/2026

02222882-0004